AN INTRODUCTION TO CONTINUUM MECHANICS

This textbook on continuum mechanics reflects the modern view that scientists and engineers should be trained to think and work in multidisciplinary environments. A course on continuum mechanics introduces the basic principles of mechanics and prepares students for advanced courses in traditional and emerging fields such as biomechanics and nanomechanics. This text introduces the main concepts of continuum mechanics simply with rich supporting examples but does not compromise mathematically in providing the invariant form as well as component form of the basic equations and their applications to problems in elasticity, fluid mechanics, and heat transfer. The book is ideal for advanced undergraduate and beginning graduate students. The book features: derivations of the basic equations of mechanics in invariant (vector and tensor) form and specializations of the governing equations to various coordinate systems; numerous illustrative examples; chapter-end summaries; and exercise problems to test and extend the understanding of concepts presented.

J. N. Reddy is a University Distinguished Professor and the holder of the Oscar S. Wyatt Endowed Chair in the Department of Mechanical Engineering at Texas A&M University, College Station, Texas. Dr. Reddy is internationally known for his contributions to theoretical and applied mechanics and computational mechanics. He is the author of over 350 journal papers and 15 books, including *Introduction to the Finite Element Method*, Third Edition; *Energy Principles and Variational Methods in Applied Mechanics*, Second Edition; *Theory and Analysis of Elastic Plates and Shells*, Second Edition; *Mechanics of Laminated Plates and Shells: Theory and Analysis*, Second Edition; and *An Introduction to Nonlinear Finite Element Analysis*. Professor Reddy is the recipient of numerous awards, including the Walter L. Huber Civil Engineering Research Prize of the American Society of Civil Engineers (ASCE), the Worcester Reed Warner Medal and the Charles Russ Richards Memorial Award of the American Society of Mechanical Engineers (ASME), the 1997 Archie Higdon Distinguished Educator Award from the American Society of Engineering Education (ASEE), the 1998 Nathan M. Newmark Medal from the ASCE, the 2000 Excellence in the Field of Composites from the American Society of Composites (ASC), the 2003 Bush Excellence Award for Faculty in International Research from Texas A&M University, and the 2003 Computational Solid Mechanics Award from the U.S. Association of Computational Mechanics (USACM).

Professor Reddy is a Fellow of the American Institute of Aeronautics and Astronautics (AIAA), the ASME, the ASCE, the American Academy of Mechanics (AAM), the ASC, the USACM, the International Association of Computational Mechanics (IACM), and the Aeronautical Society of India (ASI). Professor Reddy is the Editor-in-Chief of *Mechanics of Advanced Materials and Structures, International Journal of Computational Methods in Engineering Science and Mechanics*, and *International Journal of Structural Stability and Dynamics*; he also serves on the editorial boards of over two dozen other journals, including the *International Journal for Numerical Methods in Engineering, Computer Methods in Applied Mechanics and Engineering*, and *International Journal of Non-Linear Mechanics*.

An Introduction to Continuum Mechanics

WITH APPLICATIONS

J. N. Reddy

Texas A&M University

CAMBRIDGE UNIVERSITY PRESS
Cambridge, New York, Melbourne, Madrid, Cape Town, Singapore,
São Paulo, Delhi, Dubai, Tokyo, Mexico City

Cambridge University Press
32 Avenue of the Americas, New York, NY 10013-2473, USA

www.cambridge.org
Information on this title: www.cambridge.org/9780521870443

First published 2008
Reprinted 2008, 2010

A catalog record for this publication is available from the British Library.

Library of Congress Cataloging in Publication Data

Reddy, J. N. (Junuthula Narasimha), 1945–
An introduction to continuum mechanics : with applications / J. N. Reddy.
  p.   cm.
Includes bibliographical references and index.
ISBN 978-0-521-87044-3 (hardback)
1. Continuum mechanics – Textbooks. I. Title.
QA808.2.R43 2007
531–dc22   2007025254

ISBN 978-0-521-87044-3 Hardback
I

'Tis the good reader that makes the good book; in every book he finds passages which seem confidences or asides hidden from all else and unmistakenly meant for his ear; the profit of books is according to the sensibility of the reader; the profoundest thought or passion sleeps as in a mine, until it is discovered by an equal mind and heart.

Ralph Waldo Emerson

You cannot teach a man anything, you can only help him find it within himself.

Galileo Galilei

Contents

Preface

> If I have been able to see further, it was only because I stood on the shoulders of giants.
>
> Isaac Newton

Many of the mathematical models of natural phenomena are based on fundamental scientific laws of physics or otherwise are extracted from centuries of research on the behavior of physical systems under the action of natural forces. Today this subject is referred to simply as *mechanics* – a phrase that encompasses broad fields of science concerned with the behavior of fluids, solids, and complex materials. Mechanics is vitally important to virtually every area of technology and remains an intellectually rich subject taught in all major universities. It is also the focus of research in departments of aerospace, chemical, civil, and mechanical engineering, in engineering science and mechanics, and in applied mathematics and physics. The past several decades have witnessed a great deal of research in continuum mechanics and its application to a variety of problems. As most modern technologies are no longer discipline-specific but involve multidisciplinary approaches, scientists and engineers should be trained to think and work in such environments. Therefore, it is necessary to introduce the subject of mechanics to senior undergraduate and beginning graduate students so that they have a strong background in the basic principles common to all major engineering fields. A first course on *continuum mechanics* or *elasticity* is the one that provides the basic principles of mechanics and prepares engineers and scientists for advanced courses in traditional as well as emerging fields such as biomechanics and nanomechanics.

There are many books on mechanics of continua. These books fall into two major categories: those that present the subject as highly mathematical and abstract and those that are too elementary to be of use for those who will pursue further work in fluid dynamics, elasticity, plates and shells, viscoelasticity, plasticity, and interdisciplinary areas such as geomechanics, biomechanics, mechanobiology, and nanoscience. As is the case with all other books written (solely) by the author, the objective is to facilitate an easy understanding of the topics covered. While the author is fully aware that he is not an authority on the subject of this book, he feels that he understands the concepts well and feels confident that he can explain them to others. It is hoped that the book, which is simple in presenting the main concepts, will be mathematically rigorous enough in providing the invariant form as well as component form of the governing equations for analysis of practical problems of engineering. In particular, the book contains formulations and applications to specific problems from heat transfer, fluid mechanics, and solid mechanics.

The motivation and encouragement that led to the writing of this book came from the experience of teaching a course on continuum mechanics at Virginia Polytechnic Institute and State University and Texas A&M University. A course on continuum mechanics takes different forms – abstract to very applied – when taught by different people. The primary objective of the course taught by the author is two-fold: (1) formulation of equations that describe the motion and thermomechanical response of materials and (2) solution of these equations for specific problems from elasticity, fluid flows, and heat transfer. This book is a formal presentation of the author's notes developed for such a course over past two-and-a-half decades.

After a brief discussion of the concept of a continuum in Chapter 1, a review of vectors and tensors is presented in Chapter 2. Since the language of mechanics is mathematics, it is necessary for all readers to familiarize themselves with the notation and operations of vectors and tensors. The subject of kinematics is discussed in Chapter 3. Various measures of strain are introduced here. In this chapter the deformation gradient, Cauchy–Green deformation, Green–Lagrange strain, Cauchy and Euler strain, rate of deformation, and vorticity tensors are introduced, and the polar decomposition theorem is discussed. In Chapter 4, various measures of stress – Cauchy stress and Piola–Kirchhoff stress measures – are introduced, and stress equilibrium equations are presented.

Chapter 5 is dedicated to the derivation of the field equations of continuum mechanics, which forms the heart of the book. The field equations are derived using the principles of conservation of mass, momenta, and energy. Constitutive relations that connect the kinematic variables (e.g., density, temperature, deformation) to the kinetic variables (e.g., internal energy, heat flux, and stresses) are discussed in Chapter 6 for elastic materials, viscous and viscoelastic fluids, and heat transfer.

Chapters 7 and 8 are devoted to the application of both the field equations derived in Chapter 5 and the constitutive models of Chapter 6 to problems of linearized elasticity, and fluid mechanics and heat transfer, respectively. Simple boundary-value problems, mostly linear, are formulated and their solutions are discussed. The material presented in these chapters illustrates how physical problems are analytically formulated with the aid of continuum equations. Chapter 9 deals with linear viscoelastic constitutive models and their application to simple problems of solid mechanics. Since a continuum mechanics course is mostly offered by solid mechanics programs, the coverage in this book is slightly more favorable, in terms of the amount and type of material covered, to solid and structural mechanics.

The book is written keeping the undergraduate seniors and first-year graduate students of engineering in mind. Therefore, it is most suitable as a textbook for adoption for a first course on continuum mechanics or elasticity. The book also serves as an excellent precursor to courses on viscoelasticity, plasticity, nonlinear elasticity, and nonlinear continuum mechanics.

The book contains so many mathematical equations that it is hardly possible not to have typographical and other kinds of errors. I wish to thank in advance those readers who are willing to draw the author's attention to typos and errors, using the following e-mail address: jnreddy@tamu.edu.

J. N. Reddy
College Station, Texas

1 Introduction

I can live with doubt and uncertainty and not knowing. I think it is much more interesting to live not knowing than to have answers that might be wrong.

Richard Feynmann

What we need is not the will to believe but the will to find out.

Bertrand Russell

1.1 Continuum Mechanics

The subject of *mechanics* deals with the study of motion and forces in solids, liquids, and gases and the deformation or flow of these materials. In such a study, we make the simplifying assumption, for analysis purposes, that the matter is distributed continuously, without gaps or empty spaces (i.e., we disregard the molecular structure of matter). Such a hypothetical continuous matter is termed a *continuum*. In essence, in a continuum all quantities such as the density, displacements, velocities, stresses, and so on vary continuously so that their spatial derivatives exist and are continuous. The continuum assumption allows us to shrink an arbitrary volume of material to a point, in much the same way as we take the limit in defining a derivative, so that we can define quantities of interest at a point. For example, density (mass per unit volume) of a material at a point is defined as the ratio of the mass Δm of the material to a small volume ΔV surrounding the point in the limit that ΔV becomes a value ϵ^3, where ϵ is small compared with the mean distance between molecules

$$\rho = \lim_{\Delta V \to \epsilon^3} \frac{\Delta m}{\Delta V}. \tag{1.1.1}$$

In fact, we take the limit $\epsilon \to 0$. A mathematical study of mechanics of such an idealized continuum is called *continuum mechanics*.

The primary objectives of this book are (1) to study the conservation principles in mechanics of continua and formulate the equations that describe the motion and mechanical behavior of materials and (2) to present the applications of these equations to simple problems associated with flows of fluids, conduction of heat, and deformation of solid bodies. While the first of these objectives is an important

topic, the reason for the formulation of the equations is to gain a quantitative understanding of the behavior of an engineering system. This quantitative understanding
is useful in the design and manufacture of better products. Typical examples of engineering problems, which are sufficiently simple to cover in this book, are described
below. At this stage of discussion, it is sufficient to rely on the reader's intuitive
understanding of concepts or background from basic courses in fluid mechanics,
heat transfer, and mechanics of materials about the meaning of the stress and strain
and what constitutes viscosity, conductivity, modulus, and so on used in the example problems below. More precise definitions of these terms will be apparent in the
chapters that follow.

PROBLEM 1 (SOLID MECHANICS)

We wish to design a diving board of given length L (which must enable the swimmer
to gain enough momentum for the swimming exercise), fixed at one end and free at
the other end (see Figure 1.1.1). The board is initially straight and horizontal and
of uniform cross section. The design process consists of selecting the material (with
Young's modulus E) and cross-sectional dimensions b and h such that the board carries the (moving) weight W of the swimmer. The design criteria are that the stresses
developed do not exceed the allowable stress values and the deflection of the free
end does not exceed a prespecified value δ. A preliminary design of such systems
is often based on mechanics of materials equations. The final design involves the
use of more sophisticated equations, such as the three-dimensional (3D) elasticity
equations. The equations of elementary beam theory may be used to find a relation
between the deflection δ of the free end in terms of the length L, cross-sectional
dimensions b and h, Young's modulus E, and weight W [see Eq. (7.6.10)]:

$$\delta = \frac{4WL^3}{Ebh^3}.$$ (1.1.2)

Given δ (allowable deflection) and load W (maximum possible weight of a swimmer), one can select the material (Young's modulus, E) and dimensions L, b, and
h (which must be restricted to the standard sizes fabricated by a manufacturer).
In addition to the deflection criterion, one must also check if the board devel-
ops stresses that exceed the allowable stresses of the material selected. Analysis
of pertinent equations provide the designer with alternatives to select the material
and dimensions of the board so as to have a cost-effective but functionally reliable
structure.

PROBLEM 2 (FLUID MECHANICS)

We wish to measure the viscosity μ of a lubricating oil used in rotating machinery to
prevent the damage of the parts in contact. Viscosity, like Young's modulus of solid
materials, is a material property that is useful in the calculation of shear stresses

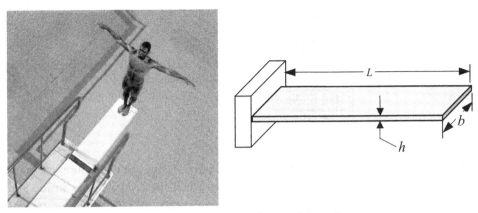

Figure 1.1.1. A diving board fixed at left end and free at right end.

developed between a fluid and solid body. A capillary tube is used to determine the viscosity of a fluid via the formula

$$\mu = \frac{\pi d^4}{128L} \frac{P_1 - P_2}{Q},$$ (1.1.3)

where d is the internal diameter and L is the length of the capillary tube, P_1 and P_2 are the pressures at the two ends of the tube (oil flows from one end to the other, as shown in Figure 1.1.2), and Q is the volume rate of flow at which the oil is discharged from the tube. Equation (1.1.3) is derived, as we shall see later in this book [see Eq. (8.2.25)], using the principles of continuum mechanics.

PROBLEM 3 (HEAT TRANSFER)

We wish to determine the heat loss through the wall of a furnace. The wall typically consists of layers of brick, cement mortar, and cinder block (see Figure 1.1.3). Each of these materials provides varying degree of thermal resistance. The Fourier heat conduction law (see Section 8.3.1)

$$q = -k\frac{dT}{dx}$$ (1.1.4)

provides a relation between the heat flux q (heat flow per unit area) and gradient of temperature T. Here k denotes thermal conductivity ($1/k$ is the thermal resistance) of the material. The negative sign in Eq. (1.1.4) indicates that heat flows from

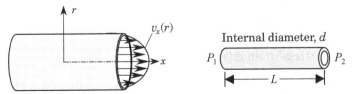

Figure 1.1.2. Measurement of viscosity of a fluid using capillary tube.

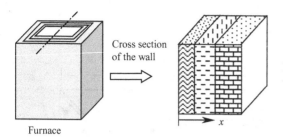

Cross section
of the wall

Furnace

x

Figure 1.1.3. Heat transfer through a composite wall of a furnace.

high temperature region to low temperature region. Using the continuum mechanics equations, one can determine the heat loss when the temperatures inside and outside of the building are known. A building designer can select the materials as well as thicknesses of various components of the wall to reduce the heat loss (while ensuring necessary structural strength – a structural analysis aspect).

The previous examples provide some indication of the need for studying the mechanical response of materials under the influence of external loads. The response of a material is consistent with the laws of physics and the constitutive behavior of the material. This book has the objective of describing the physical principles and deriving the equations governing the stress and deformation of continuous materials and then solving some simple problems from various branches of engineering to illustrate the applications of the principles discussed and equations derived.

1.2 A Look Forward

The primary objective of this book is twofold: (1) use the physical principles to derive the equations that govern the motion and thermomechanical response of materials and (2) apply these equations for the solution of specific problems of linearized elasticity, heat transfer, and fluid mechanics. The governing equations for the study of deformation and stress of a continuous material are nothing but an analytical representation of the global laws of conservation of mass, momenta, and energy and the constitutive response of the continuum. They are applicable to all materials that are treated as a continuum. Tailoring these equations to particular problems and solving them constitutes the bulk of engineering analysis and design.

The study of motion and deformation of a continuum (or a "body" consisting of continuously distributed material) can be broadly classified into four basic categories:

(1) Kinematics (strain-displacement equations)
(2) Kinetics (conservation of momenta)
(3) Thermodynamics (first and second laws of thermodynamics)
(4) Constitutive equations (stress-strain relations)

Kinematics is a study of the geometric changes or deformation in a continuum, without the consideration of forces causing the deformation. *Kinetics* is the study of the static or dynamic equilibrium of forces and moments acting on a continuum,

Table 1.2.1. *The major four topics of study, physical principles and axioms used, resulting governing equations, and variables involved*

Topic of study	Physical principle	Resulting equations	Variables involved
1. Kinematics	None – based on geometric changes	Strain–displacement relations	Displacements and strains
		Strain rate–velocity relations	Velocities and strain rates
2. Kinetics	Conservation of linear momentum	Equations of motion	Stresses, velocities, and body forces
	Conservation of angular momentum	Symmetry of stress tensor	Stresses
3. Thermodynamics	First law	Energy equation	Temperature, heat flux, stresses, heat generation, and velocities
	Second law	Clausius–Duhem inequality	Temperature, heat flux, and entropy
4. Constitutive equations (not all relations are listed)	Constitutive axioms	Hooke's law	Stresses, strains, heat flux and temperature
		Newtonian fluids	Stresses, pressure, velocities
		Fourier's law	Heat flux and temperature
		Equations of state	Density, pressure, temperature

using the principles of conservation of momenta. This study leads to equations of motion as well as the symmetry of stress tensor in the absence of body couples. *Thermodynamic principles* are concerned with the conservation of energy and relations among heat, mechanical work, and thermodynamic properties of the continuum. *Constitutive equations* describe thermomechanical behavior of the material of the continuum, and they relate the dependent variables introduced in the kinetic description to those introduced in the kinematic and thermodynamic descriptions. Table 1.2.1 provides a brief summary of the relationship between physical principles and governing equations, and physical entities involved in the equations.

1.3 Summary

In this chapter, the concept of a continuous medium is discussed, and the major objectives of the present study, namely, to use the physical principles to derive the equations governing a continuous medium and to present application of the equations in the solution of specific problems of linearized elasticity, heat transfer, and fluid mechanics, are presented. The study of physical principles is broadly divided into four topics, as outlined in Table 1.2.1. These four topics form the subject of Chapters 3 through 6, respectively. Mathematical formulation of the governing equations of a continuous medium necessarily requires the use of vectors and tensors, objects that facilitate invariant analytical formulation of the natural laws. Therefore, it is useful to study certain operational properties of vectors and tensors first. Chapter 2 is dedicated for this purpose.

While the present book is self-contained for an introduction to continuum mechanics, there are other books that may provide an advanced treatment of the subject. Interested readers may consult the titles listed in the reference list at the end of the book.

PROBLEMS

1.1 Newton's second law can be expressed as

$$\mathbf{F} = m\mathbf{a}, \tag{1}$$

where $\mathbf{F}$ is the net force acting on the body, m mass of the body, and $\mathbf{a}$ the acceleration of the body in the direction of the net force. Use Eq. (1) to determine the governing equation of a free-falling body. Consider only the forces due to gravity and the air resistance, which is assumed to be linearly proportional to the velocity of the falling body.

1.2 Consider steady-state heat transfer through a cylindrical bar of nonuniform cross section. The bar is subject to a known temperature T_0 (°C) at the left end and exposed, both on the surface and at the right end, to a medium (such as cooling fluid or air) at temperature T_∞. Assume that temperature is uniform at any section of the bar, $T = T(x)$. Use the principle of conservation of energy (which requires that the rate of change (increase) of internal energy is equal to the sum of heat gained by conduction, convection, and internal heat generation) to a typical element of the bar (see Figure P1.2) to derive the governing equations of the problem.

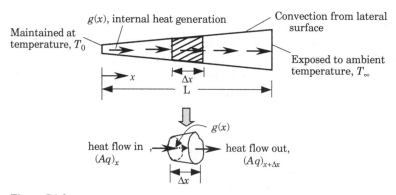

Figure P1.2.

1.3 The Euler–Bernoulli hypothesis concerning the kinematics of bending deformation of a beam assumes that straight lines perpendicular to the beam axis before deformation remain (1) straight, (2) perpendicular to the tangent line to the beam

axis, and (3) inextensible during deformation. These assumptions lead to the following displacement field:

$$u_1 = -z\frac{dw}{dx}, \quad u_2 = 0, \quad u_3 = w(x), \tag{1}$$

where (u_1, u_2, u_3) are the displacements of a point (x, y, z) along the x, y, and z
coordinates, respectively, and w is the vertical displacement of the beam at point
$(x, 0, 0)$. Suppose that the beam is subjected to distributed transverse load $q(x)$. Determine the governing equation by summing the forces and moments on an element
of the beam (see Figure P1.3). Note that the sign convention for the moment and
shear force are based on the definitions

$$V = \int_A \sigma_{xz} \, dA, \quad M = \int_A z\sigma_{xx} \, dA,$$

and it may not agree with the sign convention used in some mechanics of materials
books.

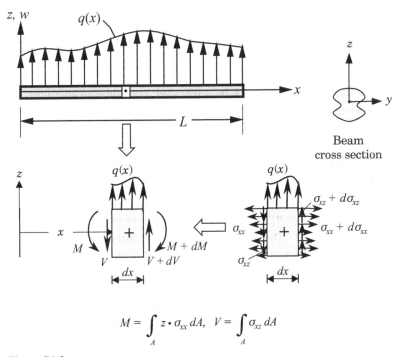

Figure P1.3.

1.4 A cylindrical storage tank of diameter D contains a liquid column of height
$h(x, t)$. Liquid is supplied to the tank at a rate of q_i (m³/day) and drained at a rate
of q_0 (m³/day). Use the principle of conservation of mass to obtain the equation
governing the flow problem.

2 Vectors and Tensors

A mathematical theory is not to be considered complete until you have made it so clear that you can explain it to the first man whom you meet on the street.

David Hilbert

2.1 Background and Overview

In the mathematical description of equations governing a continuous medium, we derive relations between various quantities that characterize the stress and deformation of the continuum by means of the laws of nature (such as Newton's laws, conservation of energy, and so on). As a means of expressing a natural law, a coordinate system in a chosen frame of reference is often introduced. The mathematical form of the law thus depends on the chosen coordinate system and may appear different in another type of coordinate system. The laws of nature, however, should be independent of the choice of a coordinate system, and we may seek to represent the law in a manner independent of a particular coordinate system. A way of doing this is provided by vector and tensor analysis. When vector notation is used, a particular coordinate system need not be introduced. Consequently, the use of vector notation in formulating natural laws leaves them *invariant* to coordinate transformations. A study of physical phenomena by means of vector equations often leads to a deeper understanding of the problem in addition to bringing simplicity and versatility into the analysis.

In basic engineering courses, the term *vector* is used often to imply a *physical* vector that has 'magnitude and direction and satisfy the parallelogram law of addition.' In mathematics, vectors are more abstract objects than physical vectors. Like physical vectors, *tensors* are more general objects that are endowed with a magnitude and multiple direction(s) and satisfy rules of tensor addition and scalar multiplication. In fact, physical vectors are often termed the *first-order tensors*. As will be shown shortly, the specification of a stress component (i.e., force per unit area) requires a magnitude and two directions – one normal to the plane on which the stress component is measured and the other is its direction – to specify it uniquely.

This chapter is dedicated to a review of algebra and calculus of physical vectors and tensors. Those who are familiar with the material covered in any of the sections may skip them and go to the next section or Chapter 3.

2.2 Vector Algebra

In this section, we present a review of the formal definition of a geometric (or physical) vector, discuss various products of vectors and physically interpret them, introduce index notation to simplify representations of vectors in terms of their components as well as vector operations, and develop transformation equations among the components of a vector expressed in two different coordinate systems. Many of these concepts, with the exception of the index notation, may be familiar to most students of engineering, physics, and mathematics and may be skipped.

2.2.1 Definition of a Vector

The quantities encountered in analytical description of physical phenomena may be classified into two groups according to the information needed to specify them completely: scalars and nonscalars. The scalars are given by a single number. Nonscalars have not only a magnitude specified but also additional information, such as direction. Nonscalars that obey certain rules (such as the parallelogram law of addition) are called *vectors*. Not all nonscalar quantities are vectors (e.g., a finite rotation is not a vector).

A physical vector is often shown as a directed line segment with an arrow head at the end of the line. The length of the line represents the magnitude of the vector and the arrow indicates the direction. In written or typed material, it is customary to place an arrow over the letter denoting the vector, such as $\vec{A}$. In printed material, the vector letter is commonly denoted by a boldface letter $\mathbf{A}$, such as used in this book. The magnitude of the vector $\mathbf{A}$ is denoted by $|\mathbf{A}|$, $\|\mathbf{A}\|$, or A. Magnitude of a vector is a scalar.

A vector of unit length is called a *unit vector*. The unit vector along $\mathbf{A}$ may be defined as follows:

$$\hat{\mathbf{e}}_A = \frac{\mathbf{A}}{A}. \tag{2.2.1}$$

We may now write

$$\mathbf{A} = A\,\hat{\mathbf{e}}_A. \tag{2.2.2}$$

Thus *any vector may be represented as a product of its magnitude and a unit vector along the vector*. A unit vector is used to designate direction. It does not have any physical dimensions. We denote a unit vector by a "hat" (caret) above the boldface letter, $\hat{\mathbf{e}}$. A vector of zero magnitude is called a *zero vector* or a *null vector*. All null vectors are considered equal to each other without consideration as to direction. Note that a light face zero, 0, is a scalar and boldface zero, $\mathbf{0}$, is the zero vector.

2.2.1.1 Vector Addition

Let **A**, **B**, and **C** be any vectors. Then there exists a vector **A** + **B**, called sum of **A** and **B**, such that

(1) $\mathbf{A} + \mathbf{B} = \mathbf{B} + \mathbf{A}$ (commutative).
(2) $(\mathbf{A} + \mathbf{B}) + \mathbf{C} = \mathbf{A} + (\mathbf{B} + \mathbf{C})$ (associative).
(3) there exists a unique vector, **0**, independent of **A** such that
     $\mathbf{A} + \mathbf{0} = \mathbf{A}$ (existence of zero vector).   (2.2.3)
(4) to every vector **A** there exists a unique vector $-\mathbf{A}$
     (that depends on **A**) such that
     $\mathbf{A} + (-\mathbf{A}) = \mathbf{0}$ (existence of negative vector).

The negative vector $-\mathbf{A}$ has the same magnitude as **A** but has the opposite *sense*. Subtraction of vectors is carried out along the same lines. To form the difference $\mathbf{A} - \mathbf{B}$, we write $\mathbf{A} + (-\mathbf{B})$ and subtraction reduces to the operation of addition.

2.2.1.2 Multiplication of Vector by Scalar

Let **A** and **B** be vectors and α and β be real numbers (scalars). To every vector **A** and every real number α, there corresponds a unique vector $\alpha\mathbf{A}$ such that

(1) $\alpha(\beta\mathbf{A}) = (\alpha\beta)\mathbf{A}$ (associative).
(2) $(\alpha + \beta)\mathbf{A} = \alpha\mathbf{A} + \beta\mathbf{A}$ (distributive scalar addition).
(3) $\alpha(\mathbf{A} + \mathbf{B}) = \alpha\mathbf{A} + \alpha\mathbf{B}$ (distributive vector addition).   (2.2.4)
(4) $1 \cdot \mathbf{A} = \mathbf{A} \cdot 1 = \mathbf{A}, \quad 0 \cdot \mathbf{A} = \mathbf{0}$.

Equations (2.2.3) and (2.2.4) clearly show that the laws that govern addition, subtraction, and scalar multiplication of vectors are identical with those governing the operations of scalar algebra.

Two vectors **A** and **B** are equal if their magnitudes are equal, $|\mathbf{A}| = |\mathbf{B}|$, *and* if their directions are equal. Consequently, a vector is not changed if it is moved parallel to itself. This means that the position of a vector in space, that is, the point from which the line segment is drawn (or the end without arrowhead), may be chosen arbitrarily. In certain applications, however, the actual point of location of a vector may be important, for instance, a moment or a force acting on a body. A vector associated with a given point is known as a *localized* or *bound vector*. A finite rotation of a rigid body is not a vector although infinitesimal rotations are. That vectors can be represented graphically is an *incidental* rather than a fundamental feature of the vector concept.

2.2.1.3 Linear Independence of Vectors

The concepts of collinear and coplanar vectors can be stated in algebraic terms. A set of n vectors is said to be *linearly dependent* if a set of n numbers $\beta_1, \beta_2, \ldots, \beta_n$ can be found such that

$$\beta_1\mathbf{A}_1 + \beta_2\mathbf{A}_2 + \cdots + \beta_n\mathbf{A}_n = \mathbf{0},$$ (2.2.5)

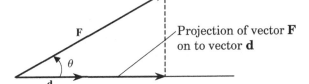

Figure 2.2.1. Representation of work.

Projection of vector **F** on to vector **d**

where $\beta_1, \beta_2, \ldots, \beta_n$ cannot all be zero. If this expression cannot be satisfied, the vectors are said to be *linearly independent*. If two vectors are linearly dependent, then they are *collinear*. If three vectors are linearly dependent, then they are *coplanar*. Four or more vectors in three-dimensional space are always linearly dependent.

2.2.2 Scalar and Vector Products

Besides addition and subtraction of vectors, and multiplication of a vector by a scalar, we also encounter product of two vectors. There are several ways the product of two vectors can be defined. We consider first the so-called scalar product.

2.2.2.1 Scalar Product

When a force **F** acts on a mass point and moves through a displacement vector **d**, the work done by the force vector is defined by the *projection* of the force in the direction of the displacement, as shown in Figure 2.2.1, times the magnitude of the displacement. Such an operation may be defined for any two vectors. Since the result of the product is a scalar, it is called the *scalar product*. We denote this product as $\mathbf{F} \cdot \mathbf{d} \equiv (\mathbf{F}, \mathbf{d})$ and it is defined as follows:

$$\mathbf{F} \cdot \mathbf{d} \equiv (\mathbf{F}, \mathbf{d}) = Fd \cos\theta, \qquad 0 \leq \theta \leq \pi. \tag{2.2.6}$$

The scalar product is also known as the *dot product* or *inner product*.

A few simple results follow from the definition in Eq. (2.2.6):

1. Since $\mathbf{A} \cdot \mathbf{B} = \mathbf{B} \cdot \mathbf{A}$, the scalar product is commutative.
2. If the vectors **A** and **B** are perpendicular to each other, then $\mathbf{A} \cdot \mathbf{B} = AB\cos(\pi/2) = 0$. Conversely, if $\mathbf{A} \cdot \mathbf{B} = 0$, then either **A** or **B** is zero or **A** is perpendicular, or *orthogonal*, to **B**.
3. If two vectors **A** and **B** are parallel and in the same direction, then $\mathbf{A} \cdot \mathbf{B} = AB\cos 0 = AB$, since $\cos 0 = 1$. Thus the scalar product of a vector with itself is equal to the square of its magnitude:

$$\mathbf{A} \cdot \mathbf{A} = AA = A^2. \tag{2.2.7}$$

4. The orthogonal projection of a vector **A** in any direction $\hat{\mathbf{e}}$ is given by $\mathbf{A} \cdot \hat{\mathbf{e}}$.
5. The scalar product follows the distributive law also:

$$\mathbf{A} \cdot (\mathbf{B} + \mathbf{C}) = (\mathbf{A} \cdot \mathbf{B}) + (\mathbf{A} \cdot \mathbf{C}). \tag{2.2.8}$$

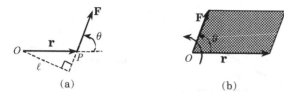

(a)   (b)

Figure 2.2.2. (a) Representation of a moment. (b) Direction of rotation.

2.2.2.2 Vector Product

To see the need for the *vector product*, consider the concept of the *moment* due to a force. Let us describe the moment about a point O of a force $\mathbf{F}$ acting at a point P, such as shown in Figure 2.2.2(a). By definition, the magnitude of the moment is given by

$$M = F\ell, \quad F = |\mathbf{F}|, \tag{2.2.9}$$

where ℓ is the perpendicular distance from the point O to the force $\mathbf{F}$ (called lever arm). If $\mathbf{r}$ denotes the vector $\mathbf{OP}$ and θ the angle between $\mathbf{r}$ and $\mathbf{F}$ as shown in Figure 2.2.2(a) such that $0 \leq \theta \leq \pi$, we have $\ell = r \sin \theta$ and thus

$$M = Fr \sin \theta. \tag{2.2.10}$$

A direction can now be assigned to the moment. Drawing the vectors $\mathbf{F}$ and $\mathbf{r}$ from the common origin O, we note that the rotation due to $\mathbf{F}$ tends to bring $\mathbf{r}$ into $\mathbf{F}$, as can be seen from Figure 2.2.2(b). We now set up an axis of rotation perpendicular to the plane formed by $\mathbf{F}$ and $\mathbf{r}$. Along this axis of rotation we set up a preferred direction as that in which a right-handed screw would advance when turned in the direction of rotation due to the moment, as can be seen from Figure 2.2.3(a). Along this axis of rotation, we draw a unit vector $\hat{\mathbf{e}}_M$ and agree that it represents the direction of the moment $\mathbf{M}$. Thus we have

$$\mathbf{M} = Fr \sin \theta \, \hat{\mathbf{e}}_M = \mathbf{r} \times \mathbf{F}. \tag{2.2.11}$$

According to this expression, $\mathbf{M}$ may be looked upon as resulting from a special operation between the two vectors $\mathbf{F}$ and $\mathbf{r}$. It is thus the basis for defining a product between any two vectors. Since the result of such a product is a vector, it may be called the *vector product*.

The product of two vectors $\mathbf{A}$ and $\mathbf{B}$ is a vector $\mathbf{C}$ whose magnitude is equal to the product of the magnitude of $\mathbf{A}$ and $\mathbf{B}$ times the sine of the angle measured from

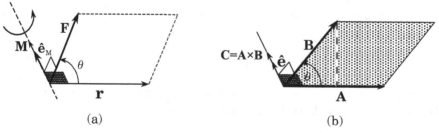

(a)   (b)

Figure 2.2.3. (a) Axis of rotation. (b) Representation of the vector product.

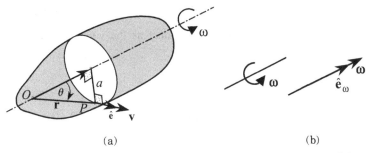

(a)   (b)

Figure 2.2.4. (a) Velocity at a point in a rotating rigid body. (b) Angular velocity as a vector.

$\mathbf{A}$ to $\mathbf{B}$ such that $0 \leq \theta \leq \pi$, and whose direction is specified by the condition that $\mathbf{C}$ be perpendicular to the plane of the vectors $\mathbf{A}$ and $\mathbf{B}$ and points in the direction in which a right-handed screw advances when turned so as to bring $\mathbf{A}$ into $\mathbf{B}$, as shown in Figure 2.2.3(b). The vector product is usually denoted by

$$\mathbf{C} = \mathbf{A} \times \mathbf{B} = AB \sin(\mathbf{A}, \mathbf{B}) \, \hat{\mathbf{e}} = AB \sin\theta \, \hat{\mathbf{e}}, \qquad (2.2.12)$$

where $\sin(\mathbf{A}, \mathbf{B})$ denotes the sine of the angle between vectors $\mathbf{A}$ and $\mathbf{B}$. This product is called the *cross product, skew product*, and also *outer product*, as well as the vector product. When $\mathbf{A} = a \, \hat{\mathbf{e}}_A$ and $\mathbf{B} = b \, \hat{\mathbf{e}}_B$ are the vectors representing the sides of a parallelogram, with a and b denoting the lengths of the sides, then the vector product $\mathbf{A} \times \mathbf{B}$ represents the area of the parallelogram, $AB \sin\theta$. The unit vector $\hat{\mathbf{e}} = \hat{\mathbf{e}}_A \times \hat{\mathbf{e}}_B$ denotes the normal to the plane area. Thus, an area can be represented as a vector (see Section 2.2.3 for additional discussion).

The description of the velocity of a point of a rotating rigid body is an important example of geometrical and physical applications of vectors. Suppose a rigid body is rotating with an angular velocity ω about an axis, and we wish to describe the velocity of some point P of the body, as shown in Figure 2.2.4(a). Let $\mathbf{v}$ denote the velocity at the point. Each point of the body describes a circle that lies in a plane perpendicular to the axis with its center on the axis. The radius of the circle, a, is the perpendicular distance from the axis to the point of interest. The magnitude of the velocity is equal to ωa. The direction of $\mathbf{v}$ is perpendicular to a and to the axis of rotation. We denote the direction of the velocity by the unit vector $\hat{\mathbf{e}}$. Thus we can write

$$\mathbf{v} = \omega a \, \hat{\mathbf{e}}. \qquad (2.2.13)$$

Let O be a reference point on the axis of revolution, and let $\mathbf{OP} = \mathbf{r}$. We then have $a = r\sin\theta$, so that

$$\mathbf{v} = \omega r \, \sin\theta \, \hat{\mathbf{e}}. \qquad (2.2.14)$$

The angular velocity is a vector since it has an assigned direction, magnitude, and obeys the parallelogram law of addition. We denote it by ω and represent its

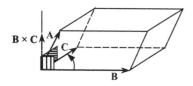

Figure 2.2.5. Scalar triple product as the volume of a parallelepiped.

direction in the sense of a right-handed screw, as shown in Figure 2.2.4(b). If we further let $\hat{\mathbf{e}}_r$ be a unit vector in the direction of $\mathbf{r}$, we see that

$$\hat{\mathbf{e}}_\omega \times \hat{\mathbf{e}}_r = \hat{\mathbf{e}} \sin\theta. \tag{2.2.15}$$

With these relations, we have

$$\mathbf{v} = \omega \times \mathbf{r}. \tag{2.2.16}$$

Thus the velocity of a point of a rigid body rotating about an axis is given by the vector product of ω and a position vector $\mathbf{r}$ drawn from any reference point on the axis of revolution.

From the definition of vector (cross) product, a few simple results follow:

1. The products $\mathbf{A} \times \mathbf{B}$ and $\mathbf{B} \times \mathbf{A}$ are not equal. In fact, we have

$$\mathbf{A} \times \mathbf{B} \equiv -\mathbf{B} \times \mathbf{A}. \tag{2.2.17}$$

Thus the vector product *does not commute*. We must therefore preserve the order of the vectors when vector products are involved.
2. If two vectors $\mathbf{A}$ and $\mathbf{B}$ are parallel to each other, then $\theta = \pi, 0$ and $\sin\theta = 0$. Thus

$$\mathbf{A} \times \mathbf{B} = \mathbf{0}.$$

Conversely, if $\mathbf{A} \times \mathbf{B} = \mathbf{0}$, then either $\mathbf{A}$ or $\mathbf{B}$ is zero, or they are parallel vectors. It follows that the vector product of a vector with itself is zero; that is, $\mathbf{A} \times \mathbf{A} = \mathbf{0}$.
3. The distributive law still holds, but the order of the factors must be maintained:

$$(\mathbf{A} + \mathbf{B}) \times \mathbf{C} = (\mathbf{A} \times \mathbf{C}) + (\mathbf{B} \times \mathbf{C}). \tag{2.2.18}$$

2.2.2.3 Triple Products of Vectors

Now consider the various products of three vectors:

$$\mathbf{A}(\mathbf{B} \cdot \mathbf{C}), \quad \mathbf{A} \cdot (\mathbf{B} \times \mathbf{C}), \quad \mathbf{A} \times (\mathbf{B} \times \mathbf{C}). \tag{2.2.19}$$

The product $\mathbf{A}(\mathbf{B} \cdot \mathbf{C})$ is merely a multiplication of the vector $\mathbf{A}$ by the scalar $\mathbf{B} \cdot \mathbf{C}$. The product $\mathbf{A} \cdot (\mathbf{B} \times \mathbf{C})$ is a scalar and it is termed *the scalar triple product*. It can be seen that the product $\mathbf{A} \cdot (\mathbf{B} \times \mathbf{C})$, except for the algebraic sign, is the volume of the parallelepiped formed by the vectors $\mathbf{A}, \mathbf{B}$, and $\mathbf{C}$, as shown in Figure 2.2.5.

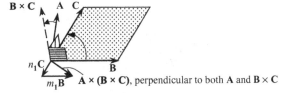

Figure 2.2.6. The vector triple product.

We also note the following properties:

1. The dot and cross can be interchanged without changing the value:

$$\mathbf{A} \cdot \mathbf{B} \times \mathbf{C} = \mathbf{A} \times \mathbf{B} \cdot \mathbf{C} \equiv [\mathbf{ABC}].\tag{2.2.20}$$

2. A cyclical permutation of the order of the vectors leaves the result unchanged:

$$\mathbf{A} \cdot \mathbf{B} \times \mathbf{C} = \mathbf{C} \cdot \mathbf{A} \times \mathbf{B} = \mathbf{B} \cdot \mathbf{C} \times \mathbf{A} \equiv [\mathbf{ABC}].\tag{2.2.21}$$

3. If the cyclic order is changed, the sign changes:

$$\mathbf{A} \cdot \mathbf{B} \times \mathbf{C} = -\mathbf{A} \cdot \mathbf{C} \times \mathbf{B} = -\mathbf{C} \cdot \mathbf{B} \times \mathbf{A} = -\mathbf{B} \cdot \mathbf{A} \times \mathbf{C}.\tag{2.2.22}$$

4. A necessary and sufficient condition for any three vectors, $\mathbf{A}, \mathbf{B}, \mathbf{C}$ to be coplanar is that $\mathbf{A} \cdot (\mathbf{B} \times \mathbf{C}) = 0$. Note also that the scalar triple product is zero when any two vectors are the same.

The *vector triple product* $\mathbf{A} \times (\mathbf{B} \times \mathbf{C})$ is a vector normal to the plane formed by $\mathbf{A}$ and $(\mathbf{B} \times \mathbf{C})$. The vector $(\mathbf{B} \times \mathbf{C})$, however, is perpendicular to the plane formed by $\mathbf{B}$ and $\mathbf{C}$. This means that $\mathbf{A} \times (\mathbf{B} \times \mathbf{C})$ lies in the plane formed by $\mathbf{B}$ and $\mathbf{C}$ and is perpendicular to $\mathbf{A}$, as shown in Figure 2.2.6. Thus $\mathbf{A} \times (\mathbf{B} \times \mathbf{C})$ can be expressed as a linear combination of $\mathbf{B}$ and $\mathbf{C}$:

$$\mathbf{A} \times (\mathbf{B} \times \mathbf{C}) = m_1\mathbf{B} + n_1\mathbf{C}.\tag{2.2.23}$$

Likewise, we would find that

$$(\mathbf{A} \times \mathbf{B}) \times \mathbf{C} = m_2\mathbf{A} + n_2\mathbf{B}.\tag{2.2.24}$$

Thus, the parentheses *cannot* be interchanged or removed. It can be shown that

$$m_1 = \mathbf{A} \cdot \mathbf{C}, \qquad n_1 = -\mathbf{A} \cdot \mathbf{B},$$

and hence that

$$\mathbf{A} \times (\mathbf{B} \times \mathbf{C}) = (\mathbf{A} \cdot \mathbf{C})\mathbf{B} - (\mathbf{A} \cdot \mathbf{B})\mathbf{C}.\tag{2.2.25}$$

The example below illustrates the use of the vector triple product.

EXAMPLE 2.2.1: Let $\mathbf{A}$ and $\mathbf{B}$ be any two vectors in space. Express vector $\mathbf{A}$ in terms of its components along (i.e., parallel) and perpendicular to vector $\mathbf{B}$.

SOLUTION: The component of $\mathbf{A}$ along $\mathbf{B}$ is given by $(\mathbf{A} \cdot \hat{\mathbf{e}}_B)$, where $\hat{\mathbf{e}}_B = \mathbf{B}/B$ is the unit vector in the direction of $\mathbf{B}$. The component of $\mathbf{A}$ perpendicular to $\mathbf{B}$

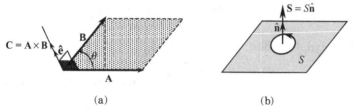

Figure 2.2.7. (a) Plane area as a vector. (b) Unit normal vector and sense of travel.

and in the plane of $\mathbf{A}$ and $\mathbf{B}$ is given by the vector triple product $\hat{\mathbf{e}}_B \times (\mathbf{A} \times \hat{\mathbf{e}}_B)$. Thus,

$$\mathbf{A} = (\mathbf{A} \cdot \hat{\mathbf{e}}_B)\hat{\mathbf{e}}_B + \hat{\mathbf{e}}_B \times (\mathbf{A} \times \hat{\mathbf{e}}_B). \qquad (2.2.26)$$

Alternatively, using Eq. (2.2.25) with $\mathbf{A} = \mathbf{C} = \hat{\mathbf{e}}_B$ and $\mathbf{B} = \mathbf{A}$, we obtain

$$\hat{\mathbf{e}}_B \times (\mathbf{A} \times \hat{\mathbf{e}}_B) = \mathbf{A} - (\mathbf{A} \cdot \hat{\mathbf{e}}_B)\hat{\mathbf{e}}_B$$

or

$$\mathbf{A} = (\mathbf{A} \cdot \hat{\mathbf{e}}_B)\hat{\mathbf{e}}_B + \hat{\mathbf{e}}_B \times (\mathbf{A} \times \hat{\mathbf{e}}_B).$$

2.2.3 Plane Area as a Vector

The magnitude of the vector $\mathbf{C} = \mathbf{A} \times \mathbf{B}$ is equal to the area of the parallelogram formed by the vectors $\mathbf{A}$ and $\mathbf{B}$, as shown in Figure 2.2.7(a). In fact, the vector $\mathbf{C}$ may be considered to represent *both* the magnitude and the direction of the product $\mathbf{A}$ and $\mathbf{B}$. Thus, a plane area may be looked upon as possessing a direction in addition to a magnitude, the directional character arising out of the need to specify an orientation of the plane in space.

It is customary to denote the direction of a plane area by means of a unit vector drawn normal to that plane. To fix the direction of the normal, we assign a *sense of travel* along the contour of the boundary of the plane area in question. The direction of the normal is taken by convention as that in which a right-handed screw advances as it is rotated according to the sense of travel along the boundary curve or contour, as shown in Figure 2.2.7(b). Let the unit normal vector be given by $\hat{\mathbf{n}}$. Then the area can be denoted by $\mathbf{S} = S\hat{\mathbf{n}}$.

Representation of a plane as a vector has many uses. The vector can be used to determine the area of an inclined plane in terms of its projected area, as illustrated in the next example.

EXAMPLE 2.2.2:

(1) Determine the plane area of the surface obtained by cutting a cylinder of cross-sectional area S_0 with an inclined plane whose normal is $\hat{\mathbf{n}}$, as shown in Fig 2.2.8(a).

(2) Consider a cube (or a prism) cut by an inclined plane whose normal is $\hat{\mathbf{n}}$, as shown in Figure 2.2.8(b). Express the areas of the sides of the resulting tetrahedron in terms of the area S of the inclined surface.

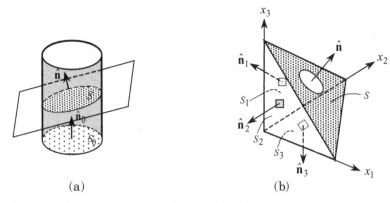

Figure 2.2.8. Vector representation of an inclined plane area.

SOLUTION:

(1) Let the plane area of the inclined surface be S, as shown in Fig 2.2.8(a). First, we express the areas as vectors

$$\mathbf{S}_0 = S_0\,\hat{\mathbf{n}}_0 \quad \text{and} \quad \mathbf{S} = S\hat{\mathbf{n}}. \tag{2.2.27}$$

Since S_0 is the projection of $\mathbf{S}$ along $\hat{\mathbf{n}}_0$ (if the angle between $\hat{\mathbf{n}}$ and $\hat{\mathbf{n}}_0$ is acute; otherwise the negative of it),

$$S_0 = \mathbf{S} \cdot \hat{\mathbf{n}}_0 = S\hat{\mathbf{n}} \cdot \hat{\mathbf{n}}_0. \tag{2.2.28}$$

The scalar product $\hat{\mathbf{n}} \cdot \hat{\mathbf{n}}_0$ is the cosine of the angle between the two unit normal vectors.

(2) For reference purposes we label the sides of the cube by 1, 2, and 3 and the normals and surface areas by $(\hat{\mathbf{n}}_1, S_1)$, $(\hat{\mathbf{n}}_2, S_2)$, and $(\hat{\mathbf{n}}_3, S_3)$, respectively (i.e., S_i is the surface area of the plane perpendicular to the ith line or $\hat{\mathbf{n}}_i$ vector), as shown in Figure 2.2.8(b). Then we have

$$S_1 = S\hat{\mathbf{n}} \cdot \hat{\mathbf{n}}_1, \quad S_2 = S\hat{\mathbf{n}} \cdot \hat{\mathbf{n}}_2, \quad S_3 = S\hat{\mathbf{n}} \cdot \hat{\mathbf{n}}_3. \tag{2.2.29}$$

2.2.4 Components of a Vector

So far we have considered a geometrical description of a vector. We now embark on an analytical description based on the notion of its components of a vector. In following discussion, we shall consider a three-dimensional space, and the extensions to n dimensions will be evident. In a three-dimensional space, a set of no more than three linearly independent vectors can be found. Let us choose any set and denote it as $\mathbf{e}_1, \mathbf{e}_2, \mathbf{e}_3$. This set is called a *basis*. We can represent any vector in three-dimensional space as a linear combination of the basis vectors

$$\mathbf{A} = A_1\mathbf{e}_1 + A_2\mathbf{e}_2 + A_3\mathbf{e}_3. \tag{2.2.30}$$

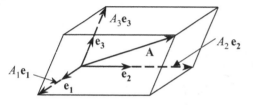

Figure 2.2.9. Components of a vector.

The vectors $A_1\mathbf{e}_1$, $A_2\mathbf{e}_2$, and $A_3\mathbf{e}_3$ are called the *vector components* of $\mathbf{A}$, and A_1, A_2, and A_3 are called *scalar components* of $\mathbf{A}$ associated with the basis $(\mathbf{e}_1, \mathbf{e}_2, \mathbf{e}_3)$, as indicated in Figure 2.2.9.

2.2.5 Summation Convention

The equations governing a continuous medium contains, especially in three dimensions, long expressions with many additive terms. Often these terms have similar structure because they represent components of a tensor. For example, consider the component form of vector $\mathbf{A}$:

$$\mathbf{A} = A_1\mathbf{e}_1 + A_2\mathbf{e}_2 + A_3\mathbf{e}_3, \tag{2.2.31}$$

which can be abbreviated as

$$\mathbf{A} = \sum_{i=1}^{3} A_i\mathbf{e}_i, \quad \text{or} \quad \mathbf{A} = \sum_{j=1}^{3} A_j\mathbf{e}_j. \tag{2.2.32}$$

The summation index i or j is arbitrary as long as the same index is used for both A and $\hat{\mathbf{e}}$. The expression can be further shortened by omitting the summation sign and having the understanding that a repeated index means summation over all values of that index. Thus, the three-term expression $A_1\mathbf{e}_1 + A_2\mathbf{e}_2 + A_3\mathbf{e}_3$ can be simply written as

$$\mathbf{A} = A_i\mathbf{e}_i. \tag{2.2.33}$$

This notation is called the *summation convention*.

2.2.5.1 Dummy Index
The repeated index is called a *dummy index* because it can be replaced by *any other symbol that has not already been used* in that expression. Thus, the expression in Eq. (2.2.33) can also be written as

$$\mathbf{A} = A_i\mathbf{e}_i = A_j\mathbf{e}_j = A_m\mathbf{e}_m, \tag{2.2.34}$$

and so on. As a rule, no index must appear more than twice in an expression. For example, $A_i B_i C_i$ is not a valid expression because the index i appears more than twice. Other examples of dummy indices are

$$F_i = A_i B_j C_j, \quad G_k = H_k(2 - 3A_i B_i) + P_j Q_j F_k. \tag{2.2.35}$$

The first equation above expresses three equations when the range of i and j is 1 to 3. We have

$$F_1 = A_1(B_1 C_1 + B_2 C_2 + B_3 C_3),$$

$$F_2 = A_2(B_1 C_1 + B_2 C_2 + B_3 C_3),$$

$$F_3 = A_3(B_1 C_1 + B_2 C_2 + B_3 C_3).$$

This amply illustrates the usefulness of the summation convention in shortening long and multiple expressions into a single expression.

2.2.5.2 Free Index

A *free index* is one that appears in every expression of an equation, except for expressions that contain real numbers (scalars) only. Index i in the equation $F_i = A_i B_j C_j$ and k in the equation $G_k = H_k(2 - 3A_i B_i) + P_j Q_j F_k$ above are free indices. Another example is

$$A_i = 2 + B_i + C_i + D_i + (F_j G_j - H_j P_j) E_i.$$

The above expression contains three equations ($i = 1, 2, 3$). The expressions $A_i = B_j C_k$, $A_i = B_j$, and $F_k = A_i B_j C_k$ do not make sense and should not arise because the indices on the two sides of the equal sign do not match.

2.2.5.3 Physical Components

For an orthonormal basis, the vectors $\mathbf{A}$ and $\mathbf{B}$ can be written as

$$\mathbf{A} = A_1 \hat{\mathbf{e}}_1 + A_2 \hat{\mathbf{e}}_2 + A_3 \hat{\mathbf{e}}_3 = A_i \hat{\mathbf{e}}_i,$$

$$\mathbf{B} = B_1 \hat{\mathbf{e}}_1 + B_2 \hat{\mathbf{e}}_2 + B_3 \hat{\mathbf{e}}_3 = B_i \hat{\mathbf{e}}_i,$$

where $(\hat{\mathbf{e}}_1, \hat{\mathbf{e}}_2, \hat{\mathbf{e}}_3)$ is the orthonormal basis and A_i and B_i are the corresponding *physical components* of the vector $\mathbf{A}$; that is, the components have the same physical dimensions or units as the vector.

2.2.5.4 Kronecker Delta and Permutation Symbols

It is convenient to introduce the Kronecker delta δ_{ij} and alternating symbol e_{ijk} because they allow simple representation of the dot product (or scalar product) and cross product, respectively, of orthonormal vectors in a right-handed basis system. We define the dot product $\hat{\mathbf{e}}_i \cdot \hat{\mathbf{e}}_j$ as

$$\hat{\mathbf{e}}_i \cdot \hat{\mathbf{e}}_j = \delta_{ij}, \tag{2.2.36}$$

where

$$\delta_{ij} = \begin{cases} 1, & \text{if } i = j \\ 0, & \text{if } i \neq j. \end{cases} \tag{2.2.37}$$

The Kronecker delta δ_{ij} modifies (or contracts) the subscripts in the coefficients of an expression in which it appears:

$$A_i \delta_{ij} = A_j, \quad A_i B_j \delta_{ij} = A_i B_i = A_j B_j, \quad \delta_{ij} \delta_{ik} = \delta_{jk}.$$

As we shall see shortly, δ_{ij} denote the components of a second-order unit tensor, $\mathbf{I} = \delta_{ij}\,\hat{\mathbf{e}}_i\hat{\mathbf{e}}_j = \hat{\mathbf{e}}_i\hat{\mathbf{e}}_i$.

We define the cross product $\hat{\mathbf{e}}_i \times \hat{\mathbf{e}}_j$ as

$$\hat{\mathbf{e}}_i \times \hat{\mathbf{e}}_j \equiv e_{ijk}\hat{\mathbf{e}}_k, \qquad (2.2.38)$$

where

$$e_{ijk} = \begin{cases} 1, & \text{if } i, j, k \text{ are in cyclic order} \\ & \text{and not repeated } (i \neq j \neq k), \\ -1, & \text{if } i, j, k \text{ are not in cyclic order} \\ & \text{and not repeated } (i \neq j \neq k), \\ 0, & \text{if any of } i, j, k \text{ are repeated.} \end{cases} \qquad (2.2.39)$$

The symbol e_{ijk} is called the *alternating symbol* or *permutation symbol*. By definition, the subscripts of the permutation symbol can be permuted without changing its value; an interchange of any two subscripts will change the sign (hence, interchange of two subscripts twice keeps the value unchanged):

$$e_{ijk} = e_{kij} = e_{jki}, \quad e_{ijk} = -e_{jik} = e_{jki} = -e_{kji}.$$

In an orthonormal basis, the scalar and vector products can be expressed in the index form using the Kronecker delta and the alternating symbols:

$$\mathbf{A} \cdot \mathbf{B} = (A_i\hat{\mathbf{e}}_i) \cdot (B_j\hat{\mathbf{e}}_j) = A_i B_j \delta_{ij} = A_i B_i,$$
$$\mathbf{A} \times \mathbf{B} = (A_i\hat{\mathbf{e}}_i) \times (B_j\hat{\mathbf{e}}_j) = A_i B_j e_{ijk}\hat{\mathbf{e}}_k. \qquad (2.2.40)$$

Note that the components of a vector in an orthonormal coordinate system can be expressed as

$$A_i = \mathbf{A} \cdot \hat{\mathbf{e}}_i, \qquad (2.2.41)$$

and therefore we can express vector $\mathbf{A}$ as

$$\mathbf{A} = A_i\hat{\mathbf{e}}_i = (\mathbf{A} \cdot \hat{\mathbf{e}}_i)\hat{\mathbf{e}}_i. \qquad (2.2.42)$$

Further, the Kronecker delta and the permutation symbol are related by the identity, known as the *e-δ identity* [see Problem 2.5(d)],

$$e_{ijk}e_{imn} = \delta_{jm}\delta_{kn} - \delta_{jn}\delta_{km}. \qquad (2.2.43)$$

The permutation symbol and the Kronecker delta prove to be very useful in proving vector identities. Since a vector form of any identity is invariant (i.e., valid in any coordinate system), it suffices to prove it in one coordinate system. In particular, an orthonormal system is very convenient because we can use the index notation, permutation symbol, and the Kronecker delta. The following examples contain several cases of incorrect and correct use of index notation and illustrate some of the uses of δ_{ij} and e_{ijk}.

EXAMPLE 2.2.3: Discuss the validity of the following expressions:
1. $a_m b_s = c_m(d_r - f_r)$.
2. $a_m b_s = c_m(d_s - f_s)$.

3. $a_i = b_j c_i d_i$.

4. $x_i x_i = r^2$.

5. $a_i b_j c_j = 3$.

SOLUTION:

1. Not a valid expression because the free indices r and s do not match.

2. Valid; both m and s are free indices. There are nine equations ($m, s = 1, 2, 3$).

3. Not a valid expression because the free index j is not matched on both sides of the equality, and index i is a dummy index in one expression and a free index in the other; i cannot be used both as a free and dummy index in the same equation. The equation would have been valid if i on the left side of the equation is replaced with j; then there will be three equations.

4. A valid expression, containing one equation: $x_1^2 + x_2^2 + x_3^2 = r^2$.

5. A valid expression; it contains three equations ($i = 1, 2, 3$): $a_1 b_1 c_1 + a_1 b_2 c_2 + a_1 b_3 c_3 = 3$, $\quad a_2 b_1 c_1 + a_2 b_2 c_2 + a_2 b_3 c_3 = 3$, $\quad$ and $\quad a_3 b_1 c_1 + a_3 b_2 c_2 + a_3 b_3 c_3 = 3$.

EXAMPLE 2.2.4: Simplify the following expressions:

1. $\delta_{ij} \delta_{jk} \delta_{kp} \delta_{pi}$.

2. $\varepsilon_{mjk} \varepsilon_{njk}$.

3. $(\mathbf{A} \times \mathbf{B}) \cdot (\mathbf{C} \times \mathbf{D})$.

SOLUTION:

1. Successive contraction of subscripts yield the result:

$$\delta_{ij} \delta_{jk} \delta_{kp} \delta_{pi} = \delta_{ij} \delta_{jk} \delta_{ki} = \delta_{ij} \delta_{ji} = \delta_{ii} = 3.$$

2. Expand using the e-δ identity

$$\varepsilon_{mjk} \varepsilon_{njk} = \delta_{mn} \delta_{jj} - \delta_{mj} \delta_{nj} = 3\delta_{mn} - \delta_{mn} = 2\delta_{mn}.$$

In particular, the expression $\varepsilon_{ijk} \varepsilon_{ijk}$ is equal to $2\delta_{ii} = 6$.

3. Expanding the expression using the index notation, we obtain

$$(\mathbf{A} \times \mathbf{B}) \cdot (\mathbf{C} \times \mathbf{D}) = (A_i B_j e_{ijk} \hat{\mathbf{e}}_k) \cdot (C_m D_n e_{mnp} \hat{\mathbf{e}}_p)$$

$$= A_i B_j C_m D_n e_{ijk} e_{mnp} \delta_{kp}$$

$$= A_i B_j C_m D_n e_{ijk} e_{mnk}$$

$$= A_i B_j C_m D_n (\delta_{im} \delta_{jn} - \delta_{in} \delta_{jm})$$

$$= A_i B_j C_m D_n \delta_{im} \delta_{jn} - A_i B_j C_m D_n \delta_{in} \delta_{jm}$$

$$= A_i B_j C_i D_j - A_i B_j C_j D_i = A_i C_i B_j D_j - A_i D_i B_j C_j$$

$$= (\mathbf{A} \cdot \mathbf{C})(\mathbf{B} \cdot \mathbf{D}) - (\mathbf{A} \cdot \mathbf{D})(\mathbf{B} \cdot \mathbf{C}),$$

where we have used the e-δ identity (2.2.43).

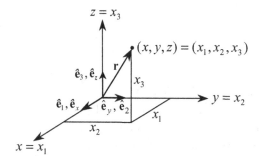

Figure 2.2.10. Rectangular Cartesian coordinates.

Although the above vector identity is established in an orthonormal coordinate system, it holds in a general coordinate system. That is, the above vector identity is invariant.

EXAMPLE 2.2.5: Rewrite the expression $e_{mni} A_i B_j C_m D_n \hat{\mathbf{e}}_j$ in vector form.

SOLUTION: We note that $B_j \hat{\mathbf{e}}_j = \mathbf{B}$. Examining the indices in the permutation symbol and the remaining coefficients, it is clear that vectors $\mathbf{C}$ and $\mathbf{D}$ must have a cross product between them and the resulting vector must have a dot product with vector $\mathbf{A}$. Thus we have

$$e_{mni} A_i B_j C_m D_n \hat{\mathbf{e}}_j = [(\mathbf{C} \times \mathbf{D}) \cdot \mathbf{A}]\mathbf{B} = (\mathbf{C} \times \mathbf{D} \cdot \mathbf{A})\,\mathbf{B}.$$

2.2.6 Transformation Law for Different Bases

When the basis vectors are constant, that is, with fixed lengths (with the same units) and directions, the basis is called *Cartesian*. The general Cartesian system is oblique. When the basis vectors are unit and orthogonal (orthonormal), the basis system is called *rectangular Cartesian* or simply *Cartesian*. In much of our study, we shall deal with Cartesian bases.

Let us denote an orthonormal Cartesian basis by

$$\{\hat{\mathbf{e}}_x, \hat{\mathbf{e}}_y, \hat{\mathbf{e}}_z\} \qquad \text{or} \qquad \{\hat{\mathbf{e}}_1, \hat{\mathbf{e}}_2, \hat{\mathbf{e}}_3\}.$$

The Cartesian coordinates are denoted by (x, y, z) or (x_1, x_2, x_3). The familiar rectangular Cartesian coordinate system is shown in Figure 2.2.10. We shall always use right-handed coordinate systems.

A position vector to an arbitrary point (x, y, z) or (x_1, x_2, x_3), measured from the origin, is given by

$$\mathbf{r} = x\hat{\mathbf{e}}_x + y\hat{\mathbf{e}}_y + z\hat{\mathbf{e}}_z$$

$$= x_1\hat{\mathbf{e}}_1 + x_2\hat{\mathbf{e}}_2 + x_3\hat{\mathbf{e}}_3, \tag{2.2.44}$$

or, in summation notation, by

$$\mathbf{r} = x_j\hat{\mathbf{e}}_j, \qquad \mathbf{r} \cdot \mathbf{r} = r^2 = x_i x_i. \tag{2.2.45}$$

We shall also use the symbol $\mathbf{x}$ for the position vector $\mathbf{r} = \mathbf{x}$. The length of a line element $d\mathbf{r} = d\mathbf{x}$ is given by

$$d\mathbf{r} \cdot d\mathbf{r} = (ds)^2 = dx_j dx_j = (dx)^2 + (dy)^2 + (dz)^2. \qquad (2.2.46)$$

Here we discuss the relationship between the components of two different orthonormal coordinate systems. Consider the first coordinate basis

$$\{\hat{\mathbf{e}}_1, \hat{\mathbf{e}}_2, \hat{\mathbf{e}}_3\}$$

and the second coordinate basis

$$\{\hat{\bar{\mathbf{e}}}_1, \hat{\bar{\mathbf{e}}}_2, \hat{\bar{\mathbf{e}}}_3\}.$$

Now we can express the same vector in the coordinate system without bars (referred as "unbarred") and also in the coordinate system with bars (referred as "barred"):

$$\begin{aligned}
\mathbf{A} &= A_i \hat{\mathbf{e}}_i = (\mathbf{A} \cdot \hat{\mathbf{e}}_i)\hat{\mathbf{e}}_i \\
&= \bar{A}_j \hat{\bar{\mathbf{e}}}_j = (\mathbf{A} \cdot \hat{\bar{\mathbf{e}}}_i)\hat{\bar{\mathbf{e}}}_i.
\end{aligned} \qquad (2.2.47)$$

From Eq. (2.2.42), we have

$$\bar{A}_j = \mathbf{A} \cdot \hat{\bar{\mathbf{e}}}_j = A_i (\hat{\mathbf{e}}_i \cdot \hat{\bar{\mathbf{e}}}_j) \equiv \ell_{ji} A_i, \qquad (2.2.48)$$

where

$$\ell_{ij} = \hat{\bar{\mathbf{e}}}_i \cdot \hat{\mathbf{e}}_j. \qquad (2.2.49)$$

Equation (2.2.48) gives the relationship between the components $(\bar{A}_1, \bar{A}_2, \bar{A}_3)$ and (A_1, A_2, A_3), and it is called the *transformation rule* between the barred and unbarred components in the two coordinate systems. The coefficients ℓ_{ij} can be interpreted as the direction cosines of the barred coordinate system with respect to the unbarred coordinate system:

$$\ell_{ij} = \text{cosine of the angle between } \hat{\bar{\mathbf{e}}}_i \text{ and } \hat{\mathbf{e}}_j. \qquad (2.2.50)$$

Note that the first subscript of ℓ_{ij} comes from the barred coordinate system and the second subscript from the unbarred system. Obviously, ℓ_{ij} is not symmetric (i.e., $\ell_{ij} \neq \ell_{ji}$). The rectangular array of these components is called a *matrix*, which is the topic of the next section. The next example illustrates the computation of direction cosines.

EXAMPLE 2.2.6: Let $\hat{\mathbf{e}}_i$ $(i = 1, 2, 3)$ be a set of orthonormal base vectors, and define a new right-handed coordinate basis by (note that $\hat{\bar{\mathbf{e}}}_1.\hat{\bar{\mathbf{e}}}_2 = 0$)

$$\hat{\bar{\mathbf{e}}}_1 = \frac{1}{3}(2\hat{\mathbf{e}}_1 + 2\hat{\mathbf{e}}_2 + \hat{\mathbf{e}}_3), \quad \hat{\bar{\mathbf{e}}}_2 = \frac{1}{\sqrt{2}}(\hat{\mathbf{e}}_1 - \hat{\mathbf{e}}_2),$$

$$\hat{\bar{\mathbf{e}}}_3 = \hat{\bar{\mathbf{e}}}_1 \times \hat{\bar{\mathbf{e}}}_2 = \frac{1}{3\sqrt{2}}(\hat{\mathbf{e}}_1 + \hat{\mathbf{e}}_2 - 4\hat{\mathbf{e}}_3).$$

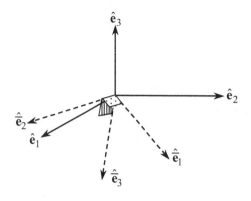

Figure 2.2.11. The original and transformed coordinate systems defined in Example 2.2.6.

The original and new coordinate systems are depicted in Figure 2.2.11. Determine the direction cosines ℓ_{ij} of the transformation and display them in a rectangular array.

SOLUTION: From Eq. (2.2.49) we have

$$\ell_{11} = \hat{\bar{\mathbf{e}}}_1 \cdot \hat{\mathbf{e}}_1 = \frac{2}{3}, \qquad \ell_{12} = \hat{\bar{\mathbf{e}}}_1 \cdot \hat{\mathbf{e}}_2 = \frac{2}{3}, \qquad \ell_{13} = \hat{\bar{\mathbf{e}}}_1 \cdot \hat{\mathbf{e}}_3 = \frac{1}{3},$$

$$\ell_{21} = \hat{\bar{\mathbf{e}}}_2 \cdot \hat{\mathbf{e}}_1 = \frac{1}{\sqrt{2}}, \qquad \ell_{22} = \hat{\bar{\mathbf{e}}}_2 \cdot \hat{\mathbf{e}}_2 = -\frac{1}{\sqrt{2}}, \qquad \ell_{23} = \hat{\bar{\mathbf{e}}}_2 \cdot \hat{\mathbf{e}}_3 = 0,$$

$$\ell_{31} = \hat{\bar{\mathbf{e}}}_3 \cdot \hat{\mathbf{e}}_1 = \frac{1}{3\sqrt{2}}, \qquad \ell_{32} = \hat{\bar{\mathbf{e}}}_3 \cdot \hat{\mathbf{e}}_2 = \frac{1}{3\sqrt{2}}, \qquad \ell_{33} = \hat{\bar{\mathbf{e}}}_3 \cdot \hat{\mathbf{e}}_3 = -\frac{4}{3\sqrt{2}}.$$

The rectangular array of these components is denoted by **L** and has the form

$$\mathbf{L} = \frac{1}{3\sqrt{2}} \begin{bmatrix} 2\sqrt{2} & 2\sqrt{2} & \sqrt{2} \\ 3 & -3 & 0 \\ 1 & 1 & -4 \end{bmatrix}.$$

2.3 Theory of Matrices

2.3.1 Definition

In the preceding sections, we studied the algebra of ordinary vectors and the transformation of vector components from one coordinate system to another. For example, the transformation equation (2.2.48) relates the components of a vector in the barred coordinate system to unbarred coordinate system. Writing Eq. (2.2.48) in expanded form,

$$\bar{A}_1 = \ell_{11} A_1 + \ell_{12} A_2 + \ell_{13} A_3,$$
$$\bar{A}_2 = \ell_{21} A_1 + \ell_{22} A_2 + \ell_{23} A_3, \qquad\qquad (2.3.1)$$
$$\bar{A}_3 = \ell_{31} A_1 + \ell_{32} A_2 + \ell_{33} A_3,$$

we see that there are nine coefficients relating the components A_i to $\bar{A}_i$. The form of these linear equations suggests writing down the scalars of ℓ_{ij} (jth components in the ith equation) in the rectangular array

$$\mathbf{L} = \begin{bmatrix} \ell_{11} & \ell_{12} & \ell_{13} \\ \ell_{21} & \ell_{22} & \ell_{23} \\ \ell_{31} & \ell_{32} & \ell_{33} \end{bmatrix}.$$

This rectangular array $\mathbf{L}$ of scalars ℓ_{ij} is called a *matrix*, and the quantities ℓ_{ij} are called the *elements* of $\mathbf{L}$.[1]

If a matrix has m rows and n columns, we will say that it is an m by n ($m \times n$) matrix, the number of rows always being listed first. The element in the ith row and jth column of a matrix $\mathbf{A}$ is generally denoted by a_{ij}, and we will sometimes designate a matrix by $\mathbf{A} = [A] = [a_{ij}]$. A square matrix is one that has the same number of rows as columns. An $n \times n$ matrix is said to be of *order n*. The elements of a square matrix for which the row number and the column number are the same (i.e., a_{ij} for $i = j$) are called *diagonal elements* or simply the *diagonal*. A square matrix is said to be a *diagonal matrix* if all of the off-diagonal elements are zero. An *identity matrix*, denoted by $\mathbf{I} = [I]$, is a diagonal matrix whose elements are all 1's. Examples of a diagonal and an identity matrix are given below:

$$\begin{bmatrix} 5 & 0 & 0 & 0 \\ 0 & -2 & 0 & 0 \\ 0 & 0 & 1 & 0 \\ 0 & 0 & 0 & 3 \end{bmatrix}, \quad \mathbf{I} = \begin{bmatrix} 1 & 0 & 0 & 0 \\ 0 & 1 & 0 & 0 \\ 0 & 0 & 1 & 0 \\ 0 & 0 & 0 & 1 \end{bmatrix}.$$

The sum of the diagonal elements is called the *trace* of the matrix.

If the matrix has only one row or one column, we will normally use only a single subscript to designate its elements. For example,

$$\mathbf{X} = \begin{Bmatrix} x_1 \\ x_2 \\ x_3 \end{Bmatrix}, \quad \mathbf{Y} = \{y_1 \ y_2 \ y_3\}$$

denote a column matrix and a row matrix, respectively. Row and column matrices can be used to denote the components of a vector.

2.3.2 Matrix Addition and Multiplication of a Matrix by a Scalar

The *sum* of two matrices of the same size is defined to be a matrix of the same size obtained by simply adding the corresponding elements. If $\mathbf{A}$ is an $m \times n$ matrix and $\mathbf{B}$ is an $m \times n$ matrix, their sum is an $m \times n$ matrix, $\mathbf{C}$, with

$$c_{ij} = a_{ij} + b_{ij} \quad \text{for all} \quad i, j. \tag{2.3.2}$$

---

[1] The word "matrix" was first used in 1850 by James Sylvester (1814–1897), an English algebraist. However, Arthur Caley (1821–1895), professor of mathematics at Cambridge, was the first to explore properties of matrices. Significant contributions in the early years were made by Charles Hermite, Georg Frobenius, and Camille Jordan, among others.

A constant multiple of a matrix is equal to the matrix obtained by multiplying all of the elements by the constant. That is, the multiple of a matrix $\mathbf{A}$ by a scalar α, $\alpha\mathbf{A}$, is the matrix obtained by multiplying each of its elements with α:

$$
\mathbf{A} = \begin{bmatrix} a_{11} & a_{12} & \cdots & a_{1n} \\ a_{21} & a_{22} & \cdots & a_{2n} \\ \vdots & \vdots & \cdots & \vdots \\ a_{m1} & a_{m2} & \cdots & a_{mn} \end{bmatrix}, \quad \alpha\mathbf{A} = \begin{bmatrix} \alpha a_{11} & \alpha a_{12} & \cdots & \alpha a_{1n} \\ \alpha a_{21} & \alpha a_{22} & \cdots & \alpha a_{2n} \\ \cdots & \cdots & \cdots & \cdots \\ \alpha a_{m1} & \alpha a_{m2} & \cdots & \alpha a_{mn} \end{bmatrix}.
$$

Matrix addition has the following properties:

1. Addition is commutative: $\mathbf{A} + \mathbf{B} = \mathbf{B} + \mathbf{A}$.
2. Addition is associative: $\mathbf{A} + (\mathbf{B} + \mathbf{C}) = (\mathbf{A} + \mathbf{B}) + \mathbf{C}$.
3. There exists a unique matrix $\mathbf{0}$, such that $\mathbf{A} + \mathbf{0} = \mathbf{0} + \mathbf{A} = \mathbf{A}$. The matrix $\mathbf{0}$ is called *zero matrix*; all elements of it are zeros.
4. For each matrix $\mathbf{A}$, there exists a unique matrix $-\mathbf{A}$ such that $\mathbf{A} + (-\mathbf{A}) = \mathbf{0}$.
5. Addition is distributive with respect to scalar multiplication: $\alpha(\mathbf{A} + \mathbf{B}) = \alpha\mathbf{A} + \alpha\mathbf{B}$.
6. Addition is distributive with respect to matrix multiplication, which will be discussed shortly (note the order):

$$(\mathbf{A} + \mathbf{B})\mathbf{C} = \mathbf{A}\mathbf{C} + \mathbf{B}\mathbf{C}.$$

2.3.3 Matrix Transpose and Symmetric Matrix

If $\mathbf{A}$ is an $m \times n$ matrix, then the $n \times m$ matrix obtained by interchanging its rows and columns is called the *transpose* of $\mathbf{A}$ and is denoted by $\mathbf{A}^{\mathrm{T}}$. For example, consider the matrices

$$
\mathbf{A} = \begin{bmatrix} 5 & -2 & 1 \\ 8 & 7 & 6 \\ 2 & 4 & 3 \\ -1 & 9 & 0 \end{bmatrix}, \quad \mathbf{B} = \begin{bmatrix} 3 & -1 & 2 & 4 \\ -6 & 3 & 5 & 7 \\ 9 & 6 & -2 & 1 \end{bmatrix}. \tag{2.3.3}
$$

The transposes of $\mathbf{A}$ and $\mathbf{B}$ are

$$
\mathbf{A}^{\mathrm{T}} = \begin{bmatrix} 5 & 8 & 2 & -1 \\ -2 & 7 & 4 & 9 \\ 1 & 6 & 3 & 0 \end{bmatrix}, \quad \mathbf{B}^{\mathrm{T}} = \begin{bmatrix} 3 & -6 & 9 \\ -1 & 3 & 6 \\ 2 & 5 & -2 \\ 4 & 7 & 1 \end{bmatrix}.
$$

The following basic properties of a transpose should be noted:

1. $(\mathbf{A}^{\mathrm{T}})^{\mathrm{T}} = \mathbf{A}$.
2. $(\mathbf{A} + \mathbf{B})^{\mathrm{T}} = \mathbf{A}^{\mathrm{T}} + \mathbf{B}^{\mathrm{T}}$.

A square matrix $\mathbf{A}$ of real numbers is said to be *symmetric* if $\mathbf{A}^{\mathrm{T}} = \mathbf{A}$. It is said to be *skew symmetric* if $\mathbf{A}^{\mathrm{T}} = -\mathbf{A}$. In terms of the elements of $\mathbf{A}$, these definitions imply that $\mathbf{A}$ is symmetric if and only if $a_{ij} = a_{ji}$, and it is skew symmetric if and only if $a_{ij} = -a_{ji}$. Note that the diagonal elements of a skew symmetric matrix are always zero since $a_{ij} = -a_{ij}$ implies $a_{ij} = 0$ for $i = j$. Examples of symmetric and skew symmetric matrices, respectively, are

$$
\begin{bmatrix}
5 & -2 & 12 & 21 \\
-2 & 2 & 16 & -3 \\
12 & 16 & 13 & 8 \\
21 & -3 & 8 & 19
\end{bmatrix},
\qquad
\begin{bmatrix}
0 & -11 & 32 & 4 \\
11 & 0 & 25 & 7 \\
-32 & -25 & 0 & 15 \\
-4 & -7 & -15 & 0
\end{bmatrix}.
$$

2.3.4 Matrix Multiplication

Consider a vector $\mathbf{A} = a_1\hat{\mathbf{e}}_1 + a_2\hat{\mathbf{e}}_2 + a_3\hat{\mathbf{e}}_3$ in a Cartesian system. We can represent $\mathbf{A}$ as a *product* of a row matrix with a column matrix,

$$
\mathbf{A} = \{a_1 \; a_2 \; a_3\}
\begin{Bmatrix}
\hat{\mathbf{e}}_1 \\
\hat{\mathbf{e}}_2 \\
\hat{\mathbf{e}}_3
\end{Bmatrix}.
$$

The vector $\mathbf{A}$ is obtained by multiplying the ith element in the row matrix with the ith element in the column matrix and adding them. This gives us a strong motivation for defining the product of two matrices.

Let $\mathbf{x}$ and $\mathbf{y}$ be the vectors (matrices with one column)

$$
\mathbf{x} =
\begin{Bmatrix}
x_1 \\
x_2 \\
\vdots \\
x_m
\end{Bmatrix},
\qquad
\mathbf{y} =
\begin{Bmatrix}
y_1 \\
y_2 \\
\vdots \\
y_m
\end{Bmatrix}.
$$

We define the product $\mathbf{x}^T\mathbf{y}$ to be the scalar

$$
\mathbf{x}^T\mathbf{y} = \{x_1, x_2, \ldots, x_m\}
\begin{Bmatrix}
y_1 \\
y_2 \\
\vdots \\
y_m
\end{Bmatrix}
$$

$$
= x_1 y_1 + x_2 y_2 + \cdots + x_m y_m = \sum_{i=1}^{m} x_i y_i. \tag{2.3.4}
$$

It follows from Eq. (2.3.4) that $\mathbf{x}^T\mathbf{y} = \mathbf{y}^T\mathbf{x}$. More generally, let $\mathbf{A} = [a_{ij}]$ be $m \times n$ and $\mathbf{B} = [b_{ij}]$ be $n \times p$ matrices. The product $\mathbf{AB}$ is defined to be the $m \times p$ matrix $\mathbf{C} = [c_{ij}]$ with

$$
c_{ij} = \{i \text{ th row of } [\mathbf{A}]\}
\begin{Bmatrix}
j\text{th col.} \\
\text{of} \\
\mathbf{B}
\end{Bmatrix}
= \{a_{i1}, a_{i2}, \ldots, a_{in}\}
\begin{Bmatrix}
b_{1j} \\
b_{2j} \\
\vdots \\
b_{nj}
\end{Bmatrix}
$$

$$
= a_{i1}b_{1j} + a_{i2}b_{2j} + \cdots + a_{in}b_{nj} = \sum_{k=1}^{n} a_{ik}b_{kj}. \tag{2.3.5}
$$

The next example illustrates the computation of the product of a square matrix with a column matrix.

The following comments are in order on the matrix multiplication, wherein **A** denotes an $m \times n$ matrix and **B** denotes a $p \times q$ matrix:

1. The product **AB** is defined only if the number of columns n in **A** is equal to the number of rows p in **B**. Similarly, the product **BA** is defined only if $q = m$.
2. If **AB** is defined, **BA** may or may not be defined. If both **AB** and **BA** are defined, it is not necessary that they be of the same size.
3. The products **AB** and **BA** are of the same size if and only if both **A** and **B** are square matrices of the same size.
4. The products **AB** and **BA** are, in general, not equal $\mathbf{AB} \neq \mathbf{BA}$ (even if they are of equal size); that is, the matrix multiplication is not *commutative*.
5. For any real square matrix **A**, **A** is said to be *normal* if $\mathbf{AA}^\mathrm{T} = \mathbf{A}^\mathrm{T}\mathbf{A}$; **A** is said to be *orthogonal* if $\mathbf{AA}^\mathrm{T} = \mathbf{A}^\mathrm{T}\mathbf{A} = \mathbf{I}$.
6. If **A** is a square matrix, the powers of **A** are defined by $\mathbf{A}^2 = \mathbf{AA}$, $\mathbf{A}^3 = \mathbf{AA}^2 = \mathbf{A}^2\mathbf{A}$, and so on.
7. Matrix multiplication is associative: $(\mathbf{AB})\mathbf{C} = \mathbf{A}(\mathbf{BC})$.
8. The product of any square matrix with the identity matrix is the matrix itself.
9. The transpose of the product is $(\mathbf{AB})^\mathrm{T} = \mathbf{B}^\mathrm{T}\mathbf{A}^\mathrm{T}$ (note the order).

The next example illustrates computation of the product of two matrices and verifies Property 9.

EXAMPLE 2.3.1: Verify Property 3 using the matrices $[A]$ and $[B]$ in Eq. (2.3.3). The product of matrix **A** and **B** is

$$\mathbf{AB} = \begin{bmatrix} 5 & -2 & 1 \\ 8 & 7 & 6 \\ 2 & 4 & 3 \\ -1 & 9 & 0 \end{bmatrix} \begin{bmatrix} 3 & -1 & 2 & 4 \\ -6 & 3 & 5 & 7 \\ 9 & 6 & -2 & 1 \end{bmatrix}$$

$$= \begin{bmatrix} 36 & -5 & -2 & 7 \\ 36 & 49 & 39 & 87 \\ 9 & 28 & 18 & 39 \\ -57 & 28 & 43 & 59 \end{bmatrix},$$

and

$$(\mathbf{AB})^\mathrm{T} = \begin{bmatrix} 36 & 36 & 9 & -57 \\ -5 & 49 & 28 & 28 \\ -2 & 39 & 18 & 43 \\ 7 & 87 & 39 & 59 \end{bmatrix}.$$

Now compute the product

$$\mathbf{B}^T\mathbf{A}^T = \begin{bmatrix} 3 & -6 & 9 \\ -1 & 3 & 6 \\ 2 & 5 & -2 \\ 4 & 7 & 1 \end{bmatrix} \begin{bmatrix} 5 & 8 & 2 & -1 \\ -2 & 7 & 4 & 9 \\ 1 & 6 & 3 & 0 \end{bmatrix}$$

$$= \begin{bmatrix} 36 & 36 & 9 & -57 \\ -5 & 49 & 28 & 28 \\ -2 & 39 & 18 & 43 \\ 7 & 87 & 39 & 59 \end{bmatrix}.$$

Thus, $(\mathbf{AB})^T = \mathbf{B}^T\mathbf{A}^T$ is verified.

2.3.5 Inverse and Determinant of a Matrix

If $\mathbf{A}$ is an $n \times n$ matrix and $\mathbf{B}$ is any $n \times n$ matrix such that $\mathbf{AB} = \mathbf{BA} = \mathbf{I}$, then $\mathbf{B}$ is called an *inverse* of $\mathbf{A}$. If it exists, the inverse of a matrix is unique (a consequence of the associative law). If both $\mathbf{B}$ and $\mathbf{C}$ are inverses for $\mathbf{A}$, then by definition,

$$\mathbf{AB} = \mathbf{BA} = \mathbf{AC} = \mathbf{CA} = \mathbf{I}.$$

Since matrix multiplication is associative, we have

$$\mathbf{BAC} = (\mathbf{BA})\mathbf{C} = \mathbf{IC} = \mathbf{C}$$

$$= \mathbf{B}(\mathbf{AC}) = \mathbf{BI} = \mathbf{B}.$$

This shows that $\mathbf{B} = \mathbf{C}$, and the inverse is unique. The inverse of $\mathbf{A}$ is denoted by $\mathbf{A}^{-1}$. A matrix is said to be *singular* if it does not have an inverse. If $\mathbf{A}$ is nonsingular, then the transpose of the inverse is equal to the inverse of the transpose: $(\mathbf{A}^{-1})^T = (\mathbf{A}^T)^{-1}$.

Let $\mathbf{A} = [a_{ij}]$ be an $n \times n$ matrix. We wish to associate with $\mathbf{A}$ a scalar that in some sense measures the "size" of $\mathbf{A}$ and indicates whether $\mathbf{A}$ is nonsingular. The *determinant* of the matrix $\mathbf{A} = [a_{ij}]$ is defined to be the scalar $\det \mathbf{A} = |A|$ computed according to the rule

$$\det\mathbf{A} = |a_{ij}| = \sum_{i=1}^{n}(-1)^{i+1}a_{i1}|A_{i1}|, \tag{2.3.6}$$

where $|A_{ij}|$ is the determinant of the $(n-1) \times (n-1)$ matrix that remains on deleting out the ith row and the first column of $\mathbf{A}$. For convenience, we define the determinant of a zeroth-order matrix to be unity. For 1×1 matrices, the determinant is defined according to $|a_{11}| = a_{11}$. For a 2×2 matrix $\mathbf{A}$, the determinant is defined by

$$\mathbf{A} = \begin{bmatrix} a_{11} & a_{12} \\ a_{21} & a_{22} \end{bmatrix}, \quad |A| = \begin{vmatrix} a_{11} & a_{12} \\ a_{21} & a_{22} \end{vmatrix} = a_{11}a_{22} - a_{12}a_{21}.$$

In the previous definition, special attention is given to the first column of the matrix $\mathbf{A}$. We call it the expansion of $|A|$ according to the first column of $\mathbf{A}$. One can expand

$|A|$ according to any column or row:

$$|A| = \sum_{i=1}^{n}(-1)^{i+j}a_{ij}|A_{ij}|, \tag{2.3.7}$$

where $|A_{ij}|$ is the determinant of the matrix obtained by deleting the ith row and jth column of matrix $\mathbf{A}$. A numerical example of the calculation of determinant is presented next.

EXAMPLE 2.3.2: Compute the determinant of the matrix

$$\mathbf{A} = \begin{bmatrix} 2 & 5 & -1 \\ 1 & 4 & 3 \\ 2 & -3 & 5 \end{bmatrix}.$$

SOLUTION: Using the definition (2.3.7) and expanding by the first column, we have

$$|A| = \sum_{i=1}^{3}(-1)^{i+1}a_{i1}|A_{i1}|$$

$$= (-1)^2 a_{11}\begin{vmatrix} 4 & 3 \\ -3 & 5 \end{vmatrix} + (-1)^3 a_{21}\begin{vmatrix} 5 & -1 \\ -3 & 5 \end{vmatrix} + (-1)^4 a_{31}\begin{vmatrix} 5 & -1 \\ 4 & 3 \end{vmatrix}$$

$$= 2\big[(4)(5) - (3)(-3)\big] + (-1)\big[(5)(5) - (-1)(-3)\big] + 2\big[(5)(3) - (-1)(4)\big]$$

$$= 2(20 + 9) - (25 - 3) + 2(15 + 4) = 74.$$

The cross product of two vectors $\mathbf{A}$ and $\mathbf{B}$ can be expressed as the value of the determinant

$$\mathbf{A} \times \mathbf{B} \equiv \begin{vmatrix} \hat{\mathbf{e}}_1 & \hat{\mathbf{e}}_2 & \hat{\mathbf{e}}_3 \\ \hat{A}_1 & \hat{A}_2 & \hat{A}_3 \\ \hat{B}_1 & \hat{B}_2 & \hat{B}_3 \end{vmatrix}, \tag{2.3.8}$$

and the scalar triple product can be expressed as the value of a determinant

$$\mathbf{A} \cdot (\mathbf{B} \times \mathbf{C}) \equiv \begin{vmatrix} \hat{A}_1 & \hat{A}_2 & \hat{A}_3 \\ \hat{B}_1 & \hat{B}_2 & \hat{B}_3 \\ \hat{C}_1 & \hat{C}_2 & \hat{C}_3 \end{vmatrix}. \tag{2.3.9}$$

In general, the determinant of a 3×3 matrix $\mathbf{A}$ can be expressed in the form

$$|A| = e_{ijk}a_{1i}a_{2j}a_{3k}, \tag{2.3.10}$$

where a_{ij} is the element occupying the ith row and the jth column of the matrix. The verification of these results is left as an exercise for the reader (Problem 2.6 is designed to prove some of them).

We note the following properties of determinants:

1. $\det(\mathbf{AB}) = \det\mathbf{A} \cdot \det\mathbf{B}$.
2. $\det\mathbf{A}^{\mathrm{T}} = \det\mathbf{A}$.

3. $\det(\alpha \, \mathbf{A}) = \alpha^n \det \mathbf{A}$, where α is a scalar and n is the order of $\mathbf{A}$.

4. If $\mathbf{A}'$ is a matrix obtained from $\mathbf{A}$ by multiplying a row (or column) of $\mathbf{A}$ by a scalar α, then $\det \mathbf{A}' = \alpha \det \mathbf{A}$.

5. If $\mathbf{A}'$ is the matrix obtained from $\mathbf{A}$ by interchanging any two rows (or columns) of $\mathbf{A}$, then $\det \mathbf{A}' = -\det \mathbf{A}$.

6. If $\mathbf{A}$ has two rows (or columns) one of which is a scalar multiple of another (i.e., linearly dependent), $\det \mathbf{A} = 0$.

7. If $\mathbf{A}'$ is the matrix obtained from $\mathbf{A}$ by adding a multiple of one row (or column) to another, then $\det \mathbf{A}' = \det \mathbf{A}$.

We define (in fact, the definition given earlier is an indirect definition) singular matrices in terms of their determinants. A matrix is said to be *singular* if and only if its determinant is zero. By Property 6 mentioned earlier the determinant of a matrix is zero if it has linearly dependent rows (or columns).

For an $n \times n$ matrix $\mathbf{A}$, the determinant of the $(n-1) \times (n-1)$ sub-matrix, of $\mathbf{A}$ obtained by deleting row i and column j of $\mathbf{A}$ is called *minor* of a_{ij} and is denoted by $M_{ij}(\mathbf{A})$. The quantity $\mathrm{cof}_{ij}(\mathbf{A}) \equiv (-1)^{i+j} M_{ij}(\mathbf{A})$ is called the *cofactor* of a_{ij}. The determinant of $\mathbf{A}$ can be cast in terms of the minor and cofactor of a_{ij}

$$\det \mathbf{A} = \sum_{i=1}^{n} a_{ij} \, \mathrm{cof}_{ij}(\mathbf{A}) \tag{2.3.11}$$

for any value of j.

The *adjunct* (also called *adjoint*) of a matrix $\mathbf{A}$ is the transpose of the matrix obtained from $\mathbf{A}$ by replacing each element by its cofactor. The adjunct of $\mathbf{A}$ is denoted by $\mathrm{Adj}\mathbf{A}$.

Now we have the essential tools to compute the inverse of a matrix. If $\mathbf{A}$ is nonsingular (i.e., $\det \mathbf{A} \neq 0$), the inverse $\mathbf{A}^{-1}$ of $\mathbf{A}$ can be computed according to

$$\mathbf{A}^{-1} = \frac{1}{\det \mathbf{A}} \mathrm{Adj}\mathbf{A}. \tag{2.3.12}$$

The next example illustrates the computation of an inverse of a matrix.

EXAMPLE 2.3.3: Determine the inverse of the matrix $[A]$ of Example 2.3.2.

SOLUTION: The determinant is given by (expanding by the first row)

$$|A| = (2)(29) + (-)(5)(-1) + (-1)(-11) = 74.$$

The we compute M_{ij}

$$M_{11}(\mathbf{A}) = \begin{vmatrix} 4 & 3 \\ -3 & 5 \end{vmatrix}, \qquad M_{12}(\mathbf{A}) = \begin{vmatrix} 1 & 3 \\ 2 & 5 \end{vmatrix}, \qquad M_{13}(\mathbf{A}) = \begin{vmatrix} 1 & 4 \\ 2 & -3 \end{vmatrix}$$

$$\mathrm{cof}_{11}(\mathbf{A}) = (-1)^2 M_{11}(\mathbf{A}) = 4 \times 5 - (-3)3 = 29$$

$$\mathrm{cof}_{12}(\mathbf{A}) = (-1)^3 M_{12}(\mathbf{A}) = -(1 \times 5 - 3 \times 2) = 1$$

$$\mathrm{cof}_{13}(\mathbf{A}) = (-1)^4 M_{13}(\mathbf{A}) = 1 \times (-3) - 2 \times 4 = -11.$$

The Adj($\mathbf{A}$) is given by

$$Adj(\mathbf{A}) = \begin{bmatrix} cof_{11}(\mathbf{A}) & cof_{12}(\mathbf{A}) & cof_{13}(\mathbf{A}) \\ cof_{21}(\mathbf{A}) & cof_{22}(\mathbf{A}) & cof_{23}(\mathbf{A}) \\ cof_{31}(\mathbf{A}) & cof_{32}(\mathbf{A}) & cof_{33}(\mathbf{A}) \end{bmatrix}^T$$

$$= \begin{bmatrix} 29 & -22 & 19 \\ 1 & 12 & -7 \\ -11 & 16 & 3 \end{bmatrix}.$$

The inverse of $\mathbf{A}$ can be now computed using Eq. (2.3.12),

$$\mathbf{A}^{-1} = \frac{1}{74} \begin{bmatrix} 29 & -22 & 19 \\ 1 & 12 & -7 \\ -11 & 16 & 3 \end{bmatrix}.$$

It can be easily verified that $\mathbf{A}\mathbf{A}^{-1} = \mathbf{I}$.

2.4 Vector Calculus

2.4.1 Derivative of a Scalar Function of a Vector

The basic notions of vector and scalar calculus, especially with regard to physical applications, are closely related to the rate of change of a scalar field (such as the velocity potential or temperature) with distance. Let us denote a scalar field by $\phi = \phi(\mathbf{x})$, $\mathbf{x}$ being the position vector, as shown in Figure 2.4.1.

In general coordinates, we can write $\phi = \phi(q^1, q^2, q^3)$. The coordinate system (q^1, q^2, q^3) is referred to as the *unitary system*. We now define the unitary basis $(\mathbf{e}_1, \mathbf{e}_2, \mathbf{e}_3)$ as follows:

$$\mathbf{e}_1 \equiv \frac{\partial \mathbf{x}}{\partial q^1}, \qquad \mathbf{e}_2 \equiv \frac{\partial \mathbf{x}}{\partial q^2}, \qquad \mathbf{e}_3 \equiv \frac{\partial \mathbf{x}}{\partial q^3}. \tag{2.4.1}$$

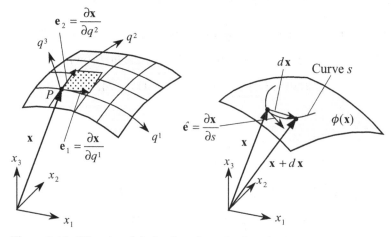

Figure 2.4.1. Directional derivative of a scalar function.

Note that $(\mathbf{e}_1, \mathbf{e}_2, \mathbf{e}_3)$ is not necessarily an orthogonal or unit basis. Hence, an arbitrary vector $\mathbf{A}$ is expressed as

$$\mathbf{A} = A^1\mathbf{e}_1 + A^2\mathbf{e}_2 + A^3\mathbf{e}_3 = A^i\mathbf{e}_i, \tag{2.4.2}$$

and a differential distance is denoted by

$$d\mathbf{x} = dq^1\mathbf{e}_1 + dq^2\mathbf{e}_2 + dq^3\mathbf{e}_3 = dq^i\mathbf{e}_i. \tag{2.4.3}$$

Observe that the A's and dq's have superscripts, whereas the unitary basis $(\mathbf{e}_1, \mathbf{e}_2, \mathbf{e}_3)$ has subscripts. The dq^i are referred to as the *contravariant components* of the differential vector $d\mathbf{x}$, and A^i are the contravariant components of vector $\mathbf{A}$. The unitary basis can be described in terms of the rectangular Cartesian basis $(\hat{\mathbf{e}}_x, \hat{\mathbf{e}}_y, \hat{\mathbf{e}}_z) = (\hat{\mathbf{e}}_1, \hat{\mathbf{e}}_2, \hat{\mathbf{e}}_3)$ as follows:

$$\mathbf{e}_1 = \frac{\partial\mathbf{x}}{\partial q^1} = \frac{\partial x}{\partial q^1}\hat{\mathbf{e}}_x + \frac{\partial y}{\partial q^1}\hat{\mathbf{e}}_y + \frac{\partial z}{\partial q^1}\hat{\mathbf{e}}_z,$$

$$\mathbf{e}_2 = \frac{\partial\mathbf{x}}{\partial q^2} = \frac{\partial x}{\partial q^2}\hat{\mathbf{e}}_x + \frac{\partial y}{\partial q^2}\hat{\mathbf{e}}_y + \frac{\partial z}{\partial q^2}\hat{\mathbf{e}}_z, \tag{2.4.4}$$

$$\mathbf{e}_3 = \frac{\partial\mathbf{x}}{\partial q^3} = \frac{\partial x}{\partial q^3}\hat{\mathbf{e}}_x + \frac{\partial y}{\partial q^3}\hat{\mathbf{e}}_y + \frac{\partial z}{\partial q^3}\hat{\mathbf{e}}_z.$$

In the summation convection, we have

$$\mathbf{e}_i \equiv \frac{\partial\mathbf{x}}{\partial q^i} = \frac{\partial x^j}{\partial q^i}\hat{\mathbf{e}}_j, \qquad i = 1, 2, 3. \tag{2.4.5}$$

Associated with any arbitrary basis is another basis that can be derived from it. We can construct this basis in the following way: Taking the scalar product of the vector $\mathbf{A}$ in Eq. (2.4.2) with the cross product $\mathbf{e}_1 \times \mathbf{e}_2$ and noting that since $\mathbf{e}_1 \times \mathbf{e}_2$ is perpendicular to both $\mathbf{e}_1$ and $\mathbf{e}_2$, we obtain

$$\mathbf{A} \cdot (\mathbf{e}_1 \times \mathbf{e}_2) = A^3\mathbf{e}_3 \cdot (\mathbf{e}_1 \times \mathbf{e}_2).$$

Of course, in the evaluation of the cross products, we shall always use the right-hand rule. Solving for A^3 gives

$$A^3 = \mathbf{A} \cdot \frac{\mathbf{e}_1 \times \mathbf{e}_2}{\mathbf{e}_3 \cdot (\mathbf{e}_1 \times \mathbf{e}_2)} = \mathbf{A} \cdot \frac{\mathbf{e}_1 \times \mathbf{e}_2}{[\mathbf{e}_1\mathbf{e}_2\mathbf{e}_3]}. \tag{2.4.6}$$

In similar fashion, we can obtain the following expressions for

$$A^1 = \mathbf{A} \cdot \frac{\mathbf{e}_2 \times \mathbf{e}_3}{[\mathbf{e}_1\mathbf{e}_2\mathbf{e}_3]}, \qquad A^2 = \mathbf{A} \cdot \frac{\mathbf{e}_3 \times \mathbf{e}_1}{[\mathbf{e}_1\mathbf{e}_2\mathbf{e}_3]}. \tag{2.4.7}$$

Thus, we observe that we can obtain the components A^1, A^2, and A^3 by taking the scalar product of the vector $\mathbf{A}$ with special vectors, which we denote as follows:

$$\mathbf{e}^1 = \frac{\mathbf{e}_2 \times \mathbf{e}_3}{[\mathbf{e}_1\mathbf{e}_2\mathbf{e}_3]}, \qquad \mathbf{e}^2 = \frac{\mathbf{e}_3 \times \mathbf{e}_1}{[\mathbf{e}_1\mathbf{e}_2\mathbf{e}_3]}, \qquad \mathbf{e}^3 = \frac{\mathbf{e}_1 \times \mathbf{e}_2}{[\mathbf{e}_1\mathbf{e}_2\mathbf{e}_3]}. \tag{2.4.8}$$

The set of vectors $(\mathbf{e}^1, \mathbf{e}^2, \mathbf{e}^3)$ is called the *dual basis* or *reciprocal basis*. Notice from the basic definitions that we have the following relations:

$$\mathbf{e}^i \cdot \mathbf{e}_j = \delta^i_j = \begin{cases} 1, & i = j \\ 0, & i \neq j \end{cases}. \tag{2.4.9}$$

It is possible, since the dual basis is linearly independent (the reader should verify this), to express a vector $\mathbf{A}$ in terms of the dual basis [cf. Eq. (2.4.2)]:

$$\mathbf{A} = A_1 \mathbf{e}^1 + A_2 \mathbf{e}^2 + A_3 \mathbf{e}^3 = A_i \mathbf{e}^i. \tag{2.4.10}$$

Notice now that the components associated with the dual basis have subscripts, and A_i are the *covariant components* of $\mathbf{A}$.

By an analogous process as that just described, we can show that the original basis can be expressed in terms of the dual basis in the following way:

$$\mathbf{e}_1 = \frac{\mathbf{e}^2 \times \mathbf{e}^3}{[\mathbf{e}^1 \mathbf{e}^2 \mathbf{e}^3]}, \quad \mathbf{e}_2 = \frac{\mathbf{e}^3 \times \mathbf{e}^1}{[\mathbf{e}^1 \mathbf{e}^2 \mathbf{e}^3]}, \quad \mathbf{e}_3 = \frac{\mathbf{e}^1 \times \mathbf{e}^2}{[\mathbf{e}^1 \mathbf{e}^2 \mathbf{e}^3]}. \tag{2.4.11}$$

It follows from Eqs. (2.4.2) and (2.4.10), in view of the orthogonality property in Eq. (2.4.9), that

$$A^i = \mathbf{A} \cdot \mathbf{e}^i, \quad A_i = \mathbf{A} \cdot \mathbf{e}_i,$$

$$A^i = g^{ij} \mathbf{e}_j, \quad A_i = g_{ij} \mathbf{e}^j, \tag{2.4.12}$$

$$g^{ij} = \mathbf{e}^i \cdot \mathbf{e}^j, \quad g_{ij} = \mathbf{e}_i \cdot \mathbf{e}_j.$$

Returning to the scalar field ϕ, the differential change is given by

$$d\phi = \frac{\partial \phi}{\partial q^1} dq^1 + \frac{\partial \phi}{\partial q^2} dq^2 + \frac{\partial \phi}{\partial q^3} dq^3. \tag{2.4.13}$$

The differentials dq^1, dq^2, dq^3 are components of $d\mathbf{x}$ [see Eq. (2.4.3)]. We would now like to write $d\phi$ in such a way that we elucidate *the direction* as well as the magnitude of $d\mathbf{x}$. Since $\mathbf{e}^1 \cdot \mathbf{e}_1 = 1$, $\mathbf{e}^2 \cdot \mathbf{e}_2 = 1$, and $\mathbf{e}^3 \cdot \mathbf{e}_3 = 1$, we can write

$$d\phi = \mathbf{e}^1 \frac{\partial \phi}{\partial q^1} \cdot \mathbf{e}_1 dq^1 + \mathbf{e}^2 \frac{\partial \phi}{\partial q^2} \cdot \mathbf{e}_2 dq^2 + \mathbf{e}^3 \frac{\partial \phi}{\partial q^3} \cdot \mathbf{e}_3 dq^3$$

$$= (dq^1 \mathbf{e}_1 + dq^2 \mathbf{e}_2 + dq^3 \mathbf{e}_3) \cdot \left(\mathbf{e}^1 \frac{\partial \phi}{\partial q^1} + \mathbf{e}^2 \frac{\partial \phi}{\partial q^2} + \mathbf{e}^3 \frac{\partial \phi}{\partial q^3} \right)$$

$$= d\mathbf{x} \cdot \left(\mathbf{e}^1 \frac{\partial \phi}{\partial q^1} + \mathbf{e}^2 \frac{\partial \phi}{\partial q^2} + \mathbf{e}^3 \frac{\partial \phi}{\partial q^3} \right). \tag{2.4.14}$$

Let us now denote the magnitude of $d\mathbf{x}$ by $ds \equiv |d\mathbf{x}|$. Then $\hat{\mathbf{e}} = d\mathbf{x}/ds$ is a unit vector in the direction of $d\mathbf{x}$, and we have

$$\left(\frac{d\phi}{ds} \right)_{\hat{\mathbf{e}}} = \hat{\mathbf{e}} \cdot \left(\mathbf{e}^1 \frac{\partial \phi}{\partial q^1} + \mathbf{e}^2 \frac{\partial \phi}{\partial q^2} + \mathbf{e}^3 \frac{\partial \phi}{\partial q^3} \right). \tag{2.4.15}$$

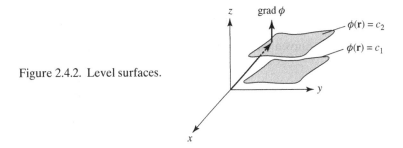

Figure 2.4.2. Level surfaces.

The derivative $(d\phi/ds)_{\hat{e}}$ is called the *directional derivative* of ϕ. We see that it is the *rate of change* of ϕ with respect to distance and that it depends on the direction $\hat{e}$ in which the distance is taken.

The vector in Eq. (2.4.15) that is scalar multiplied by $\hat{e}$ can be obtained immediately whenever the scalar field ϕ is given. Because the magnitude of this vector is equal to the maximum value of the directional derivative, it is called the *gradient vector* and is denoted by grad ϕ:

$$\text{grad } \phi \equiv \mathbf{e}^1 \frac{\partial\phi}{\partial q^1} + \mathbf{e}^2 \frac{\partial\phi}{\partial q^2} + \mathbf{e}^3 \frac{\partial\phi}{\partial q^3}. \tag{2.4.16}$$

From this representation, it can be seen that

$$\frac{\partial\phi}{\partial q^1}, \qquad \frac{\partial\phi}{\partial q^2}, \qquad \frac{\partial\phi}{\partial q^3}$$

are the *covariant components* of the gradient vector.

When the scalar function $\phi(\mathbf{x})$ is set equal to a constant, $\phi(\mathbf{x}) = \text{constant}$, a family of surfaces is generated. A different surface is designated by different values of the constant, and each surface is called a *level surface*, as shown in Figure 2.4.2. The unit vector $\hat{e}$ is tangent to a level surface. If the direction in which the directional derivative is taken lies within a level surface, then $d\phi/ds$ is zero, since ϕ is a constant on a level surface. It follows, therefore, that if $d\phi/ds$ is zero, then grad ϕ must be perpendicular to $\hat{e}$ and, hence, *perpendicular to a level surface*. Thus, if any surface is defined by $\phi(\mathbf{x}) = \text{constant}$, the unit normal to the surface is determined from

$$\hat{\mathbf{n}} = \pm \frac{\text{grad } \phi}{|\text{grad } \phi|}. \tag{2.4.17}$$

In general, the normal vector is a function of position $\mathbf{x}$; $\hat{\mathbf{n}}$ is independent of $\mathbf{x}$ only when ϕ is a plane (i.e., linear function of $\mathbf{x}$). The plus or minus sign appears in Eq. (2.4.17) because the direction of $\hat{\mathbf{n}}$ may point in either direction away from the surface. If the surface is closed, the usual convention is to take $\hat{\mathbf{n}}$ pointing outward from the surface.

2.4.2 The del Operator

It is convenient to write the gradient vector as

$$\text{grad } \phi \equiv \left(\mathbf{e}^1 \frac{\partial}{\partial q^1} + \mathbf{e}^2 \frac{\partial}{\partial q^2} + \mathbf{e}^3 \frac{\partial}{\partial q^3} \right) \phi \tag{2.4.18}$$

and interpret grad ϕ as some operator operating on ϕ, that is, grad $\phi \equiv \nabla \phi$. This operator is denoted by

$$\nabla \equiv \mathbf{e}^1 \frac{\partial}{\partial q^1} + \mathbf{e}^2 \frac{\partial}{\partial q^2} + \mathbf{e}^3 \frac{\partial}{\partial q^3} \tag{2.4.19}$$

and is called the *del operator*. The del operator is a *vector differential* operator, and the "components" $\partial/\partial q^1$, $\partial/\partial q^2$, and ∂/q^3 appear as covariant components.

Whereas the del operator has some of the properties of a vector, it does not have them all because it is an operator. For instance $\nabla \cdot \mathbf{A}$ is a scalar (called the divergence of $\mathbf{A}$), whereas $\mathbf{A} \cdot \nabla$ is a scalar *differential operator*. Thus the del operator does not commute in this sense.

In Cartesian systems, we have the simple form

$$\nabla \equiv \hat{\mathbf{e}}_x \frac{\partial}{\partial x} + \hat{\mathbf{e}}_y \frac{\partial}{\partial y} + \hat{\mathbf{e}}_z \frac{\partial}{\partial z}, \tag{2.4.20}$$

or, in the summation convection, we have

$$\nabla \equiv \hat{\mathbf{e}}_i \frac{\partial}{\partial x_i}. \tag{2.4.21}$$

2.4.3 Divergence and Curl of a Vector

The dot product of a del operator with a vector is called the *divergence of a vector* and denoted by

$$\nabla \cdot \mathbf{A} \equiv \text{div} \mathbf{A}. \tag{2.4.22}$$

If we take the divergence of the gradient vector, we have

$$\text{div}(\text{grad } \phi) \equiv \nabla \cdot \nabla \phi = (\nabla \cdot \nabla)\phi = \nabla^2 \phi. \tag{2.4.23}$$

The notation $\nabla^2 = \nabla \cdot \nabla$ is called the *Laplacian operator*. In Cartesian systems, this reduces to the simple form

$$\nabla^2 \phi = \frac{\partial^2 \phi}{\partial x^2} + \frac{\partial^2 \phi}{\partial y^2} + \frac{\partial^2 \phi}{\partial z^2} = \frac{\partial^2 \phi}{\partial x_i \partial x_i}. \tag{2.4.24}$$

The Laplacian of a scalar appears frequently in the partial differential equations governing physical phenomena (see Section 8.3.3).

The *curl of a vector* is defined as the del operator operating on a vector by means of the cross product:

$$\text{curl } \mathbf{A} = \nabla \times \mathbf{A} = e_{ijk} \hat{\mathbf{e}}_i \frac{\partial A_k}{\partial x_j}. \tag{2.4.25}$$

The quantity $\hat{\mathbf{n}} \cdot \text{grad } \phi$ of a function ϕ is called the *normal derivative* of ϕ, and it is denoted by

$$\frac{\partial \phi}{\partial n} \equiv \hat{\mathbf{n}} \cdot \text{grad } \phi = \hat{\mathbf{n}} \cdot \nabla \phi. \tag{2.4.26}$$

In a Cartesian system, this becomes

$$\frac{\partial \phi}{\partial n} = \frac{\partial \phi}{\partial x} n_x + \frac{\partial \phi}{\partial y} n_y + \frac{\partial \phi}{\partial z} n_z, \tag{2.4.27}$$

where n_x, n_y, and n_z are the direction cosines of the unit normal

$$\hat{\mathbf{n}} = n_x \hat{\mathbf{e}}_x + n_y \hat{\mathbf{e}}_y + n_z \hat{\mathbf{e}}_z. \tag{2.4.28}$$

Next, we present several examples to illustrate the use of index notation to prove certain identities involving vector calculus.

EXAMPLE 2.4.1: Establish the following identities using the index notation:
1. $\nabla(r) = \frac{\mathbf{r}}{r}$.
2. $\nabla(r^n) = nr^{n-2}\mathbf{r}$.
3. $\nabla \times (\nabla F) = \mathbf{0}$.
4. $\nabla \cdot (\nabla F \times \nabla G) = 0$.
5. $\nabla \times (\nabla \times \mathbf{v}) = \nabla(\nabla \cdot \mathbf{v}) - \nabla^2 \mathbf{v}$.
6. $\text{div}(\mathbf{A} \times \mathbf{B}) = \nabla \times \mathbf{A} \cdot \mathbf{B} - \nabla \times \mathbf{B} \cdot \mathbf{A}$.

SOLUTION:
1. Consider

$$\nabla(r) = \hat{\mathbf{e}}_i \frac{\partial r}{\partial x_i} = \hat{\mathbf{e}}_i \frac{\partial}{\partial x_i} (x_j x_j)^{\frac{1}{2}}$$

$$= \hat{\mathbf{e}}_i \frac{1}{2} (x_j x_j)^{\frac{1}{2}-1} 2x_i = \hat{\mathbf{e}}_i x_i (x_j x_j)^{-\frac{1}{2}} = \frac{\mathbf{r}}{r} = \frac{\mathbf{x}}{r}, \tag{a}$$

from which we note the identity

$$\frac{\partial r}{\partial x_i} = \frac{x_i}{r}. \tag{b}$$

2. Similar to **1**, we have

$$\nabla(r^n) = \hat{\mathbf{e}}_i \frac{\partial}{\partial x_i} (r^n) = nr^{n-1} \hat{\mathbf{e}}_i \frac{\partial r}{\partial x_i} = nr^{n-2} x_i \hat{\mathbf{e}}_i = nr^{n-2}\mathbf{r}.$$

3. Consider the expression

$$\nabla \times (\nabla F) = \left(\hat{\mathbf{e}}_i \frac{\partial}{\partial x_i}\right) \times \left(\hat{\mathbf{e}}_j \frac{\partial F}{\partial x_j}\right) = e_{ijk} \hat{\mathbf{e}}_k \frac{\partial^2 F}{\partial x_i \partial x_j}.$$

Note that $\frac{\partial^2 F}{\partial x_i \partial x_j}$ is symmetric in i and j. Consider the kth component of the above vector

$$e_{ijk} \frac{\partial^2 F}{\partial x_i \partial x_j} = -e_{jik} \frac{\partial^2 F}{\partial x_i \partial x_j} \quad \text{(interchanged } i \text{ and } j)$$

$$= -e_{ijk} \frac{\partial^2 F}{\partial x_j \partial x_i} \quad \text{(renamed } i \text{ as } j \text{ and } j \text{ as } i)$$

$$= -e_{ijk} \frac{\partial^2 F}{\partial x_i \partial x_j} \quad \left(\text{used the symmetry of } \tfrac{\partial^2 F}{\partial x_i \partial x_j}\right).$$

Thus, the expression is equal to its own negative. Obviously, the only parameter that is equal to its own negative is zero. Hence, we have $\nabla \times (\nabla F) = \mathbf{0}$. It also follows that $e_{ijk} F_{ij} = 0$ whenever $F_{ij} = F_{ji}$, that is, F_{ij} is symmetric.

4. We have

$$\nabla \cdot (\nabla F \times \nabla G) = \left(\hat{\mathbf{e}}_i \frac{\partial}{\partial x_i}\right) \cdot \left(\hat{\mathbf{e}}_j \frac{\partial F}{\partial x_j} \times \hat{\mathbf{e}}_k \frac{\partial G}{\partial x_k}\right)$$

$$= e_{jk\ell}(\hat{\mathbf{e}}_i \cdot \hat{\mathbf{e}}_\ell) \left(\frac{\partial^2 F}{\partial x_i \partial x_j} \frac{\partial G}{\partial x_k} + \frac{\partial F}{\partial x_j} \frac{\partial^2 G}{\partial x_i \partial x_k}\right)$$

$$= e_{ijk} \left(\frac{\partial^2 F}{\partial x_i \partial x_j} \frac{\partial G}{\partial x_k} + \frac{\partial F}{\partial x_j} \frac{\partial^2 G}{\partial x_i \partial x_k}\right) = 0,$$

where we have used the result from **3**.

5. Observe that

$$\nabla \times (\nabla \times \mathbf{v}) = \hat{\mathbf{e}}_i \frac{\partial}{\partial x_i} \times \left(\hat{\mathbf{e}}_j \frac{\partial}{\partial x_j} \times v_k \hat{\mathbf{e}}_k\right)$$

$$= \hat{\mathbf{e}}_i \frac{\partial}{\partial x_i} \times \left(e_{jk\ell} \frac{\partial v_k}{\partial x_j} \hat{\mathbf{e}}_\ell\right) = e_{i\ell m} e_{jk\ell} \frac{\partial^2 v_k}{\partial x_i \partial x_j} \hat{\mathbf{e}}_m.$$

Using the e-δ identity, we obtain

$$\nabla \times (\nabla \times \mathbf{v}) \equiv (\delta_{mj}\delta_{ik} - \delta_{mk}\delta_{ij}) \frac{\partial^2 v_k}{\partial x_i \partial x_j} \hat{\mathbf{e}}_m = \frac{\partial^2 v_i}{\partial x_i \partial x_j} \hat{\mathbf{e}}_j - \frac{\partial^2 v_k}{\partial x_i \partial x_i} \hat{\mathbf{e}}_k$$

$$= \hat{\mathbf{e}}_j \frac{\partial}{\partial x_j} \left(\frac{\partial v_i}{\partial x_i}\right) - \frac{\partial^2}{\partial x_i \partial x_i} (v_k \hat{\mathbf{e}}_k) = \nabla (\nabla \cdot \mathbf{v}) - \nabla^2 \mathbf{v}.$$

This result is sometimes used as the definition of the Laplacian of a vector; that is,

$$\nabla^2 \mathbf{v} = \text{grad}(\text{div } \mathbf{v}) - \text{curl curl } \mathbf{v}.$$

6. Expanding the vector expression

$$\text{div} (\mathbf{A} \times \mathbf{B}) = \hat{\mathbf{e}}_i \cdot \frac{\partial}{\partial x_i} (e_{jk\ell} A_j B_k \hat{\mathbf{e}}_\ell) = e_{ijk} \left(\frac{\partial A_j}{\partial x_i} B_k + A_j \frac{\partial B_k}{\partial x_i}\right)$$

$$= \nabla \times \mathbf{A} \cdot \mathbf{B} - \nabla \times \mathbf{B} \cdot \mathbf{A}.$$

Table 2.4.1. *Vector expressions and their Cartesian component forms (**A**, **B**, and **C**) are vector functions, and U is a scalar function; (ê₁, ê₂, ê₃) are the Cartesian unit vectors*

No.	Vector form	Component form
1.	$\mathbf{A}$	$A_i \hat{\mathbf{e}}_i$
2.	$\mathbf{A} \cdot \mathbf{B}$	$A_i B_i$
3.	$\mathbf{A} \times \mathbf{B}$	$e_{ijk} A_i B_j \hat{\mathbf{e}}_k$
4.	$\mathbf{A} \cdot (\mathbf{B} \times \mathbf{C})$	$e_{ijk} A_i B_j C_k$
5.	$\mathbf{A} \times (\mathbf{B} \times \mathbf{C}) = \mathbf{B}(\mathbf{A} \cdot \mathbf{C}) - \mathbf{C}(\mathbf{A} \cdot \mathbf{B})$	$e_{ijk} e_{klm} A_j B_l C_m \hat{\mathbf{e}}_i$
6.	∇U	$\frac{\partial U}{\partial x_i} \hat{\mathbf{e}}_i$
7.	$\nabla \mathbf{A}$	$\frac{\partial A_j}{\partial x_i} \hat{\mathbf{e}}_i \hat{\mathbf{e}}_j$
8.	$\nabla \cdot \mathbf{A}$	$\frac{\partial A_i}{\partial x_i}$
9.	$\nabla \times \mathbf{A}$	$e_{ijk} \frac{\partial A_j}{\partial x_i} \hat{\mathbf{e}}_k$
10.	$\nabla \cdot (\mathbf{A} \times \mathbf{B}) = \mathbf{B} \cdot (\nabla \times \mathbf{A}) - \mathbf{A} \cdot (\nabla \times \mathbf{B})$	$e_{ijk} \frac{\partial}{\partial x_i}(A_j B_k)$
11.	$\nabla \cdot (U\mathbf{A}) = U\nabla \cdot \mathbf{A} + \nabla U \cdot \mathbf{A}$	$\frac{\partial}{\partial x_i}(U A_i)$
12.	$\nabla \times (U\mathbf{A}) = \nabla U \times \mathbf{A} + U\nabla \times \mathbf{A}$	$e_{ijk} \frac{\partial}{\partial x_j}(U A_k)\hat{\mathbf{e}}_i$
13.	$\nabla(U\mathbf{A}) = \nabla U\mathbf{A} + U\nabla\mathbf{A}$	$\hat{\mathbf{e}}_j \frac{\partial}{\partial x_j}(U A_k \hat{\mathbf{e}}_k)$
14.	$\nabla \times (\mathbf{A} \times \mathbf{B}) = \mathbf{A}(\nabla \cdot \mathbf{B}) - \mathbf{B}(\nabla \cdot \mathbf{A})$ $+ \mathbf{B} \cdot \nabla\mathbf{A} - \mathbf{A} \cdot \nabla\mathbf{B}$	$e_{ijk} e_{mkl} \frac{\partial}{\partial x_m}(A_i B_j)\hat{\mathbf{e}}_l$
15.	$(\nabla \times \mathbf{A}) \times \mathbf{B} = \mathbf{B} \cdot [\nabla\mathbf{A} - (\nabla\mathbf{A})^{\mathrm{T}}]$	$e_{ijk} e_{klm} B_l \frac{\partial A_j}{\partial x_i} \hat{\mathbf{e}}_m$
16.	$\nabla \cdot (\nabla U) = \nabla^2 U$	$\frac{\partial^2 U}{\partial x_i \partial x_i}$
17.	$\nabla \cdot (\nabla \mathbf{A}) = \nabla^2 \mathbf{A}$	$\frac{\partial^2 A_j}{\partial x_i \partial x_i} \hat{\mathbf{e}}_j$
18.	$\nabla \times \nabla \times \mathbf{A} = \nabla(\nabla \cdot \mathbf{A}) - (\nabla \cdot \nabla)\mathbf{A}$	$e_{mil} e_{jkl} \frac{\partial^2 A_k}{\partial x_i \partial x_j} \hat{\mathbf{e}}_m$
19.	$(\mathbf{A} \cdot \nabla)\mathbf{B}$	$A_j \frac{\partial B_i}{\partial x_j} \hat{\mathbf{e}}_i$
20.	$\mathbf{A}(\nabla \cdot \mathbf{B})$	$A_i \hat{\mathbf{e}}_i \frac{\partial B_j}{\partial x_j}$

The examples presented illustrate the convenience of index notation in establishing vector identities and simplifying vector expressions. The difficult step in these proofs is recognizing vector operations from index notation. A list of vector operations in both vector notation and in Cartesian component form is presented in Table 2.4.1.

2.4.4 Cylindrical and Spherical Coordinate Systems

Two commonly used orthogonal curvilinear coordinate systems are *cylindrical* coordinate system [see Figure 2.4.3(a)] and *spherical* coordinate system [see Figure 2.4.3(b)]. Table 2.4.2 contains a summary of the basic information for the two coordinate systems. It is clear from Table 2.4.2 that the matrix of direction cosines between the orthogonal rectangular Cartesian system (x, y, z) and the orthogonal

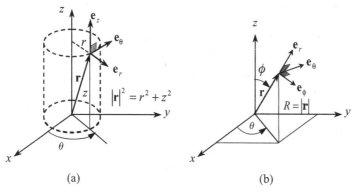

Figure 2.4.3. (a) Cylindrical coordinate system. (b) Spherical coordinate system.

curvilinear systems (r, θ, z) and (R, ϕ, θ), respectively, are as given in Eqs. (2.4.29)–(2.4.32)

Cylindrical coordinates

$$
\begin{Bmatrix} \hat{\mathbf{e}}_r \\ \hat{\mathbf{e}}_\theta \\ \hat{\mathbf{e}}_z \end{Bmatrix} = \begin{bmatrix} \cos\theta & \sin\theta & 0 \\ -\sin\theta & \cos\theta & 0 \\ 0 & 0 & 1 \end{bmatrix} \begin{Bmatrix} \hat{\mathbf{e}}_x \\ \hat{\mathbf{e}}_y \\ \hat{\mathbf{e}}_x \end{Bmatrix},
\tag{2.4.29}
$$

$$
\begin{Bmatrix} \hat{\mathbf{e}}_x \\ \hat{\mathbf{e}}_y \\ \hat{\mathbf{e}}_z \end{Bmatrix} = \begin{bmatrix} \cos\theta & -\sin\theta & 0 \\ \sin\theta & \cos\theta & 0 \\ 0 & 0 & 1 \end{bmatrix} \begin{Bmatrix} \hat{\mathbf{e}}_r \\ \hat{\mathbf{e}}_\theta \\ \hat{\mathbf{e}}_z \end{Bmatrix}.
\tag{2.4.30}
$$

Spherical coordinates

$$
\begin{Bmatrix} \hat{\mathbf{e}}_R \\ \hat{\mathbf{e}}_\phi \\ \hat{\mathbf{e}}_\theta \end{Bmatrix} = \begin{bmatrix} \sin\phi\cos\theta & \sin\phi\sin\theta & \cos\phi \\ \cos\phi\cos\theta & \cos\phi\sin\theta & -\sin\phi \\ -\sin\theta & \cos\theta & 0 \end{bmatrix} \begin{Bmatrix} \hat{\mathbf{e}}_x \\ \hat{\mathbf{e}}_y \\ \hat{\mathbf{e}}_x \end{Bmatrix},
\tag{2.4.31}
$$

$$
\begin{Bmatrix} \hat{\mathbf{e}}_x \\ \hat{\mathbf{e}}_y \\ \hat{\mathbf{e}}_z \end{Bmatrix} = \begin{bmatrix} \sin\phi\cos\theta & \cos\phi\cos\theta & -\sin\theta \\ \sin\phi\sin\theta & \cos\phi\sin\theta & \cos\theta \\ \cos\phi & -\sin\phi & 0 \end{bmatrix} \begin{Bmatrix} \hat{\mathbf{e}}_R \\ \hat{\mathbf{e}}_\phi \\ \hat{\mathbf{e}}_\theta \end{Bmatrix}.
\tag{2.4.32}
$$

2.4.5 Gradient, Divergence, and Curl Theorems

Integral identities involving the gradient of a vector, divergence of a vector, and curl of a vector can be established from integral relations between volume integrals and surface integrals. These identities will be useful in later chapters when we derive the equations of a continuous medium.

Let Ω denote a region in $\Re^3$ bounded by the closed surface Γ. Let ds be a differential element of surface and $\hat{\mathbf{n}}$ the unit outward normal, and let $d\mathbf{x}$ be a differential

Table 2.4.2. *Base vectors and del and Laplace operators in cylindrical and spherical coordinate systems*

Cylindrical coordinate system (r, θ, z)

$x = r\cos\theta,\ y = r\sin\theta,\ z = z,\ \mathbf{r} = r\hat{\mathbf{e}}_r + z\hat{\mathbf{e}}_z$

$\mathbf{A} = A_r\hat{\mathbf{e}}_r + A_\theta\hat{\mathbf{e}}_\theta + A_z\hat{\mathbf{e}}_z$ (typical vector)

$\hat{\mathbf{e}}_r = \cos\theta\,\hat{\mathbf{e}}_x + \sin\theta\,\hat{\mathbf{e}}_y,\ \hat{\mathbf{e}}_\theta = -\sin\theta\,\hat{\mathbf{e}}_x + \cos\theta\,\hat{\mathbf{e}}_y,\ \hat{\mathbf{e}}_z = \hat{\mathbf{e}}_z$

$\frac{\partial\hat{\mathbf{e}}_r}{\partial\theta} = -\sin\theta\,\hat{\mathbf{e}}_x + \cos\theta\,\hat{\mathbf{e}}_y = \hat{\mathbf{e}}_\theta,\ \frac{\partial\hat{\mathbf{e}}_\theta}{\partial\theta} = -\cos\theta\,\hat{\mathbf{e}}_x - \sin\theta\,\hat{\mathbf{e}}_y = -\hat{\mathbf{e}}_r$

All other derivatives of the base vectors are zero.

$\nabla = \hat{\mathbf{e}}_r\frac{\partial}{\partial r} + \frac{1}{r}\hat{\mathbf{e}}_\theta\frac{\partial}{\partial\theta} + \hat{\mathbf{e}}_z\frac{\partial}{\partial z},$

$\nabla^2 = \frac{1}{r}\left[\frac{\partial}{\partial r}\left(r\frac{\partial}{\partial r}\right) + \frac{1}{r}\frac{\partial^2}{\partial\theta^2} + r\frac{\partial^2}{\partial z^2}\right]$

$\nabla\cdot\mathbf{A} = \frac{1}{r}\left[\frac{\partial(rA_r)}{\partial r} + \frac{\partial A_\theta}{\partial\theta} + r\frac{\partial A_z}{\partial z}\right]$

$\nabla\times\mathbf{A} = \left(\frac{1}{r}\frac{\partial A_z}{\partial\theta} - \frac{\partial A_\theta}{\partial z}\right)\hat{\mathbf{e}}_r + \left(\frac{\partial A_r}{\partial z} - \frac{\partial A_z}{\partial r}\right)\hat{\mathbf{e}}_\theta + \frac{1}{r}\left[\frac{\partial(rA_\theta)}{\partial r} - \frac{\partial A_r}{\partial\theta}\right]\hat{\mathbf{e}}_z$

Spherical coordinate system (R, ϕ, θ)

$x = R\sin\phi\cos\theta,\ y = R\sin\phi\sin\theta,\ z = R\cos\phi,\ \mathbf{r} = R\hat{\mathbf{e}}_R$

$\mathbf{A} = A_R\hat{\mathbf{e}}_R + A_\phi\hat{\mathbf{e}}_\phi + A_\theta\hat{\mathbf{e}}_\theta$ (typical vector)

$\hat{\mathbf{e}}_R = \sin\phi\cos\theta\,\hat{\mathbf{e}}_x + \sin\phi\sin\theta\,\hat{\mathbf{e}}_y + \cos\phi\,\hat{\mathbf{e}}_z$

$\hat{\mathbf{e}}_\phi = \cos\phi\cos\theta\,\hat{\mathbf{e}}_x + \cos\phi\sin\theta\,\hat{\mathbf{e}}_y - \sin\phi\,\hat{\mathbf{e}}_z$

$\hat{\mathbf{e}}_\theta = -\sin\theta\,\hat{\mathbf{e}}_x + \cos\theta\,\hat{\mathbf{e}}_y$

$\hat{\mathbf{e}}_x = \sin\phi\cos\theta\,\hat{\mathbf{e}}_R + \cos\phi\cos\theta\,\hat{\mathbf{e}}_\phi - \sin\theta\,\hat{\mathbf{e}}_\theta$

$\hat{\mathbf{e}}_y = \sin\phi\sin\theta\,\hat{\mathbf{e}}_R + \cos\phi\sin\theta\,\hat{\mathbf{e}}_\phi + \cos\theta\,\hat{\mathbf{e}}_\theta$

$\hat{\mathbf{e}}_z = \cos\phi\,\hat{\mathbf{e}}_R - \sin\phi\,\hat{\mathbf{e}}_\phi$

$\frac{\partial\hat{\mathbf{e}}_R}{\partial\phi} = \hat{\mathbf{e}}_\phi,\ \frac{\partial\hat{\mathbf{e}}_R}{\partial\theta} = \sin\phi\,\hat{\mathbf{e}}_\theta,\ \frac{\partial\hat{\mathbf{e}}_\phi}{\partial\phi} = -\hat{\mathbf{e}}_R$

$\frac{\partial\hat{\mathbf{e}}_\phi}{\partial\theta} = \cos\phi\,\hat{\mathbf{e}}_\theta,\ \frac{\partial\hat{\mathbf{e}}_\theta}{\partial\theta} = -\sin\phi\,\hat{\mathbf{e}}_R - \cos\phi\,\hat{\mathbf{e}}_\phi$

All other derivatives of the base vectors are zero.

$\nabla = \hat{\mathbf{e}}_R\frac{\partial}{\partial R} + \frac{1}{R}\hat{\mathbf{e}}_\phi\frac{\partial}{\partial\phi} + \frac{1}{R\sin\phi}\hat{\mathbf{e}}_\theta\frac{\partial}{\partial\theta},$

$\nabla^2 = \frac{1}{R^2}\frac{\partial}{\partial R}\left(R^2\frac{\partial}{\partial R}\right) + \frac{1}{R^2\sin\phi}\frac{\partial}{\partial\phi}\left(\sin\phi\frac{\partial}{\partial\phi}\right) + \frac{1}{R^2\sin^2\phi}\frac{\partial^2}{\partial\theta^2}$

$\nabla\cdot\mathbf{A} = 2\frac{A_R}{R} + \frac{\partial A_R}{\partial R} + \frac{1}{R\sin\phi}\frac{\partial(A_\phi\sin\phi)}{\partial\phi} + \frac{1}{R\sin\phi}\frac{\partial A_\theta}{\partial\theta}$

$\nabla\times\mathbf{A} = \frac{1}{R\sin\phi}\left[\frac{\partial(\sin\phi A_\theta)}{\partial\phi} - \frac{\partial A_\phi}{\partial\theta}\right]\hat{\mathbf{e}}_R + \left[\frac{1}{R\sin\phi}\frac{\partial A_R}{\partial\theta} - \frac{1}{R}\frac{\partial(RA_\theta)}{\partial R}\right]\hat{\mathbf{e}}_\phi$

$\quad + \frac{1}{R}\left[\frac{\partial(RA_\phi)}{\partial R} - \frac{\partial A_R}{\partial\phi}\right]\hat{\mathbf{e}}_\theta$

volume element in Ω. The following relations, known from advanced calculus, hold:

$$\int_\Omega \nabla\phi \, d\mathbf{x} = \oint_\Gamma \hat{\mathbf{n}}\phi \, ds \quad \text{(Gradient theorem).} \tag{2.4.33}$$

$$\int_\Omega \nabla \cdot \mathbf{A} \, d\mathbf{x} = \oint_\Gamma \hat{\mathbf{n}} \cdot \mathbf{A} \, ds \quad \text{(Divergence theorem).} \tag{2.4.34}$$

$$\int_\Omega \nabla \times \mathbf{A} \, d\mathbf{x} = \oint_\Gamma \hat{\mathbf{n}} \times \mathbf{A} \, ds \quad \text{(Curl theorem).} \tag{2.4.35}$$

These forms are known as the *invariant forms* since they do not depend in any way upon defined coordinate systems.

The combination $\hat{\mathbf{n}} \cdot \mathbf{A} \, ds$ is called the *outflow of $\mathbf{A}$ through the differential surface ds*. The integral is called the total or net outflow through the surrounding surface Δs. This is easiest to see if one imagines that $\mathbf{A}$ is a velocity vector and the outflow is an amount of fluid flow. In the limit as the region shrinks to a point, the net outflow per unit volume is associated therefore with the *divergence of the vector field*.

2.5 Tensors

2.5.1 Dyads and Polyads

As stated earlier, the surface force acting on a small element of area in a continuous medium depends not only on the magnitude of the area but also on the orientation of the area. The stress, which is force per unit area, not only depends on the magnitude of the force and orientation of the plane but also on the direction of the force. Thus, specification of stress at a point requires two vectors, one perpendicular to the plane on which the force is acting and the other in the direction of the force. Such an object is known as a *dyad*, or what we shall call a *second-order tensor*. Because of its utilization in physical applications, a dyad is defined as two vectors standing side by side and acting as a unit. A linear combination of dyads is called a *dyadic*. Let $\mathbf{A}_1, \mathbf{A}_2, \ldots, \mathbf{A}_n$ and $\mathbf{B}_1, \mathbf{B}_2, \ldots, \mathbf{B}_n$ be arbitrary vectors. Then we can represent a dyadic as

$$\boldsymbol{\Phi} = \mathbf{A}_1\mathbf{B}_1 + \mathbf{A}_2\mathbf{B}_2 + \cdots + \mathbf{A}_n\mathbf{B}_n. \tag{2.5.1}$$

The transpose of a dyadic is defined as the result obtained by the interchange of the two vectors in each of the dyads. For example, the transpose of the dyadic in Eq. (2.5.1) is

$$\boldsymbol{\Phi}^{\mathrm{T}} = \mathbf{B}_1\mathbf{A}_1 + \mathbf{B}_2\mathbf{A}_2 + \cdots + \mathbf{B}_n\mathbf{A}_n.$$

One of the properties of a dyadic is defined by the dot product with a vector, say $\mathbf{V}$:

$$\boldsymbol{\Phi} \cdot \mathbf{V} = \mathbf{A}_1(\mathbf{B}_1 \cdot \mathbf{V}) + \mathbf{A}_2(\mathbf{B}_2 \cdot \mathbf{V}) + \cdots + \mathbf{A}_n(\mathbf{B}_n \cdot \mathbf{V}),$$
$$\mathbf{V} \cdot \boldsymbol{\Phi} = (\mathbf{V} \cdot \mathbf{A}_1)\mathbf{B}_1 + (\mathbf{V} \cdot \mathbf{A}_2)\mathbf{B}_2 + \cdots + (\mathbf{V} \cdot \mathbf{A}_n)\mathbf{B}_n. \tag{2.5.2}$$

The dot operation with a vector produces another vector. In the first case, the dyad acts as a *prefactor* and in the second case as a *postfactor*. The two operations in general produce different vectors. The expressions in Eq. (2.5.2) can also be written in alternative form using the definition of the transpose of a dyad as

$$\mathbf{V} \cdot \mathbf{\Phi} = \mathbf{\Phi}^{\mathrm{T}} \cdot \mathbf{V}, \quad \mathbf{\Phi} \cdot \mathbf{V} = \mathbf{V} \cdot \mathbf{\Phi}^{\mathrm{T}}. \tag{2.5.3}$$

In general, one can show (see Problem 2.25) that the transpose of the product of tensors (of any order) follows the rule

$$(\mathbf{\Phi} \cdot \mathbf{\Psi})^{\mathrm{T}} = \mathbf{\Psi}^{\mathrm{T}} \cdot \mathbf{\Phi}^{\mathrm{T}}, \quad (\mathbf{\Phi} \cdot \mathbf{\Psi} \cdot \mathbf{V})^{\mathrm{T}} = \mathbf{V}^{\mathrm{T}} \cdot \mathbf{\Psi}^{\mathrm{T}} \cdot \mathbf{\Phi}^{\mathrm{T}}. \tag{2.5.4}$$

The dot product of a dyadic with itself is a dyadic, and it is denoted by

$$\mathbf{\Phi} \cdot \mathbf{\Phi} = \mathbf{\Phi}^2. \tag{2.5.5}$$

In general, we have

$$\mathbf{\Phi}^n = \mathbf{\Phi}^{n-1} \cdot \mathbf{\Phi}. \tag{2.5.6}$$

2.5.2 Nonion Form of a Dyadic

Let each of the vectors in the dyadic (2.5.1) be represented in a given basis system. In Cartesian system, we have

$$\mathbf{A}_i = A_{ij}\mathbf{e}_j, \quad \mathbf{B}_i = B_{ik}\mathbf{e}_k.$$

The summations on j and k are implied by the repeated indices.

We can display all of the components of a dyadic $\mathbf{\Phi}$ by letting the k index run to the right and the j index run downward:

$$\mathbf{\Phi} = \phi_{11}\hat{\mathbf{e}}_1\hat{\mathbf{e}}_1 + \phi_{12}\hat{\mathbf{e}}_1\hat{\mathbf{e}}_2 + \phi_{13}\hat{\mathbf{e}}_1\hat{\mathbf{e}}_3$$

$$+ \phi_{21}\hat{\mathbf{e}}_2\hat{\mathbf{e}}_1 + \phi_{22}\hat{\mathbf{e}}_2\hat{\mathbf{e}}_2 + \phi_{23}\hat{\mathbf{e}}_2\hat{\mathbf{e}}_3$$

$$+ \phi_{31}\hat{\mathbf{e}}_3\hat{\mathbf{e}}_1 + \phi_{32}\hat{\mathbf{e}}_3\hat{\mathbf{e}}_2 + \phi_{33}\hat{\mathbf{e}}_3\hat{\mathbf{e}}_3. \tag{2.5.7}$$

This form is called the *nonion* form of a dyadic. Equation (2.5.7) illustrates that a dyad in three-dimensional space has nine independent components in general, each component associated with a certain dyad pair. The components are thus said to be ordered. When the ordering is understood, such as suggested by the nonion form (2.5.7), the explicit writing of the dyads can be suppressed and the dyadic written as an array:

$$[\mathbf{\Phi}] = \begin{bmatrix} \phi_{11} & \phi_{12} & \phi_{13} \\ \phi_{21} & \phi_{22} & \phi_{23} \\ \phi_{31} & \phi_{32} & \phi_{33} \end{bmatrix} \quad \text{and} \quad \mathbf{\Phi} = \begin{Bmatrix} \hat{\mathbf{e}}_1 \\ \hat{\mathbf{e}}_2 \\ \hat{\mathbf{e}}_3 \end{Bmatrix}^{\mathrm{T}} [\mathbf{\Phi}] \begin{Bmatrix} \hat{\mathbf{e}}_1 \\ \hat{\mathbf{e}}_2 \\ \hat{\mathbf{e}}_3 \end{Bmatrix}. \tag{2.5.8}$$

This representation is simpler than Eq. (2.5.7), but it is taken to mean the same.

The unit dyad is defined as

$$\mathbf{I} = \hat{\mathbf{e}}_i\hat{\mathbf{e}}_i. \tag{2.5.9}$$

It is clear that the unit second-order tensor is symmetric. With the help of the Kronecker delta symbol δ_{ij}, the unit dyadic in an orthogonal Cartesian coordinate system can be written alternatively as

$$\mathbf{I} = \delta_{ij}\,\hat{\mathbf{e}}_i\hat{\mathbf{e}}_j, \quad \mathbf{I} = \left\{\begin{array}{c}\hat{\mathbf{e}}_1\\\hat{\mathbf{e}}_2\\\hat{\mathbf{e}}_3\end{array}\right\}^{\mathrm{T}}[I]\left\{\begin{array}{c}\hat{\mathbf{e}}_1\\\hat{\mathbf{e}}_2\\\hat{\mathbf{e}}_3\end{array}\right\}, \quad [\mathbf{I}] = \begin{bmatrix}1 & 0 & 0\\0 & 1 & 0\\0 & 0 & 1\end{bmatrix}. \tag{2.5.10}$$

The permutation symbol e_{ijk} can be viewed as the Cartesian components of a third-order tensor of a special kind.

The "double-dot product" between two dyads is useful in the sequel. The double-dot product between a dyad $(\mathbf{AB})$ and another dyad $(\mathbf{CD})$ is defined as the scalar

$$(\mathbf{AB}) : (\mathbf{CD}) \equiv (\mathbf{B}\cdot\mathbf{C})(\mathbf{A}\cdot\mathbf{D}). \tag{2.5.11}$$

The double-dot product, by this definition, is commutative. The double-dot product between two dyads in a rectangular Cartesian system is given by

$$\begin{aligned}\boldsymbol{\Phi} : \boldsymbol{\Psi} &= (\phi_{ij}\hat{\mathbf{e}}_i\hat{\mathbf{e}}_j) : (\psi_{mn}\hat{\mathbf{e}}_m\hat{\mathbf{e}}_n)\\&= \phi_{ij}\psi_{mn}(\hat{\mathbf{e}}_i\cdot\hat{\mathbf{e}}_n)(\hat{\mathbf{e}}_j\cdot\hat{\mathbf{e}}_m)\\&= \phi_{ij}\psi_{mn}\delta_{in}\delta_{jm}\\&= \phi_{ij}\psi_{ji}.\end{aligned} \tag{2.5.12}$$

The *trace* of a dyad is defined to be the double-dot product of the dyad with the unit dyad

$$\mathrm{tr}\,\boldsymbol{\Phi} = \boldsymbol{\Phi} : \mathbf{I}. \tag{2.5.13}$$

The trace of a tensor is *invariant*, called the *first principal invariant*, and it is denoted by I_1; that is, it is invariant under coordinate transformations ($\phi_{ii} = \bar{\phi}_{ii}$). The first, second, and third *principal invariants of a dyadic* are defined to be

$$I_1 = \mathrm{tr}\,\boldsymbol{\Phi}, \quad I_2 = \frac{1}{2}\left[(\mathrm{tr}\,\boldsymbol{\Phi})^2 - \mathrm{tr}\left(\boldsymbol{\Phi}^2\right)\right], \quad I_3 = \det\boldsymbol{\Phi}. \tag{2.5.14}$$

In terms of the rectangular Cartesian components, the three invariants have the form

$$I_1 = \phi_{ii}, \quad I_2 = \frac{1}{2}\left(\phi_{ii}\phi_{jj} - \phi_{ij}\phi_{ji}\right), \quad I_3 = |\phi|. \tag{2.5.15}$$

In the general scheme that is developed, scalars are the *zeroth-order tensors*, vectors are *first-order tensors*, and dyads are *second-order tensors*. The *third-order tensors* can be viewed as those derived from *triads*, or three vectors standing side by side.

2.5.3 Transformation of Components of a Dyadic

A second-order Cartesian tensor $\mathbf{\Phi}$ may be represented in barred and unbarred coordinate systems as

$$\mathbf{\Phi} = \phi_{ij}\hat{\mathbf{e}}_i\hat{\mathbf{e}}_j$$

$$= \bar{\phi}_{kl}\hat{\bar{\mathbf{e}}}_k\hat{\bar{\mathbf{e}}}_l.$$

The unit base vectors in the barred and unbarred systems are related by

$$\hat{\mathbf{e}}_i = \frac{\partial \bar{x}_j}{\partial x_i}\hat{\bar{\mathbf{e}}}_j \equiv \ell_{ji}\hat{\bar{\mathbf{e}}}_j \quad \text{or} \quad \ell_{ij} = \hat{\bar{\mathbf{e}}}_i \cdot \hat{\mathbf{e}}_j, \tag{2.5.16}$$

where ℓ_{ij} denote the direction cosines between barred and unbarred systems [see Eqs. (2.2.48)–(2.2.50)]. Thus the components of a second-order tensor transform according to

$$\bar{\phi}_{k\ell} = \ell_{ki}\ell_{\ell j}\phi_{ij} \quad \text{or} \quad \bar{\mathbf{\Phi}} = \mathbf{L}\,\mathbf{\Phi}\,\mathbf{L}^{\mathsf{T}}. \tag{2.5.17}$$

In some books, a second-order tensor is defined to be one whose components transform according to Eq. (2.5.17). In orthogonal coordinate systems, the determinant of the matrix of direction cosines is unity and its inverse is equal to the transpose:

$$\mathbf{L}^{-1} = \mathbf{L}^{\mathsf{T}} \quad \text{or} \quad \mathbf{L}\mathbf{L}^{\mathsf{T}} = \mathbf{I}. \tag{2.5.18}$$

Tensors $\mathbf{L}$ that satisfy the property (2.5.18) are called *orthogonal tensors*.

Tensors of various orders, especially the zeroth-, first-, and second-order appear in the study of a continuous medium. As we shall see in Chapter 6, the tensor that characterizes the material constitution is a fourth-order tensor. Tensors whose components are the same in all coordinate systems, that is, the components are invariant under coordinate transformations, are known as *isotropic* tensors. By definition, all zeroth- order tensors (i.e., scalars) are isotropic and the only isotropic tensor of order 1 is the zero vector $\mathbf{0}$. Every isotropic tensor $\mathbf{T}$ of order 2 can be written as $\mathbf{T} = \lambda\mathbf{I}$, and the components $C_{ijk\ell}$ of every fourth-order tensor $\mathbf{C}$ can be expressed as

$$C_{ijk\ell} = \lambda\delta_{ij}\delta_{k\ell} + \mu(\delta_{ik}\delta_{j\ell} + \delta_{i\ell}\delta_{jk}) + \kappa(\delta_{ik}\delta_{j\ell} - \delta_{i\ell}\delta_{jk}), \tag{2.5.19}$$

where λ, μ, and κ are scalars. The proof of the above statements are left as exercises to the reader.

2.5.4 Tensor Calculus

We note that the gradient of a vector is a second-order tensor

$$\nabla\mathbf{A} = \hat{\mathbf{e}}_i\frac{\partial}{\partial x_i}(A_j\hat{\mathbf{e}}_j) = \frac{\partial A_j}{\partial x_i}\hat{\mathbf{e}}_i\hat{\mathbf{e}}_j.$$

Note that the order of the base vectors is kept intact (i.e., not switched from the order in which they appear). It can be expressed as the sum of symmetric and antisymmetric parts by adding and subtracting $(1/2)(\partial A_i/\partial x_j)$:

$$\nabla \mathbf{A} = \frac{1}{2}\left(\frac{\partial A_j}{\partial x_i} + \frac{\partial A_i}{\partial x_j}\right)\hat{\mathbf{e}}_i\hat{\mathbf{e}}_j + \frac{1}{2}\left(\frac{\partial A_j}{\partial x_i} - \frac{\partial A_i}{\partial x_j}\right)\hat{\mathbf{e}}_i\hat{\mathbf{e}}_j. \tag{2.5.20}$$

Analogously to the divergence of a vector, the divergence of a second-order Cartesian tensor is defined as

$$\text{div } \boldsymbol{\Phi} = \nabla \cdot \boldsymbol{\Phi} = \hat{\mathbf{e}}_i \frac{\partial}{\partial x_i} \cdot (\phi_{mn}\hat{\mathbf{e}}_m\hat{\mathbf{e}}_n)$$

$$= \frac{\partial \phi_{mn}}{\partial x_i}(\hat{\mathbf{e}}_i \cdot \hat{\mathbf{e}}_m)\hat{\mathbf{e}}_n = \frac{\partial \phi_{in}}{\partial x_i}\hat{\mathbf{e}}_n. \tag{2.5.21}$$

Thus, the divergence of a second-order tensor is a vector. The integral theorems of vectors presented in Section 2.4.5 are also valid for tensors (second order and higher), but it is important that the order of the operations be observed.

The gradient and divergence of a tensor can be expressed in cylindrical and spherical coordinate systems by writing the del operator and the tensor in component form (see Table 2.4.2). For example, the gradient of a vector $\mathbf{u}$ in the cylindrical coordinate system can be obtained by writing

$$\mathbf{u} = u_r\hat{\mathbf{e}}_r + u_\theta\hat{\mathbf{e}}_\theta + u_z\hat{\mathbf{e}}_z, \tag{2.5.22}$$

$$\nabla = \hat{\mathbf{e}}_r\frac{\partial}{\partial r} + \frac{1}{r}\hat{\mathbf{e}}_\theta\frac{\partial}{\partial \theta} + \hat{\mathbf{e}}_z\frac{\partial}{\partial z}. \tag{2.5.23}$$

Then we have

$$\nabla\mathbf{u} = \left(\hat{\mathbf{e}}_r\frac{\partial}{\partial r} + \hat{\mathbf{e}}_\theta\frac{1}{r}\frac{\partial}{\partial \theta} + \hat{\mathbf{e}}_z\frac{\partial}{\partial z}\right)(u_r\hat{\mathbf{e}}_r + u_\theta\hat{\mathbf{e}}_\theta + u_z\hat{\mathbf{e}}_z)$$

$$= \hat{\mathbf{e}}_r\hat{\mathbf{e}}_r\frac{\partial u_r}{\partial r} + \hat{\mathbf{e}}_r\hat{\mathbf{e}}_\theta\frac{\partial u_\theta}{\partial r} + \hat{\mathbf{e}}_r\hat{\mathbf{e}}_z\frac{\partial u_z}{\partial r} + \frac{1}{r}\hat{\mathbf{e}}_\theta\hat{\mathbf{e}}_r\frac{\partial u_r}{\partial \theta}$$

$$+ \frac{u_r}{r}\hat{\mathbf{e}}_\theta\frac{\partial \hat{\mathbf{e}}_r}{\partial \theta} + \frac{1}{r}\hat{\mathbf{e}}_\theta\hat{\mathbf{e}}_\theta\frac{\partial u_\theta}{\partial \theta} + \frac{u_\theta}{r}\hat{\mathbf{e}}_\theta\frac{\partial \hat{\mathbf{e}}_\theta}{\partial \theta} + \frac{1}{r}\hat{\mathbf{e}}_\theta\hat{\mathbf{e}}_z\frac{\partial u_z}{\partial \theta}$$

$$+ \hat{\mathbf{e}}_z\hat{\mathbf{e}}_r\frac{\partial u_r}{\partial z} + \hat{\mathbf{e}}_z\hat{\mathbf{e}}_\theta\frac{\partial u_\theta}{\partial z} + \hat{\mathbf{e}}_z\hat{\mathbf{e}}_z\frac{\partial u_z}{\partial z}$$

$$= \hat{\mathbf{e}}_r\hat{\mathbf{e}}_r\frac{\partial u_r}{\partial r} + \hat{\mathbf{e}}_r\hat{\mathbf{e}}_\theta\frac{\partial u_\theta}{\partial r} + \hat{\mathbf{e}}_\theta\hat{\mathbf{e}}_r\frac{1}{r}\left(\frac{\partial u_r}{\partial \theta} - u_\theta\right)$$

$$+ \hat{\mathbf{e}}_r\hat{\mathbf{e}}_z\frac{\partial u_z}{\partial r} + \hat{\mathbf{e}}_z\hat{\mathbf{e}}_r\frac{\partial u_r}{\partial z} + \hat{\mathbf{e}}_\theta\hat{\mathbf{e}}_\theta\frac{1}{r}\left(u_r + \frac{\partial u_\theta}{\partial \theta}\right)$$

$$+ \frac{1}{r}\hat{\mathbf{e}}_\theta\hat{\mathbf{e}}_z\frac{\partial u_z}{\partial \theta} + \hat{\mathbf{e}}_z\hat{\mathbf{e}}_\theta\frac{\partial u_\theta}{\partial z} + \hat{\mathbf{e}}_z\hat{\mathbf{e}}_z\frac{\partial u_z}{\partial z}, \tag{2.5.24}$$

where the following derivatives of the base vectors are used:

$$\frac{\partial \hat{\mathbf{e}}_r}{\partial \theta} = \hat{\mathbf{e}}_\theta, \qquad \frac{\partial \hat{\mathbf{e}}_\theta}{\partial \theta} = -\hat{\mathbf{e}}_r. \tag{2.5.25}$$

Similarly, one can compute the curl and divergence of a tensor. The following example illustrates the procedure (also see Problems 2.26–2.28).

EXAMPLE 2.5.1: Suppose that the second-order tensor **E** is of the form (i.e., other components are zero)

$$\mathbf{E} = E_{rr}(r, z)\hat{\mathbf{e}}_r \hat{\mathbf{e}}_r + E_{\theta\theta}(r, z)\hat{\mathbf{e}}_\theta \hat{\mathbf{e}}_\theta$$

in the cylindrical coordinate system. Determine the curl and divergence of the tensor **E**.

SOLUTION: We note that $\partial(\cdot)/\partial\theta = 0$ because E_{rr} and $E_{\theta\theta}$ are not functions of θ. Using the del operator in Eq. (2.5.23), we can write $\nabla \times \mathbf{E}$ as

$$\nabla \times \mathbf{E} = \left(\hat{\mathbf{e}}_r \frac{\partial}{\partial r} + \frac{\hat{\mathbf{e}}_\theta}{r} \frac{\partial}{\partial \theta} + \hat{\mathbf{e}}_z \frac{\partial}{\partial z} \right) \times (E_{rr}\hat{\mathbf{e}}_r \hat{\mathbf{e}}_r + E_{\theta\theta}\hat{\mathbf{e}}_\theta \hat{\mathbf{e}}_\theta)$$

$$= \hat{\mathbf{e}}_r \times \frac{\partial}{\partial r} (E_{rr}\hat{\mathbf{e}}_r \hat{\mathbf{e}}_r + E_{\theta\theta}\hat{\mathbf{e}}_\theta \hat{\mathbf{e}}_\theta) + \frac{1}{r}\hat{\mathbf{e}}_\theta \times \frac{\partial}{\partial \theta} (E_{rr}\hat{\mathbf{e}}_r \hat{\mathbf{e}}_r + E_{\theta\theta}\hat{\mathbf{e}}_\theta \hat{\mathbf{e}}_\theta)$$

$$+ \hat{\mathbf{e}}_z \times \frac{\partial}{\partial z} (E_{rr}\hat{\mathbf{e}}_r \hat{\mathbf{e}}_r + E_{\theta\theta}\hat{\mathbf{e}}_\theta \hat{\mathbf{e}}_\theta)$$

$$= \hat{\mathbf{e}}_r \times \left(\frac{\partial E_{\theta\theta}}{\partial r} \hat{\mathbf{e}}_\theta \hat{\mathbf{e}}_\theta \right) + \frac{1}{r}\hat{\mathbf{e}}_\theta \times \left(E_{rr}\hat{\mathbf{e}}_r \frac{\partial \hat{\mathbf{e}}_r}{\partial \theta} + E_{\theta\theta} \frac{\partial \hat{\mathbf{e}}_\theta}{\partial \theta} \hat{\mathbf{e}}_\theta \right)$$

$$+ \hat{\mathbf{e}}_z \times \left(\frac{\partial E_{rr}}{\partial z} \hat{\mathbf{e}}_r \hat{\mathbf{e}}_r + \frac{\partial E_{\theta\theta}}{\partial z} \hat{\mathbf{e}}_\theta \hat{\mathbf{e}}_\theta \right)$$

$$= \frac{\partial E_{\theta\theta}}{\partial r} (\hat{\mathbf{e}}_r \times \hat{\mathbf{e}}_\theta) \hat{\mathbf{e}}_\theta + \frac{1}{r} E_{rr} (\hat{\mathbf{e}}_\theta \times \hat{\mathbf{e}}_r) \frac{\partial \hat{\mathbf{e}}_r}{\partial \theta}$$

$$+ \frac{1}{r} E_{\theta\theta} \left(\hat{\mathbf{e}}_\theta \times \frac{\partial \hat{\mathbf{e}}_\theta}{\partial \theta} \right) \hat{\mathbf{e}}_\theta + \frac{\partial E_{rr}}{\partial z} (\hat{\mathbf{e}}_z \times \hat{\mathbf{e}}_r) \hat{\mathbf{e}}_r + \frac{\partial E_{\theta\theta}}{\partial z} (\hat{\mathbf{e}}_z \times \hat{\mathbf{e}}_\theta) \hat{\mathbf{e}}_\theta$$

$$= \frac{\partial E_{\theta\theta}}{\partial r} \hat{\mathbf{e}}_z \hat{\mathbf{e}}_\theta - \frac{E_{rr}}{r} \hat{\mathbf{e}}_z \hat{\mathbf{e}}_\theta + \frac{1}{r} E_{\theta\theta} \hat{\mathbf{e}}_z \hat{\mathbf{e}}_\theta + \frac{\partial E_{rr}}{\partial z} \hat{\mathbf{e}}_\theta \hat{\mathbf{e}}_r - \frac{\partial E_{\theta\theta}}{\partial z} \hat{\mathbf{e}}_r \hat{\mathbf{e}}_\theta. \quad (2.5.26)$$

Similarly, we compute the divergence of **E** as

$$\nabla \cdot \mathbf{E} = \left(\hat{\mathbf{e}}_r \frac{\partial}{\partial r} + \frac{\hat{\mathbf{e}}_\theta}{r} \frac{\partial}{\partial \theta} + \hat{\mathbf{e}}_z \frac{\partial}{\partial z} \right) \cdot (E_{rr}\hat{\mathbf{e}}_r \hat{\mathbf{e}}_r + E_{\theta\theta}\hat{\mathbf{e}}_\theta \hat{\mathbf{e}}_\theta)$$

$$= \hat{\mathbf{e}}_r \cdot \frac{\partial}{\partial r} (E_{rr}\hat{\mathbf{e}}_r \hat{\mathbf{e}}_r + E_{\theta\theta}\hat{\mathbf{e}}_\theta \hat{\mathbf{e}}_\theta) + \frac{1}{r}\hat{\mathbf{e}}_\theta \cdot \frac{\partial}{\partial \theta} (E_{rr}\hat{\mathbf{e}}_r \hat{\mathbf{e}}_r + E_{\theta\theta}\hat{\mathbf{e}}_\theta \hat{\mathbf{e}}_\theta)$$

$$+ \hat{\mathbf{e}}_z \cdot \frac{\partial}{\partial z} (E_{rr}\hat{\mathbf{e}}_r \hat{\mathbf{e}}_r + E_{\theta\theta}\hat{\mathbf{e}}_\theta \hat{\mathbf{e}}_\theta)$$

$$= \hat{\mathbf{e}}_r \cdot \left(\frac{\partial E_{rr}}{\partial r} \hat{\mathbf{e}}_r \hat{\mathbf{e}}_r \right) + \frac{1}{r}\hat{\mathbf{e}}_\theta \cdot \left(E_{rr} \frac{\partial \hat{\mathbf{e}}_r}{\partial \theta} \hat{\mathbf{e}}_r + E_{\theta\theta}\hat{\mathbf{e}}_\theta \frac{\partial \hat{\mathbf{e}}_\theta}{\partial \theta} \right)$$

$$= \frac{\partial E_{rr}}{\partial r} \hat{\mathbf{e}}_r + \frac{1}{r} (E_{rr} - E_{\theta\theta}) \hat{\mathbf{e}}_r. \quad (2.5.27)$$

2.5.5 Eigenvalues and Eigenvectors of Tensors

It is conceptually useful to regard a tensor as an operator that changes a vector
into another vector (by means of the dot product). In this regard, it is of interest to
inquire whether there are certain vectors that have only their lengths, and not their
orientation, changed when operated upon by a given tensor (i.e., seek vectors that
are transformed into multiples of themselves). If such vectors exist, they must satisfy
the equation

$$\mathbf{A} \cdot \mathbf{x} = \lambda \mathbf{x}. \tag{2.5.28}$$

Such vectors $\mathbf{x}$ are called *characteristic vectors*, *principal planes*, or *eigenvectors* associated with $\mathbf{A}$. The parameter λ is called an *characteristic value*, *principal value*, or
eigenvalue, and it characterizes the change in length of the eigenvector $\mathbf{x}$ after it has
been operated upon by $\mathbf{A}$.

Since $\mathbf{x}$ can be expressed as $\mathbf{x} = \mathbf{I} \cdot \mathbf{x}$, Eq. (2.5.28) can also be written as

$$(\mathbf{A} - \lambda \mathbf{I}) \cdot \mathbf{x} = \mathbf{0}. \tag{2.5.29}$$

Because this is a homogeneous set of equations for $\mathbf{x}$, a nontrivial solution (i.e., vector with at least one component of $\mathbf{x}$ is nonzero) will not exist unless the determinant
of the matrix $[\mathbf{A} - \lambda \mathbf{I}]$ vanishes:

$$\det(\mathbf{A} - \lambda \mathbf{I}) = 0. \tag{2.5.30}$$

The vanishing of this determinant yields an algebraic equation of degree n, called
the *characteristic equation*, for λ when $\mathbf{A}$ is a $n \times n$ matrix.

For a second-order tensor $\boldsymbol{\Phi}$, which is of interest in the present study, the characteristic equation yields three eigenvalues λ_1, λ_2, and λ_3. The character of these
eigenvalues depends on the character of the dyadic $\boldsymbol{\Phi}$. At least one of the eigenvalues must be real. The other two may be real and distinct, real and repeated,
or complex conjugates. The vanishing of the determinant assures that three eigenvectors are not unique to within a multiplicative constant, however, and an infinite
number of solutions exist having at least 3 different orientations. Since only orientation is important, it is, therefore, useful to represent the three eigenvectors by three
unit eigenvectors $\hat{\mathbf{e}}_1^*, \hat{\mathbf{e}}_2^*, \hat{\mathbf{e}}_3^*$, denoting three different orientations, each associated
with a particular eigenvalue.

In a Cartesian system, the characteristic equation associated with a second-
order tensor can be expressed in the form

$$\lambda^3 - I_1 \lambda^2 + I_2 \lambda - I_3 = 0, \tag{2.5.31}$$

where I_1, I_2, and I_3 are the invariants of $\boldsymbol{\Phi}$ as defined in Eq. (2.5.15). The invariants
can also be expressed in terms of the eigenvalues,

$$I_1 = \lambda_1 + \lambda_2 + \lambda_3, \quad I_2 = (\lambda_1 \lambda_2 + \lambda_2 \lambda_3 + \lambda_3 \lambda_1), \quad I_3 = \lambda_1 \lambda_2 \lambda_3. \tag{2.5.32}$$

Finding the roots of the cubic Eq. (2.5.31) is not always easy. However, when the matrix under consideration is either a 2×2 matrix or 3×3 matrix but of the special form

$$\begin{bmatrix} \phi_{11} & 0 & 0 \\ 0 & \phi_{22} & \phi_{23} \\ 0 & \phi_{32} & \phi_{33} \end{bmatrix},$$

one of the roots is $\lambda_1 = \phi_{11}$, and the remaining two roots can be found from the quadratic equation

$$\begin{vmatrix} \phi_{22} - \lambda & \phi_{23} \\ \phi_{32} & \phi_{33} - \lambda \end{vmatrix} = (\phi_{22} - \lambda)(\phi_{33} - \lambda) - \phi_{23}\phi_{32} = 0.$$

That is,

$$\lambda_{2,3} = \frac{\phi_{22} + \phi_{33}}{2} \pm \frac{1}{2}\sqrt{(\phi_{22} + \phi_{33})^2 - 4(\phi_{22}\phi_{33} - \phi_{23}\phi_{32})}. \tag{2.5.33}$$

A computational example of finding eigenvalues and eigenvectors is presented next.

EXAMPLE 2.5.2: Determine the eigenvalues and eigenvectors of the following matrix:

$$[A] = \begin{bmatrix} 5 & -1 \\ 3 & 1 \end{bmatrix}.$$

SOLUTION: The eigenvalue problem associated with the matrix $\mathbf{A}$ is

$$\mathbf{Ax} = \lambda\mathbf{x} \quad \rightarrow \quad \begin{bmatrix} 5 - \lambda & -1 \\ 3 & 1 - \lambda \end{bmatrix} \begin{Bmatrix} x_1 \\ x_2 \end{Bmatrix} = \begin{Bmatrix} 0 \\ 0 \end{Bmatrix} \tag{a}$$

or

$$\det(\mathbf{A} - \lambda\mathbf{I}) = \begin{vmatrix} 5 - \lambda & -1 \\ 3 & 1 - \lambda \end{vmatrix} = (5 - \lambda)(1 - \lambda) + 3 = 0.$$

The two roots of the resulting quadratic equation, $\lambda^2 - 6\lambda + 8 = 0$, are the eigenvalues $\lambda_1 = 2$ and $\lambda_2 = 4$.

To find the eigenvectors, we return to Eq. (a) and substitute for λ each of the eigenvalues and solve the resulting algebraic equations for (x_1, x_2). For $\lambda = 2$, we have

$$\begin{bmatrix} 5 - 2 & -1 \\ 3 & 1 - 2 \end{bmatrix} \begin{Bmatrix} x_1^{(1)} \\ x_2^{(1)} \end{Bmatrix} = \begin{Bmatrix} 0 \\ 0 \end{Bmatrix}. \tag{b}$$

Each row of the above matrix equation yields the same condition $3x_1^{(1)} - x_2^{(1)} = 0$ or $x_2^{(1)} = 3x_1^{(1)}$. The eigenvector $\mathbf{x}^{(1)}$ is given by

$$\mathbf{x}^{(1)} = \begin{Bmatrix} 1 \\ 3 \end{Bmatrix} x_1^{(1)}, \quad x_1^{(1)} \neq 0, \text{ arbitrary.}$$

Usually, we take $x_1^{(1)} = 1$, as we are interested in the direction of the vector $\mathbf{x}^{(1)}$ rather than in its magnitude. One may also normalize the eigenvector by using the condition

$$(x_1^{(1)})^2 + (x_2^{(1)})^2 = 1. \tag{c}$$

Then we obtain the following normalized eigenvector:

$$\mathbf{x}_n^{(1)} = \pm \frac{1}{\sqrt{10}} \begin{Bmatrix} 1 \\ 3 \end{Bmatrix}. \tag{d}$$

The second eigenvector is found using the same procedure. Substituting for $\lambda = 4$ into Eq. (a)

$$\begin{bmatrix} 5 - 4 & -1 \\ 3 & 1 - 4 \end{bmatrix} \begin{Bmatrix} x_1^{(2)} \\ x_2^{(2)} \end{Bmatrix} = \begin{Bmatrix} 0 \\ 0 \end{Bmatrix} \tag{e}$$

we obtain the condition $x_1^{(2)} - x_2^{(2)} = 0$ or $x_2^{(2)} = x_1^{(2)}$. The eigenvector $\mathbf{x}^{(2)}$ is

$$\mathbf{x}^{(2)} = \begin{Bmatrix} 1 \\ 1 \end{Bmatrix} \quad \text{or} \quad \mathbf{x}_n^{(2)} = \pm \frac{1}{\sqrt{2}} \begin{Bmatrix} 1 \\ 1 \end{Bmatrix}. \tag{f}$$

When the matrix $[A]$ is a full 3×3 matrix, we use a method that facilitates the computation of eigenvalues. In the alternative method, we seek the eigenvalues of the so-called *deviatoric tensor* associated with the tensor $\mathbf{A}$:

$$a_{ij}' \equiv a_{ij} - \frac{1}{3} a_{kk} \delta_{ij}. \tag{2.5.34}$$

Note that

$$a_{ii}' = a_{ii} - a_{kk} = 0. \tag{2.5.35}$$

That is, the first invariant I_1' of the deviatoric tensor is zero. As a result, the characteristic equation associated with the deviatoric tensor is of the form

$$(\lambda')^3 + I_2' \lambda' - I_3' = 0, \tag{2.5.36}$$

where λ' is the eigenvalue of the deviatoric tensor. The eigenvalues associated with a_{ij} itself can be computed from

$$\lambda = \lambda' + \frac{1}{3} a_{kk}. \tag{2.5.37}$$

The cubic equation in Eq. (2.5.36) is of a special form that allows a direct computation of its roots. Equation (2.5.36) can be solved explicitly by introducing the transformation

$$\lambda' = 2 \left(-\frac{1}{3} I_2' \right)^{1/2} \cos \alpha, \tag{2.5.38}$$

which transforms Eq. (2.5.36) into

$$2 \left(-\frac{1}{3} I_2' \right)^{3/2} [4 \cos^3 \alpha - 3 \cos \alpha] = I_3'. \tag{2.5.39}$$

The expression in square brackets is equal to $\cos 3\alpha$. Hence

$$\cos 3\alpha = \frac{I_3'}{2}\left(-\frac{3}{I_2'}\right)^{3/2}. \qquad (2.5.40)$$

If α_1 is the angle satisfying $0 \le 3\alpha_1 \le \pi$ whose cosine is given by Eq. (2.5.40), then $3\alpha_1$, $3\alpha_1 + 2\pi$, and $3\alpha_1 - 2\pi$ all have the same cosine and furnish three independent roots of Eq. (2.5.36),

$$\lambda_i' = 2\left(-\frac{1}{3}I_2'\right)^{1/2}\cos\alpha_i, \quad i = 1, 2, 3, \qquad (2.5.41)$$

where

$$\alpha_1 = \frac{1}{3}\left\{\cos^{-1}\left[\frac{I_3'}{2}\left(-\frac{3}{I_2'}\right)^{3/2}\right]\right\}, \quad \alpha_2 = \alpha_1 + \frac{2\pi}{3}, \quad \alpha_3 = \alpha_1 - \frac{2\pi}{3}. \qquad (2.5.42)$$

Finally, we can compute λ_i from Eq. (2.5.37).

An example of application of the previous procedures is presented next.

EXAMPLE 2.5.3: Determine the eigenvalues and eigenvectors of the following matrix:

$$\mathbf{A} = \begin{bmatrix} 2 & 1 & 0 \\ 1 & 4 & 1 \\ 0 & 1 & 2 \end{bmatrix}.$$

SOLUTION: The characteristic equation is obtained by setting $\det(\mathbf{A} - \lambda\mathbf{I}) = 0$:

$$\begin{vmatrix} 2 - \lambda & 1 & 0 \\ 1 & 4 - \lambda & 1 \\ 0 & 1 & 2 - \lambda \end{vmatrix} = (2 - \lambda)[(4 - \lambda)(2 - \lambda) - 1] - 1 \cdot (2 - \lambda) = 0,$$

or

$$(2 - \lambda)[(4 - \lambda)(2 - \lambda) - 2] = 0.$$

Hence

$$\lambda_1 = 3 + \sqrt{3} = 4.7321, \quad \lambda_2 = 3 - \sqrt{3} = 1.2679, \quad \lambda_3 = 2.$$

Alternatively, using Eqs. (2.5.34)–(2.5.42), we have

$$\mathbf{A}' = \begin{bmatrix} 2 - \frac{8}{3} & 1 & 0 \\ 1 & 4 - \frac{8}{3} & 1 \\ 0 & 1 & 2 - \frac{8}{3} \end{bmatrix}$$

$$I_2' = \frac{1}{2}\left(a_{ii}'a_{jj}' - a_{ij}'a_{ij}'\right) = -\frac{1}{2}a_{ij}'a_{ij}'$$

$$= -\frac{1}{2}\left[\left(-\frac{2}{3}\right)^2 + \left(-\frac{2}{3}\right)^2 + \left(\frac{4}{3}\right)^2 + 2 + 2\right] = -\frac{10}{3}$$

$$I_3' = \det(a_{ij}') = \frac{52}{27}.$$

From Eq. (2.5.42),

$$\alpha_1 = \frac{1}{3}\left\{\cos^{-1}\left[\frac{52}{54}\left(\frac{9}{10}\right)^{3/2}\right]\right\} = 11.565°, \quad \alpha_2 = 131.565°, \quad \alpha_3 = -108.435°,$$

and from Eq. (2.5.41),

$$\lambda_1' = 2.065384, \quad \lambda_2' = -1.3987, \quad \lambda_3' = -0.66667.$$

Finally, using Eq. (2.5.37), we obtain the eigenvalues

$$\lambda_1 = 4.7321, \quad \lambda_2 = 1.2679, \quad \lambda_3 = 2.00.$$

The eigenvector corresponding to $\lambda_3 = 2$, for example, is calculated as follows. From $(a_{ij} - \lambda_3\delta_{ij})x_j = 0$, we have

$$\begin{bmatrix} 2-2 & 1 & 0 \\ 1 & 4-2 & 1 \\ 0 & 1 & 2-2 \end{bmatrix}\begin{Bmatrix} x_1 \\ x_2 \\ x_3 \end{Bmatrix} = \begin{Bmatrix} 0 \\ 0 \\ 0 \end{Bmatrix}.$$

This gives $x_2 = 0$ and $x_1 = -x_3$, and the eigenvector associated with $\lambda_3 = 2$ is

$$\mathbf{x}^{(3)} = \begin{Bmatrix} 1 \\ 0 \\ -1 \end{Bmatrix} \quad \text{or} \quad \mathbf{x}_n^{(3)} = \pm\frac{1}{\sqrt{2}}\begin{Bmatrix} 1 \\ 0 \\ -1 \end{Bmatrix}.$$

Similarly, the normalized eigenvectors corresponding to $\lambda_{1,2} = 3 \pm \sqrt{3}$ are given by

$$\mathbf{x}_n^{(1)} = \pm\frac{(3-\sqrt{3})}{12}\left\{\begin{pmatrix} 1 \\ 1+\sqrt{3} \\ 1 \end{pmatrix}\right\}, \quad \mathbf{x}_n^{(2)} = \pm\frac{(3+\sqrt{3})}{12}\left\{\begin{pmatrix} 1 \\ (1-\sqrt{3}) \\ 1 \end{pmatrix}\right\}.$$

When $\mathbf{A}$ in Eq. (2.5.28) is an nth order tensor, Eq. (2.5.29) is a polynomial of degree n in λ, and therefore, there are n eigenvalues $\lambda_1, \lambda_2, \ldots, \lambda_n$, some of which may be repeated. In general, if an eigenvalue appears m times as a root of Eq. (2.5.29), then that eigenvalue is said to have *algebraic multiplicity m*. An eigenvalue of algebraic multiplicity m may have r linearly independent eigenvectors. The number r is called the *geometric multiplicity* of the eigenvalue, and r lies (not shown here) in the range $1 \leq r \leq m$. Thus, a matrix $\mathbf{A}$ of order n may have fewer than n linearly independent eigenvectors. The example below illustrates the calculation of eigenvectors of a matrix when it has repeated eigenvalues.

EXAMPLE 2.5.4: Determine the eigenvalues and eigenvectors of the following matrix:

$$\mathbf{A} = \begin{bmatrix} 0 & 1 & 1 \\ 1 & 0 & 1 \\ 1 & 1 & 0 \end{bmatrix}.$$

SOLUTION: The condition $\det(\mathbf{A} - \lambda \mathbf{x}) = 0$ gives

$$\begin{vmatrix} -\lambda & 1 & 1 \\ 1 & -\lambda & 1 \\ 1 & 1 & -\lambda \end{vmatrix} = -\lambda^3 + 3\lambda + 2 = 0.$$

The three roots are

$$\lambda_1 = 2, \quad \lambda_2 = -1, \quad \lambda_3 = -1.$$

Thus, $\lambda = -1$ is an eigenvalue with algebraic multiplicity of 2.

The eigenvector associated with $\lambda = 2$ is obtained from

$$\begin{bmatrix} -2 & 1 & 1 \\ 1 & -2 & 1 \\ 1 & 1 & -2 \end{bmatrix} \begin{Bmatrix} x_1 \\ x_2 \\ x_3 \end{Bmatrix} = \begin{Bmatrix} 0 \\ 0 \\ 0 \end{Bmatrix}$$

from which we have

$$-2x_1 + x_2 + x_3 = 0, \quad x_1 - 2x_2 + x_3 = 0, \quad x_1 + x_2 - 2x_3 = 0.$$

Eliminating x_3 from the first two (or the last two) equations, we obtain $x_2 = x_1$. Then the last equation gives $x_3 = x_2$. Thus the eigenvector associated with $\lambda_1 = 2$ is the vector

$$\mathbf{x}^{(1)} = \begin{Bmatrix} 1 \\ 1 \\ 1 \end{Bmatrix} x_1 \quad \text{or} \quad \mathbf{x}_n^{(1)} = \pm \frac{1}{\sqrt{3}} \begin{Bmatrix} 1 \\ 1 \\ 1 \end{Bmatrix}.$$

The eigenvector associated with $\lambda = -1$ is obtained from

$$\begin{bmatrix} 1 & 1 & 1 \\ 1 & 1 & 1 \\ 1 & 1 & 1 \end{bmatrix} \begin{Bmatrix} x_1 \\ x_2 \\ x_3 \end{Bmatrix} = \begin{Bmatrix} 0 \\ 0 \\ 0 \end{Bmatrix}.$$

All three equations yield the same equation $x_1 + x_2 + x_3 = 0$. Thus, values of two of the three components (x_1, x_2, x_3) can be chosen arbitrarily. For the choice of $x_3 = 1$ and $x_2 = 0$, we obtain the vector (or any nonzero multiples of it)

$$\mathbf{x}^{(2)} = \begin{Bmatrix} -1 \\ 0 \\ 1 \end{Bmatrix} x_1 \quad \text{or} \quad \mathbf{x}_n^{(2)} = \mp \frac{1}{\sqrt{2}} \begin{Bmatrix} 1 \\ 0 \\ -1 \end{Bmatrix}.$$

A second independent vector can be found by choosing $x_3 = 0$ and $x_2 = 1$. We obtain

$$\mathbf{x}^{(3)} = \begin{Bmatrix} -1 \\ 1 \\ 0 \end{Bmatrix} x_1 \quad \text{or} \quad \mathbf{x}_n^{(3)} = \mp \frac{1}{\sqrt{2}} \begin{Bmatrix} 1 \\ -1 \\ 0 \end{Bmatrix}.$$

Thus, in the present case, there exist two linearly independent eigenvectors associated with the double eigenvalue.

A real symmetric matrix $\mathbf{A}$ of order n has some desirable consequences as for the eigenvalues and eigenvectors are concerned. These are

1. All eigenvalues of $\mathbf{A}$ are real.
2. $\mathbf{A}$ always has n linearly independent eigenvectors, regardless of the algebraic ·multiplicities of the eigenvalues.
3. Eigenvectors $\mathbf{x}^{(1)}$ and $\mathbf{x}^{(2)}$ associated with two distinct eigenvalues λ_1 and λ_2 are orthogonal: $\mathbf{x}^{(1)} \cdot \mathbf{x}^{(2)} = 0$. If all eigenvalues are distinct, then the associated eigenvectors are all orthogonal to each other.
4. For an eigenvalue of algebraic multiplicity m, it is possible to choose m eigenvectors that are mutually orthogonal. Hence, the set of n vectors can always be chosen to be linearly independent.

We note that the matrix $\mathbf{A}$ considered in Example 2.5.4 is symmetric. Clearly, Properties 1 through 3 listed above are satisfied. However, Property 4 was not illustrated there. It is possible to choose the values of the two of the three components (x_1, x_2, x_3) to have a set of linearly independent eigenvectors that are orthogonal. The second vector associated with $\lambda = -1$ could have been chosen by setting $x_1 = x_3 = 1$. We obtain

$$\mathbf{x}^{(3)} = \left\{ \begin{array}{c} 1 \\ -2 \\ 1 \end{array} \right\} x_1 \ \text{ or } \ \mathbf{x}_n^{(3)} = \pm \frac{1}{\sqrt{6}} \left\{ \begin{array}{c} 1 \\ -2 \\ 1 \end{array} \right\}.$$

Next, we prove Properties 1 and 2 of a symmetric matrix. The vanishing of the determinant $|\mathbf{A} - \lambda \mathbf{I}| = 0$ assures that n eigenvectors exist, $\mathbf{x}^{(1)}, \mathbf{x}^{(2)}, \ldots, \mathbf{x}^{(n)}$, each corresponding to an eigenvalue. The eigenvectors are not unique to within a multiplicative constant, however, and an infinite number of solutions exist having at least n different orientations. Since only orientation is important, it is thus useful to represent the n eigenvectors by n unit eigenvectors $\hat{\mathbf{e}}_1^*, \hat{\mathbf{e}}_2^*, \ldots, \hat{\mathbf{e}}_n^*$, denoting n different orientations, each associated with a particular eigenvalue λ_*.

Suppose now that λ_1 and λ_2 are two distinct eigenvalues and $\mathbf{x}^{(1)}$ and $\mathbf{x}^{(2)}$ are their corresponding eigenvectors:

$$\mathbf{A} \cdot \mathbf{x}^{(1)} = \lambda_1 \mathbf{x}^{(1)}, \quad \mathbf{A} \cdot \mathbf{x}^{(2)} = \lambda_2 \mathbf{x}^{(2)}. \tag{2.5.43}$$

Scalar product of the first equation by $\mathbf{x}^{(2)}$ and the second by $\mathbf{x}^{(1)}$, and then subtraction, yields

$$\mathbf{x}^{(2)} \cdot \mathbf{A} \cdot \mathbf{x}^{(1)} - \mathbf{x}^{(1)} \cdot \mathbf{A} \cdot \mathbf{x}^{(2)} = (\lambda_1 - \lambda_2) \mathbf{x}^{(1)} \cdot \mathbf{x}^{(2)}. \tag{2.5.44}$$

Since $\mathbf{A}$ is symmetric, one can establish that the left-hand side of this equation vanishes. Thus

$$0 = (\lambda_1 - \lambda_2) \mathbf{x}^{(1)} \cdot \mathbf{x}^{(2)}. \tag{2.5.45}$$

Now suppose that λ_1 and λ_2 are complex conjugates such that $\lambda_1 - \lambda_2 = 2i\lambda_{1I}$, where $i = \sqrt{-1}$ and λ_{1I} is the imaginary part of λ_1. Then $\mathbf{x}^{(1)} \cdot \mathbf{x}^{(2)}$ is always positive since $\mathbf{x}^{(1)}$ and $\mathbf{x}^{(2)}$ are complex conjugate vectors associated with λ_1 and λ_2. It then follows from Eq. (2.5.45) that $\lambda_{1I} = 0$ and hence *that the n eigenvalues associated with a symmetric matrix are all real.*

Next, assume that λ_1 and λ_2 are real and distinct such that $\lambda_1 - \lambda_2$ is not zero. It then follows from Eq. (2.5.45) that $\mathbf{x}^{(1)} \cdot \mathbf{x}^{(2)} = 0$. Thus the *eigenvectors associated with distinct eigenvalues of a symmetric dyadic are orthogonal.* If the three eigenvalues are all distinct, then the three eigenvectors are mutually orthogonal.

If an eigenvalue is repeated, say $\lambda_3 = \lambda_2$, then $\mathbf{x}^{(3)}$ must also be perpendicular to $\mathbf{x}^{(i)}$, $i \neq 2$ as deducted by an argument similar to that for $\mathbf{x}^{(2)}$ stemming from Eq. (2.5.45). Neither $\mathbf{x}^{(2)}$ nor $\mathbf{x}^{(3)}$ is preferred, and they are both arbitrary, except insofar as they are both perpendicular to $\mathbf{x}^{(1)}$. It is useful, however, to select $\mathbf{x}^{(3)}$ such that it is perpendicular to both $\mathbf{x}^{(1)}$ and $\mathbf{x}^{(2)}$. We do this by choosing $\mathbf{x}^{(3)} = \mathbf{x}^{(1)} \times \mathbf{x}^{(2)}$ and thus establishing a mutually orthogonal set of eigenvectors.

Cayley–Hamilton Theorem

Consider a square matrix $\mathbf{A}$ of order n. The characteristic equation $\phi(\lambda) = 0$ is obtained by setting $\phi(\lambda) \equiv \det|\mathbf{A} - \lambda\mathbf{I}| = 0$. Then the Cayley–Hamilton theorem states that

$$\phi(\mathbf{A}) = (\mathbf{A} - \lambda_1\mathbf{I})(\mathbf{A} - \lambda_2\mathbf{I}) \cdots (\mathbf{A} - \lambda_n\mathbf{I}) = 0. \tag{2.5.49}$$

The proof of the theorem can be found in any book on matrix theory; see, for example, Gantmacher (1959).

2.6 Summary

In this chapter, mathematical preliminaries for this course are reviewed. In particular, the notion of geometric vector, vector algebra, vector calculus, theory of matrices, and tensors and tensor calculus are thoroughly reviewed, and a number of examples are presented to review the ideas introduced. Readers familiar with these topics may skip this chapter.

PROBLEMS

2.1 Find the equation of a line (or a set of lines) passing through the terminal point of a vector $\mathbf{A}$ and in the direction of vector $\mathbf{B}$.

2.2 Find the equation of a plane connecting the terminal points of vectors $\mathbf{A}$, $\mathbf{B}$, and $\mathbf{C}$. Assume that all three vectors are referred to a common origin.

2.3 Prove the following vector identity without the use of a coordinate system

$$\mathbf{A} \times (\mathbf{B} \times \mathbf{C}) = (\mathbf{A} \cdot \mathbf{C})\mathbf{B} - (\mathbf{A} \cdot \mathbf{B})\mathbf{C}.$$

2.4 If $\hat{\mathbf{e}}$ is any unit vector and $\mathbf{A}$ an arbitrary vector, show that

$$\mathbf{A} = (\mathbf{A} \cdot \hat{\mathbf{e}})\hat{\mathbf{e}} + \hat{\mathbf{e}} \times (\mathbf{A} \times \hat{\mathbf{e}}).$$

This identity shows that a vector can resolved into a component parallel to and one perpendicular to an arbitrary direction $\hat{\mathbf{e}}$.

2.5 Establish the following identities for a second-order tensor $\mathbf{A}$:

(a) $|A| = e_{ijk} A_{1i} A_{2j} A_{3k}.$

(b) $|A| = \dfrac{1}{6} A_{ir} A_{js} A_{kt} e_{rst} e_{ijk}.$

(c) $e_{lmn}|A| = e_{ijk} A_{il} A_{jm} A_{kn}.$

(d) $e_{ijk} e_{mnk} = \delta_{im}\delta_{jn} - \delta_{in}\delta_{jm}.$

(e) $e_{ijk} = \begin{vmatrix} \delta_{i1} & \delta_{i2} & \delta_{i3} \\ \delta_{j1} & \delta_{j2} & \delta_{j3} \\ \delta_{k1} & \delta_{k2} & \delta_{k3} \end{vmatrix}.$

(f) $e_{ijk} e_{pqr} = \begin{vmatrix} \delta_{ip} & \delta_{iq} & \delta_{ir} \\ \delta_{jp} & \delta_{jq} & \delta_{jr} \\ \delta_{kp} & \delta_{kq} & \delta_{kr} \end{vmatrix}.$

2.6 Given the following components

$$\mathbf{A} = \begin{Bmatrix} 2 \\ -1 \\ 4 \end{Bmatrix}, \quad \mathbf{S} = \begin{bmatrix} -1 & 0 & 5 \\ 3 & 7 & 4 \\ 9 & 8 & 6 \end{bmatrix}, \quad \mathbf{T} = \begin{bmatrix} 8 & -1 & 6 \\ 5 & 4 & 9 \\ -7 & 8 & -2 \end{bmatrix},$$

determine

(a) $\text{tr}(\mathbf{S}).$

(b) $\mathbf{S} : \mathbf{S}.$

(c) $\mathbf{S} : \mathbf{S}^{\mathrm{T}}.$

(d) $\mathbf{A} \cdot \mathbf{S}.$

(e) $\mathbf{S} \cdot \mathbf{A}.$

(f) $\mathbf{S} \cdot \mathbf{T} \cdot \mathbf{A}.$

2.7 Using the index notation prove the identities

(a) $(\mathbf{A} \times \mathbf{B}) \cdot (\mathbf{B} \times \mathbf{C}) \times (\mathbf{C} \times \mathbf{A}) = (\mathbf{A} \cdot (\mathbf{B} \times \mathbf{C}))^2.$

(b) $(\mathbf{A} \times \mathbf{B}) \times (\mathbf{C} \times \mathbf{D}) = [\mathbf{A} \cdot (\mathbf{C} \times \mathbf{D})]\mathbf{B} - [\mathbf{B} \cdot (\mathbf{C} \times \mathbf{D}]\mathbf{A}.$

2.8 Determine whether the following set of vectors is linearly independent:

$$\mathbf{A} = 2\hat{\mathbf{e}}_1 - \hat{\mathbf{e}}_2 + \hat{\mathbf{e}}_3, \quad \mathbf{B} = -\hat{\mathbf{e}}_2 - \hat{\mathbf{e}}_3, \quad \mathbf{C} = -\hat{\mathbf{e}}_1 + \hat{\mathbf{e}}_2.$$

Here $\hat{\mathbf{e}}_i$ are orthonormal unit base vectors in $\Re^3$.

2.9 Consider two rectangular Cartesian coordinate systems that are translated and rotated with respect to each other. The transformation between the two coordinate systems is given by

$$\bar{\mathbf{x}} = \mathbf{c} + \mathbf{Lx},$$

where $\mathbf{c}$ is a constant vector and $\mathbf{L} = [\ell_{ij}]$ is the matrix of direction cosines

$$\ell_{ij} \equiv \hat{\mathbf{e}}_i \cdot \hat{\mathbf{e}}_j.$$

Deduce that the following orthogonality conditions hold:

$$\mathbf{L} \cdot \mathbf{L}^{\mathrm{T}} = \mathbf{I}.$$

That is, $\mathbf{L}$ is an orthogonal matrix.

2.10 Determine the transformation matrix relating the orthonormal basis vectors $(\hat{\mathbf{e}}_1, \hat{\mathbf{e}}_2, \hat{\mathbf{e}}_3)$ and $(\hat{\mathbf{e}}'_1, \hat{\mathbf{e}}'_2, \hat{\mathbf{e}}'_3)$, when $\hat{\mathbf{e}}'_i$ are given by

(a) $\hat{\mathbf{e}}'_1$ is along the vector $\hat{\mathbf{e}}_1 - \hat{\mathbf{e}}_2 + \hat{\mathbf{e}}_3$ and $\hat{\mathbf{e}}'_2$ is perpendicular to the plane $2x_1 + 3x_2 + x_3 - 5 = 0.$

(b) $\hat{\mathbf{e}}'_1$ is along the line segment connecting point $(1, -1, 3)$ to $(2, -2, 4)$ and $\hat{\mathbf{e}}'_3 = (-\hat{\mathbf{e}}_1 + \hat{\mathbf{e}}_2 + 2\hat{\mathbf{e}}_3)/\sqrt{6}.$

2.11 The angles between the barred and unbarred coordinate lines are given by

	$\hat{e}_1$	$\hat{e}_2$	$\hat{e}_3$
$\hat{e}_1$	60°	30°	90°
$\hat{e}_2$	150°	60°	90°
$\hat{e}_3$	90°	90°	0°

Determine the direction cosines of the transformation.

2.12 The angles between the barred and unbarred coordinate lines are given by

	x_1	x_2	x_3
$\bar{x}_1$	45°	90°	45°
$\bar{x}_2$	60°	45°	120°
$\bar{x}_3$	120°	45°	60°

Determine the transformation matrix.

2.13 Show that the following expressions for the components of an arbitrary second-order tensor $\mathbf{S} = [s_{ij}]$ are invariant:
(a) s_{ii}, (b) $s_{ij}s_{ij}$, and (c) $s_{ij}s_{jk}s_{ki}$.

2.14 Let $\mathbf{r}$ denote a position vector $\mathbf{r} = x_i \hat{e}_i$ $(r^2 = x_i x_i)$ and $\mathbf{A}$ an arbitrary constant vector. Show that:

(a) $\nabla^2(r^n) = n(n+1)r^{n-2}$.   (b) grad $(\mathbf{r} \cdot \mathbf{A}) = \mathbf{A}$.

(c) div $(\mathbf{r} \times \mathbf{A}) = 0$.   (d) curl$(\mathbf{r} \times \mathbf{A}) = -2\mathbf{A}$.

(e) div $(r\mathbf{A}) = \dfrac{1}{r}(\mathbf{r} \cdot \mathbf{A})$.   (f) curl $(r\mathbf{A}) = \dfrac{1}{r}(\mathbf{r} \times \mathbf{A})$.

2.15 Let $\mathbf{A}$ and $\mathbf{B}$ be continuous vector functions of the position vector $\mathbf{x}$ with continuous first derivatives, and let F and G be continuous scalar functions of position $\mathbf{x}$ with continuous first and second derivatives. Show that:

(a) div(curl $\mathbf{A}$) = 0.

(b) div(grad $F \times$ grad G) = 0.

(c) grad$(\mathbf{A} \cdot \mathbf{x}) = \mathbf{A} +$ grad $\mathbf{A} \cdot \mathbf{x}$.

(d) div$(F\mathbf{A}) = \mathbf{A} \cdot$ grad$F + F$div$\mathbf{A}$.

(e) curl$(F\mathbf{A}) = F$ curl$\mathbf{A}$ - $\mathbf{A} \times$ grad F.

(f) grad$(\mathbf{A} \cdot \mathbf{B}) = \mathbf{A} \cdot$ grad $\mathbf{B} + \mathbf{B} \cdot$ grad$\mathbf{A} + \mathbf{A} \times$ curl $\mathbf{B} + \mathbf{B} \times$ curl $\mathbf{A}$.

(g) div $(\mathbf{A} \times \mathbf{B}) =$ curl $\mathbf{A} \cdot \mathbf{B} -$ curl $\mathbf{B} \cdot \mathbf{A}$.

(h) curl $(\mathbf{A} \times \mathbf{B}) = \mathbf{B} \cdot \nabla\mathbf{A} - \mathbf{A} \cdot \nabla\mathbf{B} + \mathbf{A}$ div$\mathbf{B} - \mathbf{B}$ div$\mathbf{A}$.

(i) $(\nabla \times \mathbf{A}) \times \mathbf{A} = \mathbf{A} \cdot \nabla\mathbf{A} - \nabla\mathbf{A} \cdot \mathbf{A}$.

(j) $\nabla^2(FG) = F \nabla^2 G + 2\nabla F \cdot \nabla G + G \nabla^2 F$.

(k) $\nabla^2(F\mathbf{x}) = 2\nabla F + \mathbf{x} \nabla^2 F$.

(l) $\mathbf{A} \cdot$ grad $\mathbf{A} =$ grad $(\frac{1}{2}\mathbf{A} \cdot \mathbf{A}) - \mathbf{A} \times$ curl $\mathbf{A}$.

2.16 Show that the vector area of a closed surface is zero, that is,

$$\oint_s \hat{\mathbf{n}}\, ds = \mathbf{0}.$$

2.17 Show that the volume of the region Ω enclosed by a boundary surface Γ is

$$\text{volume} = \frac{1}{6} \oint_\Gamma \text{grad}(r^2) \cdot \hat{\mathbf{n}}\, ds = \frac{1}{3} \oint_\Gamma \mathbf{r} \cdot \hat{\mathbf{n}}\, ds.$$

2.18 Let $\phi(\mathbf{r})$ be a scalar field. Show that

$$\int_\Omega \nabla^2\phi\, d\mathbf{x} = \oint_\Gamma \frac{\partial \phi}{\partial n}\, ds.$$

2.19 In the divergence theorem (2.4.34), set $\mathbf{A} = \phi\, \text{grad}\psi$ and $\mathbf{A} = \psi\, \text{grad}\phi$ successively and obtain the integral forms

(a) $$\int_\Omega \left[\phi\nabla^2\psi + \nabla\phi \cdot \nabla\psi \right] d\mathbf{x} = \oint_\Gamma \phi\frac{\partial \psi}{\partial n}\, ds,$$

(b) $$\int_\Omega \left[\phi\nabla^2\psi - \psi\nabla^2\phi \right] d\mathbf{x} = \oint_\Gamma \left[\phi\frac{\partial \psi}{\partial n} - \psi\frac{\partial \phi}{\partial n} \right] ds,$$

(c) $$\int_\Omega \left[\phi\nabla^4\psi - \nabla^2\phi\nabla^2\psi \right] d\mathbf{x} = \oint_\Gamma \left[\phi\frac{\partial}{\partial n}(\nabla^2\psi) - \nabla^2\psi\frac{\partial \phi}{\partial n} \right] ds,$$

where Ω denotes a (two-dimensional or three-dimensional) region with bounding surface Γ. The first two identities are sometimes called *Green's first and second theorems*.

2.20 Determine the rotation transformation matrix such that the new base vector $\hat{\mathbf{e}}_1$ is along $\hat{\mathbf{e}}_1 - \hat{\mathbf{e}}_2 + \hat{\mathbf{e}}_3$, and $\hat{\mathbf{e}}_2$ is along the normal to the plane $2x_1 + 3x_2 + x_3 = 5$. If $\mathbf{S}$ is the dyadic whose components in the unbarred system are given by $s_{11} = 1$, $s_{12} = s_{21} = 0$, $s_{13} = s_{31} = -1$, $s_{22} = 3$, $s_{23} = s_{32} = -2$, and $s_{33} = 0$, find the components in the barred coordinates.

2.21 Suppose that the new axes $\bar{x}_i$ are obtained by rotating x_i through a $60°$ about the x_2-axis. Determine the components $\bar{A}_i$ of a vector $\mathbf{A}$ whose components with respect to the x_i coordinates are $(2, 1, 3)$.

2.22 If $\mathbf{A}$ and $\mathbf{B}$ are arbitrary vectors and $\mathbf{S}$ and $\mathbf{T}$ are arbitrary dyads, verify that

(a) $(\mathbf{A} \cdot \mathbf{S}) \cdot \mathbf{B} = \mathbf{A} \cdot (\mathbf{S} \cdot \mathbf{B})$. (b) $(\mathbf{S} \cdot \mathbf{T}) \cdot \mathbf{A} = \mathbf{S} \cdot (\mathbf{T} \cdot \mathbf{A})$.

(c) $\mathbf{A} \cdot (\mathbf{S} \cdot \mathbf{T}) = (\mathbf{A} \cdot \mathbf{S}) \cdot \mathbf{T}$. (d) $(\mathbf{S} \cdot \mathbf{A}) \cdot (\mathbf{T} \cdot \mathbf{B}) = \mathbf{A} \cdot (\mathbf{S}^T \cdot \mathbf{T}) \cdot \mathbf{B}$.

2.23 If $\mathbf{A}$ is an arbitrary vector and $\boldsymbol{\Phi}$ and $\boldsymbol{\Psi}$ are arbitrary dyads, verify that

(a) $(\mathbf{I} \times \mathbf{A}) \cdot \boldsymbol{\Phi} = \mathbf{A} \times \boldsymbol{\Phi}$.   (b) $(\mathbf{A} \times \mathbf{I}) \cdot \boldsymbol{\Phi} = \mathbf{A} \times \boldsymbol{\Phi}$.

(c) $(\boldsymbol{\Phi} \times \mathbf{A})^T = -\mathbf{A} \times \boldsymbol{\Phi}^T$.   (d) $(\boldsymbol{\Phi} \cdot \boldsymbol{\Psi})^T = \boldsymbol{\Psi}^T \cdot \boldsymbol{\Phi}^T$.

2.24 The determinant of a dyadic is also defined by the expression

$$|\mathbf{S}| = \frac{[(\mathbf{S} \cdot \mathbf{A}) \times (\mathbf{S} \cdot \mathbf{B})] \cdot (\mathbf{S} \cdot \mathbf{C})}{\mathbf{A} \times \mathbf{B} \cdot \mathbf{C}},$$

where **A**, **B**, and **C** are arbitrary vectors. Verify the definition in an orthonormal basis $\{\hat{\mathbf{e}}_i\}$.

2.25 For an arbitrary second-order tensor **S**, show that $\nabla \cdot \mathbf{S}$ in the cylindrical coordinate system is given by

$$
\nabla \cdot \mathbf{S} = \left[\frac{\partial S_{rr}}{\partial r} + \frac{1}{r}\frac{\partial S_{\theta r}}{\partial \theta} + \frac{\partial S_{zr}}{\partial z} + \frac{1}{r}(S_{rr} - S_{\theta\theta}) \right] \hat{\mathbf{e}}_r
$$

$$
+ \left[\frac{\partial S_{r\theta}}{\partial r} + \frac{1}{r}\frac{\partial S_{\theta\theta}}{\partial \theta} + \frac{\partial S_{z\theta}}{\partial z} + \frac{1}{r}(S_{r\theta} + S_{\theta r}) \right] \hat{\mathbf{e}}_\theta
$$

$$
+ \left[\frac{\partial S_{rz}}{\partial r} + \frac{1}{r}\frac{\partial S_{\theta z}}{\partial \theta} + \frac{\partial S_{zz}}{\partial z} + \frac{1}{r}S_{rz} \right] \hat{\mathbf{e}}_z.
$$

2.26 For an arbitrary second-order tensor **S**, show that $\nabla \times \mathbf{S}$ in the cylindrical coordinate system is given by

$$
\nabla \times \mathbf{S} = \hat{\mathbf{e}}_r\hat{\mathbf{e}}_r \left(\frac{1}{r}\frac{\partial S_{zr}}{\partial \theta} - \frac{\partial S_{\theta r}}{\partial z} - \frac{1}{r}S_{z\theta} \right) + \hat{\mathbf{e}}_\theta\hat{\mathbf{e}}_\theta \left(\frac{\partial S_{r\theta}}{\partial z} - \frac{\partial S_{z\theta}}{\partial r} \right)
$$

$$
+ \hat{\mathbf{e}}_z\hat{\mathbf{e}}_z \left(\frac{1}{r}S_{\theta z} - \frac{1}{r}\frac{\partial S_{rz}}{\partial \theta} + \frac{\partial S_{\theta z}}{\partial r} \right) + \hat{\mathbf{e}}_r\hat{\mathbf{e}}_\theta \left(\frac{1}{r}\frac{\partial S_{z\theta}}{\partial \theta} - \frac{\partial S_{\theta\theta}}{\partial z} + \frac{1}{r}S_{zr} \right)
$$

$$
+ \hat{\mathbf{e}}_\theta\hat{\mathbf{e}}_r \left(\frac{\partial S_{rr}}{\partial z} - \frac{\partial S_{zr}}{\partial r} \right) + \hat{\mathbf{e}}_r\hat{\mathbf{e}}_z \left(\frac{1}{r}\frac{\partial S_{zz}}{\partial \theta} - \frac{\partial S_{\theta z}}{\partial z} \right)
$$

$$
+ \hat{\mathbf{e}}_z\hat{\mathbf{e}}_r \left(\frac{\partial S_{\theta r}}{\partial r} - \frac{1}{r}\frac{\partial S_{rr}}{\partial \theta} + \frac{1}{r}S_{r\theta} + \frac{1}{r}S_{\theta r} \right) + \hat{\mathbf{e}}_\theta\hat{\mathbf{e}}_z \left(\frac{\partial S_{rz}}{\partial z} - \frac{\partial S_{zz}}{\partial r} \right)
$$

$$
+ \hat{\mathbf{e}}_z\hat{\mathbf{e}}_\theta \left(\frac{\partial S_{\theta\theta}}{\partial r} + \frac{1}{r}S_{\theta\theta} - \frac{1}{r}S_{rr} - \frac{1}{r}\frac{\partial S_{r\theta}}{\partial \theta} \right).
$$

2.27 For an arbitrary second-order tensor **S**, show that $\nabla \cdot \mathbf{S}$ in the spherical coordinate system is given by

$$
\nabla \cdot \mathbf{S} = \left\{ \frac{\partial S_{RR}}{\partial R} + \frac{1}{R}\frac{\partial S_{\phi R}}{\partial \phi} + \frac{1}{R\sin\phi}\frac{\partial S_{\theta R}}{\partial \theta} \right.
$$

$$
\left. + \frac{1}{R}[2S_{RR} - S_{\phi\phi} - S_{\theta\theta} + S_{\phi R}\cot\phi] \right\} \hat{\mathbf{e}}_R
$$

$$
+ \left\{ \frac{\partial S_{R\phi}}{\partial R} + \frac{1}{R}\frac{\partial S_{\phi\phi}}{\partial \phi} + \frac{1}{R\sin\phi}\frac{\partial S_{\theta\phi}}{\partial \theta} \right.
$$

$$
\left. + \frac{1}{R}[(S_{\phi\phi} - S_{\theta\theta})\cot\phi + S_{\phi R} + 2S_{R\phi}] \right\} \hat{\mathbf{e}}_\phi
$$

$$
+ \left\{ \frac{\partial S_{R\theta}}{\partial R} + \frac{1}{R}\frac{\partial S_{\phi\theta}}{\partial \phi} + \frac{1}{R\sin\phi}\frac{\partial S_{\theta\theta}}{\partial \theta} \right.
$$

$$
\left. + \frac{1}{R}[(S_{\phi\theta} + S_{\theta\phi})\cot\phi + 2S_{R\theta} + S_{\theta R}] \right\} \hat{\mathbf{e}}_\theta.
$$

2.28 Show that $\nabla \mathbf{u}$ in the spherical coordinate system is given by

$$\nabla \mathbf{u} = \frac{\partial u_R}{\partial R}\hat{\mathbf{e}}_R\hat{\mathbf{e}}_R + \frac{\partial u_\phi}{\partial R}\hat{\mathbf{e}}_R\hat{\mathbf{e}}_\phi + \frac{\partial u_\theta}{\partial R}\hat{\mathbf{e}}_R\hat{\mathbf{e}}_\theta$$

$$+ \frac{1}{R}\left(\frac{\partial u_R}{\partial \phi} - u_\phi\right)\hat{\mathbf{e}}_\phi\hat{\mathbf{e}}_R + \frac{1}{R}\left(\frac{\partial u_\phi}{\partial \phi} + u_R\right)\hat{\mathbf{e}}_\phi\hat{\mathbf{e}}_\phi + \frac{1}{R}\frac{\partial u_\theta}{\partial \phi}\hat{\mathbf{e}}_\phi\hat{\mathbf{e}}_\theta$$

$$+ \frac{1}{R\sin\phi}\left[\left(\frac{\partial u_R}{\partial \theta} - u_\theta\sin\phi\right)\hat{\mathbf{e}}_\theta\hat{\mathbf{e}}_R + \left(\frac{\partial u_\phi}{\partial \theta} - u_\theta\cos\phi\right)\hat{\mathbf{e}}_\theta\hat{\mathbf{e}}_\phi\right.$$

$$\left. + \left(\frac{\partial u_\theta}{\partial \theta} + u_R\sin\phi + u_\phi\cos\phi\right)\hat{\mathbf{e}}_\theta\hat{\mathbf{e}}_\theta\right].$$

2.29 Show that the characteristic equation for a symmetric second-order tensor $\mathbf{\Phi}$ can be expressed as

$$\lambda^3 - I_1\lambda^2 + I_2\lambda - I_3 = 0,$$

where

$$I_1 = \phi_{kk}, \quad I_2 = \frac{1}{2}(\phi_{ii}\phi_{jj} - \phi_{ij}\phi_{ji}),$$

$$I_3 = \frac{1}{6}(2\phi_{ij}\phi_{jk}\phi_{ki} - 3\phi_{ij}\phi_{ji}\phi_{kk} + \phi_{ii}\phi_{jj}\phi_{kk}) = \det(\phi_{ij}),$$

are the three invariants of $\mathbf{\Phi}$.

2.30 Find the eigenvalues and eigenvectors of the following matrices:

(a) $\begin{bmatrix} 4 & -4 & 0 \\ -4 & 0 & 0 \\ 0 & 0 & 3 \end{bmatrix}.$ (b) $\begin{bmatrix} 2 & -\sqrt{3} & 0 \\ -\sqrt{3} & 4 & 0 \\ 0 & 0 & 4 \end{bmatrix}.$

(c) $\begin{bmatrix} 1 & 0 & 0 \\ 0 & 3 & -1 \\ 0 & -1 & 3 \end{bmatrix}.$ (d) $\begin{bmatrix} 2 & -1 & 1 \\ -1 & 0 & 1 \\ 1 & 1 & 2 \end{bmatrix}.$

(e) $\begin{bmatrix} 3 & 5 & 8 \\ 5 & 1 & 0 \\ 8 & 0 & 2 \end{bmatrix}.$ (f) $\begin{bmatrix} 1 & -1 & 0 \\ -1 & 2 & -1 \\ 0 & -1 & 2 \end{bmatrix}.$

2.31 Consider the matrix in Example 2.5.3

$$\mathbf{A} = \begin{bmatrix} 2 & 1 & 0 \\ 1 & 4 & 1 \\ 0 & 1 & 2 \end{bmatrix}.$$

Verify the Cayley–Hamilton theorem and use it to compute the inverse of $[A]$.

3 Kinematics of Continua

The man who cannot occasionally imagine events and conditions of existence that are contrary to the causal principle as he knows it will never enrich his science by the addition of a new idea.

Max Planck

It is through science that we prove, but through intuition that we discover.

H. Poincaré

3.1 Introduction

Material or matter is composed of discrete molecules, which in turn are made up of atoms. An atom consists of negatively charged electrons, positively charged protons, and neutrons. Electrons form chemical bonds. The study of matter at molecular or atomistic levels is very useful for understanding a variety of phenomena, but studies at these scales are not useful to solve common engineering problems. *Continuum mechanics* is concerned with a study of various forms of matter at macroscopic level. Central to this study is the assumption that the discrete nature of matter can be overlooked, provided the length scales of interest are large compared with the length scales of discrete molecular structure. Thus, matter at sufficiently large length scales can be treated as a *continuum* in which all physical quantities of interest, including density, are continuously differentiable.

Engineers and scientists undertake the study of continuous systems to understand their behavior under "working conditions," so that the systems can be designed to function properly and produced economically. For example, if we were to repair or replace a damaged artery in human body, we must understand the function of the original artery and the conditions that lead to its damage. An artery carries blood from the heart to different parts of the body. Conditions like high blood pressure and increase in cholesterol content in the blood may lead to deposition of particles in the arterial wall, as shown in Figure 3.1.1. With time, accumulation of these particles in the arterial wall hardens and constricts the passage, leading to cardiovascular diseases. A possible remedy for such diseases is to repair or replace

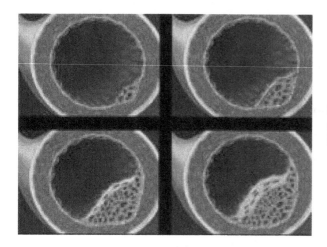

Figure 3.1.1. Progressive damage of artery due to the deposition of particles in the arterial wall.

the damaged portion of the artery. This in turn requires an understanding of the deformation and stresses caused in the arterial wall by the flow of blood. The understanding is then used to design the vascular prosthesis (i.e., artificial artery).

The present chapter is devoted to the study of geometric changes in a continuous medium (such as the artery) that is in static or dynamic equilibrium. In the subsequent chapters, we will study stresses and physical principles that govern the mechanical response of a continuous medium. The study of geometric changes in a continuum without regard to the forces causing the changes is known as *kinematics*.

3.2 Descriptions of Motion

3.2.1 Configurations of a Continuous Medium

Consider a body B of known geometry, constitution, and loading in a three-dimensional Euclidean space $\Re^3$; B may be viewed as a set of particles, each particle representing a large collection of molecules, having a continuous distribution of matter in space and time. Examples of the body B are provided by the diving board. For a given geometry and loading, the body B will undergo macroscopic geometric changes within the body, which are termed *deformation*. The geometric changes are accompanied by stresses that are induced in the body. If the applied loads are time dependent, the deformation of the body will be a function of time, that is, the geometry of the body B will change continuously with time. If the loads are applied slowly so that the deformation is only dependent on the loads, the body will occupy a continuous sequence of geometrical regions. The region occupied by the continuum at a given time t is termed a *configuration* and denoted by κ. Thus, the simultaneous positions occupied in space $\Re^3$ by all material points of the continuum B at different instants of time are called *configurations*.

Suppose that the continuum initially occupies a configuration κ_0, in which a particle X occupies the position $\mathbf{X}$, referred to a rectangular Cartesian system (X_1, X_2, X_3). Note that X (lightface letter) is the name of the particle that occupies

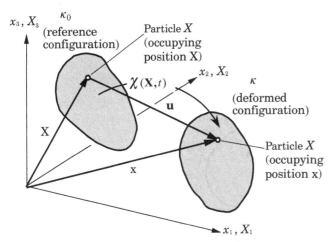

Figure 3.2.1. Reference and deformed configurations of a body.

the location $\mathbf{X}$ (boldface letter) in configuration κ_0, and therefore (X_1, X_2, X_3) are called the *material coordinates*. After the application of the loads, the continuum changes its geometric shape and thus assumes a new configuration κ, called the *current* or *deformed configuration*. The particle X now occupies the position $\mathbf{x}$ in the deformed configuration κ, as shown in Figure 3.2.1. The mapping $\chi : \mathcal{B}_{\kappa_0} \to \mathcal{B}_\kappa$ is called the *deformation mapping* of the body $\mathcal{B}$ from κ_0 to κ. The deformation mapping $\chi(\mathbf{X})$ takes the position vector $\mathbf{X}$ from the reference configuration and places the same point in the deformed configuration as $\mathbf{x} = \chi(\mathbf{X})$.

A frame of reference is chosen, explicitly or implicitly, to describe the deformation. We shall use the same reference frame for reference and current configurations. The components X_i and x_i of vectors $\mathbf{X} = X_i \, \hat{\mathbf{E}}_i$ and $\mathbf{x} = x_i \, \hat{\mathbf{e}}_i$ are along the coordinates used. We assume that the origins of the basis vectors $\hat{\mathbf{E}}_i$ and $\hat{\mathbf{e}}_i$ coincide.

The mathematical description of the deformation of a continuous body follows one of the two approaches: (1) the material description and (2) spatial description. The material description is also known as the *Lagrangian description*, and the spatial description is known as the *Eulerian description*. These descriptions are discussed next.

3.2.2 Material Description

In the material description, the motion of the body is referred to a reference configuration κ_R, which is often chosen to be the undeformed configuration, $\kappa_R = \kappa_0$. Thus, in the Lagrangian description, the current coordinates $(\mathbf{x} \in \kappa)$ are expressed in terms of the reference coordinates $(\mathbf{X} \in \kappa_0)$:

$$\mathbf{x} = \chi(\mathbf{X}, t), \quad \chi(\mathbf{X}, 0) = \mathbf{X}, \tag{3.2.1}$$

and the variation of a typical variable ϕ over the body is described with respect to the material coordinates $\mathbf{X}$ and time t:

$$\phi = \phi(\mathbf{X}, t). \tag{3.2.2}$$

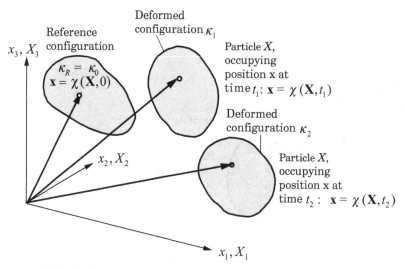

Figure 3.2.2. Reference configuration and deformed configurations at two different times in material description.

For a fixed value of $\mathbf{X} \in \kappa_0$, $\phi(\mathbf{X}, t)$ gives the value of ϕ at time t associated with the fixed material point X whose position in the reference configuration is $\mathbf{X}$, as shown in Figure 3.2.2. Thus, a change in time t implies that the *same* material particle X, occupying position $\mathbf{X}$ in κ_0, has a different value ϕ. Thus the attention is focused on the material particles X of the continuum.

3.2.3 Spatial Description

In the spatial description, the motion is referred to the current configuration κ occupied by the body B, and ϕ is described with respect to the current position ($\mathbf{x} \in \kappa$) in space, currently occupied by material particle X:

$$\phi = \phi(\mathbf{x}, t), \quad \mathbf{X} = \mathbf{X}(\mathbf{x}, t). \tag{3.2.3}$$

The coordinates ($\mathbf{x}$) are termed the *spatial coordinates*. For a fixed value of $\mathbf{x} \in \kappa$, $\phi(\mathbf{x}, t)$ gives the value of ϕ associated with a fixed point $\mathbf{x}$ in space, which will be the value of ϕ associated with different material points at different times, because different material points occupy the position $\mathbf{x} \in \kappa$ at different times, as shown in Figure 3.2.3. Thus, a change in time t implies that a different value ϕ is observed at the *same* spatial location $\mathbf{x} \in \kappa$, now probably occupied by a different material particle X. Hence, attention is focused on a spatial position $\mathbf{x} \in \kappa$.

When ϕ is known in the material description, $\phi = \phi(\mathbf{X}, t)$, its time derivative is simply the partial derivative with respect to time because the material coordinates $\mathbf{X}$ do not change with time:

$$\frac{d}{dt}[\phi(\mathbf{X}, t)] = \frac{\partial}{\partial t}[\phi(\mathbf{X}, t)]\bigg|_{\mathbf{x} \text{ fixed}} = \frac{\partial \phi}{\partial t}. \tag{3.2.4}$$

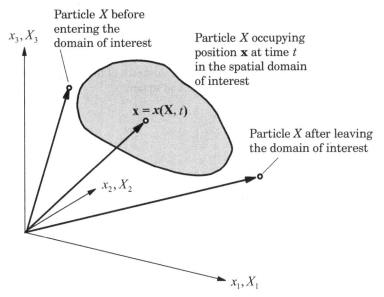

Figure 3.2.3. Material points within and outside the spatial domain of interest in spatial description.

However, when ϕ is known in the spatial description, $\phi = \phi(\mathbf{x}, t)$, its time derivative, known as the *material derivative*,[1] is

$$\frac{d}{dt}[\phi(\mathbf{x}, t)] = \frac{\partial}{\partial t}[\phi(\mathbf{x}, t)] + \frac{\partial}{\partial x_i}[\phi(\mathbf{x}, t)]\frac{dx_i}{dt}$$

$$= \frac{\partial \phi}{\partial t} + v_i \frac{\partial \phi}{\partial x_i} = \frac{\partial \phi}{\partial t} + \mathbf{v} \cdot \nabla \phi, \tag{3.2.5}$$

where $\mathbf{v}$ is the velocity $\mathbf{v} = d\mathbf{x}/dt = \dot{\mathbf{x}}$. For example, the acceleration of a particle is given by

$$\mathbf{a} = \frac{d\mathbf{v}}{dt} = \frac{\partial \mathbf{v}}{\partial t} + \mathbf{v} \cdot \nabla \mathbf{v}, \quad \left(a_i = \frac{\partial v_i}{\partial t} + v_j \frac{\partial v_i}{\partial x_j}\right). \tag{3.2.6}$$

The next example illustrates the determination of the inverse of a given mapping and computation of the material time derivative of a given function.

EXAMPLE 3.2.1: Suppose that the motion of a continuous medium $\mathcal{B}$ is described by the mapping $\chi : \kappa_0 \to \kappa$:

$$\chi(\mathbf{X}, t) = (X_1 + At\,X_2)\hat{\mathbf{e}}_1 + (X_2 - At\,X_1)\hat{\mathbf{e}}_2 + X_3\,\hat{\mathbf{e}}_3,$$

and that the temperature θ in the continuum in the spatial description is given by

$$\theta(\mathbf{x}, t) = x_1 + tx_2.$$

---

[1] Stokes's notation for material derivative is D/Dt.

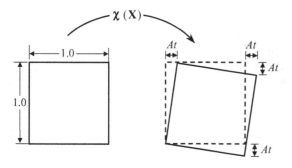

Figure 3.2.4. A sketch of the mapping as applied to a unit square.

Determine (a) inverse of the mapping, (b) the velocity components, and (c) the time derivatives of θ in the two descriptions.

SOLUTION: The mapping implies that a unit square is mapped into a rectangle that is rotated in clockwise direction, as shown in Figure 3.2.4.

(a) The inverse mapping is given by $\chi^{-1} : \kappa \to \kappa_0$:

$$\chi^{-1}(\mathbf{x}, t) = \left(\frac{x_1 - At x_2}{1 + A^2 t^2} \right) \hat{\mathbf{E}}_1 + \left(\frac{x_2 + At x_1}{1 + A^2 t^2} \right) \hat{\mathbf{E}}_2 + x_3 \, \hat{\mathbf{E}}_3.$$

(b) The velocity vector is given by $\mathbf{v} = v_1 \hat{\mathbf{E}}_1 + v_2 \hat{\mathbf{E}}_2$, with

$$v_1 = \frac{dx_1}{dt} = A X_2, \quad v_2 = \frac{dx_2}{dt} = -A X_1.$$

(c) The time rate of change of temperature of a material particle in $\mathcal{B}$ is simply

$$\frac{d}{dt}[\theta(\mathbf{X}, t)] = \frac{\partial}{\partial t}[\theta(\mathbf{X}, t)]\Big|_{\mathbf{x} \text{ fixed}} = -2At X_1 + (1 + A) X_2.$$

On the other hand, the time rate of change of temperature at point $\mathbf{x}$, which is now occupied by particle X, is

$$\frac{d}{dt}[\theta(\mathbf{x}, t)] = \frac{\partial \theta}{\partial t} + v_i \frac{\partial \theta}{\partial x_i} = x_2 + v_1 \cdot 1 + v_2 \cdot t$$

$$= -2At X_1 + (1 + A) X_2.$$

In the study of solid bodies, the Eulerian description is less useful since the configuration κ is unknown. On the other hand, it is the preferred description for the study of motion of fluids because the configuration is known and remains unchanged, and we wish to determine the changes in the fluid velocities, pressure, density and so on. Thus, in the Eulerian description, attention is focused on a given region of space instead of a given body of matter.

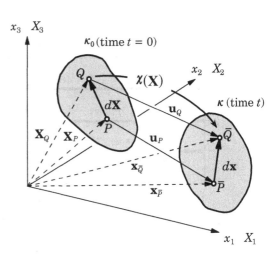

Figure 3.2.5. Points P and Q separated by a distance $d\mathbf{X}$ in the undeformed configuration κ_0 take up positions $\bar{P}$ and $\bar{Q}$, respectively, in the deformed configuration κ, where they are separated by distance $d\mathbf{x}$.

3.2.4 Displacement Field

The phrase deformation of a continuum refers to relative displacements and changes in the geometry experienced by the continuum $\mathcal{B}$ under the influence of a force system. The displacement of the particle X is given, as can be seen from Figure 3.2.5, by

$$\mathbf{u} = \mathbf{x} - \mathbf{X}. \tag{3.2.7}$$

In the Lagrangian description, the displacements are expressed in terms of the material coordinates X_i

$$\mathbf{u}(\mathbf{X}, t) = \mathbf{x}(\mathbf{X}, t) - \mathbf{X}. \tag{3.2.8}$$

If the displacement of every particle in the body $\mathcal{B}$ is known, we can construct the current configuration κ from the reference configuration κ_0, $\chi(\mathbf{X}) = \mathbf{X} + \mathbf{u}(\mathbf{X})$. However, in the Eulerian description the displacements are expressed in terms of the spatial coordinates x_i

$$\mathbf{u}(\mathbf{x}, t) = \mathbf{x} - \mathbf{X}(\mathbf{x}, t). \tag{3.2.9}$$

A rigid-body motion is one in which all material particles of the continuum $\mathcal{B}$ undergo the same linear and angular displacements. However, a deformable body is one in which the material particles can move relative to each other. Then the deformation of a continuum can be determined only by considering the change of distance between any two arbitrary but infinitesimally close points of the continuum.

To illustrate the difference between the two descriptions further, consider the one-dimensional mapping $x = X(1 + 0.5t)$ defining the motion of a rod of initial length two units. The rod experiences a temperature distribution T given by the material description $T = 2Xt^2$ or by the spatial description $T = xt^2/(1 + 0.5t)$, as shown in Figure 3.2.6 [see Bonet and Wood (1997)].

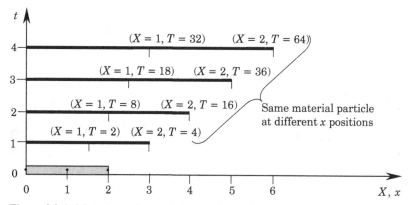

Figure 3.2.6. Material and spatial descriptions of motion.

From Figure 3.2.6, we see that the particle's material coordinate (label) X remains associated with the particle while its spatial position x changes. The temperature at a given time can be found in one of the two ways: for example, at time $t = 3$, the temperature of the particle labeled $X = 2$ is $T = 2 \times 2(3)^2 = 36$; alternatively, the temperature of the same particle which at $t = 3$ is at a spatial position $x = 2(1 + 0.5 \times 3) = 5$ is $T = 2 \times 5(3)^2/(1 + 0.5 \times 3) = 36$. The displacement of a material point occupying position X in κ_0 is

$$u(X, t) = x - X = X(1 + 0.5t) - X = 0.5Xt.$$

3.3 Analysis of Deformation

3.3.1 Deformation Gradient Tensor

One of the key quantities in deformation analysis is the *deformation gradient* of κ relative to the reference configuration κ_0, denoted $\mathbf{F}_\kappa$, which gives the relationship of a material line $d\mathbf{X}$ before deformation to the line $d\mathbf{x}$ (consisting of the same material as $d\mathbf{X}$) after deformation. It is defined as (in the interest of brevity, the subscript κ on $\mathbf{F}$ is dropped)

$$d\mathbf{x} = \mathbf{F} \cdot d\mathbf{X} = d\mathbf{X} \cdot \mathbf{F}^{\mathrm{T}}, \tag{3.3.1}$$

$$\mathbf{F} = \left(\frac{\partial \mathbf{X}}{\partial \mathbf{X}}\right)^{\mathrm{T}} = \left(\frac{\partial \mathbf{x}}{\partial \mathbf{X}}\right)^{\mathrm{T}} \equiv (\nabla_0 \mathbf{x})^{\mathrm{T}}, \tag{3.3.2}$$

and ∇_0 is the gradient operator with respect to $\mathbf{X}$. By definition, $\mathbf{F}$ is a second-order tensor. The inverse relations are given by

$$d\mathbf{X} = \mathbf{F}^{-1} \cdot d\mathbf{x} = d\mathbf{x} \cdot \mathbf{F}^{-\mathrm{T}}, \quad \text{where } \mathbf{F}^{-\mathrm{T}} = \frac{\partial \mathbf{X}}{\partial \mathbf{x}} \equiv \nabla \mathbf{X}, \tag{3.3.3}$$

and ∇ is the gradient operator with respect to $\mathbf{x}$. In indicial notation, Eqs. (3.3.2) and (3.3.3) can be written as

$$\mathbf{F} = F_{iJ}\hat{\mathbf{e}}_i\hat{\mathbf{E}}_J, \quad F_{iJ} = \frac{\partial x_i}{\partial X_J},$$

$$\mathbf{F}^{-1} = F_{Ji}^{-1}\hat{\mathbf{E}}_J\hat{\mathbf{e}}_i, \quad F_{Ji}^{-1} = \frac{\partial X_J}{\partial x_i}.$$

(3.3.4)

More explicitly, we have

$$[F] = \begin{bmatrix} \frac{\partial x_1}{\partial X_1} & \frac{\partial x_1}{\partial X_2} & \frac{\partial x_1}{\partial X_3} \\ \frac{\partial x_2}{\partial X_1} & \frac{\partial x_2}{\partial X_2} & \frac{\partial x_2}{\partial X_3} \\ \frac{\partial x_3}{\partial X_1} & \frac{\partial x_3}{\partial X_2} & \frac{\partial x_3}{\partial X_3} \end{bmatrix}, \quad [F]^{-1} = \begin{bmatrix} \frac{\partial X_1}{\partial x_1} & \frac{\partial X_1}{\partial x_2} & \frac{\partial X_1}{\partial x_3} \\ \frac{\partial X_2}{\partial x_1} & \frac{\partial X_2}{\partial x_2} & \frac{\partial X_2}{\partial x_3} \\ \frac{\partial X_3}{\partial x_1} & \frac{\partial X_3}{\partial x_2} & \frac{\partial X_3}{\partial x_3} \end{bmatrix}. \quad (3.3.5)$$

In Eqs. (3.3.3) and (3.3.4), the lowercase indices refer to the current (spatial) Cartesian coordinates, whereas uppercase indices refer to the reference (material) Cartesian coordinates. The determinant of $\mathbf{F}$ is called the *Jacobian of the motion*, and it is denoted by $J = \det \mathbf{F}$. The equation $\mathbf{F} \cdot d\mathbf{X} = 0$ for $d\mathbf{X} \neq 0$ implies that a material line in the reference configuration is reduced to zero by the deformation. Since this is physically not realistic, we conclude that $\mathbf{F} \cdot d\mathbf{X} \neq 0$ for $d\mathbf{X} \neq 0$. That is, $\mathbf{F}$ is a nonsingular tensor, $J \neq 0$. Hence, $\mathbf{F}$ has an inverse $\mathbf{F}^{-1}$. The deformation gradient can be expressed in terms of the displacement vector as

$$\mathbf{F} = (\nabla_0\mathbf{x})^{\mathrm{T}} = (\nabla_0\mathbf{u} + \mathbf{I})^{\mathrm{T}} \quad \text{or} \quad \mathbf{F}^{-1} = (\nabla\mathbf{X})^{\mathrm{T}} = (\mathbf{I} - \nabla\mathbf{u})^{\mathrm{T}}. \quad (3.3.6)$$

Example 3.3.1 illustrates the computation of the components of the deformation gradient tensor from known mapping of motion.

EXAMPLE 3.3.1: Consider the uniform deformation of a square block of side two units and initially centered at $\mathbf{X} = (0, 0)$. The deformation is defined by the mapping

$$\chi(\mathbf{X}) = (3.5 + X_1 + 0.5X_2)\,\hat{\mathbf{e}}_1 + (4 + X_2)\,\hat{\mathbf{e}}_2 + X_3\,\hat{\mathbf{e}}_3.$$

Determine deformation gradient tensor $\mathbf{F}$, sketch the deformation, and compute the displacements.

SOLUTION: From the given mapping, we have

$$x_1 = 3.5 + X_1 + 0.5X_2, \quad x_2 = 4 + X_2, \quad x_3 = X_3.$$

The above relations can be inverted to obtain

$$X_1 = -1.5 + x_1 - 0.5x_2, \quad X_2 = -4 + x_2, \quad X_3 = x_3.$$

Hence, the inverse mapping is given by

$$\chi^{-1}(\mathbf{x}) = (-1.5 + x_1 - 0.5x_2)\,\hat{\mathbf{E}}_1 + (-4 + x_2)\,\hat{\mathbf{E}}_2 + x_3\,\hat{\mathbf{E}}_3,$$

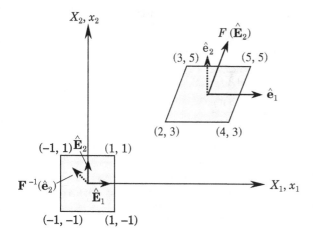

Figure 3.3.1. Uniform deformation of a square.

which produces the deformed shape shown in Figure 3.3.1. This type of deformation is known as *simple shear*, in which there exist a set of line elements (in the present case, lines parallel to the X_1-axis) whose orientation is such that they are unchanged in length and orientation by the deformation. The components of the deformation gradient tensor and its inverse can be expressed in matrix form as

$$[F] = \begin{bmatrix} \frac{\partial x_1}{\partial X_1} & \frac{\partial x_1}{\partial X_2} & \frac{\partial x_1}{\partial X_3} \\ \frac{\partial x_2}{\partial X_1} & \frac{\partial x_2}{\partial X_2} & \frac{\partial x_2}{\partial X_3} \\ \frac{\partial x_3}{\partial X_1} & \frac{\partial x_3}{\partial X_2} & \frac{\partial x_3}{\partial X_3} \end{bmatrix} = \begin{bmatrix} 1.0 & 0.5 & 0.0 \\ 0.0 & 1.0 & 0.0 \\ 0.0 & 0.0 & 1.0 \end{bmatrix},$$

$$[F]^{-1} = \begin{bmatrix} \frac{\partial X_1}{\partial x_1} & \frac{\partial X_1}{\partial x_2} & \frac{\partial X_1}{\partial x_3} \\ \frac{\partial X_2}{\partial x_1} & \frac{\partial X_2}{\partial x_2} & \frac{\partial X_2}{\partial x_3} \\ \frac{\partial X_3}{\partial x_1} & \frac{\partial X_3}{\partial x_2} & \frac{\partial X_3}{\partial x_3} \end{bmatrix} = \begin{bmatrix} 1.0 & -0.5 & 0.0 \\ 0.0 & 1.0 & 0.0 \\ 0.0 & 0.0 & 1.0 \end{bmatrix}.$$

The displacement vector is given by

$$\mathbf{u} = (3.5 + 0.5X_2)\hat{\mathbf{e}}_1 + 4\,\hat{\mathbf{e}}_2.$$

The unit vectors $\hat{\mathbf{E}}_1$ and $\hat{\mathbf{E}}_2$ in the initial configuration deform to the vectors

$$\begin{bmatrix} 1.0 & 0.5 & 0.0 \\ 0.0 & 1.0 & 0.0 \\ 0.0 & 0.0 & 1.0 \end{bmatrix} \begin{Bmatrix} 1 \\ 0 \\ 0 \end{Bmatrix} = \begin{Bmatrix} 1 \\ 0 \\ 0 \end{Bmatrix}, \quad \begin{bmatrix} 1.0 & 0.5 & 0.0 \\ 0.0 & 1.0 & 0.0 \\ 0.0 & 0.0 & 1.0 \end{bmatrix} \begin{Bmatrix} 0 \\ 1 \\ 0 \end{Bmatrix} = \begin{Bmatrix} 0.5 \\ 1.0 \\ 0.0 \end{Bmatrix}.$$

The unit vectors $\hat{\mathbf{e}}_1$ and $\hat{\mathbf{e}}_2$ in the current configuration are deformed from the vectors

$$\begin{bmatrix} 1.0 & -0.5 & 0.0 \\ 0.0 & 1.0 & 0.0 \\ 0.0 & 0.0 & 1.0 \end{bmatrix} \begin{Bmatrix} 1 \\ 0 \\ 0 \end{Bmatrix} = \begin{Bmatrix} 1 \\ 0 \\ 0 \end{Bmatrix}, \quad \begin{bmatrix} 1.0 & -0.5 & 0.0 \\ 0.0 & 1.0 & 0.0 \\ 0.0 & 0.0 & 1.0 \end{bmatrix} \begin{Bmatrix} 0 \\ 1 \\ 0 \end{Bmatrix} = \begin{Bmatrix} -0.5 \\ 1.0 \\ 0.0 \end{Bmatrix}.$$

3.3.2 Isochoric, Homogeneous, and Inhomogeneous Deformations

3.3.2.1 Isochoric Deformation

If the Jacobian is unity $J = 1$, then the deformation is a rigid rotation or the current and reference configurations coincide. If volume does not change locally (i.e., volume preserving) during the deformation, the deformation is said to be *isochoric* at $\mathbf{X}$. If $J = 1$ everywhere in the body $\mathcal{B}$, then the deformation of the body is isochoric.

3.3.2.2 Homogeneous Deformation

In general, the deformation gradient $\mathbf{F}$ is a function of $\mathbf{X}$. If $\mathbf{F} = \mathbf{I}$ everywhere in the body, then the body is not rotated and is undeformed. If $\mathbf{F}$ has the same value at every material point in a body (i.e., $\mathbf{F}$ is independent of $\mathbf{X}$), then the mapping $\mathbf{x} = \mathbf{x}(\mathbf{X}, t)$ is said to be a *homogeneous motion* of the body and the deformation is said to be homogeneous. In general, at any given time $t > 0$, a mapping $\mathbf{x} = \mathbf{x}(\mathbf{X}, t)$ is said to be a homogeneous motion if and only if it can be expressed as (so that $\mathbf{F}$ is a constant)

$$\mathbf{x} = \mathbf{A} \cdot \mathbf{X} + \mathbf{c}, \tag{3.3.7}$$

where the second-order tensor $\mathbf{A}$ and vector $\mathbf{c}$ are constants; $\mathbf{c}$ represents a rigid-body translation. For a homogeneous motion, we have $\mathbf{F} = \mathbf{A}$. Clearly, the motion described by the mapping of Example 3.3.1 is homogeneous and isochoric. Next, we consider several simple forms of homogeneous deformations.

PURE DILATATION. If a cube of material has edges of length L and ℓ in the reference and current configurations, respectively, then the deformation mapping has the form

$$\chi(\mathbf{X}) = \lambda X_1 \, \hat{\mathbf{e}}_1 + \lambda X_2 \, \hat{\mathbf{e}}_2 + \lambda X_3 \, \hat{\mathbf{e}}_3, \quad \lambda = \frac{L}{\ell}, \tag{3.3.8}$$

and $\mathbf{F}$ has the matrix representation

$$[F] = \begin{bmatrix} \lambda & 0 & 0 \\ 0 & \lambda & 0 \\ 0 & 0 & \lambda \end{bmatrix}. \tag{3.3.9}$$

This deformation is known as *pure dilatation*, or *pure stretch*, and it is isochoric if and only if $\lambda = 1$ (λ is called the *principal stretch*), as shown in Fig. 3.3.2.

SIMPLE EXTENSION. An example of homogeneous extension in the X_1-direction is shown in Fig. 3.3.3. The deformation mapping for this case is given by

$$\chi(\mathbf{X}) = (1 + \alpha) X_1 \, \hat{\mathbf{e}}_1 + X_2 \, \mathbf{e}_2 + X_3 \, \hat{\mathbf{e}}_3. \tag{3.3.10}$$

The components of the deformation gradient are given by

$$[F] = \begin{bmatrix} 1+\alpha & 0 & 0 \\ 0 & 1 & 0 \\ 0 & 0 & 1 \end{bmatrix}. \tag{3.3.11}$$

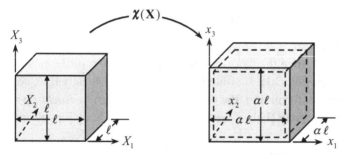

Figure 3.3.2. A deformation mapping of pure dilatation.

For example, a line $X_2 = a + b X_1$ in the undeformed configuration transforms under the mapping to [because $x_1 = (1 + \alpha) X_1$, $x_2 = X_2$, and $x_3 = X_3$]

$$x_2 = a + \frac{b}{1 + \alpha} x_1 .$$

SIMPLE SHEAR. This deformation, as discussed in Example 3.3.1, is defined to be one in which there exists a set of line elements whose lengths and orientations are unchanged, as shown in Fig. 3.3.4. The deformation mapping in this case is

$$\chi(\mathbf{X}) = (X_1 + \gamma X_2)\hat{\mathbf{e}}_1 + X_2 \, \mathbf{e}_2 + X_3 \, \hat{\mathbf{e}}_3. \qquad (3.3.12)$$

The matrix representation of the deformation gradient is given by

$$[F] = \begin{bmatrix} 1 & \gamma & 0 \\ 0 & 1 & 0 \\ 0 & 0 & 1 \end{bmatrix}, \qquad (3.3.13)$$

where γ denotes the amount of shear.

3.3.2.3 Nonhomogeneous Deformation
A nonhomogeneous deformation is one in which the deformation gradient $\mathbf{F}$ is a function of $\mathbf{X}$. An example of nonhomogeneous deformation mapping is provided,

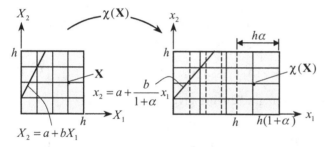

Figure 3.3.3. A deformation mapping of simple extension.

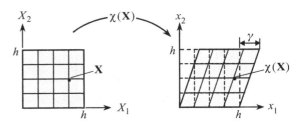

Figure 3.3.4. A deformation mapping of simple shear.

as shown in Fig. 3.3.5, by

$$\chi(\mathbf{X}) = X_1(1 + \gamma_1 X_2)\hat{\mathbf{e}}_1 + X_2(1 + \gamma_2 X_1)\mathbf{e}_2 + X_3\,\hat{\mathbf{e}}_3. \qquad (3.3.14)$$

The matrix representation of the deformation gradient is

$$[F] = \begin{bmatrix} 1 + \gamma_1 X_2 & \gamma_1 X_1 & 0 \\ \gamma_2 X_2 & 1 + \gamma_2 X_1 & 0 \\ 0 & 0 & 1 \end{bmatrix}. \qquad (3.3.15)$$

It is rather difficult to invert the mapping even for this simple nonhomogeneous deformation.

3.3.3 Change of Volume and Surface

Here we study how deformation mapping affects surface areas and volumes of a continuum. The motivation for this study comes from the need to write global equilibrium statements that involve integrals over areas and volumes.

3.3.3.1 Volume Change

We can define volume and surface elements in the reference and deformed configurations. Consider three non-coplanar line elements $d\mathbf{X}^{(1)}$, $d\mathbf{X}^{(2)}$, and $d\mathbf{X}^{(3)}$ forming the edges of a parallelepiped at point P with position vector $\mathbf{X}$ in the reference body $\mathcal{B}$, as shown in Figure 3.3.6, so that

$$d\mathbf{x}^{(i)} = \mathbf{F} \cdot d\mathbf{X}^{(i)}, \quad i = 1, 2, 3. \qquad (3.3.16)$$

The vectors $d\mathbf{x}^{(i)}$ are not necessarily parallel to or have the same length as the vectors $d\mathbf{X}^{(i)}$ because of shearing and stretching of the parallelepiped. We assume that the triad $(d\mathbf{X}^{(1)}, d\mathbf{X}^{(2)}, d\mathbf{X}^{(3)})$ is positively oriented in the sense that the triple

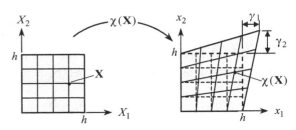

Figure 3.3.5. A deformation mapping of combined shearing and extension.

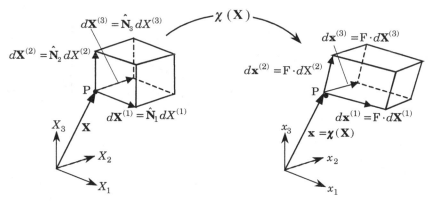

Figure 3.3.6. Transformation of a volume element under a deformation mapping.

scalar product $d\mathbf{X}^{(1)} \cdot d\mathbf{X}^{(2)} \times d\mathbf{X}^{(3)} > 0$. We denote the volume of the parallelepiped as

$$dV = d\mathbf{X}^{(1)} \cdot d\mathbf{X}^{(2)} \times d\mathbf{X}^{(3)} = \left(\hat{\mathbf{N}}_1 \cdot \hat{\mathbf{N}}_2 \times \hat{\mathbf{N}}_3 \right) dX^{(1)} dX^{(2)} dX^{(3)}$$
$$= dX^{(1)} dX^{(2)} dX^{(3)}, \tag{3.3.17}$$

where $\hat{\mathbf{N}}_i$ denote the unit vector along $d\mathbf{X}^{(i)}$. The corresponding volume in the deformed configuration is given by

$$dv = d\mathbf{x}^{(1)} \cdot d\mathbf{x}^{(2)} \times d\mathbf{x}^{(3)}$$
$$= \left(\mathbf{F} \cdot \hat{\mathbf{N}}_1 \right) \cdot \left(\mathbf{F} \cdot \hat{\mathbf{N}}_2 \right) \times \left(\mathbf{F} \cdot \hat{\mathbf{N}}_3 \right) dX^{(1)} dX^{(2)} dX^{(3)}$$
$$= \det \mathbf{F} \, dX^{(1)} dX^{(2)} dX^{(3)} = J \, dV. \tag{3.3.18}$$

We assume that the volume elements are positive so that the relative orientation
of the line elements is preserved under the deformation, that is, $J > 0$. Thus, J has
the physical meaning of being the local ratio of current to reference volume of a
material volume element.

3.3.3.2 Surface Change

Next, consider an infinitesimal vector element of material surface $d\mathbf{A}$ in a neighborhood of the point $\mathbf{X}$ in the undeformed configuration, as shown in Figure 3.3.7. The
surface vector can be expressed as $d\mathbf{A} = dA\,\hat{\mathbf{N}}$, where $\hat{\mathbf{N}}$ is the positive unit normal
to the surface in the reference configuration. Suppose that $d\mathbf{A}$ becomes $d\mathbf{a}$ in the
deformed body, where $d\mathbf{a} = da\,\hat{\mathbf{n}}$, $\hat{\mathbf{n}}$ being the positive unit normal to the surface in
the deformed configuration. The unit normals in the deformed and deformed con-
figurations can be expressed as ($\hat{\mathbf{N}}_i = \hat{\mathbf{n}}_i$):

$$\hat{\mathbf{N}} = \frac{\hat{\mathbf{N}}_1 \times \hat{\mathbf{N}}_2}{|\hat{\mathbf{N}}_1 \times \hat{\mathbf{N}}_2|}, \qquad \hat{\mathbf{n}} = \frac{\mathbf{F} \cdot \hat{\mathbf{n}}_1 \times \mathbf{F} \cdot \hat{\mathbf{n}}_2}{|\mathbf{F} \cdot \hat{\mathbf{n}}_1 \times \mathbf{F} \cdot \hat{\mathbf{n}}_2|}. \tag{3.3.19}$$

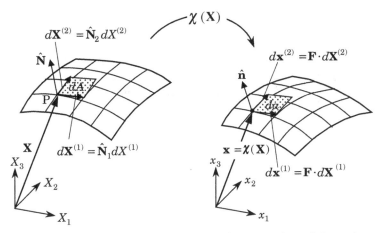

Figure 3.3.7. Transformation of a surface element under a deformation mapping.

The areas of the parallelograms in the undeformed and deformed configurations are

$$dA \equiv |\hat{\mathbf{N}}_1 \times \hat{\mathbf{N}}_2| \, dX_1 \, dX_2, \quad da \equiv |\mathbf{F} \cdot \hat{\mathbf{n}}_1 \times \mathbf{F} \cdot \hat{\mathbf{n}}_2| \, dx_1 \, dx_2. \tag{3.3.20}$$

The area vectors are

$$dA = \hat{\mathbf{N}} dA = \frac{\hat{\mathbf{N}}_1 \times \hat{\mathbf{N}}_2}{|\hat{\mathbf{N}}_1 \times \hat{\mathbf{N}}_2|} \, |\hat{\mathbf{N}}_1 \times \hat{\mathbf{N}}_2| \, dX_1 \, dX_2$$

$$= \left(\hat{\mathbf{N}}_1 \times \hat{\mathbf{N}}_2\right) dX_1 \, dX_2 = \left(\hat{\mathbf{n}}_1 \times \hat{\mathbf{n}}_2\right) dX_1 \, dX_2, \tag{3.3.21}$$

$$da = \hat{\mathbf{n}} \, da = \frac{\mathbf{F} \cdot \hat{\mathbf{n}}_1 \times \mathbf{F} \cdot \hat{\mathbf{n}}_2}{|\mathbf{F} \cdot \hat{\mathbf{n}}_1 \times \mathbf{F} \cdot \hat{\mathbf{n}}_2|} \, |\mathbf{F} \cdot \hat{\mathbf{n}}_1 \times \mathbf{F} \cdot \hat{\mathbf{n}}_2| \, dx_1 \, dx_2$$

$$= \left(\mathbf{F} \cdot \hat{\mathbf{n}}_1 \times \mathbf{F} \cdot \hat{\mathbf{n}}_2\right) dx_1 \, dx_2. \tag{3.3.22}$$

Then it can be shown that (see the result of Problem 3.10)

$$da = J\mathbf{F}^{-T} \cdot dA \quad \text{or} \quad \hat{\mathbf{n}} \, da = J\mathbf{F}^{-T} \cdot \hat{\mathbf{N}} dA. \tag{3.3.23}$$

Next we consider an example of area change under simple shear deformation [see Hjelmsted (2005) for additional examples].

EXAMPLE 3.3.2: Consider a square block with a circular hole at the center, as shown in Figure 3.3.8(a). Suppose that block is of of thickness h and plane dimensions $2b \times 2b$, and the radius of the hole is b. Determine the change in the area of the circle and the edge of the block when it is subjected to simple shear deformation mapping of Eq. (3.3.12).

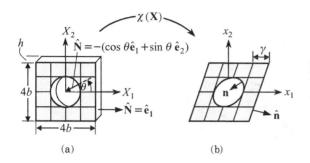

Figure 3.3.8. (a) Original geometry of the undeformed square block. (b) Deformed (simple shear) geometry of the block.

(a)   (b)

SOLUTION: The components of the deformation gradient tensor and its inverse are

$$[F] = \begin{bmatrix} 1 & \gamma & 0 \\ 0 & 1 & 0 \\ 0 & 0 & 1 \end{bmatrix}, \quad [F]^{-1} = \begin{bmatrix} 1 & -\gamma & 0 \\ 0 & 1 & 0 \\ 0 & 0 & 1 \end{bmatrix}.$$

The determinant of $\mathbf{F}$ is $\det \mathbf{F} = 1$, implying that there is no change in the volume of the block. Consider the edge with normal $\hat{\mathbf{N}} = \hat{\mathbf{E}}_1 = \hat{\mathbf{e}}_1$ in the undeformed configuration. By Eq. (3.3.23), we have

$$\hat{\mathbf{n}}\, da_1 = (\hat{\mathbf{e}}_1 - \gamma \hat{\mathbf{e}}_2)\, dX_2\, dX_3.$$

Thus, da_1 is

$$da_1 = \sqrt{(1 + \gamma^2)}\, dX_2\, dX_3.$$

The total area of the deformed edge, as shown in Fig. 3.3.8(b), is

$$\int_0^h \int_{-2b}^{2b} da_1 = 4bh\sqrt{1 + \gamma^2}.$$

The result is obvious from the deformed geometry of the edge.

Next, we determine the deformed area of the cylindrical surface of the hole. In this case, the unit vector normal to the surface is in the radial direction and it is given by

$$\hat{\mathbf{N}} = -(\cos\theta\, \hat{\mathbf{e}}_1 + \sin\theta\, \hat{\mathbf{e}}_2).$$

Hence, the components of the vector $\mathbf{F}^{-T} \cdot \hat{\mathbf{N}}$ are given by

$$\begin{bmatrix} 1 & 0 & 0 \\ -\gamma & 1 & 0 \\ 0 & 0 & 1 \end{bmatrix} \begin{Bmatrix} -\cos\theta \\ -\sin\theta \\ 0 \end{Bmatrix} = \begin{Bmatrix} -\cos\theta \\ \gamma\cos\theta - \sin\theta \\ 0 \end{Bmatrix}.$$

Using Eq. (3.3.23), we obtain

$$\hat{\mathbf{n}}\, da_n = [-\cos\theta\, \hat{\mathbf{e}}_1 + (\gamma\cos\theta - \sin\theta)\, \mathbf{e}_2]\, b\, d\theta\, dX_3.$$

Hence, the deformed surface area of the hole is

$$b \int_0^h \int_0^{2\pi} \sqrt{\cos^2 \theta + (\gamma \cos \theta - \sin \theta)^2} \, d\theta \, dX_3.$$

The integral can be evaluated for any given value of γ. In particular, we have

$$\gamma = 0: \quad a_n = 2\pi b h \quad \text{(no deformation)},$$

$$\gamma = 1: \quad a_n = bh \int_0^{2\pi} \sqrt{1.5 + 0.5 \cos 2\theta - \sin 2\theta} \, d\theta \approx 2.35\pi b h.$$

For other values of γ, the integral may be evaluated numerically.

3.4 Strain Measures

3.4.1 Cauchy–Green Deformation Tensors

The geometric changes that a continuous medium experiences can be measured in a number of ways. Here, we discuss a general measure of deformation of a continuous medium, independent of both translation and rotation.

Consider two material particles P and Q in the neighborhood of each other, separated by $d\mathbf{X}$ in the reference configuration, as shown in Figure 3.4.1. In the current (deformed) configuration, the material points P and Q occupy positions $\bar{P}$ and $\bar{Q}$, and they are separated by $d\mathbf{x}$. We wish to determine the change in the distance $d\mathbf{X}$ between the material points P and Q as the body deforms and the material points move to the new locations $\bar{P}$ and $\bar{Q}$.

The distances between points P and Q and points $\bar{P}$ and $\bar{Q}$ are given, respectively, by

$$(dS)^2 = d\mathbf{X} \cdot d\mathbf{X}, \tag{3.4.1}$$

$$(ds)^2 = d\mathbf{x} \cdot d\mathbf{x} = d\mathbf{X} \cdot (\mathbf{F}^{\mathrm{T}} \cdot \mathbf{F}) \cdot d\mathbf{X} \equiv d\mathbf{X} \cdot \mathbf{C} \cdot d\mathbf{X}, \tag{3.4.2}$$

where $\mathbf{C}$ is called the *right Cauchy–Green deformation tensor*

$$\mathbf{C} = \mathbf{F}^{\mathrm{T}} \cdot \mathbf{F}. \tag{3.4.3}$$

Figure 3.4.1. Points P and Q separated by a distance $d\mathbf{X}$ in the undeformed configuration κ_0 take up positions $\bar{P}$ and $\bar{Q}$, respectively, in the deformed configuration κ, where they are separated by distance $d\mathbf{x}$.

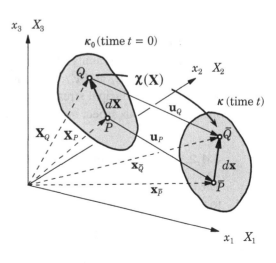

By definition, $\mathbf{C}$ is a symmetric second-order tensor. The transpose of $\mathbf{C}$ is denoted by $\mathbf{B}$ and it is called the *left Cauchy–Green deformation tensor*, or *Finger tensor*

$$\mathbf{B} = \mathbf{F} \cdot \mathbf{F}^{\mathrm{T}}. \tag{3.4.4}$$

Recall from Eq. (2.4.15) that the directional (or tangential) derivative of a field $\phi(\mathbf{X})$ is given by

$$\frac{d\phi}{dS} = \hat{\mathbf{N}} \cdot \nabla_0 \phi, \quad \hat{\mathbf{N}} = \frac{d\mathbf{X}}{|d\mathbf{X}|} = \frac{d\mathbf{X}}{dS}, \tag{3.4.5}$$

where $\hat{\mathbf{N}}$ is the unit vector in the direction of the tangent vector at point $\mathbf{X}$. Therefore, a parameterized curve in the deformed configuration is determined by the deformation mapping $\mathbf{x}(S) = \chi(\mathbf{x}(S))$, and we have ($\mathbf{F} = F_{iJ}\,\hat{\mathbf{e}}\hat{\mathbf{E}}_J$ and $\hat{\mathbf{N}} = N_K\hat{\mathbf{E}}_K$)

$$\frac{d\mathbf{x}}{dS} = \frac{d\mathbf{X}}{dS} \cdot \nabla_0 \chi(\mathbf{X}) = \mathbf{F} \cdot \frac{d\mathbf{X}}{dS}$$

$$= \mathbf{F} \cdot \hat{\mathbf{N}} = F_{iJ} N_J\,\hat{\mathbf{e}}_i. \tag{3.4.6}$$

Clearly, $d\mathbf{x}/dS = F_{iJ} N_J\,\hat{\mathbf{e}}_i$ is a vector defined in the deformed configuration.

The *stretch* of a curve at a point in the deformed configuration is defined to be the ratio of the deformed length of the curve to its original length. Let us consider an infinitesimal length dS of curve in the neighborhood of the material point $\mathbf{X}$. Then the stretch λ of the curve is simply the length of the tangent vector $\mathbf{F} \cdot \hat{\mathbf{N}}$ in the deformed configuration

$$\lambda^2(S) = (\mathbf{F} \cdot \hat{\mathbf{N}}) \cdot (\mathbf{F} \cdot \hat{\mathbf{N}}) \tag{3.4.7}$$

$$= \hat{\mathbf{N}} \cdot (\mathbf{F}^{\mathrm{T}} \cdot \mathbf{F}) \cdot \hat{\mathbf{N}}$$

$$= \hat{\mathbf{N}} \cdot \mathbf{C} \cdot \hat{\mathbf{N}} \tag{3.4.8}$$

Equation (3.4.8) holds for any arbitrary curve with $d\mathbf{X} = dS\,\hat{\mathbf{N}}$ and thus allows us to compute the stretch in any direction at a given point. In particular, the square of the stretch in the direction of the unit base vector $\hat{\mathbf{E}}_I$ is given by

$$\lambda^2(\hat{\mathbf{E}}_I) = \hat{\mathbf{E}}_I \cdot \mathbf{C} \cdot \hat{\mathbf{E}}_I = C_{II}. \tag{3.4.9}$$

That is, the diagonal terms of the left Cauchy–Green deformation tensor $\mathbf{C}$ represent the squares of the stretches in the direction of the coordinate axes (X_1, X_2, X_3). The off-diagonal elements of $\mathbf{C}$ give a measure of the angle of shearing between two base vectors $\hat{\mathbf{E}}_I$ and $\hat{\mathbf{E}}_J$, $\neq J$, under the deformation mapping χ. Further, the squares of the principal stretches at a point are equal to the eigenvalue of $\mathbf{C}$. We shall return to this aspect in Section 3.7 on polar decomposition theorem.

3.4.2 Green Strain Tensor

The change in the squared lengths that occurs as a body deforms from the reference to the current configuration can be expressed relative to the original length as

$$(ds)^2 - (dS)^2 = 2\,d\mathbf{X} \cdot \mathbf{E} \cdot d\mathbf{X}, \tag{3.4.10}$$

where $\mathbf{E}$ is called the *Green–St. Venant (Lagrangian) strain tensor* or simply the *Green strain tensor*.[2] The Green strain tensor can be expressed, in view of Eqs. (3.4.1)–(3.4.3), as

$$
\begin{aligned}
\mathbf{E} &= \frac{1}{2}\left(\mathbf{F}^{\mathrm{T}} \cdot \mathbf{F} - \mathbf{I}\right) = \frac{1}{2}\left(\mathbf{C} - \mathbf{I}\right) \\
&= \frac{1}{2}\left[(\mathbf{I} + \nabla_0 \mathbf{u}) \cdot (\mathbf{I} + \nabla_0 \mathbf{u})^{\mathrm{T}} - \mathbf{I}\right] \\
&= \frac{1}{2}\left[\nabla_0 \mathbf{u} + (\nabla_0 \mathbf{u})^{\mathrm{T}} + (\nabla_0 \mathbf{u}) \cdot (\nabla_0 \mathbf{u})^{\mathrm{T}}\right].
\end{aligned}
\tag{3.4.11}
$$

By definition, the Green strain tensor is a symmetric second-order tensor. Also, the change in the squared lengths is zero if and only if $\mathbf{E} = \mathbf{0}$.

The vector form of the Green strain tensor in Eq. (3.4.11) allows us to express it in terms of its components in any coordinate system. In particular, in rectangular Cartesian coordinate system (X_1, X_2, X_3), the components of $\mathbf{E}$ are given by

$$
E_{ij} = \frac{1}{2}\left(\frac{\partial u_i}{\partial X_j} + \frac{\partial u_j}{\partial X_i} + \frac{\partial u_k}{\partial X_i}\frac{\partial u_k}{\partial X_j}\right).
\tag{3.4.12}
$$

In expanded notation, they are given by

$$
\begin{aligned}
E_{11} &= \frac{\partial u_1}{\partial X_1} + \frac{1}{2}\left[\left(\frac{\partial u_1}{\partial X_1}\right)^2 + \left(\frac{\partial u_2}{\partial X_1}\right)^2 + \left(\frac{\partial u_3}{\partial X_1}\right)^2\right], \\
E_{22} &= \frac{\partial u_2}{\partial X_2} + \frac{1}{2}\left[\left(\frac{\partial u_1}{\partial X_2}\right)^2 + \left(\frac{\partial u_2}{\partial X_2}\right)^2 + \left(\frac{\partial u_3}{\partial X_2}\right)^2\right], \\
E_{33} &= \frac{\partial u_3}{\partial X_3} + \frac{1}{2}\left[\left(\frac{\partial u_1}{\partial X_3}\right)^2 + \left(\frac{\partial u_2}{\partial X_3}\right)^2 + \left(\frac{\partial u_3}{\partial X_3}\right)^2\right], \\
E_{12} &= \frac{1}{2}\left(\frac{\partial u_1}{\partial X_2} + \frac{\partial u_2}{\partial X_1} + \frac{\partial u_1}{\partial X_1}\frac{\partial u_1}{\partial X_2} + \frac{\partial u_2}{\partial X_1}\frac{\partial u_2}{\partial X_2} + \frac{\partial u_3}{\partial X_1}\frac{\partial u_3}{\partial X_2}\right), \\
E_{13} &= \frac{1}{2}\left(\frac{\partial u_1}{\partial X_3} + \frac{\partial u_3}{\partial X_1} + \frac{\partial u_1}{\partial X_1}\frac{\partial u_1}{\partial X_3} + \frac{\partial u_2}{\partial X_1}\frac{\partial u_2}{\partial X_3} + \frac{\partial u_3}{\partial X_1}\frac{\partial u_3}{\partial X_3}\right), \\
E_{23} &= \frac{1}{2}\left(\frac{\partial u_2}{\partial X_3} + \frac{\partial u_3}{\partial X_2} + \frac{\partial u_1}{\partial X_2}\frac{\partial u_1}{\partial X_3} + \frac{\partial u_2}{\partial X_2}\frac{\partial u_2}{\partial X_3} + \frac{\partial u_3}{\partial X_2}\frac{\partial u_3}{\partial X_3}\right).
\end{aligned}
\tag{3.4.13}
$$

The components E_{11}, E_{22}, and E_{33} are called *normal strains* and E_{12}, E_{23}, and E_{13} are called *shear strains*. The Green–Lagrange strain components in the cylindrical coordinate system are given in Problem 3.18.

---

[2] The reader should not confuse the symbol $\mathbf{E}$ used for the Lagrangian strain tensor and $\mathbf{E}_i$ used for the basis vectors in the reference configuration. One should always pay attention to different typeface and subscripts used.

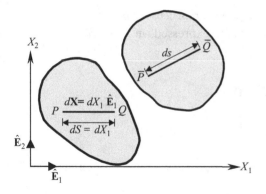

Figure 3.4.2. Physical interpretation of normal strain component E_{11}.

3.4.3 Physical Interpretation of Green Strain Components

To see the physical meaning of the normal strain component E_{11}, consider a line element initially parallel to the X_1-axis, that is, $d\mathbf{X} = dX_1 \hat{\mathbf{E}}_1$ in the undeformed body, as shown in Figure 3.4.2. Then

$$(ds)^2 - (dS)^2 = 2E_{ij}\, dX_i\, dX_j = 2E_{11}\, dX_1\, dX_1 = 2E_{11}\, (dS)^2.$$

Solving for E_{11}, we obtain

$$E_{11} = \frac{1}{2}\frac{(ds)^2 - (dS)^2}{(dS)^2} = \frac{1}{2}\left[\left(\frac{ds}{dS}\right)^2 - 1\right] = \frac{1}{2}\left(\lambda^2 - 1\right), \qquad (3.4.14)$$

where λ is the stretch

$$\lambda = \frac{ds}{dS} = \sqrt{1 + 2E_{11}} = 1 + E_{11} - \frac{1}{2}E_{11}^2 + \dots. \qquad (3.4.15)$$

In terms of the *unit extension* $\Lambda_1 = \lambda - 1$, we have (including up to the quadratic term)

$$E_{11} = \Lambda_1 + \frac{1}{2}\Lambda_1^2. \qquad (3.4.16)$$

When the unit extension is small compared with unity, the quadratic term in the last expression can be neglected in comparison with the linear term, and the strain E_{11} is approximately equal to the unit extension Λ_1. Thus, E_{11} is the ratio of the change in its length to the original length.

The shear strain components E_{ij}, $i \neq j$, can be interpreted as a measure of the change in the angle between line elements that were perpendicular to each other in the undeformed configuration. To see this, consider line elements $d\mathbf{X}^{(1)} = dX_1\hat{\mathbf{E}}_1$ and $d\mathbf{X}^{(2)} = dX_2\hat{\mathbf{E}}_2$ in the undeformed body, which are perpendicular to each other, as shown in Figure 3.4.3. The material line elements $d\mathbf{X}^{(1)}$ and $d\mathbf{X}^{(2)}$ occupy positions $d\mathbf{x}^{(1)}$ and $d\mathbf{x}^{(2)}$, respectively, in the deformed body. Then the

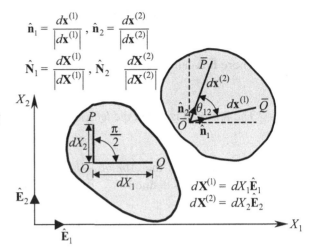

Figure 3.4.3. Physical interpretation of shear strain component E_{12}.

cosine of the angle between the line elements $\bar{O}\bar{Q}$ and $\bar{O}\bar{P}$ in the deformed body is given by [see Eq. (3.3.1)]

$$\cos\theta_{12} = \hat{\mathbf{n}}_1 \cdot \hat{\mathbf{n}}_2 = \frac{d\mathbf{x}^{(1)} \cdot d\mathbf{x}^{(2)}}{|d\mathbf{x}^{(1)}|\,|d\mathbf{x}^{(2)}|}$$

$$= \frac{[d\mathbf{X}^{(1)} \cdot \mathbf{F}^{\mathrm{T}}] \cdot [\mathbf{F} \cdot d\mathbf{X}^{(2)}]}{\sqrt{d\mathbf{X}^{(1)} \cdot \mathbf{C} \cdot d\mathbf{X}^{(1)}}\sqrt{d\mathbf{X}^{(2)} \cdot \mathbf{C} \cdot d\mathbf{X}^{(2)}}}. \tag{3.4.17}$$

Since

$$\mathbf{C} = \mathbf{F}^{\mathrm{T}} \cdot \mathbf{F}, \quad \hat{\mathbf{N}}_1 = \hat{\mathbf{E}}_1, \quad \hat{\mathbf{N}}_2 = \hat{\mathbf{E}}_2, \tag{3.4.18}$$

we have

$$\cos\theta_{12} = \frac{\hat{\mathbf{N}}_1 \cdot \mathbf{C} \cdot \mathbf{N}_2}{\sqrt{\hat{\mathbf{N}}_1 \cdot \mathbf{C} \cdot \mathbf{N}_1}\,\sqrt{\hat{\mathbf{N}}_2 \cdot \mathbf{C} \cdot \mathbf{N}_2}} = \frac{C_{12}}{\sqrt{C_{11}}\sqrt{C_{22}}}$$

or

$$\theta_{12} = \frac{C_{12}}{\lambda_1\lambda_2} = \frac{2E_{12}}{\sqrt{(1+2E_{11})}\sqrt{(1+2E_{22})}}. \tag{3.4.19}$$

Thus, $2E_{12}$ is equal to cosine of the angle between the line elements, θ_{12}, multiplied by the product of extension ratios γ_1 and γ_2. Clearly, the finite strain E_{12} not only depends on the angle θ_{12} but also on the stretches of elements involved. When the unit extensions and the angle changes are small compared with unity, we have

$$\frac{\pi}{2} - \theta_{12} \approx \sin\left(\frac{\pi}{2} - \theta_{12}\right) = \cos\theta_{12} \approx 2E_{12}. \tag{3.4.20}$$

3.4.4 Cauchy and Euler Strain Tensors

Returning to the strain measures, the change in the squared lengths that occurs as the body deforms from the initial to the current configuration can be expressed relative to the current length. First, we express dS in terms of $d\mathbf{x}$ as

$$(dS)^2 = d\mathbf{X} \cdot d\mathbf{X} = d\mathbf{x} \cdot (\mathbf{F}^{-T} \cdot \mathbf{F}^{-1}) \cdot d\mathbf{x} \equiv d\mathbf{x} \cdot \tilde{\mathbf{B}} \cdot d\mathbf{x}, \qquad (3.4.21)$$

where $\tilde{\mathbf{B}}$ is called the *Cauchy strain tensor*

$$\tilde{\mathbf{B}} = \mathbf{F}^{-T} \cdot \mathbf{F}^{-1}, \quad \tilde{\mathbf{B}}^{-1} \equiv \mathbf{B} = \mathbf{F} \cdot \mathbf{F}^T. \qquad (3.4.22)$$

The tensor $\mathbf{B}$ is called the *left Cauchy–Green tensor*, or *Finger tensor*. We can write

$$(ds)^2 - (dS)^2 = 2\, d\mathbf{x} \cdot \mathbf{e} \cdot d\mathbf{x}. \qquad (3.4.23)$$

where $\mathbf{e}$, called the *Almansi–Hamel (Eulerian) strain tensor* or simply the *Euler strain tensor*, is defined as

$$\mathbf{e} = \frac{1}{2}\left(\mathbf{I} - \mathbf{F}^{-T} \cdot \mathbf{F}^{-1}\right) = \frac{1}{2}\left(\mathbf{I} - \tilde{\mathbf{B}}\right) \qquad (3.4.24)$$

$$= \frac{1}{2}\left[\mathbf{I} - (\mathbf{I} - \nabla\mathbf{u}) \cdot (\mathbf{I} - \nabla\mathbf{u})^T\right]$$

$$= \frac{1}{2}\left[\nabla\mathbf{u} + (\nabla\mathbf{u})^T - (\nabla\mathbf{u}) \cdot (\nabla\mathbf{u})^T\right]. \qquad (3.4.25)$$

The rectangular Cartesian components of $\mathbf{C}$, $\tilde{\mathbf{B}}$, and $\mathbf{e}$ are given by

$$C_{IJ} = \frac{\partial x_k}{\partial X_I}\frac{\partial x_k}{\partial X_J}, \quad \tilde{B}_{ij} = \frac{\partial X_K}{\partial x_i}\frac{\partial X_K}{\partial x_j}, \qquad (3.4.26)$$

$$e_{ij} = \frac{1}{2}\left(\delta_{ij} - \frac{\partial X_K}{\partial x_i}\frac{\partial X_K}{\partial x_j}\right)$$

$$= \frac{1}{2}\left(\frac{\partial u_i}{\partial x_j} + \frac{\partial u_j}{\partial x_i} - \frac{\partial u_k}{\partial x_i}\frac{\partial u_k}{\partial x_j}\right). \qquad (3.4.27)$$

The next two examples illustrate the calculation of various measures of strain.

EXAMPLE 3.4.1: For the deformation given in Example 3.3.1, determine the right Cauchy–Green deformation tensor, the Cauchy strain tensor, and the components of Green and Almansi strain tensors.

SOLUTION: The right Cauchy–Green deformation tensor and the Cauchy strain tensor are, respectively,

$$[C] = \begin{bmatrix} 1.0 & 0.0 & 0.0 \\ 0.5 & 1.0 & 0.0 \\ 0.0 & 0.0 & 1.0 \end{bmatrix}\begin{bmatrix} 1.0 & 0.5 & 0.0 \\ 0.0 & 1.0 & 0.0 \\ 0.0 & 0.0 & 1.0 \end{bmatrix} = \begin{bmatrix} 1.0 & 0.5 & 0.0 \\ 0.5 & 1.25 & 0.0 \\ 0.0 & 0.0 & 1.0 \end{bmatrix},$$

$$[\tilde{B}] = \begin{bmatrix} 1.0 & 0.0 & 0.0 \\ -0.5 & 1.0 & 0.0 \\ 0.0 & 0.0 & 1.0 \end{bmatrix}\begin{bmatrix} 1.0 & -0.5 & 0.0 \\ 0.0 & 1.0 & 0.0 \\ 0.0 & 0.0 & 1.0 \end{bmatrix} = \begin{bmatrix} 1.0 & -0.5 & 0.0 \\ -0.5 & 1.25 & 0.0 \\ 0.0 & 0.0 & 1.0 \end{bmatrix}.$$

The Green and Almansi strain tensor components in matrix form are given by

$$[E] = \frac{1}{2}\begin{bmatrix} 0.0 & 0.5 & 0.0 \\ 0.5 & 0.25 & 0.0 \\ 0.0 & 0.0 & 0.0 \end{bmatrix}; \quad [e] = \frac{1}{2}\begin{bmatrix} 0.0 & 0.5 & 0.0 \\ 0.5 & -0.25 & 0.0 \\ 0.0 & 0.0 & 0.0 \end{bmatrix}.$$

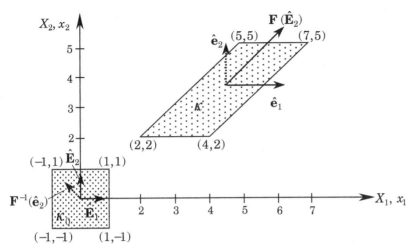

Figure 3.4.4. Undeformed (κ_0) and deformed (κ) configurations of a rectangular block, $\mathcal{B}$.

EXAMPLE 3.4.2: Consider the uniform deformation of a square block $\mathcal{B}$ of side length 2 units, initially centered at $\mathbf{X} = (0, 0)$, as shown in Figure 3.4.4. The deformation is defined by the mapping

$$\chi(\mathbf{X}) = \frac{1}{4}(18 + 4X_1 + 6X_2)\hat{\mathbf{e}}_1 + \frac{1}{4}(14 + 6X_2)\hat{\mathbf{e}}_2 + X_3\hat{\mathbf{e}}_3.$$

(a) Sketch the deformed configuration κ of the body $\mathcal{B}$.
(b) Compute the components of the deformation gradient tensor $\mathbf{F}$ and its inverse (display them in matrix form).
(c) Compute the components of the right Cauchy–Green deformation tensor $\mathbf{C}$ and Cauchy strain tensor $\tilde{\mathbf{B}}$ (display them in matrix form).
(d) Compute Green's and Almansi's strain tensor components (E_{IJ} and e_{ij}) (display them in matrix form).

SOLUTION:
(a) Sketch of the deformed configuration of the body $\mathcal{B}$ is shown in Figure 3.4.3.
(b) Note that the inverse transformation is given by ($X_3 = x_3$)

$$\begin{Bmatrix} X_1 \\ X_2 \end{Bmatrix} = 4 \begin{bmatrix} 4 & 6 \\ 0 & 6 \end{bmatrix}^{-1} \left(\begin{Bmatrix} x_1 \\ x_2 \end{Bmatrix} - \frac{1}{4} \begin{Bmatrix} 18 \\ 14 \end{Bmatrix} \right) = -\frac{1}{6} \begin{Bmatrix} 9 \\ 7 \end{Bmatrix} + \frac{1}{3} \begin{bmatrix} 3 & -3 \\ 0 & 2 \end{bmatrix} \begin{Bmatrix} x_1 \\ x_2 \end{Bmatrix}$$

or

$$\chi^{-1}(\mathbf{x}) = (-1.5 + x_1 - x_2)\hat{\mathbf{E}}_1 + \frac{1}{6}\left(7 + 4x_2\right)\hat{\mathbf{E}}_2 + x_3\,\hat{\mathbf{E}}_3.$$

The matrix form of the deformation gradient tensor and its inverse are

$$[F] = \begin{bmatrix} \frac{\partial x_1}{\partial X_1} & \frac{\partial x_1}{\partial X_2} \\ \frac{\partial x_2}{\partial X_1} & \frac{\partial x_2}{\partial X_2} \end{bmatrix} = \frac{1}{2}\begin{bmatrix} 2 & 3 \\ 0 & 3 \end{bmatrix}; \quad [F]^{-1} = \begin{bmatrix} \frac{\partial X_1}{\partial x_1} & \frac{\partial X_1}{\partial x_2} \\ \frac{\partial X_2}{\partial x_1} & \frac{\partial X_2}{\partial x_2} \end{bmatrix} = \frac{1}{3}\begin{bmatrix} 3 & -3 \\ 0 & 2 \end{bmatrix}.$$

(c) The right Cauchy–Green deformation tensor and Cauchy strain tensor are, respectively,

$$[C] = [F]^{\mathrm{T}}[F] = \frac{1}{2}\begin{bmatrix} 2 & 3 \\ 3 & 9 \end{bmatrix}, \quad [B] = [F][F]^{\mathrm{T}} = \frac{1}{4}\begin{bmatrix} 13 & 9 \\ 9 & 9 \end{bmatrix}.$$

(d) The Green and Almansi strain tensor components in matrix form are, respectively,

$$[E] = \frac{1}{2}\left([F]^{\mathrm{T}}[F] - [I]\right) = \frac{1}{2}\begin{bmatrix} 0 & 3 \\ 3 & 7 \end{bmatrix},$$

$$[e] = \frac{1}{2}\left([I] - [F]^{-\mathrm{T}}[F]^{-1}\right) = \frac{1}{18}\begin{bmatrix} 0 & 9 \\ 9 & -4 \end{bmatrix}.$$

3.4.5 Principal Strains

The tensors $\mathbf{E}$ and $\mathbf{e}$ can be expressed in any coordinate system much like any dyadic. For example, in a rectangular Cartesian system, we have

$$\mathbf{E} = E_{IJ}\hat{\mathbf{E}}_I\hat{\mathbf{E}}_J, \quad \mathbf{e} = e_{ij}\hat{\mathbf{e}}_i\hat{\mathbf{e}}_j. \tag{3.4.28}$$

Further, the components of $\mathbf{E}$ and $\mathbf{e}$ transform according to Eq. (2.5.17):

$$\bar{E}_{ij} = \ell_{ik}\,\ell_{j\ell}\,E_{k\ell}, \quad \bar{e}_{ij} = \ell_{ik}\,\ell_{j\ell}\,e_{k\ell}, \tag{3.4.29}$$

where ℓ_{ij} denotes the direction cosines between the barred and unbarred coordinate systems [see Eq. (2.2.49)].

The principal invariants of the Green–Lagrange strain tensor $\mathbf{E}$ are [see Eq. (2.5.14)]

$$J_1 = \mathrm{tr}\,\mathbf{E}, \quad J_2 = \frac{1}{2}\left[(\mathrm{tr}\mathbf{E})^2 - \mathrm{tr}(\mathbf{E}^2)\right], \quad J_3 = \det\mathbf{E}, \tag{3.4.30}$$

where the trace of $\mathbf{E}$, $\mathrm{tr}\mathbf{E}$, is defined to be the double-dot product of $\mathbf{E}$ with the unit dyad [see Eq. (2.5.13)]

$$\mathrm{tr}\,\mathbf{E} = \mathbf{E} : \mathbf{I}. \tag{3.4.31}$$

Invariant J_1 is also known as the *dilatation*.

The eigenvalue problem discussed in Section 2.5.5 for a tensor is applicable here for the strain tensors. The eigenvalues of a strain tensor are called the *principal strains*, and the corresponding eigenvectors are called the *principal directions of strain*.

EXAMPLE 3.4.3: Consider a rectangular block ($\mathcal{B}$) $ABCD$ of dimensions $a \times b \times h$, where h is thickness and it is very small compared with a and b. Suppose that the block $\mathcal{B}$ is deformed into the diamond shape $\bar{A}\bar{B}\bar{C}\bar{D}$ shown in Figure 3.4.5(a). Determine the deformation, displacements, and strains in the body.

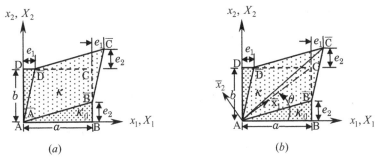

Figure 3.4.5. Undeformed (κ_0) and deformed (κ) configurations of a rectangular block, $\mathcal{B}$.

SOLUTION: By inspection, the geometry of the deformed body can be described as follows: let (X_1, X_2, X_3) denote the coordinates of a material point in the undeformed configuration, κ_0. The X_3-axis is taken out of the plane of the page and not shown in the figure. The deformation of $\mathcal{B}$ is defined by the mapping $\chi(\mathbf{x}) = x_1 \, \hat{\mathbf{e}}_1 + x_2 \, \hat{\mathbf{e}}_2 + x_3 \, \hat{\mathbf{e}}_3$, where

$$x_1 = A_0 + A_1 X_1 + A_2 X_2 + A_{12} X_1 X_2,$$

$$x_2 = B_0 + B_1 X_1 + B_2 X_2 + B_{12} X_1 X_2,$$

$$x_3 = X_3.$$

and A_i and B_i are constants, which can be determined using the deformed configuration κ. We have

$$(X_1, X_2) = (0, 0), \quad (x_1, x_2) = (0, 0) \quad \rightarrow \quad A_0 = 0, \quad B_0 = 0,$$

$$(X_1, X_2) = (a, 0), \quad (x_1, x_2) = (a, e_2) \quad \rightarrow \quad A_1 = 1, \quad B_1 = \frac{e_2}{a},$$

$$(X_1, X_2) = (0, b), \quad (x_1, x_2) = (e_1, b) \quad \rightarrow \quad A_2 = \frac{e_1}{b}, \quad B_2 = 1,$$

$$(X_1, X_2) = (a, b), \quad (x_1, x_2) = (a + e_1, b + e_2) \quad \rightarrow \quad A_{12} = 0, \quad B_{12} = 0.$$

Thus, the deformation is defined by the transformation

$$\chi(\mathbf{x}) = (X_1 + k_1 X_2)\hat{\mathbf{e}}_1 + (X_2 + k_2 X_1)\hat{\mathbf{e}}_2 + X_3 \, \hat{\mathbf{e}}_3,$$

where $k_1 = e_1/b$ and $k_2 = e_2/a$. The inverse mapping is given by

$$\chi^{-1}(\mathbf{X}) = \frac{1}{1 - k_1 k_2} (x_1 - k_1 x_2) \, \hat{\mathbf{E}}_1 + \frac{1}{1 - k_1 k_2} (-k_2 x_1 + x_2) \, \hat{\mathbf{E}}_2 + x_3 \, \hat{\mathbf{E}}_3.$$

Thus, the displacement vector of a material point in the Lagrangian description is

$$\mathbf{u} = k_1 X_2 \, \hat{\mathbf{e}}_1 + k_2 X_1 \, \hat{\mathbf{e}}_2.$$

The only nonzero Green strain tensor components are given by

$$E_{11} = \frac{1}{2} k_2^2, \quad 2 E_{12} = k_1 + k_2, \quad E_{22} = \frac{1}{2} k_1^2.$$

The deformation gradient tensor components are

$$[F] = \begin{bmatrix} 1 & k_1 & 0 \\ k_2 & 1 & 0 \\ 0 & 0 & 1 \end{bmatrix}.$$

The case in which $k_2 = 0$ is known as the *simple shear*. The Green's deformation tensor **C** is

$$\mathbf{C} = \mathbf{F}^T \cdot \mathbf{F} \quad \rightarrow \quad [C] = [F]^T[F] = \begin{bmatrix} 1 + k_1^2 & k_1 + k_2 & 0 \\ k_1 + k_2 & 1 + k_2^2 & 0 \\ 0 & 0 & 1 \end{bmatrix},$$

and $2\mathbf{E} = \mathbf{C} - \mathbf{I}$ yields the results given above.

The displacements in the spatial description are

$$u_1 = x_1 - X_1 = k_1 X_2 = \frac{k_1}{1 - k_1 k_2}(-k_2 x_1 + x_2),$$

$$u_2 = x_2 - X_2 = k_2 X_1 = \frac{k_2}{1 - k_1 k_2}(x_1 - k_1 x_2),$$

$$u_3 = x_3 - X_3 = 0.$$

The Almansi strain tensor components are

$$e_{11} = -\frac{k_1 k_2}{1 - k_1 k_2} - \frac{1}{2}\left[\left(\frac{k_1 k_2}{1 - k_1 k_2}\right)^2 + \left(\frac{k_2}{1 - k_1 k_2}\right)^2\right],$$

$$2e_{12} = \frac{k_1 + k_2}{1 - k_1 k_2} + \frac{k_1 k_2(k_1 + k_2)}{(1 - k_1 k_2)^2},$$

$$e_{22} = -\frac{k_1 k_2}{1 - k_1 k_2} - \frac{1}{2}\left[\left(\frac{k_1 k_2}{1 - k_1 k_2}\right)^2 + \left(\frac{k_1}{1 - k_1 k_2}\right)^2\right].$$

Alternatively, the same results can be obtained using the elementary mechanics of materials approach, where the strains are defined to be the ratio of the difference between the final length and original length to the original length. A line element AB in the undeformed configuration κ_0 of the body B moves to position $\bar{A}\bar{B}$. Then the Green strain in the line AB is given by

$$E_{11} = E_{AB} = \frac{\bar{A}\bar{B} - AB}{AB} = \frac{1}{a}\sqrt{a^2 + e_2^2} - 1 = \sqrt{1 + \left(\frac{e_2}{a}\right)^2} - 1$$

$$= \left[1 + \frac{1}{2}\left(\frac{e_2}{a}\right)^2 + \cdots\right] - 1 \approx \frac{1}{2}\left(\frac{e_2}{a}\right)^2 = \frac{1}{2}k_2^2.$$

Similarly,

$$E_{22} = \left[1 + \frac{1}{2}\left(\frac{e_1}{b}\right)^2 + \cdots\right] - 1 \approx \frac{1}{2}\left(\frac{e_1}{b}\right)^2 = \frac{1}{2}k_1^2.$$

The shear strain $2E_{12}$ is equal to the change in the angle between two line elements that were originally at 90°, that is, change in the angle DAB. The change is clearly equal to, as can be seen from Fig. 3.4.5(b),

$$2E_{12} = \angle DAB - \angle \bar{D}\bar{A}\bar{B} = \frac{e_1}{b} + \frac{e_2}{a} = k_1 + k_2.$$

The axial strain in line element AC is $(\bar{A} = A)$

$$E_{AC} = \frac{\bar{A}\bar{C} - AC}{AC} = \frac{1}{\sqrt{a^2 + b^2}}\sqrt{(a + e_1)^2 + (b + e_2)^2} - 1$$

$$= \frac{1}{\sqrt{a^2 + b^2}}\sqrt{a^2 + b^2 + e_1^2 + e_2^2 + 2ae_1 + 2be_2} - 1$$

$$= \left[1 + \frac{e_1^2 + e_2^2 + 2ae_1 + 2be_2}{a^2 + b^2}\right]^{\frac{1}{2}} - 1 \approx \frac{1}{2}\frac{e_1^2 + e_2^2 + 2ae_1 + 2be_2}{a^2 + b^2}$$

$$= \frac{1}{2(a^2 + b^2)}\left[a^2 k_2^2 + 2ab(k_1 + k_2) + b^2 k_1^2\right].$$

The axial strain E_{AC} can also be computed using the strain transformation equations (3.4.29). The line AC is oriented at $\theta = \tan^{-1}(b/a)$. Hence, we have

$$\beta_{11} = \cos\theta = \frac{a}{\sqrt{a^2 + b^2}}, \quad \beta_{12} = \sin\theta = \frac{b}{\sqrt{a^2 + b^2}},$$

$$\beta_{21} = -\sin\theta = -\frac{b}{\sqrt{a^2 + b^2}}, \quad \beta_{22} = \cos\theta = \frac{a}{\sqrt{a^2 + b^2}},$$

and

$$E_{AC} \equiv \bar{E}_{11} = \beta_{1i}\beta_{1j}E_{ij} = \beta_{11}\beta_{11}E_{11} + 2\beta_{11}\beta_{12}E_{12} + \beta_{12}\beta_{12}E_{22}$$

$$= \frac{1}{2(a^2 + b^2)}\left[a^2 k_2^2 + 2ab(k_1 + k_2) + b^2 k_1^2\right],$$

which is the same as that computed above.

The next example is concerned with the computation of principal strains and their directions.

EXAMPLE 3.4.4: The state of strain at a point in an elastic body is given by $(10^{-3}$ in./in.)

$$[E] = \begin{bmatrix} 4 & -4 & 0 \\ -4 & 0 & 0 \\ 0 & 0 & 3 \end{bmatrix}.$$

Determine the principal strains and principal directions of the strain.

SOLUTION: Setting $|[E] - \lambda[I]| = 0$, we obtain

$$(4 - \lambda)[(-\lambda)(3 - \lambda) - 0] + 4[-4(3 - \lambda)] = 0 \quad \rightarrow \quad [(4 - \lambda)\lambda + 16](3 - \lambda) = 0.$$

We see that $\lambda_1 = 3$ is an eigenvalue of the matrix. The remaining two eigen-values are obtained from $\lambda^2 - 4\lambda - 16 = 0$. Thus the principal strains are $(10^{-3}$ in./in.)

$$\lambda_1 = 3, \quad \lambda_2 = 2(1 + \sqrt{5}), \quad \lambda_3 = 2(1 - \sqrt{5}).$$

The eigenvector components x_i associated with $\varepsilon_1 = \lambda_1 = 3$ are calculated from

$$\begin{bmatrix} 4-3 & -4 & 0 \\ -4 & 0-3 & 0 \\ 0 & 0 & 3-3 \end{bmatrix} \begin{Bmatrix} x_1 \\ x_2 \\ x_3 \end{Bmatrix} = \begin{Bmatrix} 0 \\ 0 \\ 0 \end{Bmatrix},$$

which gives $x_1 - 4x_2 = 0$ and $-4x_1 - 3x_2 = 0$, or $x_1 = x_2 = 0$. Using the normal-ization $x_1^2 + x_2^2 + x_3^2 = 1$, we obtain $x_3 = 1$. Thus, the principal direction associ-ated with the principal strain $\varepsilon_1 = 3$ is $\hat{\mathbf{x}}^{(1)} = \pm(0, 0, 1)$.

The eigenvector components associated with principal strain $\varepsilon_2 = \lambda_2 = 2(1 + \sqrt{5})$ are calculated from

$$\begin{bmatrix} 4-\lambda_2 & -4 & 0 \\ -4 & 0-\lambda_2 & 0 \\ 0 & 0 & 3-\lambda_2 \end{bmatrix} \begin{Bmatrix} x_1 \\ x_2 \\ x_3 \end{Bmatrix} = \begin{Bmatrix} 0 \\ 0 \\ 0 \end{Bmatrix},$$

which gives

$$x_1 = -\frac{2+2\sqrt{5}}{4} x_2 = -1.618 x_2, \quad x_3 = 0, \quad \rightarrow \quad \hat{\mathbf{x}}^{(2)} = \pm(-0.851, 0.526, 0).$$

Similarly, the eigenvector components associated with principal strain $\varepsilon_3 = \lambda_3 = 2(1 - \sqrt{5})$ are obtained as

$$x_1 = \frac{2+2\sqrt{5}}{4} x_2 = 1.618 x_2, \quad x_3 = 0, \quad \rightarrow \quad \hat{\mathbf{x}}^{(3)} = \pm(0.526, 0.851, 0).$$

The principal planes of strain are shown in Figure 3.4.6.

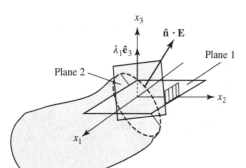

Figure 3.4.6. Principal planes 1 and 2 of strain.

3.5 Infinitesimal Strain Tensor and Rotation Tensor

3.5.1 Infinitesimal Strain Tensor

When all displacements gradients are small (or infinitesimal), that is, $|\nabla\mathbf{u}| \ll 1$, we can neglect the nonlinear terms in the definition of the Green strain tensor defined in Eq. (3.4.11). In the case of infinitesimal strains, no distinction is made between the material coordinates $\mathbf{X}$ and the spatial coordinates $\mathbf{x}$. Therefore, the linear Green–Lagrange strain tensor and the linear Eulerian strain tensor become the same. The *infinitesimal strain tensor* is denoted by ε, and it is given by

$$\varepsilon = \frac{1}{2}\left[\nabla\mathbf{u} + (\nabla\mathbf{u})^{\mathrm{T}}\right]. \tag{3.5.1}$$

The rectangular Cartesian components of the infinitesimal strain tensor are given by

$$\varepsilon_{ij} = \frac{1}{2}\left(u_{i,j} + u_{j,i}\right), \tag{3.5.2}$$

or, in expanded form,

$$\varepsilon_{11} = \frac{\partial u_1}{\partial X_1}; \qquad\qquad \varepsilon_{22} = \frac{\partial u_2}{\partial X_2};$$

$$\varepsilon_{33} = \frac{\partial u_3}{\partial X_3}; \qquad\qquad \varepsilon_{12} = \frac{1}{2}\left(\frac{\partial u_1}{\partial X_2} + \frac{\partial u_2}{\partial X_1}\right); \tag{3.5.3}$$

$$\varepsilon_{13} = \frac{1}{2}\left(\frac{\partial u_1}{\partial X_3} + \frac{\partial u_3}{\partial X_1}\right); \quad \varepsilon_{23} = \frac{1}{2}\left(\frac{\partial u_2}{\partial X_3} + \frac{\partial u_3}{\partial X_2}\right).$$

The strain components ε_{11}, ε_{22}, and ε_{33} are the infinitesimal normal strains and ε_{12}, ε_{13}, and ε_{23} are the infinitesimal shear strains. The shear strains $\gamma_{12} = 2\varepsilon_{12}$, $\gamma_{13} = 2\varepsilon_{13}$, and $\gamma_{23} = 2\varepsilon_{23}$ are called the *engineering shear strains*.

3.5.2 Physical Interpretation of Infinitesimal Strain Tensor Components

To gain insight into the physical meaning of the infinitesimal strain components, we write Eq. (3.4.10) in the form

$$(ds)^2 - (dS)^2 = 2d\mathbf{X} \cdot \varepsilon \cdot d\mathbf{X} = 2\varepsilon_{ij}dX_i dX_j$$

and dividing throughout by (dS^2), we obtain

$$\frac{(ds)^2 - (dS)^2}{dS^2} = 2\varepsilon_{ij}\frac{dX_i}{dS}\frac{dX_j}{dS}.$$

Let $d\mathbf{X}/dS = \hat{\mathbf{N}}$, the unit vector in the direction of $d\mathbf{X}$. For small deformations, we have $ds + dS = ds + dS \approx 2dS$, and therefore we have

$$\frac{ds - dS}{dS} = \hat{\mathbf{N}} \cdot \varepsilon \cdot \hat{\mathbf{N}} = \varepsilon_{ij}N_i N_j. \tag{3.5.4}$$

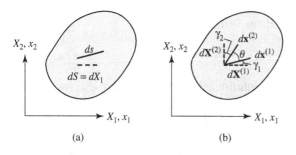

Figure 3.5.1. Physical interpretation of infinitesimal strain components.

(a)   (b)

The left side of Eq. (3.5.4) is the ratio of change in length per unit original length for a line element in the direction of $\hat{\mathbf{N}}$. For example, consider $\hat{\mathbf{N}}$ along the X_1-direction. Then we have from Figure 3.5.1(a) (also see Figure 3.4.2)

$$\frac{ds - dS}{dS} = \varepsilon_{11}.$$

Thus, the normal strain ε_{11} is the ratio of change in length of a line element that was parallel to the x_1-axis in the undeformed body to its original length. Similarly, for a line element along X_2 direction, $(ds - dS)/ds$ is the normal strain ε_{22}, and for a line element along X_3 direction, $(ds - dS)/ds$ denotes the normal strain ε_{33}.

To understand the meaning of shear components of infinitesimal strain tensor, consider line elements $d\mathbf{X}^{(1)}$ and $d\mathbf{X}^{(2)}$ at a point in the body, which deform into line elements $d\mathbf{x}^{(1)}$ and $d\mathbf{x}^{(2)}$, respectively, as shown in Figure 3.5.1(b). Then we have [see Eqs. (3.3.1), (3.4.3), and (3.4.11)]

$$d\mathbf{x}^{(1)} \cdot d\mathbf{x}^{(2)} = d\mathbf{X}^{(1)} \cdot \mathbf{F}^{\mathsf{T}} \cdot \mathbf{F} \cdot d\mathbf{X}^{(2)} = d\mathbf{X}^{(1)} \cdot \mathbf{C} \cdot d\mathbf{X}^{(2)}$$

$$= d\mathbf{X}^{(1)} \cdot (\mathbf{I} + 2\mathbf{E}) \cdot d\mathbf{X}^{(2)}$$

$$= d\mathbf{X}^{(1)} \cdot d\mathbf{X}^{(2)} + 2d\mathbf{X}^{(1)} \cdot \mathbf{E} \cdot d\mathbf{X}^{(2)}. \tag{3.5.5}$$

Now suppose that the line elements $d\mathbf{X}^{(1)}$ and $d\mathbf{X}^{(2)}$ are orthogonal to each other. Then

$$d\mathbf{x}^{(1)} \cdot d\mathbf{x}^{(2)} = 2d\mathbf{X}^{(1)} \cdot \mathbf{E} \cdot d\mathbf{X}^{(2)},$$

or

$$2d\mathbf{X}^{(1)} \cdot \mathbf{E} \cdot d\mathbf{X}^{(2)} = dx^{(1)}dx^{(2)} \cos\theta = dx^{(1)}dx^{(2)} \cos\left(\frac{\pi}{2} - \gamma_1 - \gamma_2\right)$$

$$= dx^{(1)}dx^{(2)} \sin(\gamma_1 + \gamma_2) = dx^{(1)}dx^{(2)} \sin\gamma, \tag{3.5.6}$$

where θ is the angle between the deformed line elements $dx^{(1)}$ and $dx^{(2)}$ and $\gamma = \gamma_1 + \gamma_2$ is the change in the angle from 90°, as shown in Figure 3.5.1(b) (also see Figure 3.4.2). For small deformations, we take $\sin\gamma \approx \gamma$, and obtain

$$\gamma = 2\frac{d\mathbf{X}^{(1)}}{dx^{(1)}} \cdot \mathbf{E} \cdot \frac{d\mathbf{X}^{(2)}}{dx^{(2)}} = 2\hat{\mathbf{N}}^{(1)} \cdot \boldsymbol{\varepsilon} \cdot \hat{\mathbf{N}}^{(2)}, \tag{3.5.7}$$

where $\hat{\mathbf{N}}^{(1)} = d\mathbf{X}^{(1)}/dx^{(1)}$ and $\hat{\mathbf{N}}^{(2)} = d\mathbf{X}^{(2)}/dx^{(2)}$ are the unit vectors along the line elements $d\mathbf{X}^{(1)}$ and $d\mathbf{X}^{(2)}$, respectively. If the line elements $d\mathbf{X}^{(1)}$ and $d\mathbf{X}^{(2)}$ are taken along the X_1 and X_2 coordinates, respectively, then we have $2\varepsilon_{12} = \gamma$. Thus, the engineering shear strain $\gamma_{12} = 2\varepsilon_{12}$ represents the change in angle between line elements that were perpendicular to each other in the undeformed body.

3.5.3 Infinitesimal Rotation Tensor

The displacement gradient tensor $\nabla \mathbf{u}$ (note that $\mathbf{F}$ is the deformation gradient tensor) can be expressed as the sum of a symmetric tensor and antisymmetric tensor

$$\nabla \mathbf{u} = \frac{1}{2}\left[\nabla \mathbf{u} + (\nabla \mathbf{u})^{\mathrm{T}}\right] + \frac{1}{2}\left[\nabla \mathbf{u} - (\nabla \mathbf{u})^{\mathrm{T}}\right] \equiv \varepsilon + \Omega, \tag{3.5.8}$$

where the symmetric part is clearly the infinitesimal strain tensor, and the antisymmetric part is known as the *infinitesimal rotation tensor*

$$\Omega = \frac{1}{2}\left[\nabla \mathbf{u} - (\nabla \mathbf{u})^{\mathrm{T}}\right]. \tag{3.5.9}$$

From the definition, it follows that Ω is antisymmetric (or skew-symmetric), that is, $\Omega^{\mathrm{T}} = -\Omega$. In Cartesian component form,

$$\Omega_{ij} = (u_{i,j} - u_{j,i}), \qquad \Omega_{ij} = -\Omega_{ji}. \tag{3.5.10}$$

Thus, there are only three independent components of Ω:

$$[\Omega] = \frac{1}{2}\begin{bmatrix} 0 & -\Omega_{12} & -\Omega_{13} \\ \Omega_{12} & 0 & -\Omega_{23} \\ \Omega_{13} & \Omega_{23} & 0 \end{bmatrix}. \tag{3.5.11}$$

While there is no restriction placed on the magnitude of $\nabla \mathbf{u}$ in writing (3.5.1), ε and Ω do not have the meaning of infinitesimal strain and infinitesimal rotation tensors unless the deformation is infinitesimal (i.e., $|\nabla \mathbf{u}|$ is small, $|\nabla \mathbf{u}| \ll 1$).

Since Ω has only three independent components, the three components can be used to define the components of a vector ω

$$\Omega = -\mathcal{E} \cdot \omega \quad \text{or} \quad \omega = -\frac{1}{2}\mathcal{E} : \Omega, \tag{3.5.12}$$

$$\Omega_{ij} = -e_{ijk}\omega_k \quad \text{or} \quad \omega_i = -\frac{1}{2}e_{ijk}\Omega_{jk},$$

where $\mathcal{E}$ is the permutation (alternating) tensor, $\mathcal{E} = e_{ijk}\hat{\mathbf{e}}_i\hat{\mathbf{e}}_j\hat{\mathbf{e}}_k$. In view of Eqs. (3.5.9) and (3.5.12), it follows that

$$\omega_i = \frac{1}{2}e_{ijk}u_{k,j} \quad \text{or} \quad \omega = \frac{1}{2}\mathrm{curl}\,\mathbf{u} = \frac{1}{2}\nabla \times \mathbf{u}. \tag{3.5.13}$$

In essence, infinitesimal displacements of the form $\mathbf{u} = \Omega \cdot \mathbf{x}$, where Ω is independent of the position $\mathbf{x}$, are rotations because

$$u_i = \Omega_{ij}x_j = -e_{ijk}\omega_k x_j = -(\mathbf{x} \times \omega)_i = (\omega \times \mathbf{x})_i \quad \text{or} \quad \mathbf{u} = \omega \times \mathbf{x},$$

which represents the velocity of a point $\mathbf{x}$ in a rigid material in uniform rotation about the origin.

Certain motions do not produce infinitesimal strains but they may produce finite strains. For example, consider the following deformation mapping:

$$\chi(\mathbf{X}) = (b_1 + X_1 + c_2 X_3 - c_3 X_2)\,\hat{\mathbf{e}}_1 + (b_2 + X_2 + c_3 X_1 - c_1 X_3)\,\hat{\mathbf{e}}_2$$
$$+ (b_3 + X_3 + c_1 X_2 - c_2 X_1)\,\hat{\mathbf{e}}_3, \tag{3.5.14}$$

where b_i and c_i $(i = 1, 2, 3)$ are arbitrary constants. The displacement vector is

$$\mathbf{u}(\mathbf{X}) = (b_1 + c_2 X_3 - c_3 X_2)\,\hat{\mathbf{e}}_1 + (b_2 + c_3 X_1 - c_1 X_3)\,\hat{\mathbf{e}}_2$$
$$+ (b_3 + c_1 X_2 - c_2 X_1)\,\hat{\mathbf{e}}_3. \tag{3.5.15}$$

Since

$$\frac{\partial u_1}{\partial X_1} = 0, \qquad \frac{\partial u_1}{\partial X_2} = -c_3, \qquad \frac{\partial u_1}{\partial X_3} = c_2,$$

$$\frac{\partial u_2}{\partial X_1} = c_3, \qquad \frac{\partial u_2}{\partial X_2} = 0, \qquad \frac{\partial u_2}{\partial X_3} = -c_1, \tag{3.5.16}$$

$$\frac{\partial u_3}{\partial X_1} = -c_2, \qquad \frac{\partial u_3}{\partial X_2} = c_1, \qquad \frac{\partial u_1}{\partial X_3} = 0,$$

the infinitesimal (i.e., linearized) strains are all zero.

The components of the deformation gradient tensor $\mathbf{F}$ and left Cauchy–Green deformation tensor $\mathbf{C}$ associated with the mapping are

$$[F] = \begin{bmatrix} 1 & -c_3 & c_2 \\ c_3 & 1 & -c_1 \\ -c_2 & c_1 & 1 \end{bmatrix}, \quad [C] = \begin{bmatrix} 1 + c_2^2 + c_3^2 & -c_1 c_2 & -c_1 c_3 \\ -c_1 c_2 & 1 + c_1^2 + c_3^2 & -c_2 c_3 \\ -c_1 c_3 & -c_2 c_3 & 1 + c_1^2 + c_2^2 \end{bmatrix}. \tag{3.5.17}$$

It is clear that for nonzero values of the constants c_i, the mapping produces nonzero finite strains. When all of the constants c_i are either zero or negligibly small (so that their products and squares are very small compared to unity), then $[F] = [C] = [I]$, implying that the mapping represents a rigid body rotation. Figure 3.5.2 depicts the deformation for the two-dimensional case, with $b_1 = 2$, $b_2 = 3$, and $c_3 = 1$. Thus, the finite Green strain tensor and deformation gradient tensor give true measures of

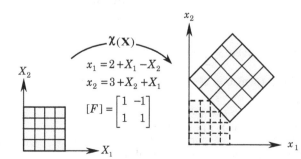

Figure 3.5.2. A mapping that produces zero infinitesimal strains but nonzero finite strains.

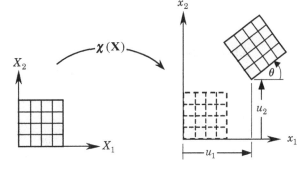

Figure 3.5.3. A mapping that produces nonzero infinitesimal strains but zero finite strains.

the deformation. The question of smallness of c_i in a given engineering application must be carefully examined before using a linearized strains.

Next, consider the mapping

$$\chi(\mathbf{X}) = (u_1 + X_1 \cos\theta - X_2 \sin\theta)\, \hat{\mathbf{e}}_1 + (u_2 + X_1 \sin\theta + X_2 \cos\theta)\, \hat{\mathbf{e}}_2$$

$$+ X_3\, \hat{\mathbf{e}}_3, \tag{3.5.18}$$

where u_1 and u_2 denote the horizontal and vertical displacements of the point $(0, 0, 0)$, as shown in Figure 3.5.3.

The components of the deformation gradient tensor $\mathbf{F}$ and left Cauchy–Green deformation tensor $\mathbf{C}$ are

$$[F] = \begin{bmatrix} \cos\theta & -\sin\theta & 0 \\ \sin\theta & \cos\theta & 0 \\ 0 & 0 & 1 \end{bmatrix}, \quad [C] = \begin{bmatrix} 1 & 0 & 0 \\ 0 & 1 & 0 \\ 0 & 0 & 1 \end{bmatrix}. \tag{3.5.19}$$

Since $\mathbf{C} = \mathbf{I}$, we have $\mathbf{E} = \mathbf{0}$, indicating that the body does not experience stretching or shearing. The mapping is a rigid-body motion (both rigid-body translation and rigid-body rotation).

If we linearize the deformation mapping by making the approximations $\cos\theta \approx 1$ and $\sin\theta \approx 0$, we obtain

$$[F] = \begin{bmatrix} 1 & -\theta & 0 \\ \theta & 1 & 0 \\ 0 & 0 & 1 \end{bmatrix}, \quad [C] = \begin{bmatrix} 1+\theta^2 & 0 & 0 \\ 0 & 1+\theta^2 & 0 \\ 0 & 0 & 1 \end{bmatrix}. \tag{3.5.20}$$

Thus, the Green strain tensor components are no longer zero. The principal stretches $\lambda_1 = \lambda_2 = 1 + \theta^2$ are not equal to 1, as required by the definition of rigid-body motion. Owing to the artificial stretch induced by the linearization of the mapping, the stretches get larger and larger as the block rotates.

3.5.4 Infinitesimal Strains in Cylindrical and Spherical Coordinate Systems

The strains defined by Eq. (3.5.1) are valid in any coordinate system. Hence, they can be expressed in component form in any given coordinate system by expanding the strain tensors in the dyadic form and the operator ∇ in that coordinate system, as given in Table 2.4.2 (also see Figure 2.4.2).

3.5.4.1 Cylindrical coordinate system
In the cylindrical coordinate system we have

$$\mathbf{u} = u_r\hat{\mathbf{e}}_r + u_\theta\hat{\mathbf{e}}_\theta + u_z\hat{\mathbf{e}}_z, \tag{3.5.16}$$

$$\nabla_0 = \hat{\mathbf{e}}_r\frac{\partial}{\partial r} + \frac{1}{r}\hat{\mathbf{e}}_\theta\frac{\partial}{\partial\theta} + \hat{\mathbf{e}}_z\frac{\partial}{\partial z}, \tag{3.5.17}$$

$$\frac{\partial\hat{\mathbf{e}}_r}{\partial\theta} = \hat{\mathbf{e}}_\theta, \quad \frac{\partial\hat{\mathbf{e}}_\theta}{\partial\theta} = -\hat{\mathbf{e}}_r. \tag{3.5.18}$$

Using Eqs. (3.5.16)–(3.5.18), we obtain [see Eq. (2.5.27)]

$$\nabla_0\mathbf{u} = \hat{\mathbf{e}}_r\hat{\mathbf{e}}_r\frac{\partial u_r}{\partial r} + \hat{\mathbf{e}}_r\hat{\mathbf{e}}_\theta\frac{\partial u_\theta}{\partial r} + \frac{1}{r}\hat{\mathbf{e}}_\theta\hat{\mathbf{e}}_r\left(\frac{\partial u_r}{\partial\theta} - u_\theta\right)$$

$$+ \hat{\mathbf{e}}_r\hat{\mathbf{e}}_z\frac{\partial u_z}{\partial r} + \hat{\mathbf{e}}_z\hat{\mathbf{e}}_r\frac{\partial u_r}{\partial z} + \frac{1}{r}\left(u_r + \frac{\partial u_\theta}{\partial\theta}\right)\hat{\mathbf{e}}_\theta\hat{\mathbf{e}}_\theta$$

$$+ \frac{1}{r}\hat{\mathbf{e}}_\theta\hat{\mathbf{e}}_z\frac{\partial u_z}{\partial\theta} + \hat{\mathbf{e}}_z\hat{\mathbf{e}}_\theta\frac{\partial u_\theta}{\partial z} + \hat{\mathbf{e}}_z\hat{\mathbf{e}}_z\frac{\partial u_z}{\partial z}, \tag{3.5.19}$$

$$(\nabla_0\mathbf{u})^{\mathrm{T}} = \hat{\mathbf{e}}_r\hat{\mathbf{e}}_r\frac{\partial u_r}{\partial r} + \hat{\mathbf{e}}_\theta\hat{\mathbf{e}}_r\frac{\partial u_\theta}{\partial r} + \frac{1}{r}\hat{\mathbf{e}}_r\hat{\mathbf{e}}_\theta\left(\frac{\partial u_r}{\partial\theta} - u_\theta\right)$$

$$+ \hat{\mathbf{e}}_z\hat{\mathbf{e}}_r\frac{\partial u_z}{\partial r} + \hat{\mathbf{e}}_r\hat{\mathbf{e}}_z\frac{\partial u_r}{\partial z} + \frac{1}{r}\hat{\mathbf{e}}_\theta\hat{\mathbf{e}}_\theta\left(u_r + \frac{\partial u_\theta}{\partial\theta}\right)$$

$$+ \frac{1}{r}\hat{\mathbf{e}}_z\hat{\mathbf{e}}_\theta\frac{\partial u_z}{\partial\theta} + \hat{\mathbf{e}}_\theta\hat{\mathbf{e}}_z\frac{\partial u_\theta}{\partial z} + \hat{\mathbf{e}}_z\hat{\mathbf{e}}_z\frac{\partial u_z}{\partial z}. \tag{3.5.20}$$

Substituting the above expressions into Eq. (3.5.1) and collecting the coefficients of various dyadics (i.e., coefficients of $\hat{\mathbf{e}}_r\hat{\mathbf{e}}_r$, $\hat{\mathbf{e}}_r\hat{\mathbf{e}}_\theta$, and so on) we obtain the infinitesimal strain tensor components

$$\varepsilon_{rr} = \frac{\partial u_r}{\partial r}, \qquad\qquad \varepsilon_{r\theta} = \frac{1}{2}\left(\frac{1}{r}\frac{\partial u_r}{\partial\theta} + \frac{\partial u_\theta}{\partial r} - \frac{u_\theta}{r}\right),$$

$$\varepsilon_{rz} = \frac{1}{2}\left(\frac{\partial u_r}{\partial z} + \frac{\partial u_z}{\partial r}\right), \qquad \varepsilon_{\theta\theta} = \frac{u_r}{r} + \frac{1}{r}\frac{\partial u_\theta}{\partial\theta}, \tag{3.5.21}$$

$$\varepsilon_{z\theta} = \frac{1}{2}\left(\frac{\partial u_\theta}{\partial z} + \frac{1}{r}\frac{\partial u_z}{\partial\theta}\right), \qquad \varepsilon_{zz} = \frac{\partial u_z}{\partial z}.$$

3.5.4.2 Spherical coordinate system
In the spherical coordinate system, we have

$$\mathbf{u} = u_R\hat{\mathbf{e}}_R + u_\phi\,\hat{\mathbf{e}}_\phi + u_\theta\,\hat{\mathbf{e}}_\theta, \tag{3.5.22}$$

$$\nabla_0 = \hat{\mathbf{e}}_R\frac{\partial}{\partial R} + \frac{1}{R}\hat{\mathbf{e}}_\phi\frac{\partial}{\partial\phi} + \frac{1}{R\sin\phi}\hat{\mathbf{e}}_\theta\frac{\partial}{\partial\theta}, \tag{3.5.23}$$

$$\frac{\partial \hat{\mathbf{e}}_R}{\partial \phi} = \hat{\mathbf{e}}_\phi, \quad \frac{\partial \hat{\mathbf{e}}_R}{\partial \theta} = \sin \phi\, \hat{\mathbf{e}}_\theta, \quad \frac{\partial \hat{\mathbf{e}}_\phi}{\partial \phi} = -\hat{\mathbf{e}}_R,$$

$$\frac{\partial \hat{\mathbf{e}}_\phi}{\partial \theta} = \cos \phi\, \hat{\mathbf{e}}_\theta, \quad \frac{\partial \hat{\mathbf{e}}_\theta}{\partial \theta} = -\sin \phi\, \hat{\mathbf{e}}_R - \cos \phi\, \hat{\mathbf{e}}_\phi . \tag{3.5.24}$$

Using Eqs. (3.5.22)–(3.5.24), we obtain (answer to Problem 2.28)

$$\begin{aligned}
\nabla_0 \mathbf{u} = {}& \frac{\partial u_R}{\partial R} \hat{\mathbf{e}}_R \hat{\mathbf{e}}_R + \frac{\partial u_\phi}{\partial R} \hat{\mathbf{e}}_R \hat{\mathbf{e}}_\phi + \frac{\partial u_\theta}{\partial R} \hat{\mathbf{e}}_R \hat{\mathbf{e}}_\theta \\
&+ \frac{1}{R}\left(\frac{\partial u_R}{\partial \phi} - u_\phi \right) \hat{\mathbf{e}}_\phi \hat{\mathbf{e}}_R + \frac{1}{R}\left(\frac{\partial u_\phi}{\partial \phi} + u_R \right) \hat{\mathbf{e}}_\phi \hat{\mathbf{e}}_\phi + \frac{1}{R}\frac{\partial u_\theta}{\partial \phi} \hat{\mathbf{e}}_\phi \hat{\mathbf{e}}_\theta \\
&+ \frac{1}{R \sin \phi}\left(\frac{\partial u_R}{\partial \theta} - u_\theta \sin \phi \right) \hat{\mathbf{e}}_\theta \hat{\mathbf{e}}_R + \frac{1}{R \sin \phi}\left(\frac{\partial u_\phi}{\partial \theta} - u_\theta \cos \phi \right) \hat{\mathbf{e}}_\theta \hat{\mathbf{e}}_\phi \\
&+ \frac{1}{R \sin \phi}\left(\frac{\partial u_\theta}{\partial \theta} + u_R \sin \phi + u_\phi \cos \phi \right) \hat{\mathbf{e}}_\theta \hat{\mathbf{e}}_\theta , \tag{3.5.25}
\end{aligned}$$

$$\begin{aligned}
(\nabla_0 \mathbf{u})^{\mathrm{T}} = {}& \frac{\partial u_R}{\partial R} \hat{\mathbf{e}}_R \hat{\mathbf{e}}_R + \frac{\partial u_\phi}{\partial R} \hat{\mathbf{e}}_\phi \hat{\mathbf{e}}_R + \frac{\partial u_\theta}{\partial R} \hat{\mathbf{e}}_\theta \hat{\mathbf{e}}_R \\
&+ \frac{1}{R}\left(\frac{\partial u_R}{\partial \phi} - u_\phi \right) \hat{\mathbf{e}}_R \hat{\mathbf{e}}_\phi + \frac{1}{R}\left(\frac{\partial u_\phi}{\partial \phi} + u_R \right) \hat{\mathbf{e}}_\phi \hat{\mathbf{e}}_\phi + \frac{1}{R}\frac{\partial u_\theta}{\partial \phi} \hat{\mathbf{e}}_\theta \hat{\mathbf{e}}_\phi \\
&+ \frac{1}{R \sin \phi}\left(\frac{\partial u_R}{\partial \theta} - u_\theta \sin \phi \right) \hat{\mathbf{e}}_R \hat{\mathbf{e}}_\theta + \frac{1}{R \sin \phi}\left(\frac{\partial u_\phi}{\partial \theta} - u_\theta \cos \phi \right) \hat{\mathbf{e}}_\phi \hat{\mathbf{e}}_\theta \\
&+ \frac{1}{R \sin \phi}\left(\frac{\partial u_\theta}{\partial \theta} + u_R \sin \phi + u_\phi \cos \phi \right) \hat{\mathbf{e}}_\theta \hat{\mathbf{e}}_\theta . \tag{3.5.26}
\end{aligned}$$

Substituting the above expressions into Eq. (3.5.1) and collecting the coefficients of various dyadics, we obtain the following infinitesimal strain tensor components in the spherical coordinate system:

$$\varepsilon_{RR} = \frac{\partial u_R}{\partial R}, \quad \varepsilon_{\phi\phi} = \frac{1}{R}\left(\frac{\partial u_\phi}{\partial \phi} + u_R \right),$$

$$\varepsilon_{R\phi} = \frac{1}{2}\left(\frac{1}{R}\frac{\partial u_R}{\partial \phi} + \frac{\partial u_\phi}{\partial R} - \frac{u_\phi}{R} \right),$$

$$\varepsilon_{R\theta} = \frac{1}{2}\left(\frac{1}{R \sin \phi}\frac{\partial u_R}{\partial \theta} + \frac{\partial u_\theta}{\partial R} - \frac{u_\theta}{R} \right),$$

$$\varepsilon_{\phi\theta} = \frac{1}{2R}\left(\frac{1}{\sin \phi}\frac{\partial u_\phi}{\partial \theta} + \frac{\partial u_\theta}{\partial \phi} - u_\theta \cot \phi \right),$$

$$\varepsilon_{\theta\theta} = \frac{1}{R \sin \phi}\left(\frac{\partial u_\theta}{\partial \theta} + u_R \sin \phi + u_\phi \cos \phi \right). \tag{3.5.27}$$

3.6 Rate of Deformation and Vorticity Tensors

3.6.1 Definitions

In fluid mechanics, velocity vector $\mathbf{v}(\mathbf{x}, t)$ is the variable of interest as opposed to the displacement vector $\mathbf{u}$ in solid mechanics. Similar to the displacement gradient tensor [see Eq. (3.5.8)], we can write the *velocity gradient tensor* $\mathbf{L} \equiv \nabla\mathbf{v}$ as the sum of symmetric and antisymmetric (or skew-symmetric) tensors

$$\mathbf{L} = \nabla\mathbf{v} = \frac{1}{2}\left[(\nabla\mathbf{v})^{\mathrm{T}} + \nabla\mathbf{v}\right] + \frac{1}{2}\left[\nabla\mathbf{v} - (\nabla\mathbf{v})^{\mathrm{T}}\right] \equiv \mathbf{D} - \mathbf{W}, \tag{3.6.1}$$

where $\mathbf{D}$ is called the *rate of deformation tensor* and $\mathbf{W}$ is called the *vorticity tensor* or *spin tensor*

$$\mathbf{D} = \frac{1}{2}\left[(\nabla\mathbf{v})^{\mathrm{T}} + \nabla\mathbf{v}\right], \quad \mathbf{W} = -\frac{1}{2}\left[\nabla\mathbf{v} - (\nabla\mathbf{v})^{\mathrm{T}}\right]. \tag{3.6.2}$$

It follows that

$$\mathbf{D} = \frac{1}{2}\left(\mathbf{L}^{\mathrm{T}} + \mathbf{L}\right), \quad \mathbf{W} = -\frac{1}{2}\left(\mathbf{L} - \mathbf{L}^{\mathrm{T}}\right). \tag{3.6.3}$$

Since $\mathbf{W}$ is skew-symmetric (i.e., $\mathbf{W}^{\mathrm{T}} = -\mathbf{W}$), it has only three independent scalar components, which can be used to define the scalar components of a vector $\mathbf{w}$, called the *axial vector of* $\mathbf{W}$, as follows:

$$W_{ij} = -e_{ijk}w_k, \quad [\mathbf{W}] = \begin{bmatrix} 0 & -w_3 & w_2 \\ w_3 & 0 & -w_1 \\ -w_2 & w_1 & 0 \end{bmatrix}. \tag{3.6.4}$$

The scalar components of $\mathbf{w}$ can be expressed in terms of the scalar components of $\mathbf{W}$ as

$$w_i = -\frac{1}{2}e_{ijk}W_{jk} = \frac{1}{2}e_{ijk}\frac{\partial v_k}{\partial x_j} \quad \text{or} \quad \mathbf{w} = \frac{1}{2}\nabla \times \mathbf{v}. \tag{3.6.5}$$

Note that div $\mathbf{w} = 0$ by virtue of the vector identity (i.e., divergence of the curl of a vector is zero). Thus, the axial vector is divergence free. As discussed in Section 3.5.3, if a velocity vector $\mathbf{v}$ is of the form $\mathbf{v} = \mathbf{W} \cdot \mathbf{x}$ for some antisymmetric tensor $\mathbf{W}$ that is independent of the position $\mathbf{x}$, then the motion is a uniform rigid body rotation about the origin with angular velocity $\mathbf{w}$.

3.6.2 Relationship between D and Ė

The rate of deformation tensor $\mathbf{D}$ is not the same as the time rate of change of the infinitesimal strain tensor ε, that is, the strain rate $\dot{\varepsilon}$, where superposed dot signifies the material time derivative. The time rate of change of Green–Lagrange strain tensor can be related to $\mathbf{D}$, as discussed next.

Taking the material time derivative of (3.4.10), we obtain

$$\frac{d}{dt}[(ds)^2] = 2d\mathbf{X} \cdot \frac{d\mathbf{E}}{dt} \cdot d\mathbf{X} = 2d\mathbf{X} \cdot \dot{\mathbf{E}} \cdot d\mathbf{X}, \tag{3.6.6}$$

where we used the fact that $d\mathbf{X}$ and dS are constants. However, from Eq. (3.4.2), we have $(ds)^2 = d\mathbf{x} \cdot d\mathbf{x}$, and the instantaneous rate of change of the squared length $(ds)^2$ is

$$\frac{d}{dt}[(ds)^2] = 2d\mathbf{x} \cdot \frac{d\mathbf{x}}{dt}. \qquad (3.6.7)$$

From Eq. (3.3.1), we have $(d\mathbf{x} = \mathbf{F} \cdot d\mathbf{X})$

$$\frac{d}{dt}(d\mathbf{x}) = \left[\frac{d}{dt}\mathbf{F}\right] \cdot d\mathbf{X} + \mathbf{F} \cdot \frac{d(d\mathbf{X})}{dt} = \dot{\mathbf{F}} \cdot d\mathbf{X}, \qquad (3.6.8)$$

where the time derivative of $d\mathbf{X}$ is zero because it does not change with time. Note that

$$\dot{\mathbf{F}} = \frac{d}{dt}(\nabla_0\mathbf{x})^{\mathrm{T}} = \left[\nabla_0\left(\frac{d\mathbf{x}}{dt}\right)\right]^{\mathrm{T}} = (\nabla_0\mathbf{v})^{\mathrm{T}}. \qquad (3.6.9)$$

Note that $\nabla_0\mathbf{v}$ is the gradient of the velocity vector $\mathbf{v}$ with respect to the material coordinates $\mathbf{X}$, and it is not the same as $\mathbf{L} = \nabla v$ (or $\mathbf{L} \cdot d\mathbf{x} = d\mathbf{v}$). From Eqs. (3.6.8) and (3.6.9), we have

$$d\mathbf{v} = (\nabla_0\mathbf{v})^{\mathrm{T}} \cdot d\mathbf{X}. \qquad (3.6.10)$$

Thus

$$(\nabla_0\mathbf{v})^{\mathrm{T}} \cdot d\mathbf{X} = \mathbf{L} \cdot d\mathbf{x}, \qquad (3.6.11)$$

and from Eqs. (3.6.7) and (3.6.11), we obtain (see Problem 3.25)

$$\frac{d}{dt}[(ds)^2] = 2d\mathbf{x} \cdot \mathbf{L} \cdot d\mathbf{x} = 2d\mathbf{x} \cdot (\mathbf{D} + \mathbf{W}) \cdot d\mathbf{x} = 2d\mathbf{x} \cdot \mathbf{D} \cdot d\mathbf{x}. \qquad (3.6.12)$$

The second term is zero because of the skew symmetry of $\mathbf{W}$. Now comparing Eqs. (3.6.6) with (3.6.12) and using Eq. (3.4.4)

$$d\mathbf{X} \cdot \frac{d\mathbf{E}}{dt} \cdot d\mathbf{X} = d\mathbf{x} \cdot \mathbf{D} \cdot d\mathbf{x}$$

$$= d\mathbf{X} \cdot \mathbf{F}^{\mathrm{T}} \cdot \mathbf{D} \cdot \mathbf{F} \cdot d\mathbf{X},$$

we arrive at the result

$$\frac{d\mathbf{E}}{dt} = \mathbf{F}^{\mathrm{T}} \cdot \mathbf{D} \cdot \mathbf{F}. \qquad (3.6.13)$$

We can also relate the velocity gradient tensor $\mathbf{L}$ to the time rate of deformation gradient tensor $\mathbf{F}$. From Eqs. (3.6.7) and (3.6.9), we have

$$\dot{\mathbf{F}} = \mathbf{L} \cdot \mathbf{F} \quad \text{or} \quad \mathbf{L} = \dot{\mathbf{F}} \cdot \mathbf{F}^{-1}. \qquad (3.6.14)$$

3.7 Polar Decomposition Theorem

Recall that the deformation gradient tensor $\mathbf{F}$ transforms a material vector $d\mathbf{X}$ at $\mathbf{X}$ into the corresponding spatial vector $d\mathbf{x}$, and it characterizes all of the deformation, stretch as well as rotation, at $\mathbf{X}$. Therefore, it forms an essential part of the definition

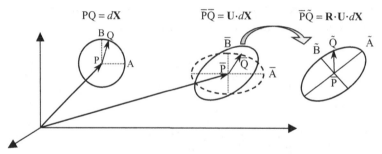

Figure 3.7.1. The roles of **U** and **R** in transforming an ellipsoidal volume of material in the neighborhood of **X**.

of any strain measure. Another role of **F** in connection with the strain measures is discussed here with the help of the polar decomposition theorem of Cauchy. The polar decomposition theorem enables one to decompose **F** into the product of an orthogonal tensor and a symmetric tensor and thereby decomposes the general deformation into pure stretch and rotation.

Suppose that **F** is nonsingular so that each line element $d\mathbf{X}$ from the reference configuration is transformed into a unique line element $d\mathbf{x}$ in the current configuration and conversely. Then the polar decomposition theorem states that **F** has a *unique* right and left decompositions of the form

$$\mathbf{F} = \mathbf{R} \cdot \mathbf{U} = \mathbf{V} \cdot \mathbf{R} \tag{3.7.1}$$

so that

$$d\mathbf{x} = \mathbf{F} \cdot d\mathbf{X} = (\mathbf{R} \cdot \mathbf{U}) \cdot d\mathbf{X} = (\mathbf{V} \cdot \mathbf{R}) \cdot d\mathbf{X}, \tag{3.7.2}$$

where **U** the symmetric *right Cauchy stretch tensor* (stretch is the ratio of the final length to the original length), **V** the symmetric *left Cauchy stretch tensor*, and **R** is the *orthogonal rotation tensor*, which satisfies the identity

$$\mathbf{R}^{\mathrm{T}} \cdot \mathbf{R} = \mathbf{I} \text{ or } \mathbf{R}^{\mathrm{T}} = \mathbf{R}^{-1}. \tag{3.7.3}$$

In Eq. (3.7.2), $\mathbf{U} \cdot d\mathbf{X}$ describes a pure stretch deformation in which there are three mutually perpendicular directions along which the material element $d\mathbf{X}$ stretches (i.e., elongates or compresses) but does not rotate. The three directions are provided by the eigenvectors of **U**. The role of **R** in $\mathbf{R} \cdot \mathbf{U} \cdot d\mathbf{X}$ is to rotate the stretched element. These ideas are illustrated in Figure 3.7.1, which shows the material occupying the spherical volume of radius $|d\mathbf{X}|$ in the undeformed configuration being mapped by the operator **U** into an ellipsoid in the deformed configuration at **x**. The role of **R** is then to rotate the ellipsoid through a rigid body rotation.

From Eqs. (3.7.1) and (3.7.3), it follows that

$$\mathbf{U} = \mathbf{R}^{-1} \cdot \mathbf{F} = \mathbf{R}^{\mathrm{T}} \cdot \mathbf{F}, \quad \mathbf{V} = \mathbf{F} \cdot \mathbf{R}^{-1} = \mathbf{F} \cdot \mathbf{R}^{\mathrm{T}} \tag{3.7.4}$$

and

$$U^2 = U \cdot U = U^T \cdot U = F^T \cdot (R \cdot R^{-1}) \cdot F = F^T \cdot F = C,$$

$$V^2 = V \cdot V = V \cdot V^T = F \cdot (R^{-1} \cdot R) \cdot F^T = F \cdot F^T = B. \tag{3.7.5}$$

We also note that

$$F = R \cdot U = (R \cdot U) \cdot (R^T \cdot R) = (R \cdot U \cdot R^T) \cdot R = V \cdot R$$

$$= V \cdot R = (R \cdot R^T) \cdot (V \cdot R) = R \cdot (R^T \cdot V \cdot R) = R \cdot U, \tag{3.7.6}$$

which show that

$$U = R^T \cdot V \cdot R, \quad V = R \cdot U \cdot R^T. \tag{3.7.7}$$

It can be shown that $\det U = \det V = \det F$. Since $F^T \cdot F$ is real and symmetric, there exists an orthogonal matrix A that transforms $F^T \cdot F$ into a diagonal matrix

$$U^2 = F^T \cdot F = C = A \begin{bmatrix} \lambda_1^2 & 0 & 0 \\ 0 & \lambda_2^2 & 0 \\ 0 & 0 & \lambda_3^2 \end{bmatrix} A^T, \tag{3.7.8}$$

where λ_i^2 are the eigenvalues of $U^2 = F^T \cdot F$, and A is the matrix of eigenvectors $\hat{N}_i$. The eigenvalues λ_i are called the *principal stretches* and the corresponding mutually orthogonal eigenvectors are called the *principal directions*. The tensors U and V have the same eigenvalues and their eigenvectors differ only by the rotation R; see Problem 3.31 for a proof. Thus

$$U = A \begin{bmatrix} \lambda_1 & 0 & 0 \\ 0 & \lambda_2 & 0 \\ 0 & 0 & \lambda_3 \end{bmatrix} A^T = \sum_{i=1}^{3} \lambda_i \hat{N}_i \hat{N}_i. \tag{3.7.9}$$

In view of Eq. (3.7.8), the eigenvalues of U and V are identical. Once the stretch tensor U is known, the rotation tensor R can be obtained from Eq. (3.7.1) as

$$R = F \cdot U^{-1}. \tag{3.7.10}$$

More details on rotation and stretch tensors can be found in the books by Malvern (1969) and Truesdell and Noll (1965). An example of the use of the polar decomposition theorem is presented next.

EXAMPLE 3.7.1: Consider the deformation given by the mapping

$$x_1 = \frac{1}{4}[4X_1 + (9 - 3X_1 - 5X_2 - X_1 X_2)t], \quad x_2 = \frac{1}{4}[4X_2 + (16 + 8X_1)t].$$

(a) For $X = (0, 0)$ and $t = 1$, determine the deformation gradient tensor F and right Cauchy–Green strain tensor C.

(b) Find the eigenvalues (stretches) λ_1 and λ_2 and the associated eigenvectors $\hat{N}_1$ and $\hat{N}_2$.

(c) Use the polar decomposition to determine the symmetric stretch tensor U and rotation tensor R.

SOLUTION:

(a) For $\mathbf{X} = (0, 0)$ and time $t = 1$, the components of the deformation gradient tensor $\mathbf{F}$ and right Cauchy–Green strain tensor $\mathbf{C}$ are

$$[F] = \frac{1}{4}\begin{bmatrix} 1 & -5 \\ 8 & 4 \end{bmatrix}, \quad [C] = \frac{1}{16}\begin{bmatrix} 65 & 27 \\ 27 & 41 \end{bmatrix}.$$

(b) The eigenvalues λ_1^2 and λ_2^2 of matrix $[C]$ are determined by setting

$$|[C] - \lambda^2[I]| = 0 \quad \rightarrow \quad \lambda_1^2 = 5.1593, \quad \lambda_2^2 = 1.4658$$

so that $\lambda_1 = 2.2714$ and $\lambda_2 = 1.2107$. The eigenvectors are (in vector component form)

$$\{N^{(1)}\} = \begin{Bmatrix} 0.8385 \\ 0.5449 \end{Bmatrix}, \quad \{N^{(2)}\} = \begin{Bmatrix} -0.5449 \\ 0.8385 \end{Bmatrix}.$$

(c) Hence, the stretch tensor can be written as

$$\mathbf{U} = \lambda_1 \hat{\mathbf{N}}^{(1)} \hat{\mathbf{N}}^{(1)} + \lambda_2 \hat{\mathbf{N}}^{(2)} \hat{\mathbf{N}}^{(2)}$$

$$= \lambda_1 \left(N_1^{(1)} \hat{\mathbf{e}}_1 + N_2^{(1)} \hat{\mathbf{e}}_2 \right) \left(N_1^{(1)} \hat{\mathbf{e}}_1 + N_2^{(1)} \hat{\mathbf{e}}_2 \right)$$

$$+ \lambda_2 \left(N_1^{(2)} \hat{\mathbf{e}}_1 + N_2^{(2)} \hat{\mathbf{e}}_2 \right) \left(N_1^{(2)} \hat{\mathbf{e}}_1 + N_2^{(2)} \hat{\mathbf{e}}_2 \right)$$

$$= \left(\lambda_1 [N_1^{(1)}]^2 + \lambda_2 [N_1^{(2)}]^2 \right) \hat{\mathbf{e}}_1 \hat{\mathbf{e}}_1 + \left(\lambda_1 [N_2^{(1)}]^2 + \lambda_2 [N_2^{(2)}]^2 \right) \hat{\mathbf{e}}_2 \hat{\mathbf{e}}_2$$

$$+ \left(\lambda_1 N_1^{(1)} N_2^{(1)} + \lambda_2 N_1^{(2)} N_2^{(2)} \right) (\hat{\mathbf{e}}_1 \hat{\mathbf{e}}_2 + \hat{\mathbf{e}}_2 \hat{\mathbf{e}}_1)$$

or in matrix form

$$[U] = \begin{bmatrix} 1.9564 & 0.4846 \\ 0.4846 & 1.5257 \end{bmatrix}.$$

Then the rotation tensor $[R]$ in matrix form is given by

$$[R] = [F][U]^{-1} = \begin{bmatrix} 0.3590 & -0.9333 \\ 0.9333 & 0.3590 \end{bmatrix}.$$

3.8 Compatibility Equations

The task of computing strains (infinitesimal or finite) from a given displacement field is a straightforward exercise. However, sometimes we face the problem of finding the displacements from a given strain field. This is not as straightforward because there are six independent partial differential equations (i.e., strain-displacement relations) for only three unknown displacements, which would in general overdetermine the solution. We will find some conditions, known as *St. Venant's compatibility equations*, that will ensure the computation of unique displacement field from a given strain field. The derivation is presented for infinitesimal strains. For finite strains, the same steps may be followed, but the process is so difficult that it is never attempted (although some general compatibility conditions may be stated to ensure integrability of the six nonlinear partial differential equations).

To understand the meaning of strain compatibility, imagine that a material body is cut up into pieces before it is strained, and then each piece is given a certain strain. The strained pieces cannot be fitted back into a single continuous body without further deformation. However, if the strain in each piece is related to or compatible with the strains in the neighboring pieces, then they can be fitted together to form a continuous body. Mathematically, the six relations that connect six strain components to the three displacement components should be consistent. To make this point clear, consider the two-dimensional case. We have three strain-displacement relations in two displacements:

$$\frac{\partial u_1}{\partial x_1} = \varepsilon_{11}, \tag{3.8.1}$$

$$\frac{\partial u_2}{\partial x_2} = \varepsilon_{22}, \tag{3.8.2}$$

$$\frac{\partial u_1}{\partial x_2} + \frac{\partial u_2}{\partial x_1} = 2\varepsilon_{12}. \tag{3.8.3}$$

If the given data $(\varepsilon_{11}, \varepsilon_{22}, \varepsilon_{12})$ is compatible (or consistent), any two of the three equations should yield the same displacement components. The compatibility of the data can be established as follows. Differentiate the first equation with respect to x_2 twice, the second equation with respect to x_1 twice, and the third equation with respect to x_1 and x_2 each to obtain

$$\frac{\partial^3 u_1}{\partial x_1 \partial x_2^2} = \frac{\partial^2 \varepsilon_{11}}{\partial x_2^2}, \tag{3.8.1'}$$

$$\frac{\partial^3 u_2}{\partial x_2 \partial x_1^2} = \frac{\partial^2 \varepsilon_{22}}{\partial x_1^2}, \tag{3.8.2'}$$

$$\frac{\partial^3 u_1}{\partial x_2^2 \partial x_1} + \frac{\partial^3 u_2}{\partial x_1^2 \partial x_2} = 2\frac{\partial^2 \varepsilon_{12}}{\partial x_1 \partial x_2}. \tag{3.8.3'}$$

Using Eqs. (3.8.1') and (3.8.2') in (3.8.3'), we arrive at the following relation between the three strains:

$$\frac{\partial^2 \varepsilon_{11}}{\partial x_2^2} + \frac{\partial^2 \varepsilon_{22}}{\partial x_1^2} = 2\frac{\partial^2 \varepsilon_{12}}{\partial x_1 \partial x_2}. \tag{3.8.4}$$

Equation (3.8.4) is called the strain compatibility condition among the three strains for a two-dimensional case.

Similar procedure can be followed to obtain the strain compatibility equations for the three-dimensional case. In addition to Eq. (3.8.4) five more such conditions can be derived:

$$\frac{\partial^2 \varepsilon_{11}}{\partial x_3^2} + \frac{\partial^2 \varepsilon_{33}}{\partial x_1^2} = 2\frac{\partial^2 \varepsilon_{13}}{\partial x_1 \partial x_3}, \tag{3.8.5}$$

$$\frac{\partial^2 \varepsilon_{22}}{\partial x_3^2} + \frac{\partial^2 \varepsilon_{33}}{\partial x_2^2} = 2\frac{\partial^2 \varepsilon_{23}}{\partial x_2 \partial x_3}, \tag{3.8.6}$$

$$\frac{\partial^2 \varepsilon_{11}}{\partial x_2 \partial x_3} + \frac{\partial^2 \varepsilon_{23}}{\partial x_1^2} = \frac{\partial^2 \varepsilon_{13}}{\partial x_1 \partial x_2} + \frac{\partial^2 \varepsilon_{12}}{\partial x_1 \partial x_3}, \tag{3.8.7}$$

$$\frac{\partial^2 \varepsilon_{22}}{\partial x_1 \partial x_3} + \frac{\partial^2 \varepsilon_{13}}{\partial x_2^2} = \frac{\partial^2 \varepsilon_{23}}{\partial x_1 \partial x_2} + \frac{\partial^2 \varepsilon_{12}}{\partial x_2 \partial x_3}, \tag{3.8.8}$$

$$\frac{\partial^2 \varepsilon_{33}}{\partial x_1 \partial x_2} + \frac{\partial^2 \varepsilon_{12}}{\partial x_3^2} = \frac{\partial^2 \varepsilon_{13}}{\partial x_2 \partial x_3} + \frac{\partial^2 \varepsilon_{23}}{\partial x_1 \partial x_3}. \tag{3.8.9}$$

The six equations in Eqs. (3.8.4)–(3.8.9) can be written as a single relation using the index notation

$$\frac{\partial^2 \varepsilon_{mn}}{\partial x_i \partial x_j} + \frac{\partial^2 \varepsilon_{ij}}{\partial x_m \partial x_n} = \frac{\partial^2 \varepsilon_{im}}{\partial x_j \partial x_n} + \frac{\partial^2 \varepsilon_{jn}}{\partial x_i \partial x_m}. \tag{3.8.10}$$

These conditions are both necessary and sufficient to determine a single-valued displacement field. Similar compatibility conditions hold for the rate of deformation tensor **D**.

Equation (3.8.10) can be derived in vector form as follows. We begin with the curl of ε:

$$\nabla \times \varepsilon = e_{ijk}\frac{\partial \varepsilon_{jr}}{\partial x_i}\hat{\mathbf{e}}_k\hat{\mathbf{e}}_r = \frac{1}{2}e_{ijk}\left(\frac{\partial^2 u_j}{\partial x_i \partial x_r} + \frac{\partial^2 u_r}{\partial x_i \partial x_j}\right)\hat{\mathbf{e}}_k\hat{\mathbf{e}}_r$$

$$= \frac{1}{2}\left(e_{ijk}\frac{\partial^2 u_j}{\partial x_i \partial x_r} + 0\right)\hat{\mathbf{e}}_k\hat{\mathbf{e}}_r. \tag{3.8.11}$$

Using Eq. (3.5.13), we have

$$\nabla \times \varepsilon = \frac{1}{2}e_{ijk}\frac{\partial^2 u_j}{\partial x_i \partial x_r}\hat{\mathbf{e}}_k\hat{\mathbf{e}}_r = \frac{\partial}{\partial x_r}\left(\frac{1}{2}e_{ijk}\frac{\partial u_j}{\partial x_i}\hat{\mathbf{e}}_k\right)\hat{\mathbf{e}}_r = \frac{\partial \omega_k}{\partial x_r}\hat{\mathbf{e}}_k\hat{\mathbf{e}}_r$$

or

$$(\nabla \times \varepsilon)^{\mathrm{T}} = \hat{\mathbf{e}}_r\frac{\partial \omega_k}{\partial x_r}\hat{\mathbf{e}}_k = \hat{\mathbf{e}}_r\frac{\partial}{\partial x_r}(\omega_k\hat{\mathbf{e}}_k) = \nabla\omega.$$

Since the curl ($\nabla \times$) of the gradient (∇) of a vector (or tensor) is zero, we take curl of the above equation and arrive at the compatibility equation in vector/tensor form:

$$\nabla \times (\nabla \times \varepsilon)^{\mathsf{T}} = \mathbf{0} \quad \text{or} \quad e_{ikr}\, e_{jls}\, \varepsilon_{ij,kl} = 0. \tag{3.8.12}$$

The next example illustrates how to determine if a given strain field is compatible.

EXAMPLE 3.8.1: Given the following two-dimensional, infinitesimal strain field:

$$\varepsilon_{11} = c_1 x_1 \left(x_1^2 + x_2^2\right), \quad \varepsilon_{22} = \frac{1}{3}c_2 x_1^3, \quad \varepsilon_{12} = c_3 x_1^2 x_2,$$

where c_1, c_2, and c_3 are constants, determine whether the strain field is compatible.

SOLUTION: Using Eq. (3.8.4), we obtain

$$\frac{\partial^2 \varepsilon_{11}}{\partial x_2^2} + \frac{\partial^2 \varepsilon_{22}}{\partial x_1^2} - 2\frac{\partial^2 \varepsilon_{12}}{\partial x_1 \partial x_2} = 2c_1 x_1 + 2c_2 x_1 - 4c_3 x_1.$$

Thus the strain field is not compatible, unless $c_1 + c_2 - 2c_3 = 0$.

The next example illustrates how to determine the displacement field from a given compatible strain field.

EXAMPLE 3.8.2: Consider the problem of the isotropic cantilever beam bent by a load P at the free end, as shown in Figure 3.8.1. From the elementary beam theory, we have the following strains:

$$\varepsilon_{11} = -\frac{Px_1x_2}{EI}, \quad \varepsilon_{22} = -\nu\varepsilon_{11} = \nu\frac{Px_1x_2}{EI}, \quad \varepsilon_{12} = -\frac{(1+\nu)P}{2EI}(h^2 - x_2^2), \tag{3.8.13}$$

where I is the second moment of area about the x_3-axis, ν is the Poisson ratio, E is Young's modulus, and $2h$ is the height of the beam. (a) Determine whether the strains are compatible, and if it is, (b) find the displacement field using the linearized strain-displacement relations, and (c) determine the constants of integration using suitable boundary conditions.

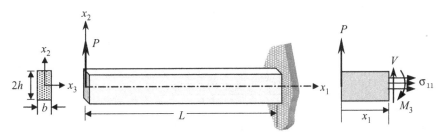

Figure 3.8.1. Cantilever beam bent by a point load, P.

SOLUTION

(a) Substituting ε_{ij} into the single compatibility Eq. (3.8.4), we obtain $0 + 0 = 0$. Thus the strains satisfy the compatibility equation in two dimensions. Although the two-dimensional strains are compatible, the three-dimensional strains are not compatible. For example, using the additional strains, $\varepsilon_{33} = -\nu\varepsilon_{11}$, $\varepsilon_{13} = \varepsilon_{23} = 0$, one can show that all of the equations except the one in Eq. (3.8.9) are satisfied.

(b) Using the strain-displacement equations and integrating the strains

$$\frac{\partial u_1}{\partial x_1} = \varepsilon_{11} = -\frac{Px_1x_2}{EI} \quad \text{or} \quad u_1 = -\frac{Px_1^2x_2}{2EI} + f(x_2), \qquad (3.8.14)$$

$$\frac{\partial u_2}{\partial x_2} = \varepsilon_{22} = \frac{\nu Px_1x_2}{EI} \quad \text{or} \quad u_2 = \frac{\nu Px_1x_2^2}{2EI} + g(x_1), \qquad (3.8.15)$$

where $f(x_2)$ and $g(x_1)$ are functions of integration. Substituting u_1 and u_2 into the definition of $2\varepsilon_{12}$, we obtain

$$2\varepsilon_{12} = \frac{\partial u_1}{\partial x_2} + \frac{\partial u_2}{\partial x_1} = -\frac{Px_1^2}{2EI} + \frac{df}{dx_2} + \frac{\nu Px_2^2}{2EI} + \frac{dg}{dx_1}. \qquad (3.8.16)$$

But this must be equal to the strain value given in Eq. (3.8.13):

$$-\frac{Px_1^2}{2EI} + \frac{df}{dx_2} + \frac{\nu Px_2^2}{2EI} + \frac{dg}{dx_1} = -\frac{(1+\nu)}{EI}P(h^2 - x_2^2).$$

Separating the x_1 and x_2 terms, we obtain

$$-\frac{dg}{dx_1} + \frac{Px_1^2}{2EI} - \frac{(1+\nu)Ph^2}{EI} = \frac{df}{dx_2} - \frac{(2+\nu)Px_2^2}{2EI}.$$

Since the left side depends only on x_1 and the right side depends only on x_2, and yet the equality must hold, it follows that both sides should be equal to a constant c_0:

$$\frac{df}{dx_2} - \frac{(2+\nu)Px_2^2}{2EI} = c_0, \quad -\frac{dg}{dx_1} + \frac{Px_1^2}{2EI} - \frac{(1+\nu)Ph^2}{EI} = c_0.$$

Integrating the expressions for f and g, we obtain

$$f(x_2) = \frac{(2+\nu)Px_2^3}{6EI} + c_0x_2 + c_1$$

$$g(x_1) = \frac{Px_1^3}{6EI} - \frac{(1+\nu)Ph^2x_1}{EI} - c_0x_1 + c_2, \qquad (3.8.17)$$

where c_1 and c_2 are constants of integration that are to be determined. The most general form of displacement field (u_1, u_2) corresponding to the strains in Eq. (3.8.13) is given by

$$u_1(x_1, x_2) = -\frac{Px_1^2 x_2}{2EI} + \frac{(2+v)Px_2^3}{6EI} + c_0 x_2 + c_1,$$

$$u_2(x_1, x_2) = -\frac{(1+v)Ph^2 x_1}{EI} + \frac{vPx_1 x_2^2}{2EI} + \frac{Px_1^3}{6EI} - c_0 x_1 + c_2.$$

(3.8.18)

(c) The three constants c_0, c_1, and c_2 are determined using boundary conditions necessary to prevent the beam from moving as a rigid body. From Figure 3.8.1, we have the following boundary conditions:

$$u_1(L, 0) = 0, \quad u_2(L, 0) = 0. \tag{3.8.19}$$

This prevents rigid body translations in the x_1- and x_2-directions. Substituting the expressions for u_1 and u_2 into the boundary conditions (3.8.19), we obtain

$$u_1(L, 0) = 0 \quad \rightarrow \quad c_1 = 0,$$

$$u_2(L, 0) = 0 \quad \rightarrow \quad c_0 L - c_2 = -\frac{(1+v)Ph^2 L}{EI} + \frac{PL^3}{6EI}.$$

(3.8.20)

To remove rigid body rotation, we assume that the rotation of the vertical edge at the point $(x_1, x_2) = (L, 0)$ is zero:

$$\left(\frac{\partial u_2}{\partial x_1}\right)_{x_1=L, x_2=0} = 0 \rightarrow c_0 = \frac{PL^2}{2EI} - \frac{(1+v)Ph^2}{EI}, \quad c_2 = \frac{PL^3}{3EI}. \tag{3.8.21}$$

The displacement field is

$$u_1(x_1, x_2) = \frac{PL^2 x_2}{6EI}\left[3\left(1 - \frac{x_1^2}{L^2}\right) + (2+v)\frac{x_2^2}{L^2} - 6(1+v)\frac{h^2}{L^2}\right],$$

$$u_2(x_1, x_2) = \frac{PL^3}{6EI}\left[2 - 3\frac{x_1}{L}\left(1 - v\frac{x_2^2}{L^2}\right) + \frac{x_1^3}{L^3}\right].$$

(3.8.22)

3.9 Change of Observer: Material Frame Indifference

In the analytical description of physical events, the following two requirements must be followed:

1. Invariance of the equations with respect to stationary coordinate frames of reference.
2. Invariance of the equations with respect to frames of reference that move in arbitrary relative motion.

The first requirement is readily met by expressing the equations in vector/tensor form, which is invariant. The assertion that an equation is in "invariant form" refers to the vector form that is independent of the choice of a coordinate system. The coordinate systems used in the present study were assumed to be relatively at rest.

The second requirement is that the invariance property holds for reference frames (or observers) moving arbitrarily with respect to each other. This requirement is dictated by the need for forces to be the same as measured by all observers irrespective of their relative motions. The concept of frames of reference should not be confused with that of coordinate systems, as they are not the same at all. A given observer is free to choose any coordinate system as may be convenient to observe or analyze the system response. Invariance with respect to changes of observer is termed *material frame indifference* or *material objectivity*.

A detailed discussion of frame indifference is outside the scope of the present study, and only a brief discussion is presented here. Functions and fields whose values are scalars, vectors, or tensors are called *frame indifferent* or *objective* if both the dependent and independent vector and tensor variables transform according to the following equations:

1. Events, $\mathbf{x}$, t: $\mathbf{x}^* = \mathbf{c}(t) + \mathbf{Q}(t) \cdot \mathbf{x}$
2. Vectors, $\mathbf{v}$: $\mathbf{v}^* = \mathbf{Q}(t) \cdot \mathbf{v}$
3. General second-order tensors, $\mathbf{S}$: $\mathbf{S}^* = \mathbf{Q}(t) \cdot \mathbf{S} \cdot \mathbf{Q}^{\mathrm{T}}(t)$
4. Deformation gradient tensor, $\mathbf{F}$: $\mathbf{F}^* = \mathbf{Q}(t) \cdot \mathbf{F}$

Here quantities without an asterisk refer to a frame of reference (or observer) $\mathcal{F}$ with origin O, and those with an asterisk (*) refer to another frame of reference $\mathcal{F}^*$ with origin O^*; $\mathbf{c}(t)$ is a constant vector from O to O^*, and $\mathbf{Q}(t)$ is the orthogonal rotation tensor that rotates frame $\mathcal{F}^*$ into frame $\mathcal{F}$. For example, $\mathbf{x}$ and $\mathbf{x}^*$ refer to the same motion, but mathematically $\mathbf{x}^*$ is the motion obtained from $\mathbf{x}$ by superposition of a rigid rotation and translation. One can show that the velocity and acceleration vectors are *not* objective.

To see the effect on the deformation gradient of a change of observer, consider the most general mapping between observer O and observer O^*

$$\mathbf{x}^* = \mathbf{c}(t) + \mathbf{Q}(t) \cdot \mathbf{x}, \tag{3.9.1}$$

where $\mathbf{c}$ is an arbitrary vector and $\mathbf{Q}$ is a second-order orthogonal tensor (i.e., $\mathbf{Q} \cdot \mathbf{Q} = \mathbf{I}$), both of which depend on time t. The mapping in Eq. (3.9.1) may be interpreted as one that takes $(\mathbf{x}, t)$ to $(\mathbf{x}^*, t^*)$ as a change of observer from O to O^*, so that the event which is observed at place $\mathbf{x}$ at time t by observer O is the *same* event as that observed at $\mathbf{x}^*$ at time t^* by observer O^*, where $t^* = t - a$, and a is a constant. Thus, a change of observer merely changes the *description* of an event.

The motion of body $\mathcal{B}$ as seen by observer O can be written as

$$\mathbf{x} = \chi(\mathbf{X}, t), \tag{3.9.2}$$

whereas observer O^* describes the *same* motion as

$$\mathbf{x}^* = \chi^*(\mathbf{X}, t^*), \tag{3.9.3}$$

where $\chi^*(\mathbf{X}, t^*)$ is defined through the observer mapping by

$$\chi^*(\mathbf{X}, t^*) = \mathbf{c}(t) + \mathbf{Q}(t) \cdot \chi(\mathbf{X}, t), \quad t^* = t - a. \tag{3.9.4}$$

The velocity and acceleration of the particle X as observed by O are

$$\frac{\partial \chi}{\partial t}, \quad \frac{\partial^2 \chi}{\partial t^2}, \tag{3.9.5}$$

respectively. The velocity and acceleration observed by O^* are

$$\frac{\partial \chi^*}{\partial t^*} = \frac{\partial \mathbf{c}}{\partial t} + \frac{\partial \mathbf{Q}}{\partial t} \cdot \chi(\mathbf{X}, t) + \mathbf{Q}(t) \cdot \frac{\partial \chi}{\partial t},$$

$$\frac{\partial^2 \chi^*}{\partial t^{*2}} = \frac{\partial^2 \mathbf{c}}{\partial t^2} + 2 \frac{\partial \mathbf{Q}}{\partial t} \cdot \frac{\partial \chi}{\partial t} + \mathbf{Q}(t) \cdot \frac{\partial^2 \chi}{\partial t^2} + \frac{\partial^2 \mathbf{Q}}{\partial t^2} \cdot \chi(\mathbf{X}, t), \tag{3.9.6}$$

respectively. We note that the two observers' view of the velocity and acceleration of a given motion are different, even though the rate of change at fixed $\mathbf{X}$ is the same in each case.

If the reference configuration is independent of the observer, the deformation gradients in the two frames of reference are $\mathbf{F}$ and $\mathbf{F}^*$, respectively, where

$$\mathbf{F}^* = \mathbf{Q}(t) \cdot \mathbf{F}(\mathbf{X}, t). \tag{3.9.7}$$

The respective Jacobians are given by

$$J = \det \mathbf{F}, \quad J^* = \det \mathbf{F}^* = \det \mathbf{F} = J, \tag{3.9.8}$$

where the fact that the determinant of $\mathbf{Q}$ is unity is used. Thus the volume change is unaffected by an observer transformation, which makes sense as the local volume ratio should be independent of the (kinematic) description of motion.

To see how the Green–Lagrange strain tensor changes under the observer transformation, consider

$$\mathbf{C}^* = (\mathbf{F}^*)^{\mathrm{T}} \cdot \mathbf{F}^* = \left(\mathbf{F}^{\mathrm{T}} \cdot \mathbf{Q}^{\mathrm{T}} \right) \cdot \left(\mathbf{Q} \cdot \mathbf{F} \right) = \mathbf{F}^{\mathrm{T}} \cdot \mathbf{F} = \mathbf{C}, \tag{3.9.9}$$

where Eq. (3.9.7) and the the property $\mathbf{Q}^{\mathrm{T}} \cdot \mathbf{Q} = \mathbf{I}$ of an orthogonal matrix $\mathbf{Q}$ is used. Hence, by definition [see Eqs. (3.4.3) and (3.4.5)], the Green strain tensor as measured by the two different observers is the same:

$$\mathbf{E} = \mathbf{E}^*. \tag{3.9.10}$$

3.10 Summary

In this chapter, the two descriptions of motion, namely, the spatial (Eulerian) and material (Lagrange), are discussed, and the deformation gradient tensor, Cauchy–Green deformation tensors, several forms of homogeneous deformations, and various measures of strain are introduced. The strain tensors discussed include the Green–Lagrange strain tensor, Cauchy strain tensor, and the Euler strain tensor. Physical interpretation of the normal and shear strain tensor components is also discussed. Compatibility conditions on strains to ensure a unique determination of displacements from a given strain field are presented. The polar decomposition theorem is also presented and its utility in determining the principal stretches was illustrated. Numerous examples are presented to illustrate the concepts introduced. Finally, the concept of frame indifference is briefly discussed.

PROBLEMS

3.1 Given the motion

$$\mathbf{x} = (1 + t)\mathbf{X},$$

determine the velocity and acceleration fields of the motion.

3.2 Show that the acceleration components in cylindrical coordinates are

$$a_r = \frac{\partial v_r}{\partial t} + v_r \frac{\partial v_r}{\partial r} + \frac{v_\theta}{r} \frac{\partial v_r}{\partial \theta} + v_z \frac{\partial v_r}{\partial z} - \frac{v_\theta^2}{r},$$

$$a_\theta = \frac{\partial v_\theta}{\partial t} + v_r \frac{\partial v_\theta}{\partial r} + \frac{v_\theta}{r} \frac{\partial v_\theta}{\partial \theta} + v_z \frac{\partial v_\theta}{\partial z} + \frac{v_r v_\theta}{r},$$

$$a_z = \frac{\partial v_z}{\partial t} + v_r \frac{\partial v_z}{\partial r} + \frac{v_\theta}{r} \frac{\partial v_z}{\partial \theta} + v_z \frac{\partial v_z}{\partial z}.$$

3.3 The motion of a body is described by the mapping

$$\chi(\mathbf{X}) = (X_1 + t^2 X_2)\,\hat{\mathbf{e}}_1 + (X_2 + t^2 X_1)\,\hat{\mathbf{e}}_2 + X_3\,\mathbf{e}_3,$$

where t denotes time. Determine

(a) the components of the deformation gradient tensor $\mathbf{F}$,

(b) the components of the displacement, velocity, and acceleration vectors, and

(c) the position (X_1, X_2, X_3) of the particle in undeformed configuration that occupies the position $(x_1, x_2, x_3) = (9, 6, 1)$ at time $t = 2$ s in the deformed configuration.

3.4 *Homogeneous stretch.* Consider a body with deformation mapping of the form

$$\chi(\mathbf{X}) = k_1 X_1\,\hat{\mathbf{e}}_1 + k_2 X_2\,\hat{\mathbf{e}}_2 + k_3 X_3\,\hat{\mathbf{e}}_3,$$

where k_i are constants. Determine the components of

(a) the deformation gradient tensor $\mathbf{F}$, and

(b) the left and right Cauchy–Green tensors $\mathbf{C}$ and $\mathbf{B}$.

3.5 *Homogeneous stretch followed by simple shear.* Consider a body with deformation mapping of the form

$$\chi(\mathbf{X}) = (k_1 X_1 + e_0 k_2 X_2)\,\hat{\mathbf{e}}_1 + k_2 X_2\,\hat{\mathbf{e}}_2 + k_3 X_3\,\hat{\mathbf{e}}_3,$$

where k_i and e_0 are constants. Determine the components of

(a) the deformation gradient tensor $\mathbf{F}$, and

(b) the left and right Cauchy–Green tensors $\mathbf{C}$ and $\mathbf{B}$.

3.6 Suppose that the motion of a continuous medium is given by

$$x_1 = X_1 \cos At + X_2 \sin At,$$

$$x_2 = -X_1 \sin At + X_2 \cos At,$$

$$x_3 = (1 + Bt)X_3,$$

where A and B are constants. Determine the components of

(a) the displacement vector in the material description,

(b) the displacement vector in the spatial description, and

(c) the Green–Lagrange and Eulerian strain tensors.

3.7 If the deformation mapping of a body is given by

$$\chi(\mathbf{X}) = (X_1 + AX_2)\,\hat{\mathbf{e}}_1 + (X_2 + BX_1)\,\hat{\mathbf{e}}_2 + X_3\,\hat{\mathbf{e}}_3,$$

where A and B are constants, determine

(a) the displacement components in the material description,

(b) the displacement components in the spatial description, and

(c) the components of the Green–Lagrange and Eulerian strain tensors.

3.8 For the deformation field is given in Problem 3.5, determine the positions (x_1, x_2, x_3) of a circle of material particles $X_1^2 + X_2^2 = a^2$.

3.9 The motion of a continuous medium is given by

$$x_1 = \frac{1}{2}(X_1 + X_2)e^t + \frac{1}{2}(X_1 - X_2)e^{-t},$$

$$x_2 = \frac{1}{2}(X_1 + X_2)e^t - \frac{1}{2}(X_1 - X_2)e^{-t},$$

$$x_3 = X_3.$$

Determine

(a) the velocity components in the material description,

(b) the velocity components in the spatial description, and

(c) the components of the rate of deformation and vorticity tensors.

3.10 *Nanson's formula.* Let the differential area in the reference configuration be dA. Then

$$\hat{\mathbf{N}}dA = d\mathbf{X}^{(1)} \times d\mathbf{X}^{(2)} \quad \text{or} \quad N_I \, dA = e_{IJK}\, dX_J^{(1)} dX_K^{(2)},$$

where $d\mathbf{X}^{(1)}$ and $d\mathbf{X}^{(2)}$ are two nonparallel differential vectors in the reference configuration. The mapping from the undeformed configuration to the deformed configuration maps $d\mathbf{X}^{(1)}$ and $d\mathbf{X}^{(2)}$ into $d\mathbf{x}^{(1)}$ and $d\mathbf{x}^{(2)}$, respectively. Then $\hat{\mathbf{n}}da = d\mathbf{x}^{(1)} \times d\mathbf{x}^{(2)}$. Show that

$$\hat{\mathbf{n}}\,da = J\mathbf{F}^{-\mathrm{T}} \cdot \hat{\mathbf{N}}\,dA.$$

3.11 Consider a rectangular block of material of thickness h and sides $3b$ and $4b$ and having a triangular hole as shown in Figure P3.11. If the material is subjected to the deformation mapping given in Eq. (3.3.12),

$$\chi(\mathbf{X}) = (X_1 + \gamma X_2)\hat{\mathbf{e}}_1 + X_2\,\mathbf{e}_2 + X_3\,\hat{\mathbf{e}}_3,$$

determine (a) the equation of the line BC in the undeformed and deformed configurations, (b) the angle ABC in the undeformed and deformed configurations, and (c) the area of the triangle ABC in the undeformed and deformed configurations.

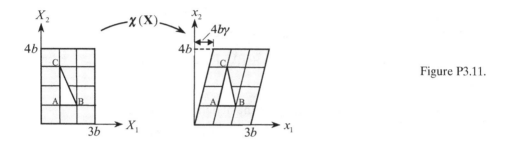

Figure P3.11.

3.12 Consider a square block of material of thickness h, as shown in Figure P3.12. If the material is subjected to the deformation mapping given in Eq. (3.3.14) with $\gamma_1 = 1$ and $\gamma_2 = 3$,

$$\chi(\mathbf{X}) = X_1(1 + X_2)\hat{\mathbf{e}}_1 + X_2(1 + 3X_1)\mathbf{e}_2 + X_3\,\hat{\mathbf{e}}_3,$$

(a) compute the components of the Cauchy–Green deformation tensor $\mathbf{C}$ and Green–Lagrange strain tensor $\mathbf{E}$ at the point $\mathbf{X} = (1, 1, 0)$, and

(b) the principal strains and directions at $\mathbf{X} = (1, 1, 0)$.

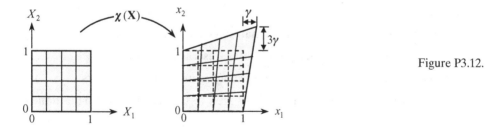

Figure P3.12.

3.13 Determine the displacements and Green–Lagrange strain tensor components for the deformed configuration shown in Figure P3.13. The undeformed configuration is shown in dashed lines.

3.14 Determine the displacements and Green–Lagrange strain components for the deformed configuration shown in Figure P3.14. The undeformed configuration is shown in dashed lines.

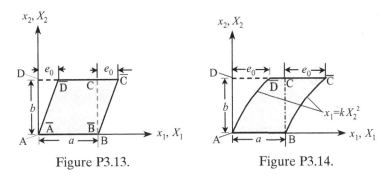

Figure P3.13.   Figure P3.14.

3.15 Determine the displacements and Green-Lagrange strains in the (x_1, x_2, x_3) system for the deformed configuration shown in Figure P3.15. The undeformed configuration is shown in dashed lines.

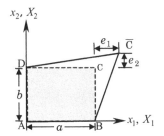

Figure P3.15.

3.16 Determine the displacements and Green–Lagrange strains for the deformed configuration shown in Figure P3.16. The undeformed configuration is shown in dashed lines.

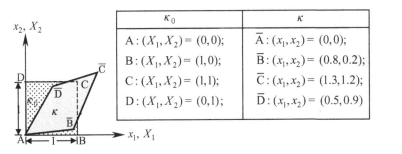

κ_0	κ
$A : (X_1, X_2) = (0,0);$	$\bar{A} : (x_1, x_2) = (0,0);$
$B : (X_1, X_2) = (1,0);$	$\bar{B} : (x_1, x_2) = (0.8, 0.2);$
$C : (X_1, X_2) = (1,1);$	$\bar{C} : (x_1, x_2) = (1.3, 1.2);$
$D : (X_1, X_2) = (0,1);$	$\bar{D} : (x_1, x_2) = (0.5, 0.9)$

Figure P3.16.

Discussion. Discuss the validity of the following comments by a reviewer of Problems 3.13–3.16: "…in these (problems) the student is asked to construct the deformation in the interior of the body from the boundary data alone. This is really quite absurd, for the answer is not kinematically determined in general. It depends on the details of material constitution, material homogeneity, whether the body is in equilibrium or not, etc."

3.17 Given the following displacement vector in a material description using a cylindrical coordinate system

$$\mathbf{u} = A r \hat{\mathbf{e}}_r + B r z \hat{\mathbf{e}}_\theta + C \sin \theta \hat{\mathbf{e}}_z,$$

where A, B, and C are constants, determine the infinitesimal strains.

3.18 Show that the components of the Green–Lagrange strain tensor in cylindrical coordinate system are given by

$$E_{rr} = \frac{\partial u_r}{\partial r} + \frac{1}{2}\left[\left(\frac{\partial u_r}{\partial r}\right)^2 + \left(\frac{\partial u_\theta}{\partial r}\right)^2 + \left(\frac{\partial u_z}{\partial r}\right)^2\right],$$

$$E_{r\theta} = \frac{1}{2}\left(\frac{1}{r}\frac{\partial u_r}{\partial \theta} + \frac{\partial u_\theta}{\partial r} - \frac{u_\theta}{r} + \frac{1}{r}\frac{\partial u_r}{\partial r}\frac{\partial u_r}{\partial \theta} + \frac{1}{r}\frac{\partial u_\theta}{\partial r}\frac{\partial u_\theta}{\partial \theta}\right.$$

$$\left. + \frac{1}{r}\frac{\partial u_z}{\partial r}\frac{\partial u_z}{\partial \theta} + \frac{u_r}{r}\frac{\partial u_\theta}{\partial r} - \frac{u_\theta}{r}\frac{\partial u_r}{\partial r}\right),$$

$$E_{rz} = \frac{1}{2}\left(\frac{\partial u_r}{\partial z} + \frac{\partial u_z}{\partial r} + \frac{\partial u_r}{\partial r}\frac{\partial u_r}{\partial z} + \frac{\partial u_\theta}{\partial r}\frac{\partial u_\theta}{\partial z} + \frac{\partial u_z}{\partial r}\frac{\partial u_z}{\partial z}\right),$$

$$E_{\theta\theta} = \frac{u_r}{r} + \frac{1}{r}\frac{\partial u_\theta}{\partial \theta} + \frac{1}{2}\left[\left(\frac{1}{r}\frac{\partial u_r}{\partial \theta}\right)^2 + \left(\frac{1}{r}\frac{\partial u_\theta}{\partial \theta}\right)^2 + \left(\frac{1}{r}\frac{\partial u_z}{\partial \theta}\right)^2\right.$$

$$\left. - \frac{2}{r^2}u_\theta\frac{\partial u_r}{\partial \theta} + \frac{2}{r^2}u_r\frac{\partial u_\theta}{\partial \theta} + \left(\frac{u_\theta}{r}\right)^2 + \left(\frac{u_r}{r}\right)^2\right],$$

$$E_{\theta z} = \frac{1}{2}\left(\frac{\partial u_\theta}{\partial z} + \frac{1}{r}\frac{\partial u_z}{\partial \theta} + \frac{1}{r}\frac{\partial u_r}{\partial \theta}\frac{\partial u_r}{\partial z} + \frac{1}{r}\frac{\partial u_\theta}{\partial \theta}\frac{\partial u_\theta}{\partial z}\right.$$

$$\left. + \frac{1}{r}\frac{\partial u_z}{\partial \theta}\frac{\partial u_z}{\partial z} - \frac{u_\theta}{r}\frac{\partial u_r}{\partial z} + \frac{u_r}{r}\frac{\partial u_\theta}{\partial z}\right),$$

$$E_{zz} = \frac{\partial u_z}{\partial z} + \frac{1}{2}\left[\left(\frac{\partial u_r}{\partial z}\right)^2 + \left(\frac{\partial u_\theta}{\partial z}\right)^2 + \left(\frac{\partial u_z}{\partial z}\right)^2\right].$$

3.19 The two-dimensional displacement field in a body is given by

$$u_1 = X_1\left[X_1^2 X_2 + c_1\left(2c_2^3 + 3c_2^2 X_2 - X_2^3\right)\right],$$

$$u_2 = -X_2\left(2c_2^3 + \frac{3}{2}c_2^2 X_2 - \frac{1}{4}X_2^3 + \frac{3}{2}c_1 X_1^2 X_2\right),$$

where c_1 and c_2 are constants. Find the linear and nonlinear Green–Lagrange strains.

3.20 Determine whether the following strain fields are possible in a continuous body:

$$\text{(a) } [\varepsilon] = \begin{bmatrix} (X_1^2 + X_2^2) & X_1 X_2 \\ X_1 X_2 & X_2^2 \end{bmatrix}, \quad \text{(b) } [\varepsilon] = \begin{bmatrix} X_3(X_1^2 + X_2^2) & 2X_1 X_2 X_3 & X_3 \\ 2X_1 X_2 X_3 & X_2^2 & X_1 \\ X_3 & X_1 & X_3^2 \end{bmatrix}.$$

3.21 Find the axial strain in the diagonal element of Problem 3.13, using (a) the basic definition of normal strain and (b) the strain transformation equations.

3.22 The biaxial state of strain at a point is given by $\varepsilon_{11} = 800 \times 10^{-6}$ in./in., $\varepsilon_{22} = 200 \times 10^{-6}$ in./in., $\varepsilon_{12} = 400 \times 10^{-6}$ in./in. Find the principal strains and their directions.

3.23 Consider the following infinitesimal strain field:

$$\varepsilon_{11} = c_1 X_2^2, \quad \varepsilon_{22} = c_1 X_1^2, \quad 2\varepsilon_{12} = c_2 X_1 X_2,$$

$$\varepsilon_{31} = \varepsilon_{32} = \varepsilon_{33} = 0,$$

where c_1 and c_2 are constants. Determine

(a) c_1 and c_2 such that there exists a continuous, single-valued displacement field that corresponds to this strain field,

(b) the most general form of the corresponding displacement field using c_1 and c_2 obtained in Part (a), and

(c) the constants of integration introduced in Part (b) for the boundary conditions $\mathbf{u} = \mathbf{0}$ and $\mathbf{\Omega} = 0$ at $\mathbf{X} = \mathbf{0}$.

3.24 Show that the invariants J_1, J_2, and J_3 of the Green–Lagrange strain tensor $\mathbf{E}$ can be expressed in terms of the principal values λ_i of $\mathbf{E}$ as

$$J_1 = \lambda_1 + \lambda_2 + \lambda_3, \quad J_2 = \lambda_1\lambda_2 + \lambda_2\lambda_3 + \lambda_3\lambda_1, \quad J_3 = \lambda_1\lambda_2\lambda_3.$$

Of course, the above result holds for any second-order tensor.

3.25 Show that

$$\frac{d}{dt}\left[(ds)^2\right] = 2d\mathbf{x} \cdot \mathbf{D} \cdot d\mathbf{x}.$$

3.26 Show that the spin tensor $\mathbf{W}$ can be written as

$$\mathbf{W} = \dot{\mathbf{R}} \cdot \mathbf{R}^{\mathrm{T}},$$

where $\mathbf{R}$ is the rotation tensor.

3.27 Verify that

$$\dot{\mathbf{v}} = \frac{\partial \mathbf{v}}{\partial t} + \frac{1}{2}\operatorname{grad}(\mathbf{v} \cdot \mathbf{v}) + 2\mathbf{W} \cdot \mathbf{v}$$

$$= \frac{\partial \mathbf{v}}{\partial t} + \frac{1}{2}\operatorname{grad}(\mathbf{v} \cdot \mathbf{v}) + 2\mathbf{w} \times \mathbf{v},$$

where $\mathbf{W}$ is the spin tensor and $\mathbf{w}$ is the vorticity vector [see Eq. (3.6.5)].

3.28 Evaluate the compatibility conditions $\nabla_0 \times (\nabla_0 \times \mathbf{E})^{\mathrm{T}} = \mathbf{0}$ in cylindrical coordinates.

3.29 Given the strain components

$$\varepsilon_{11} = f(X_2, X_3), \quad \varepsilon_{22} = \varepsilon_{33} = -\nu f(X_2, X_3), \quad \varepsilon_{12} = \varepsilon_{13} = \varepsilon_{23} = 0,$$

determine the form of $f(X_2, X_3)$ in order that the strain field is compatible.

3.30 Given the strain tensor $\mathbf{E} = E_{rr}\hat{\mathbf{e}}_r\hat{\mathbf{e}}_r + E_{\theta\theta}\hat{\mathbf{e}}_\theta\hat{\mathbf{e}}_\theta$ in an axisymmetric body (i.e., E_{rr} and $E_{\theta\theta}$ are functions of r and z only), determine the compatibility conditions on E_{rr} and $E_{\theta\theta}$.

3.31 Show that the components of the spin tensor $\mathbf{W}$ in cylindrical coordinate system are

$$W_{\theta r} = \frac{1}{2}\left(\frac{1}{r}\frac{\partial v_r}{\partial \theta} - \frac{v_\theta}{r} - \frac{\partial v_\theta}{\partial r}\right) = -W_{r\theta},$$

$$W_{zr} = \frac{1}{2}\left(\frac{\partial v_r}{\partial z} - \frac{\partial v_z}{\partial r}\right) = -W_{rz},$$

$$W_{\theta z} = \frac{1}{2}\left(\frac{1}{r}\frac{\partial v_z}{\partial \theta} - \frac{\partial v_\theta}{\partial z}\right) = -W_{z\theta}.$$

3.32 Establish the uniqueness of the decomposition $\mathbf{F} = \mathbf{R}\cdot\mathbf{U} = \mathbf{V}\cdot\mathbf{R}$. For example, if $\mathbf{F} = \mathbf{R}_1\cdot\mathbf{U}_1 = \mathbf{R}_2\cdot\mathbf{U}_2$, then show that $\mathbf{R}_1 = \mathbf{R}_2$ and $\mathbf{U}_1 = \mathbf{U}_2$.

3.33 Show that the eigenvalues of the left and right Cauchy stretch tensors $\mathbf{U}$ and $\mathbf{V}$ are the same and that the eigenvector of $\mathbf{V}$ is given by $\mathbf{R}\cdot\mathbf{n}$, where $\mathbf{n}$ is the eigenvector of $\mathbf{U}$.

3.34 Calculate the left and right Cauchy stretch tensors $\mathbf{U}$ and $\mathbf{V}$ associated with $\mathbf{F}$ of Problem 3.5.

3.35 Given that

$$[F] = \frac{1}{5}\begin{bmatrix} 2 & -5 \\ 11 & 2 \end{bmatrix},$$

determine the right and left stretch tensors.

3.36 Calculate the left and right Cauchy stretch tensors $\mathbf{U}$ and $\mathbf{V}$ associated with $\mathbf{F}$ of Problem 3.7 for the choice of $A = 2$ and $B = 0$.

Stress Measures

Most of the fundamental ideas of science are essentially simple, and may, as a rule, be expressed in a language comprehensible to everyone.

Albert Einstein

4.1 Introduction

In the beginning of Chapter 3, we have briefly discussed the need to study deformation and stresses in material systems that we may design for engineering applications. All materials have certain threshold to withstand forces, beyond which they "fail" to perform their intended function. The force per unit area, called *stress*, is a measure of the capacity of the material to carry loads, and all designs are based on the criterion that the materials used have the capacity to carry the working loads of the system. Thus, it is necessary to determine the state of stress in a material.

In the present chapter, we study the concept of stress and its various measures. For instance, stress can be measured per unit deformed area or undeformed area. As we shall see shortly, stress at a point in a three-dimensional continuum can be measured in terms of nine quantities, three per plane, on three mutually perpendicular planes at the point. These nine quantities may be viewed as the components of a second-order tensor, called *stress tensor*. Coordinate transformations and principal values associated with the stress tensor and stress equilibrium equations will also be discussed.

4.2 Cauchy Stress Tensor and Cauchy's Formula

First we introduce the true stress, that is, stress in the deformed configuration κ that is measured per unit area of the deformed configuration κ. The surface force acting on a small element of area in a continuous medium depends not only on the magnitude of the area but also upon the orientation of the area. It is customary to denote the direction of a plane area by means of a unit vector drawn normal to that plane, as discussed in Section 2.2.3. The direction of the normal is taken by convention as that in which a right-handed screw advances as it is rotated according

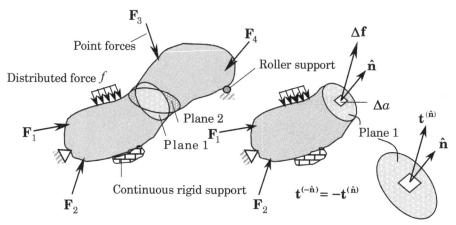

Figure 4.2.1. A material body supported at various points on the surface and subjected to a number forces. Cuts through a point by planes of different orientation. Stress vector on a plane normal to $\hat{\mathbf{n}}$.

to the sense of travel along the boundary curve or contour. Let the unit normal vector be denoted by $\hat{\mathbf{n}}$. Then the area is expressed as $\mathbf{A} = A\hat{\mathbf{n}}$.

If we denote by $d\mathbf{f}(\hat{\mathbf{n}})$ the force on a small area $\hat{\mathbf{n}} da$ located at the position $\mathbf{x}$, the *stress vector* can be defined, shown graphically in Figure 4.2.1, as

$$\mathbf{t}(\hat{\mathbf{n}}) = \lim_{\Delta a \to 0} \frac{\Delta \mathbf{f}(\hat{\mathbf{n}})}{\Delta a}. \tag{4.2.1}$$

We see that the stress vector is a point function of the unit normal $\hat{\mathbf{n}}$ which denotes the orientation of the surface Δa. The component of $\mathbf{t}$ that is in the direction of $\hat{\mathbf{n}}$ is called the *normal stress*. The component of $\mathbf{t}$ that is normal to $\hat{\mathbf{n}}$ is called the *shear* stress. Because of Newton's third law for action and reaction, we see that $\mathbf{t}(-\hat{\mathbf{n}}) = -\mathbf{t}(\hat{\mathbf{n}})$.

At a fixed point $\mathbf{x}$ for each given unit vector $\hat{\mathbf{n}}$, there is a stress vector $\mathbf{t}(\hat{\mathbf{n}})$ acting on the plane normal to $\hat{\mathbf{n}}$. Note that $\mathbf{t}(\hat{\mathbf{n}})$ is, in general, not in the direction of $\hat{\mathbf{n}}$. It is fruitful to establish a relationship between $\mathbf{t}$ and $\hat{\mathbf{n}}$.

To establish the relationship between $\mathbf{t}$ and $\hat{\mathbf{n}}$, we now set up an infinitesimal tetrahedron in Cartesian coordinates, as shown in Figure 4.2.2. If $-\mathbf{t}_1, -\mathbf{t}_2, -\mathbf{t}_3$, and $\mathbf{t}$ denote the stress vectors in the outward directions on the faces of the infinitesimal tetrahedron whose areas are $\Delta a_1, \Delta a_2, \Delta a_3$, and Δa, respectively, we have by Newton's second law for the mass inside the tetrahedron,

$$\mathbf{t}\Delta a - \mathbf{t}_1 \Delta a_1 - \mathbf{t}_2 \Delta a_2 - \mathbf{t}_3 \Delta a_3 + \rho \Delta v \mathbf{f} = \rho \Delta v \mathbf{a}, \tag{4.2.2}$$

where Δv is the volume of the tetrahedron, ρ the density, $\mathbf{f}$ the body force per unit mass, and $\mathbf{a}$ the acceleration. Since the total vector area of a closed surface is zero (use the gradient theorem), we have

$$\Delta a\, \hat{\mathbf{n}} - \Delta a_1 \hat{\mathbf{e}}_1 - \Delta a_2 \hat{\mathbf{e}}_2 - \Delta a_3 \hat{\mathbf{e}}_3 = \mathbf{0}. \tag{4.2.3}$$

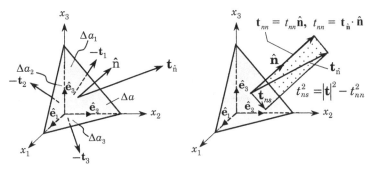

Figure 4.2.2. Tetrahedral element in Cartesian coordinates.

It follows that

$$\Delta a_1 = (\hat{\mathbf{n}} \cdot \hat{\mathbf{e}}_1)\Delta a, \quad \Delta a_2 = (\hat{\mathbf{n}} \cdot \hat{\mathbf{e}}_2)\Delta a, \quad \Delta a_3 = (\hat{\mathbf{n}} \cdot \hat{\mathbf{e}}_3)\Delta a. \tag{4.2.4}$$

The volume of the element Δv can be expressed as

$$\Delta v = \frac{\Delta h}{3}\Delta a, \tag{4.2.5}$$

where Δh is the perpendicular distance from the origin to the slant face.

Substitution of Eqs. (4.2.3) and (4.2.4) into Eq. (4.2.2) and dividing throughout by Δa, yields

$$\mathbf{t} = (\hat{\mathbf{n}} \cdot \hat{\mathbf{e}}_1)\mathbf{t}_1 + (\hat{\mathbf{n}} \cdot \hat{\mathbf{e}}_2)\mathbf{t}_2 + (\hat{\mathbf{n}} \cdot \hat{\mathbf{e}}_3)\mathbf{t}_3 + \rho\frac{\Delta h}{3}(\mathbf{a} - \mathbf{f}). \tag{4.2.6}$$

In the limit when the tetrahedron shrinks to a point, $\Delta h \to 0$, we are left with

$$\mathbf{t} = (\hat{\mathbf{n}} \cdot \hat{\mathbf{e}}_1)\mathbf{t}_1 + (\hat{\mathbf{n}} \cdot \hat{\mathbf{e}}_2)\mathbf{t}_2 + (\hat{\mathbf{n}} \cdot \hat{\mathbf{e}}_3)\mathbf{t}_3$$

$$= (\hat{\mathbf{n}} \cdot \hat{\mathbf{e}}_i)\mathbf{t}_i, \tag{4.2.7}$$

where the summation convention is used. It is now convenient to display the above equation as

$$\mathbf{t} = \hat{\mathbf{n}} \cdot (\hat{\mathbf{e}}_1\mathbf{t}_1 + \hat{\mathbf{e}}_2\mathbf{t}_2 + \hat{\mathbf{e}}_3\mathbf{t}_3). \tag{4.2.8}$$

The terms in the parenthesis are to be treated as a dyadic, called *stress dyadic* or *stress tensor* σ:

$$\sigma \equiv \hat{\mathbf{e}}_1\mathbf{t}_1 + \hat{\mathbf{e}}_2\mathbf{t}_2 + \hat{\mathbf{e}}_3\mathbf{t}_3. \tag{4.2.9}$$

The stress tensor is a property of the medium that is independent of the $\hat{\mathbf{n}}$. Thus, from Eqs. (4.2.8) and (4.2.9), we have

$$\mathbf{t}(\hat{\mathbf{n}}) = \hat{\mathbf{n}} \cdot \sigma = \sigma^{\mathrm{T}} \cdot \hat{\mathbf{n}}, \tag{4.2.10}$$

and the dependence of $\mathbf{t}$ on $\hat{\mathbf{n}}$ has been explicitly displayed.

The stress vector $\mathbf{t}$ represents the vectorial stress on a plane whose normal is $\hat{\mathbf{n}}$. Equation (4.2.10) is known as the *Cauchy stress formula*, and σ is termed the *Cauchy stress tensor*. Thus, the Cauchy stress tensor σ is defined to be the *current force* per

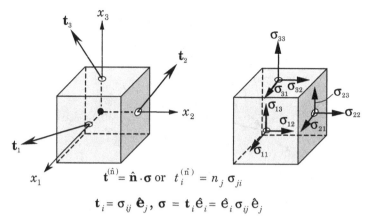

Figure 4.2.3. Display of stress components in Cartesian rectangular coordinates.

unit *deformed area*, $d\mathbf{f} = \mathbf{t}da = \boldsymbol{\sigma} \cdot d\mathbf{a}$, where Cauchy's formula, $\mathbf{t} = \hat{\mathbf{n}} \cdot \boldsymbol{\sigma} = \boldsymbol{\sigma}^{\mathrm{T}} \cdot \hat{\mathbf{n}}$, is used.

In Cartesian component form, the Cauchy formula in (4.2.10) can be written as $t_i = n_j \sigma_{ji}$. The matrix form of the Cauchy's formula (for computing purposes) in rectangular Cartesian system is given by

$$\begin{Bmatrix} t_1 \\ t_2 \\ t_3 \end{Bmatrix} = \begin{bmatrix} \sigma_{11} & \sigma_{21} & \sigma_{31} \\ \sigma_{12} & \sigma_{22} & \sigma_{32} \\ \sigma_{13} & \sigma_{23} & \sigma_{33} \end{bmatrix} \begin{Bmatrix} n_1 \\ n_2 \\ n_3 \end{Bmatrix}. \tag{4.2.11}$$

It is useful to resolve the stress vectors $\mathbf{t}_1$, $\mathbf{t}_2$, and $\mathbf{t}_3$ into their orthogonal components in a rectangular Cartesian system

$$\mathbf{t}_i = \sigma_{i1}\hat{\mathbf{e}}_1 + \sigma_{i2}\hat{\mathbf{e}}_2 + \sigma_{i3}\hat{\mathbf{e}}_3 = \sigma_{ij}\hat{\mathbf{e}}_j \tag{4.2.12}$$

for $i = 1, 2, 3$. Hence, the stress tensor can be expressed in the Cartesian component form as

$$\boldsymbol{\sigma} = \hat{\mathbf{e}}_i \mathbf{t}_i = \sigma_{ij}\hat{\mathbf{e}}_i\hat{\mathbf{e}}_j. \tag{4.2.13}$$

The component σ_{ij} represents the stress (force per unit area) on a plane perpendicular to the x_i coordinate and in the x_j coordinate direction, as shown in Figure 4.2.3.

In cylindrical coordinate system, for example, the dyadic form of the stress tensor is given by

$$\boldsymbol{\sigma} = \sigma_{rr}\hat{\mathbf{e}}_r\hat{\mathbf{e}}_r + \sigma_{r\theta}\left(\hat{\mathbf{e}}_r\hat{\mathbf{e}}_\theta + \hat{\mathbf{e}}_\theta\hat{\mathbf{e}}_r\right) + \sigma_{rz}\left(\hat{\mathbf{e}}_r\hat{\mathbf{e}}_z + \hat{\mathbf{e}}_z\hat{\mathbf{e}}_r\right)$$

$$+ \sigma_{\theta\theta}\hat{\mathbf{e}}_\theta\hat{\mathbf{e}}_\theta + \sigma_{\theta z}\left(\hat{\mathbf{e}}_\theta\hat{\mathbf{e}}_z + \hat{\mathbf{e}}_z\hat{\mathbf{e}}_\theta\right) + \sigma_{zz}\hat{\mathbf{e}}_z\hat{\mathbf{e}}_z. \tag{4.2.14}$$

By Example 2.2.1, the stress vector $\mathbf{t}$ can be represented as the sum of vectors along and perpendicular to the unit normal vector $\hat{\mathbf{n}}$

$$\mathbf{t} = (\mathbf{t} \cdot \hat{\mathbf{n}})\hat{\mathbf{n}} + \hat{\mathbf{n}} \times (\mathbf{t} \times \hat{\mathbf{n}}). \tag{4.2.15}$$

The magnitudes of the component of the stress vector $\mathbf{t}$ normal to the plane is given by

$$t_{nn} = \mathbf{t} \cdot \hat{\mathbf{n}} = t_i n_i = n_j \sigma_{ji} n_i, \tag{4.2.16}$$

and the component of $\mathbf{t}$ perpendicular to $\hat{\mathbf{n}}$, as depicted in Figure 4.2.2, is

$$t_{ns} = \sqrt{|\mathbf{t}|^2 - t_{nn}^2}. \tag{4.2.17}$$

The tangential component lies in the $\hat{\mathbf{n}}$-$\mathbf{t}$ plane but perpendicular to $\hat{\mathbf{n}}$. The next example illustrates the ideas presented here.

EXAMPLE 4.2.1: With reference to a rectangular Cartesian system (x_1, x_2, x_3), the components of the stress dyadic at a certain point of a continuous medium $\mathcal{B}$ are given by

$$[\sigma] = \begin{bmatrix} 200 & 400 & 300 \\ 400 & 0 & 0 \\ 300 & 0 & -100 \end{bmatrix} \text{ psi.}$$

Determine the stress vector $\mathbf{t}$ and its normal and tangential components at the point on the plane, $\phi(x_1, x_2) \equiv x_1 + 2x_2 + 2x_3 = $ constant, passing through the point.

SOLUTION: First, we should find the unit normal to the plane on which we are required to find the stress vector. The unit normal to the plane defined by $\phi(x_1, x_2, x_3) = $ constant is given by

$$\hat{\mathbf{n}} = \frac{\nabla\phi}{|\nabla\phi|} = \frac{1}{3}(\hat{\mathbf{e}}_1 + 2\hat{\mathbf{e}}_2 + 2\hat{\mathbf{e}}_3).$$

The components of the stress vector are

$$\begin{Bmatrix} t_1 \\ t_2 \\ t_3 \end{Bmatrix} = \begin{bmatrix} 200 & 400 & 300 \\ 400 & 0 & 0 \\ 300 & 0 & -100 \end{bmatrix} \frac{1}{3} \begin{Bmatrix} 1 \\ 2 \\ 2 \end{Bmatrix} = \frac{1}{3} \begin{Bmatrix} 1600 \\ 400 \\ 100 \end{Bmatrix} \text{ psi,}$$

or

$$\mathbf{t}(\hat{\mathbf{n}}) = \frac{1}{3}(1600\hat{\mathbf{e}}_1 + 400\hat{\mathbf{e}}_2 + 100\hat{\mathbf{e}}_3) \text{ psi.}$$

The normal component t_{nn} of the stress vector $\mathbf{t}$ on the plane is given by

$$t_{nn} = \mathbf{t}(\hat{\mathbf{n}}) \cdot \hat{\mathbf{n}} = \frac{2600}{9} \text{ psi,}$$

and the tangential component is given by (the Pythagorean theorem)

$$t_{ns} = \sqrt{|\mathbf{t}|^2 - t_{nn}^2} = \frac{10^2}{9}\sqrt{(256 + 16 + 1)9 - 26 \times 26} \text{ psi} = 468.9 \text{ psi.}$$

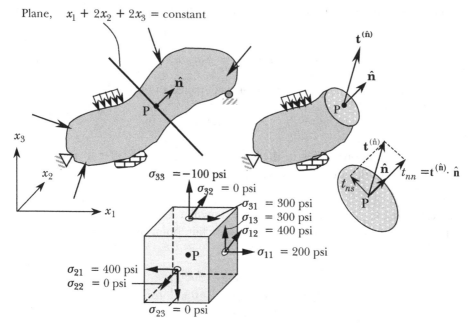

Dimensionless point cube at P

Figure 4.2.4. Stress vector and its normal and shear components.

4.3 Transformations of Stress Components and Principal Stresses

4.3.1 Transformation of Stress Components

Since the Cauchy stress is a second-order tensor, we can define its invariants, transformation laws, and eigenvalues and eigenvectors. The invariants of stress tensor σ are defined by [see Eq. (2.5.14)]

$$I_1 = \operatorname{tr} \sigma, \quad I_2 = \frac{1}{2}\left[(\operatorname{tr} \sigma)^2 - \operatorname{tr}(\sigma^2)\right], \quad I_3 = \det \sigma. \tag{4.3.1}$$

Further, the components of a stress tensor σ in one rectangular Cartesian coordinate system are related to the components in another rectangular Cartesian system according to the transformation law in Eq. (2.5.17):

$$\bar{\sigma}_{ij} = \ell_{ik}\,\ell_{j\ell}\,\sigma_{k\ell} \ \text{ or } \ [\bar{\sigma}] = [L][\sigma][L]^{\mathrm{T}}, \tag{4.3.2}$$

where ℓ_{ij} are the direction cosines

$$\ell_{ij} = \hat{\bar{\mathbf{e}}}_i \cdot \hat{\mathbf{e}}_j. \tag{4.3.3}$$

The principal stresses (eigenvalues of a stress tensor) and principal directions will be discussed in the next section.

The next two examples show how the stress transformation equations in Eq. (4.3.2) can be derived for a specific problem and used in the calculation of stresses in the new coordinate system.

EXAMPLE 4.3.1: Consider the unidirectional fiber-reinforced composite layer shown in Figure 4.3.1. If the rectangular coordinates (x, y, z) are taken such that the z-coordinate is normal to the plane of the layer and the x- and y-coordinates are in the plane of the layer, as shown in Figure 4.3.1. Now suppose we define another rectangular coordinate system (x_1, x_2, x_3) such that the x_3-coordinate coincides with the z-coordinate, but the x_1- and x_2-coordinates are oriented at an angle of θ to the x- and y-coordinates, respectively, so that the x_1-axis is along the fiber direction. Determine the transformation relationships between the stress components $\sigma_{xx}, \sigma_{yy}, \sigma_{xy}, \cdots$ referred to the (x, y, z) system and σ_{11}, $\sigma_{22}, \sigma_{12}, \cdots$ referred to the coordinates system (x_1, x_2, x_3).

SOLUTION: First, we note that the $x_1 x_2$-plane and the xy-plane are parallel, but rotated by an angle θ counterclockwise (when looking down on the lamina) from the x-axis about the z- or x_3-axis. The coordinates of a material point in the two coordinate systems are related as follows ($z = x_3$):

$$\begin{Bmatrix} x_1 \\ x_2 \\ x_3 \end{Bmatrix} = \begin{bmatrix} \cos\theta & \sin\theta & 0 \\ -\sin\theta & \cos\theta & 0 \\ 0 & 0 & 1 \end{bmatrix} \begin{Bmatrix} x \\ y \\ z \end{Bmatrix} = [L] \begin{Bmatrix} x \\ y \\ z \end{Bmatrix}. \tag{4.3.4}$$

The inverse of Eq. (4.3.4) is

$$\begin{Bmatrix} x \\ y \\ z \end{Bmatrix} = \begin{bmatrix} \cos\theta & -\sin\theta & 0 \\ \sin\theta & \cos\theta & 0 \\ 0 & 0 & 1 \end{bmatrix} \begin{Bmatrix} x_1 \\ x_2 \\ x_3 \end{Bmatrix} = [L]^{\mathrm{T}} \begin{Bmatrix} x_1 \\ x_2 \\ x_3 \end{Bmatrix}. \tag{4.3.5}$$

The inverse of $[L]$ is equal to its transpose: $[L]^{-1} = [L]^{\mathrm{T}}$. That is, $\mathbf{L}$ is an orthogonal tensor.

Next, we establish the relationship between the components of stress in the (x, y, z) and (x_1, x_2, x_3) coordinate systems. Let $\bar{\sigma}_{ij}$ be the components of the stress tensor σ in the coordinates (x, y, z), that is, $\sigma_{xx} = \bar{\sigma}_{11}$, $\sigma_{xy} = \bar{\sigma}_{12}$, and so on, and ℓ_{ij} are the direction cosines defined by

$$\ell_{ij} = \hat{\mathbf{e}}_i \cdot \hat{\mathbf{e}}_j,$$

Figure 4.3.1. Stress components in a fiber-reinforced layer referred to different rectangular Cartesian coordinate systems: (x, y, z) are parallel to the sides of the rectangular lamina, while (x_1, x_2, x_3) are taken along and transverse to the fiber.

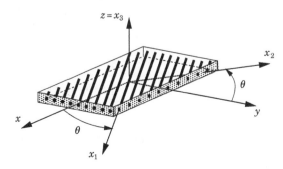

where $\hat{\bar{e}}_i$ and $\hat{e}_i$ are the orthonormal basis vectors in coordinate systems $(x = \bar{x}_1, y = \bar{x}_2, z = \bar{x}_3)$ and (x_1, x_2, x_3), respectively. Then using Eq. (4.3.2), we can write

$$[\bar{\sigma}] = [L][\sigma][L]^T, \quad [\sigma] = [L]^T[\bar{\sigma}][L], \tag{4.3.6}$$

where $[L]$ is the 3×3 matrix of direction cosines defined in (4.3.4). Carrying out the indicated matrix multiplications in Eq. (4.3.6) and rearranging the equations in terms of the column vectors of stress components, we obtain

$$\begin{Bmatrix} \sigma_{xx} \\ \sigma_{yy} \\ \sigma_{zz} \\ \sigma_{yz} \\ \sigma_{xz} \\ \sigma_{xy} \end{Bmatrix} = \begin{bmatrix} \cos^2\theta & \sin^2\theta & 0 & 0 & 0 & -\sin 2\theta \\ \sin^2\theta & \cos^2\theta & 0 & 0 & 0 & \sin 2\theta \\ 0 & 0 & 1 & 0 & 0 & 0 \\ 0 & 0 & 0 & \cos\theta & \sin\theta & 0 \\ 0 & 0 & 0 & -\sin\theta & \cos\theta & 0 \\ \frac{1}{2}\sin 2\theta & -\frac{1}{2}\sin 2\theta & 0 & 0 & 0 & \cos 2\theta \end{bmatrix} \begin{Bmatrix} \sigma_{11} \\ \sigma_{22} \\ \sigma_{33} \\ \sigma_{23} \\ \sigma_{13} \\ \sigma_{12} \end{Bmatrix}, \tag{4.3.7}$$

and

$$\begin{Bmatrix} \sigma_{11} \\ \sigma_{22} \\ \sigma_{33} \\ \sigma_{23} \\ \sigma_{13} \\ \sigma_{12} \end{Bmatrix} = \begin{bmatrix} \cos^2\theta & \sin^2\theta & 0 & 0 & 0 & \sin 2\theta \\ \sin^2\theta & \cos^2\theta & 0 & 0 & 0 & -\sin 2\theta \\ 0 & 0 & 1 & 0 & 0 & 0 \\ 0 & 0 & 0 & \cos\theta & -\sin\theta & 0 \\ 0 & 0 & 0 & \sin\theta & \cos\theta & 0 \\ -\frac{1}{2}\sin 2\theta & \frac{1}{2}\sin 2\theta & 0 & 0 & 0 & \cos 2\theta \end{bmatrix} \begin{Bmatrix} \sigma_{xx} \\ \sigma_{yy} \\ \sigma_{zz} \\ \sigma_{yz} \\ \sigma_{xz} \\ \sigma_{xy} \end{Bmatrix}. \tag{4.3.8}$$

The result in Eq. (4.3.8) can also be obtained from Eq. (4.3.7) by replacing θ with $-\theta$. The stress transformation relations can also be derived using equilibrium of forces. (Problem 4.7 illustrates this.)

EXAMPLE 4.3.2: Consider a thin, closed, filament-wound cylindrical pressure vessel shown in Figure 4.3.2. The vessel is of 63.5 cm (25 in.) internal diameter, and it is pressurized to 1.379 MPa (200 psi). If the filament winding angle is $\theta = 53.125°$ from the longitudinal axis of the pressure vessel, determine the shear and normal forces per unit length of filament winding. Assume that the material used is graphite epoxy with the following material properties [see Reddy (2004)]:

$$E_1 = 140 \text{ MPa (20.3 Msi)}, \quad E_2 = 10 \text{ MPa (1.45 Msi)},$$
$$G_{12} = 7 \text{ MPa (1.02 Msi)}, \quad \nu_{12} = 0.3, \tag{4.3.9}$$

where MPa denotes mega (10^6) Pascal (Pa) and Pa $= \text{N/m}^2$ (1 psi $= 6,894.76$ Pa). The material properties are not needed to answer the question.

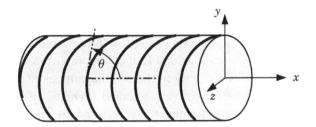

Figure 4.3.2. A filament-wound cylindrical
pressure vessel.

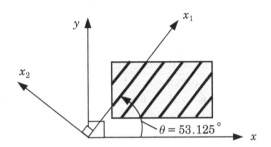

SOLUTION: First, we compute the stresses in the pressure vessel using the for-
mulas from a course on elementary mechanics of materials. The longitudinal
(σ_{xx}) and circumferential (σ_{yy}) stresses are given by

$$\sigma_{xx} = \frac{pD_i}{4h}, \quad \sigma_{yy} = \frac{pD_i}{2h}, \tag{4.3.10}$$

where p is internal pressure, D_i is internal diameter, and h is thickness of the
pressure vessel. The stresses are independent of the material properties and
only depend on the geometry and load (pressure). Using the values of various
parameters, we calculate the stresses as

$$\sigma_{xx} = \frac{1.379 \times 0.635}{4h} = \frac{0.2189}{h} \text{ MPa}, \quad \sigma_{yy} = \frac{1.379 \times 0.635}{2h} = \frac{0.4378}{h} \text{ MPa}.$$

The shear stress σ_{xy} is zero.

Next, we determine the shear stress σ_{12} along the fiber and the normal stress
σ_{11} in the fiber using the transformation equations (4.3.8)

$$\sigma_{11} = \frac{0.2189}{h}(0.6)^2 + \frac{0.4378}{h}(0.8)^2 = \frac{0.3590}{h} \text{ MPa},$$

$$\sigma_{22} = \frac{0.2189}{h}(0.8)^2 + \frac{0.4378}{h}(0.6)^2 = \frac{0.2977}{h} \text{ MPa},$$

$$\sigma_{12} = \left(\frac{0.4378}{h} - \frac{0.2189}{h}\right) \times 0.6 \times 0.8 = \frac{0.1051}{h} \text{ MPa}.$$

Thus, the normal and shear forces per unit length along the fiber-matrix interface are $F_{22} = 0.2977$ MN and $F_{12} = 0.1051$ MN, whereas the force per unit
length in the fiber direction is $F_{11} = 0.359$ MN.

4.3.2 Principal Stresses and Principal Planes

For a given state of stress, the determination of maximum normal stresses and shear stresses at a point is of considerable interest in the design of structures because failures occur when the the magnitudes of stresses exceed the allowable (normal or shear) stress values, called *strengths*, of the material. In this regard, it is of interest to determine the values and the planes on which the stresses are the maximum. Thus, we must determine the eigenvalues and eigenvectors associated with the stress tensor (see Section 2.5.5 for more detailed discussion of the eigenvalues and eigenvectors of a tensor).

It is clear from Figures 4.2.2 and 4.2.3 that the normal component of a stress vector is the largest when $\mathbf{t}$ is parallel to the normal $\hat{\mathbf{n}}$. If we denote the value of the normal stress by λ, then we can write $\mathbf{t} = \lambda\hat{\mathbf{n}}$. However, by Cauchy's formula, $\mathbf{t} = \hat{\mathbf{n}} \cdot \boldsymbol{\sigma} = \boldsymbol{\sigma} \cdot \hat{\mathbf{n}}$ (due to the symmetry of the stress tensor). Thus, we have

$$\mathbf{t} = \boldsymbol{\sigma} \cdot \hat{\mathbf{n}} = \lambda\hat{\mathbf{n}} \quad \text{or} \quad (\boldsymbol{\sigma} - \lambda\mathbf{I}) \cdot \hat{\mathbf{n}} = \mathbf{0}. \tag{4.3.11}$$

Because this is a homogeneous set of equations for the components of vector $\hat{\mathbf{n}}$, a nontrivial solution will not exist unless the determinant of the matrix for $\boldsymbol{\sigma} - \lambda\mathbf{I}$ vanishes. The vanishing of this determinant yields a cubic equation for λ, called the *characteristic equation,* the solution of which yields three values of λ. The eigenvalues λ of $\boldsymbol{\sigma}$ are called the *principal stresses* and the associated eigenvectors are called the *principal planes.* That is, for a given state of stress at a given point in the body $\mathcal{B}$, there exists a set of planes $\hat{\mathbf{n}}$ on which the stress vector is normal to the planes (i.e., there is no shear component on the planes).

The computation of the eigenvalues of the stress tensor is made easy by seeking the eigenvalues of the deviatoric stress tensor [see Eq. (2.5.34)]. Let σ_m denote the mean normal stress

$$\sigma_m = \frac{1}{3}\text{tr }\boldsymbol{\sigma} = \frac{1}{3}I_1 \quad \left(\sigma_m = \frac{1}{3}\sigma_{kk}\right). \tag{4.3.12}$$

Then the stress tensor can be expressed as the sum of *spherical* or *hydrostatic* stress tensor and *deviatoric* stress tensor

$$\boldsymbol{\sigma} = \sigma\mathbf{I} + \boldsymbol{\sigma}'. \tag{4.3.13}$$

Thus, the deviatoric stress tensor is defined by

$$\boldsymbol{\sigma}' = \boldsymbol{\sigma} - \frac{1}{3}I_1\mathbf{I} \quad \left(\sigma_{ij}' = \sigma_{ij} - \frac{1}{3}\delta_{ij}\sigma_{kk}\right). \tag{4.3.14}$$

The invariants I_1', I_2', and I_3' of the deviatoric stress tensor are

$$I_1' = 0, \quad I_2' = \frac{1}{2}\sigma_{ij}'\sigma_{ij}', \quad I_3' = \frac{1}{3}\sigma_{ij}'\sigma_{jk}'\sigma_{ki}'. \tag{4.3.15}$$

The deviatoric stress invariants are particularly important in the determination of the principal stresses as discussed in Section 2.5.5. The next example illustrates the computation of principal stresses and principal planes.

EXAMPLE 4.3.3: The components of a stress dyadic at a point, referred to the (x_1, x_2, x_3) system, are

$$[\sigma] = \begin{bmatrix} 12 & 9 & 0 \\ 9 & -12 & 0 \\ 0 & 0 & 6 \end{bmatrix} \text{MPa}.$$

Find the principal stresses and the principal plane associated with the maximum stress.

SOLUTION: Setting $|\sigma - \lambda \mathbf{I}| = 0$, we obtain

$$(6 - \lambda)[(12 - \lambda)(-12 - \lambda) - 81] = 0 \quad \rightarrow \quad [-(144 - \lambda^2) - 81](6 - \lambda) = 0.$$

Clearly, $\lambda_2 = 6$ is an eigenvalue of the matrix. The remaining two eigenvalues are obtained from $\lambda^2 - 225 = 0 \rightarrow \lambda_1 = 15$ and $\lambda_3 = -15$; thus the principal stresses are

$$\sigma_1 = \lambda_1 = 15 \, \text{MPa}, \quad \sigma_2 = \lambda_2 = 6 \, \text{MPa}, \quad \sigma_3 = \lambda_3 = -15 \, \text{MPa}.$$

The plane associated with the maximum principal stress $\lambda_1 = 15$ MPa can be calculated from

$$\begin{bmatrix} 12 - 15 & 9 & 0 \\ 9 & -12 - 15 & 0 \\ 0 & 0 & 6 - 15 \end{bmatrix} \begin{Bmatrix} n_1 \\ n_2 \\ n_3 \end{Bmatrix} = \begin{Bmatrix} 0 \\ 0 \\ 0 \end{Bmatrix},$$

which gives

$$-3n_1 + 9n_2 = 0, \quad 9n_1 - 27n_2 = 0, \quad -9n_3 = 0 \quad \rightarrow \quad n_3 = 0, \quad n_1 = 3n_2$$

or

$$\mathbf{n}^{(1)} = 3\hat{\mathbf{e}}_1 + \hat{\mathbf{e}}_2 \text{ or } \hat{\mathbf{n}}^{(1)} = \frac{1}{\sqrt{10}}(3\hat{\mathbf{e}}_1 + \hat{\mathbf{e}}_2).$$

The eigenvector associated with $\lambda_2 = 6$ MPa is $\mathbf{n}^{(2)} = \hat{\mathbf{e}}_3$. Finally, the eigenvector associated with $\lambda_3 = -15$ MPa is

$$\mathbf{n}^{(3)} = \pm(\hat{\mathbf{e}}_1 - 3\hat{\mathbf{e}}_2) \text{ or } \hat{\mathbf{n}}^{(3)} = \pm\frac{1}{\sqrt{10}}(\hat{\mathbf{e}}_1 - 3\hat{\mathbf{e}}_2).$$

The principal plane 1 is depicted in Figure 4.3.3.

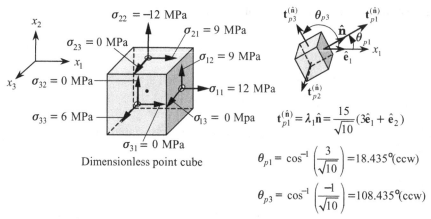

Figure 4.3.3. Stresses on a point cube at the point of interest and orientation of the first principal plane.

4.3.3 Maximum Shear Stress

In the previous section, we studied the procedure to determine the maximum normal stresses at a point. The eigenvalues of the stress tensor at the point are the maximum normal stresses on three perpendicular planes (whose normals are the eigenvectors), and the largest of these three stresses is the true maximum normal stress. Recall that the shear stresses are zero on the principal planes. In this section, we wish to determine the maximum shear stresses and their planes.

Let λ_1, λ_2, and λ_3 denote the principal (normal) stresses and $\hat{\mathbf{n}}$ be an arbitrary unit normal vector. Then the stress vector is $\mathbf{t} = \lambda_1 n_1 \hat{\mathbf{e}}_1 + \lambda_2 n_2 \hat{\mathbf{e}}_2 + \lambda_3 n_3 \hat{\mathbf{e}}_3$ and $t_{nn} = t_i n_i = \lambda_1 n_1^2 + \lambda_2 n_2^2 + \lambda_3 n_3^2$. The square of the magnitude of the shear stress on the plane with unit normal $\hat{\mathbf{n}}$ is given by Eq. (4.2.16)

$$t_{ns}^2(\hat{\mathbf{n}}) = |\mathbf{t}|^2 - t_{nn}^2 = \lambda_1^2 n_1^2 + \lambda_2^2 n_2^2 + \lambda_3^2 n_3^2 - \left(\lambda_1 n_1^2 + \lambda_2 n_2^2 + \lambda_3 n_3^2\right)^2. \tag{4.3.16}$$

We wish to determine the plane $\hat{\mathbf{n}}$ on which t_{ns} is the maximum. Thus, we seek the maximum of the function $F(n_1, n_2, n_3) = t_{ns}^2(n_1, n_2, n_3)$ subject to the constraint

$$n_1^2 + n_2^2 + n_3^2 - 1 = 0. \tag{4.3.17}$$

One way to determine the extremum of a function subjected to a constraint is to use the Lagrange multiplier method, in which we seek the stationary value of the modified function

$$F_L(n_1, n_2, n_3) = t_{ns}^2(n_1, n_2, n_3) + \lambda \left(n_1^2 + n_2^2 + n_3^2 - 1\right), \tag{4.3.18}$$

where λ is the Lagrange multiplier, which is to be determined along with n_1, n_2, and n_3. The necessary condition for the stationarity of F_L is

$$0 = dF_L = \frac{\partial F_L}{\partial n_1} dn_1 + \frac{\partial F_L}{\partial n_2} dn_2 + \frac{\partial F_L}{\partial n_3} dn_3 + \frac{\partial F_L}{\partial \lambda} d\lambda,$$

or, because the increments dn_1, dn_2, dn_3, and $d\lambda$ are linearly independent of each other, we have

$$\frac{\partial F_L}{\partial n_1} = 0, \quad \frac{\partial F_L}{\partial n_2} = 0, \quad \frac{\partial F_L}{\partial n_3} = 0, \quad \frac{\partial F_L}{\partial \lambda} = 0. \tag{4.3.19}$$

The last of the four relations in Eq. (4.3.19) is the same as that in Eq. (4.3.17). The remaining three equations in Eq. (4.3.19) yield the following two sets of solutions (not derived here)

$$(n_1, n_2, n_3) = (1, 0, 0), \quad (0, 1, 0), \quad (0, 0, 1), \tag{4.3.20}$$

$$(n_1, n_2, n_3) = \left(\frac{1}{\sqrt{2}}, \pm\frac{1}{\sqrt{2}}, 0\right), \quad \left(\frac{1}{\sqrt{2}}, 0, \pm\frac{1}{\sqrt{2}}\right), \quad \left(0, \frac{1}{\sqrt{2}}, \pm\frac{1}{\sqrt{2}}\right). \tag{4.3.21}$$

The first set of solutions corresponds to the principal planes, on which the shear stresses are the minimum, namely, zero. The second set of solutions corresponds to the maximum shear stress planes. The maximum shear stresses on the planes are given by

$$t_{ns}^2 = \frac{1}{4}(\lambda_1 - \lambda_2)^2 \text{ for } \hat{\mathbf{n}} = \frac{1}{\sqrt{2}}(\hat{\mathbf{e}}_1 \pm \hat{\mathbf{e}}_2),$$

$$t_{ns}^2 = \frac{1}{4}(\lambda_1 - \lambda_3)^2 \text{ for } \hat{\mathbf{n}} = \frac{1}{\sqrt{2}}(\hat{\mathbf{e}}_1 \pm \hat{\mathbf{e}}_3), \tag{4.3.22}$$

$$t_{ns}^2 = \frac{1}{4}(\lambda_2 - \lambda_3)^2 \text{ for } \hat{\mathbf{n}} = \frac{1}{\sqrt{2}}(\hat{\mathbf{e}}_2 \pm \hat{\mathbf{e}}_3).$$

The largest shear stress is given by the largest of the three values given in Eq. (4.3.22). Thus, we have

$$(t_{ns})_{\max} = \frac{1}{2}(\lambda_{\max} - \lambda_{\min}), \tag{4.3.23}$$

where $\lambda_{\max}$ and $\lambda_{\min}$ are the maximum and minimum principal values of stress, respectively. Clearly, the plane of the maximum shear stress lies between the planes of the maximum and minimum principal stresses (i.e., oriented at $\pm 45°$ to both planes).

EXAMPLE 4.3.4: For the state of stress given in Example 4.3.3, determine the maximum shear stress.

SOLUTION: From Example 4.3.3, the principal stresses are (ordered from the minimum to the maximum):

$$\lambda_1 = -15 \text{ MPa}, \quad \lambda_2 = 6 \text{ MPa}, \quad \lambda_3 = 15 \text{ MPa}.$$

Hence, the maximum shear stress is given by

$$(t_{ns})_{\max} = \frac{1}{2}(\lambda_3 - \lambda_1) = \frac{1}{2}[15 - (-15)] = 15 \text{ MPa}.$$

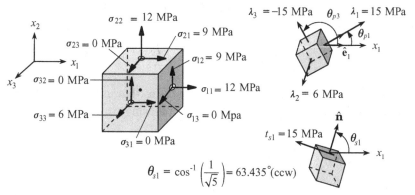

Figure 4.3.4. Stresses on a point cube at the point of interest and orientation of the maximum shear stress plane.

The planes of the maximum principal stress ($\lambda_1 = 15$ MPa) and the minimum principal stress ($\lambda_3 = -15$ MPa) are given by their normal vectors (not unit vectors):

$$\mathbf{n}^{(1)} = 3\hat{\mathbf{e}}_1 + \hat{\mathbf{e}}_2, \quad \mathbf{n}^{(3)} = \hat{\mathbf{e}}_1 - 3\hat{\mathbf{e}}_2.$$

Then the plane of the maximum shear stress is given by the vector

$$\mathbf{n}_s = \left(\mathbf{n}^{(1)} - \mathbf{n}^{(3)}\right) = 2\hat{\mathbf{e}}_1 + 4\hat{\mathbf{e}}_2 \ \text{ or } \ \hat{\mathbf{n}}_s = \frac{1}{\sqrt{5}}(\hat{\mathbf{e}}_1 - 2\hat{\mathbf{e}}_2).$$

4.4 Other Stress Measures

4.4.1 Preliminary Comments

The Cauchy stress tensor is the most natural and physical measure of the state of stress at a point in the deformed configuration and measured per unit area of the deformed configuration. It is the quantity most commonly used in spatial description of problems in fluid mechanics. The equations of motion or equilibrium of a material body in the Lagrange description must be derived for the deformed configuration of the body at time t. However, since the geometry of the deformed configuration is not known, the equations must be written in terms of the known reference configuration. In doing so, we introduce various measures of stress. They emerge in a natural way as we transform volumes and areas from the deformed configuration to undeformed (or reference) configuration. These measures are purely mathematical but facilitate analysis. These are discussed next.

4.4.2 First Piola–Kirchhoff Stress Tensor

Consider a continuum $\mathcal{B}$ subjected to a deformation mapping χ that results in the deformed configuration κ, as shown in Figure 4.4.1. Let the force vector on an

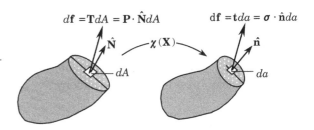

Figure 4.4.1. Definition of the first Piola–
Kirchhoff stress tensor.

elemental area da with normal $\hat{\mathbf{n}}$ in the deformed configuration be $d\mathbf{f}$. Suppose that
the area element in the undeformed configuration that corresponds to da is dA. The
force $d\mathbf{f}$ can be expressed in terms of a stress vector $\mathbf{t}$ times the deformed area da as

$$d\mathbf{f} = \mathbf{t}^{(\mathbf{n})} \, da.$$

We define a stress vector $\mathbf{T}^{(\mathbf{N})}$ over the area element dA with normal $\mathbf{N}$ in the undeformed configuration such that it results in the same total force

$$d\mathbf{f} = \mathbf{t}^{(\mathbf{n})} \, da = \mathbf{T}^{(\mathbf{N})} dA. \tag{4.4.1}$$

Clearly, both stress vectors have the same direction but different magnitudes owing
to the different areas. The stress vector $\mathbf{T}^{(\mathbf{N})}$ is measured per unit undeformed area,
while the stress vector $\mathbf{t}^{(\mathbf{n})}$ is measured per unit deformed area.

From Cauchy's formula, we have $\mathbf{t}^{(\mathbf{n})} = \boldsymbol{\sigma} \cdot \hat{\mathbf{n}}$, where $\boldsymbol{\sigma}$ is the Cauchy stress
tensor. In a similar fashion, we introduce a stress tensor $\mathbf{P}$, called the *first Piola–
Kirchhoff stress tensor*, such that $\mathbf{T}^{(\mathbf{N})} = \mathbf{P} \cdot \hat{\mathbf{N}}$. Then using Eq. (4.4.1) we can write

$$\boldsymbol{\sigma} \cdot \hat{\mathbf{n}} \, da = \mathbf{P} \cdot \hat{\mathbf{N}} \, dA \quad \text{or} \quad \boldsymbol{\sigma} \cdot d\mathbf{a} = \mathbf{P} \cdot d\mathbf{A}, \tag{4.4.2}$$

where

$$d\mathbf{a} = da \, \hat{\mathbf{n}}, \quad d\mathbf{A} = dA \, \hat{\mathbf{N}}. \tag{4.4.3}$$

The first Piola–Kirchhoff stress tensor, also referred to as the *nominal stress tensor*,
or *Lagrangian stress tensor*, gives the *current force* per unit *undeformed area*.

The stress vector $\mathbf{T}^{(\mathbf{N})}$ is known as the *psuedo stress vector* associated with the
first Piola–Kirchhoff stress tensor. The tensor Cartesian component representation
of $\mathbf{P}$ is given by

$$\mathbf{P} = P_{iI} \, \hat{\mathbf{e}}_i \, \hat{\mathbf{E}}_I. \tag{4.4.4}$$

Clearly, the first Piola–Kirchhoff stress tensor is a mixed tensor.

To derive the relation between first Piola–Kirchhoff stress tensor and the
Cauchy stress tensor, we recall from Nanson's formula in Eq. (3.3.23) the relation
between the area elements in the deformed and undeformed configurations. From
Eqs. (4.4.2) and (3.3.23), we obtain

$$\mathbf{P} \cdot d\mathbf{A} = \boldsymbol{\sigma} \cdot d\mathbf{a}$$

$$= J\boldsymbol{\sigma} \cdot \mathbf{F}^{-\mathrm{T}} \cdot d\mathbf{A}, \tag{4.4.5}$$

where J is the Jacobian. Finally, we arrive at the relation

$$\mathbf{P} = J\boldsymbol{\sigma} \cdot \mathbf{F}^{-\text{T}}. \tag{4.4.6}$$

In general, the first Piola–Kirchhoff stress tensor $\mathbf{P}$ is unsymmetric even when the Cauchy stress tensor $\boldsymbol{\sigma}$ is symmetric.

4.4.3 Second Piola–Kirchhoff Stress Tensor

The *second Piola–Kirchhoff stress tensor* $\mathbf{S}$, which is used in the study of large deformation analysis, is introduced as the stress tensor associated with the force $d\mathcal{F}$ in the undeformed elemental area $d\mathbf{A}$ that corresponds to the the force $d\mathbf{f}$ on the deformed elemental area $d\mathbf{a}$

$$d\mathcal{F} = \mathbf{S} \cdot d\mathbf{A}. \tag{4.4.8}$$

Thus, the second Piola–Kirchhoff stress tensor gives the *transformed current force per unit undeformed area*.

Similar to the relationship between $d\mathbf{x}$ and $d\mathbf{X}$, $d\mathbf{X} = \mathbf{F}^{-1} \cdot d\mathbf{x}$, the force $d\mathbf{f}$ on the deformed elemental area $d\mathbf{a}$ is related to the force $d\mathcal{F}$ on the undeformed elemental area $d\mathbf{A}$

$$d\mathcal{F} = \mathbf{F}^{-1} \cdot d\mathbf{f}$$

$$= \mathbf{F}^{-1} \cdot (\mathbf{P} \cdot d\mathbf{A})$$

$$= \mathbf{S} \cdot d\mathbf{A}. \tag{4.4.9}$$

Hence, the second Piola–Kirchhoff stress tensor is related to the first Piola–Kirchhoff stress tensor and Cauchy stress tensor according to the equations

$$\mathbf{S} = \mathbf{F}^{-1} \cdot \mathbf{P} = J\mathbf{F}^{-1} \cdot \boldsymbol{\sigma} \cdot \mathbf{F}^{-\text{T}}. \tag{4.4.10}$$

Clearly, $\mathbf{S}$ is symmetric whenever $\boldsymbol{\sigma}$ is symmetric. Cartesian component representation of $\mathbf{S}$ is

$$\mathbf{S} = S_{IJ}\hat{\mathbf{E}}_I\hat{\mathbf{E}}_J. \tag{4.4.11}$$

We can introduce the psuedo stress vector $\tilde{\mathbf{T}}$ associated with the second Piola–Kirchhoff stress tensor by

$$d\mathcal{F} = \tilde{\mathbf{T}}\,dA = \mathbf{S} \cdot \hat{\mathbf{N}}\,dA = \mathbf{S} \cdot d\mathbf{A}. \tag{4.4.12}$$

The next two examples illustrate the meaning of the first and second Piola–Kirchhoff stress tensors and the computation of first and second Piola–Kirchhoff stress tensor components from the Cauchy stress tensor components.

EXAMPLE 4.4.1: Consider a bar of cross-sectional area A and length L. The initial configuration of the bar is such that its longitudinal axis is along the X_1 axis. If the bar is subjected to uniaxial tensile stress and deformation that stretches

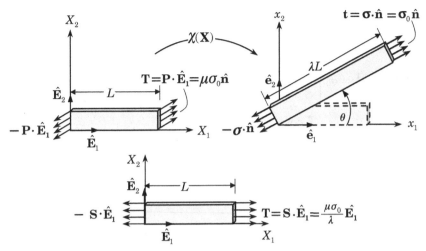

Figure 4.4.2. The undeformed and deformed bar of Example 4.4.1.

the bar by an amount λ and rotates it, without bending, by an angle θ, as shown
in Figure 4.2.2 [see Hjelmstad (2005)], the deformation mapping χ is given by

$$\chi(\mathbf{X}) = (\lambda X_1 \cos\theta - \mu X_2 \sin\theta)\,\hat{\mathbf{e}}_1 + (\lambda X_1 \sin\theta + \mu X_2 \cos\theta)\,\hat{\mathbf{e}}_2 + X_3\,\hat{\mathbf{e}}_3,$$

where λ and μ are constants; λ denotes the stretch of the bar and $\mu\lambda$ denotes
the volume change from undeformed configuration to deformed configuration.
Determine the components of the first and second Piola–Kirchhoff stress tensors.

SOLUTION: The components of the deformation gradient tensor and its inverse
are

$$[F] = \begin{bmatrix} \lambda\cos\theta & -\mu\sin\theta & 0 \\ \lambda\sin\theta & \mu\cos\theta & 0 \\ 0 & 0 & 1 \end{bmatrix}, \quad [F]^{-1} = \frac{1}{J}\begin{bmatrix} \mu\cos\theta & \mu\sin\theta & 0 \\ -\lambda\sin\theta & \lambda\cos\theta & 0 \\ 0 & 0 & \lambda\mu \end{bmatrix},$$

and the Jacobian is equal to $J = \lambda\mu$. Clearly, μ denotes the ratio of deformed
to undeformed cross-sectional area.

The unit vector normal to the undeformed cross-sectional area is $\hat{\mathbf{N}} = \hat{\mathbf{E}}_1$,
and the unit vector normal to the cross-sectional area of the deformed configuration is

$$\hat{\mathbf{n}} = \cos\theta\,\hat{\mathbf{e}}_1 + \sin\theta\,\hat{\mathbf{e}}_2.$$

The Cauchy stress tensor is $\boldsymbol{\sigma} = \sigma_0\,\hat{\mathbf{n}}\hat{\mathbf{n}}$ and associated stress vector is $\mathbf{t} = \sigma_0\,\hat{\mathbf{n}}$.
The components of the Cauchy stress tensor are

$$[\sigma] = \sigma_0 \left\{\begin{array}{c} \cos\theta \\ \sin\theta \\ 0 \end{array}\right\} \{\cos\theta \quad \sin\theta \quad 0\} = \sigma_0 \begin{bmatrix} \cos^2\theta & \cos\theta\sin\theta & 0 \\ \cos\theta\sin\theta & \cos^2\theta & 0 \\ 0 & 0 & 0 \end{bmatrix}.$$

The components of the first Piola–Kirchhoff stress tensor are given by Eq. (4.4.6)

$$[P] = J[\sigma][F]^{-T} = \sigma_0 \begin{bmatrix} \cos^2\theta & \cos\theta\sin\theta & 0 \\ \cos\theta\sin\theta & \cos^2\theta & 0 \\ 0 & 0 & 0 \end{bmatrix} \begin{bmatrix} \mu\cos\theta & -\lambda\sin\theta & 0 \\ \mu\sin\theta & \lambda\cos\theta & 0 \\ 0 & 0 & \lambda\mu \end{bmatrix}$$

$$= \mu\sigma_0 \begin{bmatrix} \cos\theta & 0 & 0 \\ \sin\theta & 0 & 0 \\ 0 & 0 & 0 \end{bmatrix}.$$

The first Piola–Kirchhoff stress tensor and the associated stress vector are

$$\mathbf{P} = \mu\sigma_0(\cos\theta\,\hat{\mathbf{e}}_1 + \sin\theta\,\hat{\mathbf{e}}_2)\,\hat{\mathbf{E}}_1, \quad \mathbf{T} = \mathbf{P}\cdot\hat{\mathbf{N}} = \mu\sigma_0(\cos\theta\,\hat{\mathbf{e}}_1 + \sin\theta\,\hat{\mathbf{e}}_2).$$

Clearly, the matrix representing $\mathbf{P}$ is not symmetric.

While the stress vector $\mathbf{t}$ satisfies both balance of linear momentum and angular momentum in the deformed configuration, the stress vector $\mathbf{T}$ satisfies only the balance of linear momentum but not the balance of angular momentum in the undeformed body (it does satisfy the balance of angular momentum in the deformed body: $\mathbf{P}\cdot\mathbf{F}^T = \mathbf{F}\cdot\mathbf{P}^T$). On the other hand, there is no reason to expect the forces occurring in the deformed configuration but measured in the undeformed configuration to satisfy the balance of momenta in the undeformed configuration.

The second Piola–Kirchhoff stress tensor is given by Eq. (4.4.10). Thus, we have

$$[S] = \frac{\mu\sigma_0}{\lambda} \begin{bmatrix} \mu\cos\theta & \mu\sin\theta & 0 \\ -\lambda\sin\theta & \lambda\cos\theta & 0 \\ 0 & 0 & \lambda\mu \end{bmatrix} \begin{bmatrix} \cos\theta & 0 & 0 \\ \sin\theta & 0 & 0 \\ 0 & 0 & 0 \end{bmatrix} = \frac{\mu\sigma_0}{\lambda} \begin{bmatrix} 1 & 0 & 0 \\ 0 & 0 & 0 \\ 0 & 0 & 0 \end{bmatrix}.$$

The second Piola–Kirchhoff stress tensor and the associated pseudo stress vector are

$$\mathbf{S} = \frac{\mu\sigma_0}{\lambda}\,\hat{\mathbf{E}}_1\hat{\mathbf{E}}_1, \quad \tilde{\mathbf{T}} = \frac{\mu\sigma_0}{\lambda}\,\hat{\mathbf{E}}_1.$$

The second Piola–Kirchhoff stress tensor does satisfy the balance equations in the undeformed body.

EXAMPLE 4.4.2: The equilibrium configuration of a deformed body is described by the mapping

$$\chi(\mathbf{X}) = A X_1\,\hat{\mathbf{e}}_1 - B X_3\,\hat{\mathbf{e}}_2 + C X_2\,\hat{\mathbf{e}}_3,$$

where A, B, and C are constants. If the Cauchy stress tensor for this body is

$$[\sigma] = \begin{bmatrix} 0 & 0 & 0 \\ 0 & 0 & 0 \\ 0 & 0 & \sigma_0 \end{bmatrix},$$

where σ_0 is a constant, determine the pseudo stress vectors associated with the first and second Piola–Kirchhoff stress tensors on the $\hat{\mathbf{e}}_3$-plane in the deformed configuration.

SOLUTION: The deformation gradient tensor and its inverse are

$$[F] = \begin{bmatrix} A & 0 & 0 \\ 0 & 0 & -B \\ 0 & C & 0 \end{bmatrix}, \quad [F]^{-1} = \begin{bmatrix} \frac{1}{A} & 0 & 0 \\ 0 & 0 & \frac{1}{C} \\ 0 & -\frac{1}{B} & 0 \end{bmatrix}, \quad J = ABC.$$

The components of the first Piola–Kirchhoff stress tensor are

$$[P] = J[\sigma][F]^{-T} = ABC \begin{bmatrix} 0 & 0 & 0 \\ 0 & 0 & 0 \\ 0 & 0 & \sigma_0 \end{bmatrix} \begin{bmatrix} \frac{1}{A} & 0 & 0 \\ 0 & 0 & -\frac{1}{B} \\ 0 & \frac{1}{C} & 0 \end{bmatrix}$$

$$= AB\sigma_0 \begin{bmatrix} 0 & 0 & 0 \\ 0 & 0 & 0 \\ 0 & 1 & 0 \end{bmatrix}.$$

The components of the second Piola–Kirchhoff stress tensor are

$$[S] = [F]^{-1}[P] = AB\sigma_0 \begin{bmatrix} \frac{1}{A} & 0 & 0 \\ 0 & 0 & \frac{1}{C} \\ 0 & -\frac{1}{B} & 0 \end{bmatrix} \begin{bmatrix} 0 & 0 & 0 \\ 0 & 0 & 0 \\ 0 & 1 & 0 \end{bmatrix}$$

$$= \frac{AB}{C}\sigma_0 \begin{bmatrix} 0 & 0 & 0 \\ 0 & 1 & 0 \\ 0 & 0 & 0 \end{bmatrix}.$$

Consider a unit area in the deformed state in the $\hat{\mathbf{e}}_3$-direction. The corresponding undeformed area $dA\,\hat{\mathbf{N}}$ is given by [see Eq. (4.4.3)]

$$dA\,\hat{\mathbf{N}} = \frac{1}{J} \mathbf{F}^T \cdot \hat{\mathbf{n}}\, da = \frac{C}{J}\hat{\mathbf{E}}_2.$$

Thus, $dA = C/J$ and $\hat{\mathbf{N}} = \hat{\mathbf{E}}_2$. The pseudo stress vector $\mathbf{T}$ associated with the first Piola–Kirchhoff stress tensor is given by Eq. (4.4.1)

$$\mathbf{T} = \mathbf{P} \cdot \hat{\mathbf{N}} = AB\sigma_0\hat{\mathbf{E}}_3.$$

The pseudo stress vector $\tilde{\mathbf{T}}$ associated with the second Piola–Kirchhoff stress tensor is given by Eq. (4.4.12)

$$\tilde{\mathbf{T}} = \mathbf{S} \cdot \hat{\mathbf{N}} = \frac{AB}{C}\sigma_0\hat{\mathbf{E}}_2.$$

4.5 Equations of Equilibrium

The principle of conservation of linear momentum, which is commonly known as Newton's second law of motion, will be discussed along with other principles of mechanics in Chapter 5. To make the present chapter on stresses complete, we derive the equations of equilibrium of a continuous medium undergoing *small deformations* using Newton's second law of motion to a volume element of the medium, as shown in Figure 4.5.1.

Consider the stresses and body forces on an infinitesimal parallelepiped element of a material body. The stresses acting on various faces of the infinitesimal parallelepiped with dimensions dx_1, dx_2, and dx_3 along coordinate lines (x_1, x_2, x_3) are shown in Figure 4.5.1. By Newton's second law of motion, the sum of the forces in the x_1-direction be zero if the body is in equilibrium. The sum of all forces in the x_1-direction is given by

$$
0 = \left(\sigma_{11} + \frac{\partial \sigma_{11}}{\partial x_1}dx_1\right)dx_2dx_3 - \sigma_{11}dx_2dx_3 + \left(\sigma_{21} + \frac{\partial \sigma_{21}}{\partial x_2}dx_2\right)dx_1dx_3
$$

$$
- \sigma_{21}dx_1dx_3 + \left(\sigma_{31} + \frac{\partial \sigma_{31}}{\partial x_3}dx_3\right)dx_1dx_2 - \sigma_{31}dx_1dx_2 + \rho f_1 dx_1dx_2dx_3
$$

$$
= \left(\frac{\partial \sigma_{11}}{\partial x_1} + \frac{\partial \sigma_{21}}{\partial x_2} + \frac{\partial \sigma_{31}}{\partial x_3} + \rho f_1\right)dx_1dx_2dx_3. \tag{4.5.1}
$$

Upon dividing throughout by $dx_1dx_2dx_3$, we obtain

$$
\frac{\partial \sigma_{11}}{\partial x_1} + \frac{\partial \sigma_{21}}{\partial x_2} + \frac{\partial \sigma_{31}}{\partial x_3} + \rho f_1 = 0. \tag{4.5.2}
$$

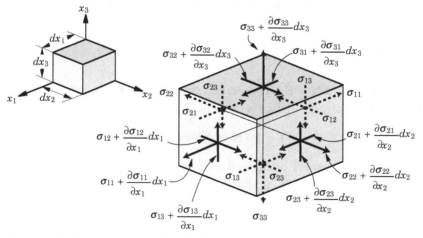

Figure 4.5.1. Stresses on a parallelepiped element.

Similarly, the application of Newton's second law in the x_2- and x_3-directions gives, respectively,

$$\frac{\partial \sigma_{12}}{\partial x_1} + \frac{\partial \sigma_{22}}{\partial x_2} + \frac{\partial \sigma_{32}}{\partial x_3} + \rho f_2 = 0, \tag{4.5.3}$$

$$\frac{\partial \sigma_{13}}{\partial x_1} + \frac{\partial \sigma_{23}}{\partial x_2} + \frac{\partial \sigma_{33}}{\partial x_3} + \rho f_3 = 0. \tag{4.5.4}$$

In index notation, the Eq. (4.5.2)–(4.5.4) can be expressed as

$$\frac{\partial \sigma_{ji}}{\partial x_j} + \rho f_i = 0. \tag{4.5.5}$$

The invariant form of the above equation is given by

$$\nabla \cdot \boldsymbol{\sigma} + \rho \mathbf{f} = \mathbf{0}. \tag{4.5.6}$$

The symmetry of the stress tensor can be established using Newton's second law for moments. Consider the moment of all forces acting on the parallelepiped about the x_3-axis (see Figure 4.5.1). Using the right-handed screw rule for positive moment, we obtain

$$\left[\left(\sigma_{12} + \frac{\partial \sigma_{12}}{\partial x_1} dx_1\right) dx_2 dx_3\right] \frac{dx_1}{2} + (\sigma_{12} dx_2 dx_3) \frac{dx_1}{2}$$

$$-\left[\left(\sigma_{21} + \frac{\partial \sigma_{21}}{\partial x_2} dx_2\right) dx_1 dx_3\right] \frac{dx_2}{2} - (\sigma_{21} dx_1 dx_3) \frac{dx_2}{2} = 0.$$

Dividing throughout by $\frac{1}{2} dx_1 dx_2 dx_3$ and taking the limit $dx_1 \to 0$ and $dx_2 \to 0$, we obtain

$$\sigma_{12} - \sigma_{21} = 0. \tag{4.5.7}$$

Similar considerations of moments about the x_1-axis and x_2-axis give, respectively, the relations

$$\sigma_{23} - \sigma_{32} = 0, \quad \sigma_{13} - \sigma_{31} = 0. \tag{4.5.8}$$

Equations (4.5.7) and (4.5.8) can be expressed in a single equation using the index notation as

$$e_{kij} \sigma_{ij} = 0 \text{ or } \sigma_{ij} - \sigma_{ji} = 0. \tag{4.5.9}$$

In Section 4.3.2, we have discussed the principal stresses and principal directions of a stress tensor. The symmetry of stress tensor with real components has the desirable properties listed in Section 2.5.5. In particular, the stress tensor has real principal values and the principal directions associated with distinct principal stresses are orthogonal.

Next, we consider couple of examples of application of the stress equilibrium equations.

EXAMPLE 4.5.1: Given the following state of stress ($\sigma_{ij} = \sigma_{ji}$) in a kinematically infinitesimal deformation,

$$\sigma_{11} = -2x_1^2, \quad \sigma_{12} = -7 + 4x_1x_2 + x_3, \quad \sigma_{13} = 1 + x_1 - 3x_2,$$

$$\sigma_{22} = 3x_1^2 - 2x_2^2 + 5x_3, \quad \sigma_{23} = 0, \quad \sigma_{33} = -5 + x_1 + 3x_2 + 3x_3,$$

determine the body force components for which the stress field describes a state of static equilibrium.

SOLUTION: The body force components are

$$\rho f_1 = -\left(\frac{\partial\sigma_{11}}{\partial x_1} + \frac{\partial\sigma_{12}}{\partial x_2} + \frac{\partial\sigma_{13}}{\partial x_3}\right) = -[(-4x_1) + (4x_1) + 0] = 0,$$

$$\rho f_2 = -\left(\frac{\partial\sigma_{12}}{\partial x_1} + \frac{\partial\sigma_{22}}{\partial x_2} + \frac{\partial\sigma_{23}}{\partial x_3}\right) = -[(4x_2) + (-4x_2) + 0] = 0,$$

$$\rho f_3 = -\left(\frac{\partial\sigma_{13}}{\partial x_1} + \frac{\partial\sigma_{23}}{\partial x_2} + \frac{\partial\sigma_{33}}{\partial x_3}\right) = -[1 + 0 + 3] = -4.$$

Thus, the body is in static equilibrium for the body force components $\rho f_1 = 0$, $\rho f_2 = 0$, and $\rho f_3 = -4$.

EXAMPLE 4.5.2: Determine whether the following stress field in a kinematically infinitesimal deformation satisfies the equations of equilibrium:

$$\sigma_{11} = x_2^2 + k\left(x_1^2 - x_2^2\right), \quad \sigma_{12} = -2kx_1x_2, \quad \sigma_{13} = 0,$$

$$\sigma_{22} = x_1^2 + k\left(x_2^2 - x_1^2\right), \quad \sigma_{23} = 0, \quad \sigma_{33} = k\left(x_1^2 + x_2^2\right).$$

SOLUTION: We have

$$\frac{\partial\sigma_{11}}{\partial x_1} + \frac{\partial\sigma_{12}}{\partial x_2} + \frac{\partial\sigma_{13}}{\partial x_3} = (2kx_1) + (-2kx_1) + 0 = 0,$$

$$\frac{\partial\sigma_{12}}{\partial x_1} + \frac{\partial\sigma_{22}}{\partial x_2} + \frac{\partial\sigma_{23}}{\partial x_3} = (-2kx_2) + (2kx_2) + 0 = 0,$$

$$\frac{\partial\sigma_{13}}{\partial x_1} + \frac{\partial\sigma_{23}}{\partial x_2} + \frac{\partial\sigma_{33}}{\partial x_3} = 0 + 0 + 0 = 0.$$

Thus the given stress field is in equilibrium in the absence of any body forces.

4.6 Summary

In this chapter, various measures of stress, namely, the Cauchy stress and Piola–Kirchhoff stress tensors, are introduced and the Cauchy formula that relates the stress tensor to the stress vector at a point on the boundary is derived. The transformation relations among stress components from two different rectangular coordinate systems are derived, the principal values and principal directions of a stress tensor are discussed, and equations of stress equilibrium are derived.

PROBLEMS

4.1 Suppose that $\mathbf{t}^{\hat{n}_1}$ and $\mathbf{t}^{\hat{n}_2}$ are stress vectors acting on planes with unit normals $\hat{n}_1$ and $\hat{n}_2$, respectively, and passing through a point with the stress state σ. Show that the component of $\mathbf{t}^{\hat{n}_1}$ along $\hat{n}_2$ is equal to the component of $\mathbf{t}^{\hat{n}_2}$ along the normal $\mathbf{t}^{\hat{n}_1}$.

4.2 Write the stress vectors on each boundary surface in terms of the given values and base vectors $\hat{\mathbf{i}}$ and $\hat{\mathbf{j}}$ for the system shown in Figure P4.2.

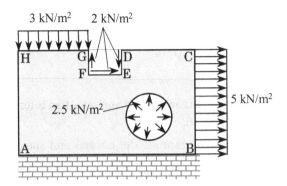

Figure P4.2.

4.3 The components of a stress dyad at a point, referred to the (x_1, x_2, x_3) system, are (in ksi = 1000 psi):

(i) $\begin{bmatrix} 12 & 9 & 0 \\ 9 & -12 & 0 \\ 0 & 0 & 6 \end{bmatrix}$, (ii) $\begin{bmatrix} 9 & 0 & 12 \\ 0 & -25 & 0 \\ 12 & 0 & 16 \end{bmatrix}$, (iii) $\begin{bmatrix} 1 & -3 & \sqrt{2} \\ -3 & 1 & -\sqrt{2} \\ \sqrt{2} & -\sqrt{2} & 4 \end{bmatrix}$.

Find the following:

(a) The stress vector acting on a plane perpendicular to the vector $2\hat{\mathbf{e}}_1 - 2\hat{\mathbf{e}}_2 + \hat{\mathbf{e}}_3$.

(b) The magnitude of the stress vector and the angle between the stress vector and the normal to the plane.

(c) The magnitudes of the normal and tangential components of the stress vector.

4.4 Consider a kinematically infinitesimal stress field whose matrix of scalar components in the vector basis $\{\hat{\mathbf{e}}_i\}$ is

$$\begin{bmatrix} 1 & 0 & 2X_2 \\ 0 & 1 & 4X_1 \\ 2X_2 & 4X_1 & 1 \end{bmatrix} \times 10^3 \ (\text{psi}),$$

where the Cartesian coordinate variables X_i are in inches (in.) and the units of stress are pounds per square inch (psi).

(a) Determine the traction vector acting at point $\mathbf{X} = \hat{\mathbf{e}}_1 + 2\hat{\mathbf{e}}_2 + 3\hat{\mathbf{e}}_3$ on the plane $X_1 + X_2 + X_3 = 6$.

(b) Determine the normal and projected shear tractions acting at this point on this plane.

(c) Determine the principal stresses and principal directions of stress at this point.

(d) Determine the maximum shear stress at the point.

4.5 The three-dimensional state of stress at a point $(1, 1, -2)$ within a body relative to the coordinate system (x_1, x_2, x_3) is

$$\begin{bmatrix} 2.0 & 3.5 & 2.5 \\ 3.5 & 0.0 & -1.5 \\ 2.5 & -1.5 & 1.0 \end{bmatrix} \times 10^6 \text{ (Pa)}.$$

Determine the normal and shear stresses at the point and on the surface of an internal sphere whose equation is $x_1^2 + (x_2 - 2)^2 + x_3^2 = 6$.

4.6 For the state of stress given in Problem 4.5, determine the normal and shear stresses on a plane intersecting the point where the plane is defined by the points $(0, 0, 0)$, $(2, -1, 3)$, and $(-2, 0, 1)$.

4.7 Use equilibrium of forces to derive the relations between the normal and shear stresses σ_n and σ_s on a plane whose normal is $\hat{n} = \cos\theta\hat{e}_1 + \sin\theta\hat{e}_2$ to the stress components σ_{11}, σ_{22}, and $\sigma_{12} = \sigma_{21}$ on the $\hat{e}_1$ and $\hat{e}_2$ planes, as shown in Figure P4.7:

$$\sigma_n = \sigma_{11}\cos^2\theta + \sigma_{22}\sin^2\theta + \sigma_{12}\sin 2\theta,$$

$$\sigma_s = -\frac{1}{2}(\sigma_{11} - \sigma_{22})\sin 2\theta + \sigma_{12}(\cos^2\theta - \sin^2\theta). \tag{1}$$

Note that θ is the angle measured from the positive x_1-axis to the normal to the inclined plane. Then show that

(a) the principal stresses at a point in a body with two-dimensional state of stress are given by

$$\sigma_{p1} = \sigma_{max} = \frac{\sigma_{11} + \sigma_{22}}{2} + \sqrt{\left(\frac{\sigma_{11} - \sigma_{22}}{2}\right)^2 + \sigma_{12}^2},$$

$$\sigma_{p2} = \sigma_{min} = \frac{\sigma_{11} + \sigma_{22}}{2} - \sqrt{\left(\frac{\sigma_{11} - \sigma_{22}}{2}\right)^2 + \sigma_{12}^2}, \tag{2}$$

and that the orientation of the principal planes is given by

$$\theta_p = \pm\frac{1}{2}\tan^{-1}\left[\frac{2\sigma_{12}}{\sigma_{11} - \sigma_{22}}\right], \tag{3}$$

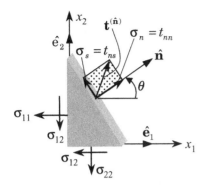

Figure P4.7.

(b) the maximum shear stress is given by

$$(\sigma_s)_{max} = \pm \frac{\sigma_{p1} - \sigma_{p2}}{2} . \tag{4}$$

Also, determine the plane on which the maximum shear stress occurs.

4.8 Determine the normal and shear stress components on the plane indicated in Figure P4.8.

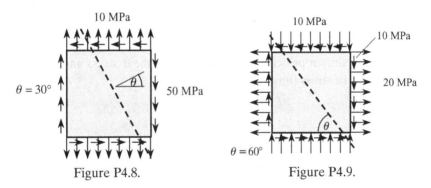

Figure P4.8.   Figure P4.9.

4.9 Determine the normal and shear stress components on the plane indicated in Figure P4.9.

4.10 Determine the normal and shear stress components on the plane indicated in Figure P4.10.

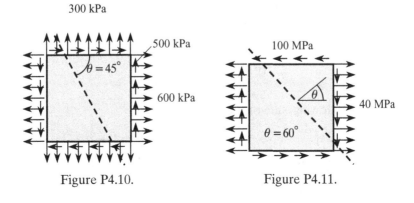

Figure P4.10.   Figure P4.11.

4.11 Determine the normal and shear stress components on the plane indicated in Figure P4.11.

4.12 Find the values of σ_s and σ_{22} for the state of stress shown in Figure P4.12.

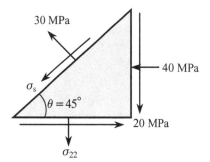

Figure P4.12.

4.13 Find the maximum and minimum normal stresses and the orientations of the principal planes for the state of stress shown in Figure P4.10. What is the maximum shear stress at the point?

4.14 Find the maximum and minimum normal stresses and the orientations of the principal planes for the state of stress shown in Figure P4.11. What is the maximum shear stress at the point?

4.15 Find the maximum principal stress, maximum shear stress and their orientations for the state of stress (units are 10^6 psi)

$$[\sigma] = \begin{bmatrix} 3 & 5 & 8 \\ 5 & 1 & 0 \\ 8 & 0 & 2 \end{bmatrix}.$$

4.16 (*Spherical and deviatoric stress tensors*) The the stress tensor can be expressed as the sum of *spherical* or *hydrostatic* stress tensor $\tilde{\sigma}$ and *deviatoric* stress tensor σ'

$$\sigma = \tilde{\sigma}\mathbf{I} + \sigma', \quad \tilde{\sigma} = \frac{1}{3}\operatorname{tr}\sigma = \frac{1}{3}I_1, \quad \sigma' = \sigma - \frac{1}{3}I_1\mathbf{I}.$$

For the state of stress given in Problem 4.15, compute the spherical and deviatoric components of the stress tensor.

4.17 Determine the invariants I_i' and the principal deviator stresses for the following state of stress (units are msi $= 10^6$ psi)

$$[\sigma] = \begin{bmatrix} 2 & -1 & 1 \\ -1 & 0 & 1 \\ 1 & 1 & 2 \end{bmatrix}.$$

4.18 Given the following state of stress ($\sigma_{ij} = \sigma_{ji}$),

$$\sigma_{11} = -2x_1^2, \quad \sigma_{12} = -7 + 4x_1x_2 + x_3, \quad \sigma_{13} = 1 + x_1 - 3x_2,$$

$$\sigma_{22} = 3x_1^2 - 2x_2^2 + 5x_3, \quad \sigma_{23} = 0, \quad \sigma_{33} = -5 + x_1 + 3x_2 + 3x_3,$$

determine (a) the stress vector at point (x_1, x_2, x_3) on the plane $x_1 + x_2 + x_3 = $ constant, (b) the normal and shearing components of the stress vector at point $(1, 1, 3)$, and (c) the principal stresses and their orientation at point $(1, 2, 1)$.

4.19 The components of a stress dyad at a point, referred to the (x_1, x_2, x_3) system, are

$$\begin{bmatrix} 25 & 0 & 0 \\ 0 & -30 & -60 \\ 0 & -60 & 5 \end{bmatrix} \text{ MPa.}$$

Determine (a) the stress vector acting on a plane perpendicular to the vector $2\hat{e}_1 + \hat{e}_2 + 2\hat{e}_3$, and (b) the magnitude of the normal and tangential components of the stress vector.

4.20 The components of a stress dyad at a point P, referred to the (x_1, x_2, x_3) system, are

$$\begin{bmatrix} 57 & 0 & 24 \\ 0 & 50 & 0 \\ 24 & 0 & 43 \end{bmatrix} \text{ MPa.}$$

Determine the principal stresses and principal directions at point P. What is the maximum shear stress at the point?

4.21 Derive the stress equilibrium equations incylindrical coordinates by considering the equilibrium of a typical volume element shown in Figure P4.21.

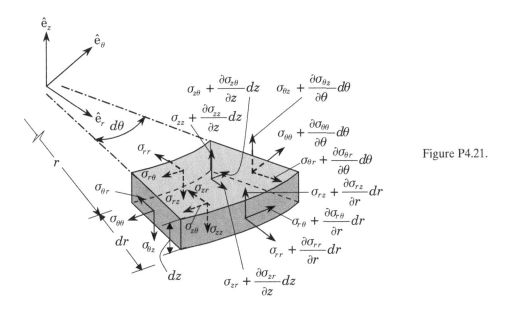

Figure P4.21.

4.22 Given the following stress field in a body in equilibrium and referred to cylindrical coordinate system:

$$\sigma_{rr} = 2A\left(r + \frac{B}{r^3} - \frac{C}{r}\right)\sin\theta,$$

$$\sigma_{\theta\theta} = 2A\left(3r - \frac{B}{r^3} - \frac{C}{r}\right)\sin\theta,$$

$$\sigma_{r\theta} = -2A\left(r + \frac{B}{r^3} - \frac{C}{r}\right)\cos\theta,$$

where A, B, and C are constants, determine whether the stress field satisfies the equilibrium equations when the body forces and all other stresses are zero.

4.23 Given the following stress field in a body in equilibrium and referred to spherical coordinate system:

$$\sigma_{rr} = -\left(A + \frac{B}{r^3}\right), \quad \sigma_{\theta\theta} = \sigma_{\phi\phi} = -\left(A + \frac{C}{r^3}\right),$$

where A, B, and C are constants, determine if the stress field satisfies the equilibrium equations when the body forces are zero and all other stresses are zero.

Conservation of Mass, Momenta, and Energy

Although to penetrate into the intimate mysteries of nature and thence to learn the true causes of phenomena is not allowed to us, nevertheless it can happen that a certain fictive hypothesis may suffice for explaining many phenomena.

Leonard Euler

Nothing is too wonderful to be true if it be consistent with the laws of nature.

Michael Faraday

5.1 Introduction

Virtually every phenomenon in nature, whether mechanical, biological, chemical, geological, or geophysical, can be described in terms of mathematical relations among various quantities of interest. Most mathematical models of physical phenomena are based on fundamental scientific laws of physics that are extracted from centuries of research on the behavior of mechanical systems subjected to the action of natural forces. What is most exciting is that the laws of physics, which are also termed *principles of mechanics*, govern biological systems as well (because of mass and energy transports). However, biological systems may require additional laws, yet to be discovered, from biology and chemistry to complete their description.

This chapter is devoted to the study of fundamental laws of physics as applied to mechanical systems. The laws of physics are expressed in analytical form with the aid of the concepts and quantities introduced in previous chapters. The laws or principles of physics that we study here are (1) the principle of conservation of mass, (2) the principle of conservation of linear momentum, (3) the principle of conservation of angular momentum, and (4) the principle of conservation of energy. These laws allow us to write mathematical relationships – algebraic, differential, or integral type – of physical quantities such as displacements, velocities, temperature, stresses, and strains in mechanical systems. The solution of these equations represents the response of the system, which aids the design and manufacturing of the system. The equations developed here will be used not only in the later chapters of this book, but they are also useful in other courses. Thus, the present chapter is the heart and soul of a course on continuum mechanics.

5.2 Conservation of Mass

5.2.1 Preliminary Discussion

It is a common knowledge that mass of a given system cannot be created or destroyed. For example, the mass flow of the blood entering a section of an artery is equal to the mass flow leaving the artery, provided that no blood is added or lost through the artery walls. Thus, mass of the blood is conserved even when the artery cross section changes along the length.

The principle of conservation of mass states that *the total mass of any part ∂B of the body B does not change in any motion.* The mathematical form of this principle is different in different descriptions of motion. Before we derive the mathematical forms of the principle, certain other identities need to be established.

5.2.2 Material Time Derivative

As discussed in Chapter 3 [see Eqs. (3.2.4) and (3.2.5)], the partial time derivative with the material coordinates $\mathbf{X}$ held constant should be distinguished from the partial time derivative with spatial coordinates $\mathbf{x}$ held constant due to the difference in the descriptions of motion. The *material time derivative*, denoted here by D/Dt, is the time derivative with the material coordinates held constant. The time derivative of a function ϕ in material description [i.e., $\phi = \phi(\mathbf{X}, t)$] with $\mathbf{X}$ held constant is nothing but the partial derivative with respect to time [see Eq. (3.2.5)]:

$$\frac{D\phi}{Dt} \equiv \left(\frac{\partial \phi}{\partial t}\right)_{\mathbf{X}=\text{const}} = \frac{\partial \phi}{\partial t}. \tag{5.2.1}$$

In particular, we have

$$\frac{D\mathbf{x}}{Dt} = \left(\frac{\partial \mathbf{x}}{\partial t}\right)_{\mathbf{X}=\text{const}} = \left(\frac{\partial \mathbf{x}}{\partial t}\right) \equiv \mathbf{v}, \tag{5.2.2}$$

where $\mathbf{v}$ is the velocity vector. Similarly, the acceleration is

$$\frac{D\mathbf{v}}{Dt} = \left(\frac{\partial \mathbf{v}}{\partial t}\right)_{\mathbf{X}=\text{const}} = \left(\frac{\partial \mathbf{v}}{\partial t}\right) \equiv \mathbf{a}, \tag{5.2.3}$$

where $\mathbf{a}$ is the acceleration vector.

In spatial description, we have $\phi = \phi(\mathbf{x}, t)$ and the partial time derivative

$$\left(\frac{\partial \phi}{\partial t}\right)_{\mathbf{X}=\text{const}} \quad \text{is different from} \quad \left(\frac{\partial \phi}{\partial t}\right)_{\mathbf{x}=\text{const}}.$$

The time derivative $\left(\frac{\partial \phi}{\partial t}\right)_{\mathbf{X}=\text{const}}$ denotes the *local rate of change* of ϕ. If $\phi = \mathbf{v}$, then it is the rate of change of $\mathbf{v}$ read by a velocity meter located at the fixed spatial location $\mathbf{x}$, which is not the same as the acceleration of the particle just passing the place $\mathbf{x}$.

To calculate the material time derivative of a function of spatial coordinates, $\phi = \phi(\mathbf{x}, t)$, we assume that there exists differentiable mapping $\mathbf{x} = \mathbf{x}(\mathbf{X}, t)$ so that we can write $\phi(\mathbf{x}, t) = \phi[\mathbf{x}(\mathbf{X}, t), t]$ and compute the derivative using chain rule of differentiation:

$$
\begin{aligned}
\frac{D\phi}{Dt} \equiv \left(\frac{\partial \phi}{\partial t}\right)_{\mathbf{X}=\mathrm{const}} &= \left(\frac{\partial \phi}{\partial t}\right)_{\mathbf{x}=\mathrm{const}} + \left(\frac{\partial x_i}{\partial t}\right)_{\mathbf{X}=\mathrm{const}} \frac{\partial \phi}{\partial x_i} \\
&= \left(\frac{\partial \phi}{\partial t}\right)_{\mathbf{x}=\mathrm{const}} + v_i \frac{\partial \phi}{\partial x_i} \\
&= \left(\frac{\partial \phi}{\partial t}\right)_{\mathbf{x}=\mathrm{const}} + \mathbf{v} \cdot \nabla \phi,
\end{aligned}
$$

where Eq. (5.2.2) is used in the second line. Thus, the material derivative operator is given by

$$
\frac{D}{Dt} = \left(\frac{\partial}{\partial t}\right)_{\mathbf{x}=\mathrm{const}} + \mathbf{v} \cdot \nabla. \tag{5.2.4}
$$

The next example illustrates the calculation of material time derivative based on material and spatial descriptions.

EXAMPLE 5.2.1: Suppose that the motion is described by the mapping, $x = (1 + t)X$. Determine (a) the velocities and accelerations in the spatial and material descriptions and (b) the time derivative of a function $\phi(X, t) = Xt^2$ in the material description.

SOLUTION: The velocity $v \equiv dx/dt$ can be expressed in the material and spatial coordinates as

$$
v(X, t) = \frac{Dx}{Dt} = \frac{\partial x}{\partial t} = X, \quad v(x, t) = \frac{x}{1+t}.
$$

The acceleration $a \equiv \frac{Dv}{Dt}$ in the two descriptions is

$$
a \equiv \frac{Dv(X, t)}{Dt} = \frac{\partial v}{\partial t} = 0,
$$

$$
a \equiv \frac{Dv(x, t)}{Dt} = \frac{\partial v}{\partial t} + v \frac{\partial v}{\partial x} = -\frac{x}{(1+t)^2} + \left(\frac{x}{1+t}\right)\left(\frac{1}{1+t}\right) = 0.
$$

The material time derivative of ϕ in the material description is simply

$$
\frac{D\phi(X, t)}{Dt} = \frac{\partial \phi(X, t)}{\partial t} = 2Xt.
$$

The material time derivative of $\phi(x, t) = xt^2/(1+t)$ in the spatial description is

$$
\frac{D\phi}{Dt} = \frac{\partial \phi}{\partial t} + v \frac{\partial \phi}{\partial x} = \frac{2xt}{1+t} - \frac{xt^2}{(1+t)^2} + \left(\frac{x}{1+t}\right)\left(\frac{t^2}{1+t}\right) = \frac{2xt}{1+t},
$$

which is the same as that calculated before, except that it is expressed in terms of the current coordinate, x.

In the later chapters of the book, we will make use of the gradient and divergence theorems [see Eqs. (2.4.33) and (2.4.34)]. For a ready reference, they are recorded here. The following relations hold for a closed region Ω bounded by surface Γ:

$$\oint_\Gamma \hat{\mathbf{n}}\Phi \, ds = \int_\Omega \nabla\Phi \, d\mathbf{x} \quad \text{(Gradient theorem)}, \tag{5.2.5}$$

$$\oint_\Gamma \hat{\mathbf{n}} \cdot \boldsymbol{\Phi} \, ds = \int_\Omega \nabla \cdot \boldsymbol{\Phi} \, d\mathbf{x} \quad \text{(Divergence theorem)}, \tag{5.2.6}$$

where $\hat{\mathbf{n}}$ denotes unit outward normal to the surface Γ.

5.2.3 Continuity Equation in Spatial Description

Let an arbitrary region in a continuous medium $\mathcal{B}$ be denoted by Ω, and the bounding closed surface of this region be continuous and denoted by Γ. Let each point on the bounding surface move with the velocity $\mathbf{v}_s$. It can be shown that the time derivative of the volume integral over some continuous function $\phi(\mathbf{x}, t)$ is given by

$$\frac{d}{dt}\int_\Omega \phi(\mathbf{x}, t) \, d\mathbf{x} \equiv \frac{\partial}{\partial t}\int_\Omega \phi \, d\mathbf{x} + \oint_\Gamma \phi\mathbf{v}_s \cdot \hat{\mathbf{n}} \, ds,$$

$$= \int_\Omega \frac{\partial \phi}{\partial t} \, d\mathbf{x} + \oint_\Gamma \phi\mathbf{v}_s \cdot \hat{\mathbf{n}} \, ds. \tag{5.2.7}$$

This expression for the differentiation of a volume integral with variable limits is sometimes known as the three-dimensional *Leibnitz rule*.

Let each element of mass in the medium move with the velocity $\mathbf{v}(\mathbf{x}, t)$ and consider a special region Ω such that the bounding surface Γ is attached to a fixed set of material elements. Then each point of this surface moves itself with the material velocity, that is, $\mathbf{v}_s = \mathbf{v}$, and the region Ω thus contains a fixed total amount of mass since no mass crosses the boundary surface Γ. To distinguish the time rate of change of an integral over this material region, we replace d/dt by D/Dt and write

$$\frac{D}{Dt}\int_\Omega \phi(\mathbf{x}, t) \, d\mathbf{x} \equiv \int_\Omega \frac{\partial \phi}{\partial t} \, d\mathbf{x} + \oint_\Gamma \phi\mathbf{v} \cdot \hat{\mathbf{n}} \, ds, \tag{5.2.8}$$

which holds for a material region, that is, a region of fixed total mass. Then the relation between the time derivative following an arbitrary region and the time derivative following a material region (fixed total mass) is

$$\frac{d}{dt}\int_\Omega \phi(\mathbf{x}, t) \, d\mathbf{x} \equiv \frac{D}{Dt}\int_\Omega \phi(\mathbf{x}, t) \, d\mathbf{x} + \oint_\Gamma \phi(\mathbf{v}_s - \mathbf{v}) \cdot \hat{\mathbf{n}} \, ds. \tag{5.2.9}$$

The velocity difference $\mathbf{v} - \mathbf{v}_s$ is the velocity of the material measured relative to the velocity of the surface. The surface integral

$$\oint_\Gamma \phi(\mathbf{v}_s - \mathbf{v}) \cdot \hat{\mathbf{n}} \, ds$$

thus measures the total *outflow* of the property ϕ from the region Ω.

Let $\rho(\mathbf{x}, t)$ denote the mass density of a continuous region. Then the principle of conservation of mass for a *material* region requires that

$$\frac{D}{Dt} \int_\Omega \rho \, d\mathbf{x} = 0.$$

Then from Eq. (5.2.9) it follows that for a *fixed region* ($\mathbf{v}_s = 0$) the principle of conservation of mass can also be stated as

$$\frac{d}{dt} \int_\Omega \rho \, d\mathbf{x} = - \oint_\Gamma \rho \mathbf{v} \cdot \hat{\mathbf{n}} \, ds. \tag{5.2.10}$$

Equation (5.2.10) is known as the control volume formulation of the conservation of mass principle. In Eq. (5.2.10), Ω denotes the *control volume* (cv) and Γ the *control surface* (cs) enclosing Ω.

Using Eq. (5.2.7) with $\phi = \rho$, Eq. (5.2.10) can be expressed as

$$\int_\Omega \frac{\partial \rho}{\partial t} \, d\mathbf{x} = - \oint_\Gamma \rho \mathbf{v} \cdot \hat{\mathbf{n}} \, ds.$$

Converting the surface integral to a volume integral by means of the divergence theorem (5.2.6), we obtain

$$\int_\Omega \left[\frac{\partial \rho}{\partial t} + \text{div}(\rho \mathbf{v}) \right] d\mathbf{x} = 0. \tag{5.2.11}$$

Equation (5.2.11) also follows directly from Eq. (5.2.8). Since this integral vanishes, for a continuous medium, for any arbitrary region Ω, we deduce that this can be true only if the integrand itself vanishes identically, giving the following local form:

$$\frac{\partial \rho}{\partial t} + \text{div}(\rho \mathbf{v}) = 0. \tag{5.2.12}$$

This equation, called the *continuity equation*, expresses local conservation of mass at any point in a continuous medium.

An alternative derivation of Eq. (5.2.12) is presented next. Consider an arbitrary control volume Ω in space where flow occurs into and out of the space. Conservation of mass in this case means that the time rate of change of mass in Ω is equal to the mass inflow through the control surface. Consider an elemental area ds with unit normal $\hat{\mathbf{n}}$ around a point P on the control surface, as shown in Figure 5.2.1. Let $\mathbf{v}$ and ρ be the velocity and density, respectively, at point P. The mass outflow (slug/s or kg/s) through the elemental surface is $\rho \mathbf{v} \cdot d\mathbf{s}$, where $d\mathbf{s} = \hat{\mathbf{n}} \, ds$. The total mass inflow through the entire surface of the control volume is

$$\oint_\Gamma (-\rho v_n) \, ds = - \oint_\Gamma \rho \mathbf{v} \cdot \hat{\mathbf{n}} \, ds = - \int_\Omega \nabla \cdot (\rho \mathbf{v}) \, d\mathbf{x}, \tag{5.2.13}$$

where the divergence theorem (5.2.6) is used in arriving at the last line. If a continuous medium of density ρ fills the region Ω at time t, the total mass in Ω is

$$M = \int_\Omega \rho(\mathbf{x}, t) \, d\mathbf{x}.$$

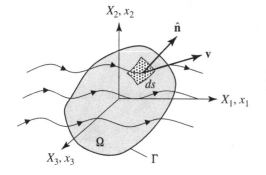

Figure 5.2.1. A control volume for the derivation of the continuity equation.

The rate of increase of mass in Ω is

$$\frac{\partial M}{\partial t} = \int_\Omega \frac{\partial \rho}{\partial t} \, d\mathbf{x}.$$

(5.2.14)

Equating Eqs. (5.2.13) and (5.2.14), we obtain

$$\int_\Omega \left[\frac{\partial \rho}{\partial t} + \nabla \cdot (\rho \mathbf{v}) \right] d\mathbf{x} = 0,$$

which is the same as Eq. (5.2.11), and hence, we obtain the continuity equation in Eq. (5.2.12).

Equation (5.2.12) can be written in an alternative form as follows:

$$0 = \frac{\partial \rho}{\partial t} + \nabla \cdot (\rho \mathbf{v}) = \frac{\partial \rho}{\partial t} + \mathbf{v} \cdot \nabla \rho + \rho \nabla \cdot \mathbf{v} = \frac{D\rho}{Dt} + \rho \nabla \cdot \mathbf{v},$$

(5.2.15)

where the definition of *material time derivative*, Eq. (5.2.4), is used in arriving at the final result.

The one-dimensional version of the local form of the continuity equation (5.2.12) can be obtained by considering flow along the x-axis (see Figure 5.2.2). The amount of mass entering (i.e., mass flow) per unit time at the left section of the elemental volume is: density × cross-sectional area × velocity of the flow $= (\rho A v_x)_x$. The mass leaving at the right section of the elemental volume is $(\rho A v_x)_{x+\Delta x}$, where v_x is the velocity along x-direction. The subscript of $(\cdot)$ denotes the distance at which the enclosed quantity is evaluated. It is assumed that the cross-sectional area A is

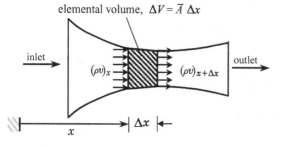

Figure 5.2.2. Derivation of the local form of the continuity equation in one dimension.

a function of position x but not of time t. The net mass inflow into the elemental volume is

$$(A\rho v_x)_x - (A\rho v_x)_{x+\Delta x}.$$

The time rate of increase of the total mass inside the elemental volume is

$$\bar{A}\Delta x \frac{(\bar{\rho})_{t+\Delta t} - (\bar{\rho})_t}{\Delta t},$$

where $\bar{\rho}$ and $\bar{A}$ are the average values of the density and cross-sectional area, respectively, inside the elemental volume.

If no mass is created or destroyed inside the elemental volume, the rate of increase of mass should be equal to the mass inflow:

$$\bar{A}\Delta x \frac{(\bar{\rho})_{t+\Delta t} - (\bar{\rho})_t}{\Delta t} = (A\rho v_x)_x - (A\rho v_x)_{x+\Delta x}.$$

Dividing throughout by Δx and taking the limits $\Delta t \to 0$ and $\Delta x \to 0$, we obtain

$$\lim_{\Delta t, \Delta x \to 0} \bar{A} \frac{(\rho)_{t+\Delta t} - (\rho)_t}{\Delta t} + \frac{(A\rho v_x)_{x+\Delta x} - (A\rho v_x)_x}{\Delta x} = 0$$

or ($\bar{\rho} \to \rho$ and $\bar{A} \to A$ as $\Delta x \to 0$)

$$A \frac{\partial \rho}{\partial t} + \frac{\partial (A\rho v_x)}{\partial x} = 0. \tag{5.2.16}$$

Equation (5.2.16) is the same as Eq. (5.2.12) when **v** is replaced with $\mathbf{v} = v_x \hat{\mathbf{e}}_x$ and A is a constant. Note that for the steady-state case, Eq. (5.2.16) reduces to

$$\frac{\partial (A\rho v_x)}{\partial x} = 0 \rightarrow A\rho v_x = \text{constant} \Rightarrow (A\rho v_x)_1 = (A\rho v_x)_2 = \cdots = (A\rho v_x)_i, \tag{5.2.17}$$

where the subscript i in $(\cdot)_i$ refers to ith section along the direction of the (one-dimensional) flow. The quantity $Q = Av_x$ is called the *flow*, whereas ρAv_x is called the *mass flow*.

The continuity equation (5.2.12) can also be expressed in orthogonal curvilinear coordinate systems as (Problems 5.4 and 5.5 are designed to obtain these results)

Cylindrical coordinate system (r, θ, z)

$$0 = \frac{\partial \rho}{\partial t} + \frac{1}{r}\left[\frac{\partial (r\rho v_r)}{\partial r} + \frac{\partial (\rho v_\theta)}{\partial \theta} + r\frac{\partial (\rho v_z)}{\partial z}\right]. \tag{5.2.18}$$

Spherical coordinate system (R, ϕ, θ)

$$0 = \frac{\partial \rho}{\partial t} + \frac{1}{R^2}\frac{\partial (\rho R^2 v_R)}{\partial R} + \frac{1}{R\sin\phi}\frac{\partial (\rho v_\phi \sin\phi)}{\partial \theta} + \frac{1}{R\sin\phi}\frac{\partial (\rho v_\theta)}{\partial \theta}. \tag{5.2.19}$$

See Table 2.4.2 for expressions of divergence of a vector in the cylindrical and spherical coordinate systems. For steady-state, we set the time derivative terms in

Eqs. (5.2.12), (5.2.18), and (5.2.19) to zero. The invariant form of continuity equation for steady-state flows is

$$\nabla \cdot (\rho \mathbf{v}) = 0, \tag{5.2.20}$$

and for materials with constant density, we set $D\rho/Dt = 0$ and obtain

$$\rho \nabla \cdot \mathbf{v} = 0 \text{ or } \nabla \cdot \mathbf{v} = 0. \tag{5.2.21}$$

Next, we consider two examples of application of the principle of conservation of mass in spatial description.

EXAMPLE 5.2.2: Consider a water hose with conical-shaped nozzle at its end, as shown in Figure 5.2.3(a). (a) Determine the pumping capacity required in order the velocity of the water (assuming incompressible for the present case) exiting the nozzle be 25 m/s. (b) If the hose is connected to a rotating sprinkler through its base, as shown in Figure 5.2.3(b), determine the average speed of the water leaving the sprinkler nozzle.

SOLUTION:
(a) The principle of conservation of mass for steady one-dimensional flow requires

$$\rho_1 A_1 v_1 = \rho_2 A_2 v_2.$$

If the exit of the nozzle is taken as the section 2, we can calculate the flow at section 1 as (for an incompressible fluid, $\rho_1 = \rho_2$)

$$Q_1 = A_1 v_1 = A_2 v_2 = \frac{\pi (20 \times 10^{-3})^2}{4} 25 = 0.0025\pi \text{ m}^3/\text{s}.$$

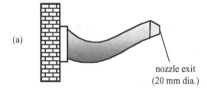

(a)

nozzle exit
(20 mm dia.)

Figure 5.2.3. (a) Water hose with a conical head. (b) Water hose connected to a sprinkler.

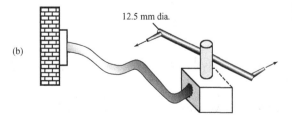

12.5 mm dia.

(b)

(b) The average speed of the water leaving the sprinkler nozzle can be calculated using the principle of conservation of mass for steady one-dimensional flow. We obtain

$$Q_1 = 2A_2 v_2 \quad \rightarrow \quad v_2 = \frac{2Q_1}{\pi d^2} = \frac{0.005}{(12.5 \times 10^{-3})^2} = 32 \text{ m/s.}$$

EXAMPLE 5.2.3: A syringe used to inoculate large animals has a cylinder, plunger, and needle combination, as shown in Figure 5.2.4. Let the internal diameter of the cylinder be d and plunger face area be A_p. If the liquid in the syringe is to be injected at a steady rate of Q_0, determine the speed of the plunger. Assume that the leakage rate past the plunger is 10% of the volume flow rate out of the needle.

SOLUTION: In this problem, the control volume (shown in dotted lines in Figure 5.2.4) is not constant. Even though there is a leakage, the plunger surface area can be taken as equal to the open cross-sectional area of the cylinder, $A_p = \pi d^2/4$. Let us consider Section 1 to be the plunger face and Section 2 to be the needle exit to apply the continuity of mass equation.

Assuming that the flow through the needle and leakage are steady, application of the global form of the continuity equation, Eq. (5.2.10), to the control volume gives

$$0 = \frac{d}{dt} \int_\Omega \rho \, d\mathbf{x} + \oint_\Gamma \rho \mathbf{v} \cdot \hat{\mathbf{n}} \, ds$$

$$= \frac{d}{dt} \int_\Omega \rho \, d\mathbf{x} + \rho Q_0 + \rho Q_{\text{leak}}. \tag{a}$$

The integral in the above equation can be evaluated as follows:

$$\frac{d}{dt} \int_\Omega \rho \, d\mathbf{x} = \frac{d}{dt} \left(\rho x A_p + \rho V_n \right) = \rho A_p \frac{dx}{dt} = -\rho A_p v_p, \tag{b}$$

where x is the distance between the plunger face and the end of the cylinder, V_n is the volume of the needle opening, and $v_p = -dx/dt$ is the speed of the plunger that we are after. Noting that $Q_{\text{leak}} = 0.1 Q_0$, we can write the continuity equation (a) as

$$-\rho A_p v_p + 1.1 \rho Q_0 = 0,$$

from which we obtain

$$v_p = 1.1 \frac{Q_0}{A_p} = \frac{4.4 Q_0}{\pi d^2}. \tag{c}$$

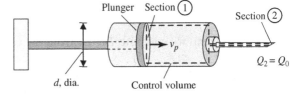

Figure 5.2.4. The syringe discussed in Example 5.2.3.

Plunger Section ①

Section ②

v_p

$Q_2 = Q_0$

d, dia.

Control volume

For $Q_0 = 250\,\text{cm}^3/\text{min}$ and $d = 25\,\text{mm}$, we obtain

$$v_p = \frac{4.4 \times (250 \times 10^3)}{\pi (25 \times 25)} = 560\,\text{mm/min}.$$

5.2.4 Continuity Equation in Material Description

Under the assumption that the mass is neither created nor destroyed during the motion, we require that the total mass of any material volume be the same at any instant during the motion. To express this in analytical terms, we consider a material body $\mathcal{B}$ that occupies configuration κ_0 with density ρ_0 and volume Ω_0 at time $t = 0$. The same material body occupies the configuration κ with volume Ω at time $t > 0$, and it has a density ρ. As per the principle of conservation of mass, we have

$$\int_{\Omega_0} \rho_0\, d\mathbf{X} = \int_{\Omega} \rho\, d\mathbf{x}. \tag{5.2.22}$$

Using the relation between $d\mathbf{X}$ and $d\mathbf{x}$, $d\mathbf{x} = J\, d\mathbf{X}$, where J is the determinant of the deformation gradient tensor $\mathbf{F}$, we arrive at

$$\int_{\Omega_0} (\rho_0 - J\rho)\, d\mathbf{X} = 0. \tag{5.2.23}$$

This is the *global form* of the *continuity equation*. Since the material volume Ω_0 we selected is arbitrarily small, as we shrink the volume to a point, we obtain the *local form* of the continuity equation

$$\rho_0 = J\rho. \tag{5.2.24}$$

The next example illustrates the use of the material time derivative in computing velocities and use of the continuity equation to compute the density in the current configuration.

EXAMPLE 5.2.4: Consider the motion of a body $\mathcal{B}$ described by the mapping

$$x_1 = \frac{X_1}{1 + t X_1}, \quad x_2 = X_2, \quad x_3 = X_3.$$

Determine the material density as a function of position $\mathbf{x}$ and time t.

SOLUTION: First, we compute the velocity components

$$\mathbf{v} = \frac{D\mathbf{x}}{Dt} = \left(\frac{\partial \mathbf{x}}{\partial t}\right)_{\mathbf{X}=\text{fixed}}; \quad v_i = \frac{Dx_i}{Dt} = \left(\frac{\partial x_i}{\partial t}\right)_{\mathbf{X}=\text{fixed}}. \tag{5.2.25}$$

Therefore, we have

$$v_1 = -\frac{X_1^2}{(1 + t X_1)^2} = -x_1^2, \quad v_2 = 0, \quad v_3 = 0.$$

Next, we compute $D\rho/Dt$ from the continuity equation (5.2.15):

$$\frac{D\rho}{Dt} = -\rho\,\text{div}\,\mathbf{v} = 2\rho x_1 = 2\rho\frac{X_1}{1 + t X_1}.$$

Integrating the above equation, we obtain

$$\int \frac{1}{\rho} D\rho = 2 \int \frac{X_1}{1 + t X_1} Dt \Rightarrow \ln \rho = 2 \ln(1 + t X_1) + \ln c,$$

where c is the constant of integration. If $\rho = \rho_0$ at time $t = 0$, we have $\ln c = \ln \rho_0$. Thus, the material density in the current configuration is

$$\rho = \rho_0 \left(1 + t X_1\right)^2 = \frac{\rho_0}{(1 - t x_1)^2}.$$

It can be verified that[1]

$$\frac{D\rho}{Dt} = \frac{\partial \rho}{\partial t} + v_1 \frac{\partial \rho}{\partial x_1} = \frac{2 \rho_0 x_1}{(1 - t x_1)^2} = 2 \rho x_1.$$

The material density in the current configuration can also be computed using the continuity equation in the material description, $\rho_0 = \rho J$. Noting that

$$dx_1 = \frac{1}{(1 + t X_1)^2} dX_1, \quad J = \frac{dx_1}{dX_1} = \frac{1}{(1 + t X_1)^2},$$

we obtain

$$\rho = \frac{1}{J} \rho_0 = \rho_0 (1 + t X_1)^2.$$

5.2.5 Reynolds Transport Theorem

The material derivative operator D/Dt corresponds to changes with respect to a fixed mass, that is, $\rho \, d\mathbf{x}$ is constant with respect to this operator. Therefore, from Eq. (5.2.8) it follows that, for $\phi = \rho Q(\mathbf{x}, t)$, the result

$$\frac{D}{Dt} \int_{\Omega} \rho Q(\mathbf{x}, t) \, d\mathbf{x} = \frac{\partial}{\partial t} \int_{\Omega} \rho Q \, d\mathbf{x} + \oint_{\Gamma} \rho Q \mathbf{v} \cdot \hat{\mathbf{n}} \, ds, \qquad (5.2.26)$$

or

$$\frac{D}{Dt} \int_{\Omega} \rho Q(\mathbf{x}, t) \, d\mathbf{x} = \int_{\Omega} \left[\rho \frac{\partial Q}{\partial t} + Q \frac{\partial \rho}{\partial t} + \mathrm{div}\,(\rho Q \mathbf{v}) \right] d\mathbf{x}$$

$$= \int_{\Omega} \left[\rho \left(\frac{\partial Q}{\partial t} + \mathbf{v} \cdot \nabla Q \right) + Q \left(\frac{\partial \rho}{\partial t} + \mathrm{div}(\rho \mathbf{v}) \right) \right] d\mathbf{x}, \qquad (5.2.27)$$

and using the continuity equation (5.2.12) and the definition of the material time derivative, we arrive at the result

$$\frac{D}{Dt} \int_{\Omega} \rho Q \, d\mathbf{x} = \int_{\Omega} \rho \frac{DQ}{Dt} \, d\mathbf{x}. \qquad (5.2.28)$$

Equation (5.2.28) is known as the *Reynolds transport theorem*.

---

[1] Note that $\rho = \rho(x_1, t)$, and $(1 + t X_1) = (1 - t x_1)^{-1}$.

5.3 Conservation of Momenta

5.3.1 Principle of Conservation of Linear Momentum

The principle of conservation of linear momentum, or Newton's second law of motion, applied to a set of particles (or rigid body) can be stated as *the time rate of change of (linear) momentum of a collection of particles equals the net force exerted on the collection.* Written in vector form, the principle implies

$$\frac{d}{dt}(m\mathbf{v}) = \mathbf{F}, \tag{5.3.1}$$

where m is the total mass, $\mathbf{v}$ the velocity, and $\mathbf{F}$ the resultant force on the collection of particles. For constant mass, Eq. (5.3.1) becomes

$$\mathbf{F} = m\frac{d\mathbf{v}}{dt} = m\mathbf{a}, \tag{5.3.2}$$

which is the familiar form of Newton's second law. Newton's second law for a control volume Ω can be expressed as

$$\mathbf{F} = \frac{\partial}{\partial t}\int_{\Omega}\rho\mathbf{v}\,d\mathbf{x} + \int_{\Gamma}\rho\mathbf{v}\mathbf{v}\cdot d\mathbf{s}, \tag{5.3.3}$$

where $\mathbf{F}$ is the resultant force and $d\mathbf{s}$ denotes the vector representing an area element of the outflow. Several simple examples that illustrate the use of Eq. (5.3.3) are presented next.

EXAMPLE 5.3.1: Suppose that a jet of fluid with area of cross-section A and mass density ρ issues from a nozzle with a velocity v and impinges against a smooth inclined flat plate, as shown in Figure 5.3.1. Assuming that there is no frictional resistance between the jet and plate, determine the distribution of the flow and the force required to keep the plate in position.

SOLUTION: Since there is no change in pressure or elevation before and after impact, the velocity of the fluid remains the same before and after impact. Let the amounts of flow to the left be Q_L and to the right be Q_R. Then the total flow $Q = vA$ of the jet is equal to the sum (by continuity equation)

$$Q = Q_L + Q_R.$$

Next, we use the principle of conservation of linear momentum to relate Q_L and Q_R. Applying Eq. (5.3.3) to the tangential direction to the plate and noting that the resultant force is zero and the first term on the right-hand side is zero by virtue of the steady-state condition, we obtain

$$0 = \int_{\Omega}\rho v_t\mathbf{v}\cdot d\mathbf{s} = \rho v(vA_L) + \rho(-v)(vA_R) + \rho v\cos\theta(-vA),$$

which, with $Q_L = A_L v$, $Q_R = A_R v$, and $Q = Av$, yields

$$Q_L - Q_R = Q\cos\theta.$$

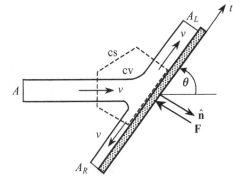

Figure 5.3.1. Jet of fluid impinging on an inclined plate.

Solving the two equations for Q_L and Q_R, we obtain

$$Q_L = \frac{1}{2}(1 + \cos\theta)\,Q, \quad Q_R = \frac{1}{2}(1 - \cos\theta)\,Q.$$

Thus, the total flow Q is divided into the left flow of Q_L and right flow of Q_R as given above.

The force exerted on the plate is normal to the plate. By applying the conservation of linear momentum in the normal direction, we obtain

$$-F_n = \int_\Omega \rho v_n \mathbf{v} \cdot d\mathbf{s} = \rho(v\sin\theta)(-vA) \;\rightarrow\; F_n = \rho\,Qv\sin\theta.$$

EXAMPLE 5.3.2: When a free jet of fluid impinges on a smooth (frictionless) curved vane with a velocity v, the jet is deflected in a tangential direction as shown in Figure 5.3.2, changing the momentum and exerting a force on the vane. Assuming that the velocity is uniform throughout the jet and there is no change in the pressure, determine the force exerted on a fixed vane.

SOLUTION: Applying Eq. (5.3.3), we obtain

$$\mathbf{F} \equiv -F_x\,\hat{\mathbf{e}}_x + F_y\,\hat{\mathbf{e}}_y = \int_{cv} \rho\mathbf{v}(\mathbf{v}\cdot d\mathbf{s})$$

$$= \rho v\,(\cos\theta\,\hat{\mathbf{e}}_x + \sin\theta\,\hat{\mathbf{e}}_y)\,vA + \rho v\,(-vA\,\hat{\mathbf{e}}_x)$$

or

$$F_x = \rho v^2 A\,(1 - \cos\theta), \quad F_y = \rho v^2 A\sin\theta.$$

When a jet of water ($\rho = 10^3$ kg/m^3) discharging 80 L/s at a velocity of 60 m/s is deflected through an angle of $\theta = 60°$, we obtain ($Q = vA$)

$$F_x = 10^3 \times 0.08 \times 60\,(1 - \cos 60°) = 2.4 \text{ kN},$$

$$F_y = 10^3 \times 0.08 \times 60\sin 60° = 4.157 \text{ kN}.$$

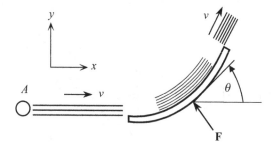

Figure 5.3.2. Jet of fluid deflected by a curved vane.

When the vane moves with a horizontal velocity of $v_0 < v$, Eq. (5.3.3) becomes

$$\mathbf{F} \equiv -F_x\,\hat{\mathbf{e}}_x + F_y\,\hat{\mathbf{e}}_y = \int_{\mathrm{cv}} \rho\mathbf{v}(\mathbf{v}\cdot d\mathbf{s})$$

$$= \rho(v - v_0)\,(\cos\theta\,\hat{\mathbf{e}}_x + \sin\theta\,\hat{\mathbf{e}}_y)\,(v - v_0)\,A$$

$$+ \rho(v - v_0)\,[-(v - v_0)A\,\hat{\mathbf{e}}_x],$$

from which we obtain

$$F_x = \rho(v - v_0)^2 A\,(1 - \cos\theta), \quad F_y = \rho(v - v_0)^2 A \sin\theta.$$

EXAMPLE 5.3.3: A chain of total length L and mass ρ per unit length slides down from the edge of a smooth table with an initial overhang x_0 to initiate motion, as shown in Figure 5.3.3. Assuming that the chain is rigid, find the equation of motion governing the chain and the tension in the chain.

SOLUTION: Let x be the amount of chain sliding down the table at any instant t. By considering the entire chain as the control volume, the linear momentum of the chain is

$$\rho(L - x)\cdot\dot{x}\,\hat{\mathbf{e}}_x - \rho x\cdot\dot{x}\,\hat{\mathbf{e}}_y.$$

The resultant force in the chain is $-\rho x g\,\hat{\mathbf{e}}_y$.
The principle of linear momentum gives

$$-\rho x g\,\hat{\mathbf{e}}_y = \frac{d}{dt}\left[\rho(L - x)\dot{x}\hat{\mathbf{e}}_x - \rho x\dot{x}\,\hat{\mathbf{e}}_y\right],$$

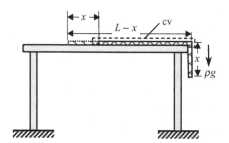

Figure 5.3.3. Chain sliding down a table.

or

$$(L - x)\ddot{x} - \dot{x}^2 = 0, \quad x\ddot{x} + \dot{x}^2 = gx.$$

Eliminating $\dot{x}^2$ from the two equations, we arrive at the equation of motion

$$\ddot{x} - \frac{g}{L}x = 0.$$

The solution of the second-order differential equation is

$$x(t) = A\cosh \lambda t + B\sinh \lambda t, \quad \text{where } \lambda = \sqrt{\frac{g}{L}}.$$

The constants of integration A and B are determined from the initial conditions

$$x(0) = x_0, \quad \dot{x}(0) = 0,$$

where x_0 denotes the initial overhang of the chain. We obtain

$$A = x_0, \quad B = 0,$$

and the solution becomes

$$x(t) = x_0 \cosh \lambda t, \quad \lambda = \sqrt{\frac{g}{L}}.$$

The tension in the chain can be computed by using the principle of linear momentum applied to the control volume of the chain on the table:

$$T = \frac{d}{dt}[\rho(L-x)\dot{x}] + \rho\dot{x}\dot{x} = \rho(L-x)\ddot{x} = \frac{\rho g}{L}(L-x)x,$$

where the term $\rho\dot{x}\dot{x}$ denotes the momentum flux.

EXAMPLE 5.3.4: Consider a chain of length L and mass density ρ per unit length which is piled on a stationary table, as shown in Figure 5.3.4. Determine the force F required to lift the chain with a constant velocity v.

SOLUTION: Let x be the height of the chain lifted off the table. Taking the control volume to be that enclosing the lifted chain and using Eq. (5.3.3), we obtain

$$F - \rho gx = \frac{\partial}{\partial t}(\rho v) + \rho vv = 0 + \rho v^2 \quad \rightarrow \quad F = \rho\left(gx + v^2\right).$$

The same result can be obtained using Newton's second law of motion:

$$F - \rho gx = \frac{d}{dt}(mv) = m\dot{v} + \dot{m}v = 0 + \dot{m}v,$$

where the rate of increase of mass m is $\dot{m} = \rho v$.

To derive the equation of motion applied to an arbitrarily fixed region in space through which material flows (i.e., control volume), we must identify the forces acting on it. Forces acting on a volume element can be classified as *internal* and *external*. The internal forces resist the tendency of one part of the region/body to be separated from another part. The internal force per unit area is termed *stress*, as defined

Figure 5.3.4. Lifting of a chain piled on a table.

in Eq. (4.2.1). The external forces are those transmitted by the body. The external forces can be further classified as *body* (or *volume*) *forces* and *surface forces*.

Body forces act on the distribution of mass inside the body. Examples of body forces are provided by the gravitational and electromagnetic forces. Body forces are usually measured per unit mass or unit volume of the body. Let **f** denote the body force per unit mass. Consider an elemental volume $d\mathbf{x}$ inside Ω. The body force of the elemental volume is equal to $\rho\, d\mathbf{x}\, \mathbf{f}$. Hence, the total body force of the control volume is

$$\int_{\Omega} \rho\, \mathbf{f}\, d\mathbf{x}. \tag{5.3.4}$$

Surface forces are contact forces acting on the boundary surface of the body. Examples of surface forces are provided by applied forces on the surface of the body. Surface forces are reckoned per unit area. Let **t** denote the surface force per unit area (or surface stress vector). The surface force on an elemental surface ds of the volume is $\mathbf{t}\, ds$. The total surface force acting on the closed surface of the region Ω is

$$\oint_{\Gamma} \mathbf{t}\, ds. \tag{5.3.5}$$

Since the stress vector **t** on the surface is related to the (internal) stress tensor σ by Cauchy's formula [see Eq. (4.2.10)]

$$\mathbf{t} = \hat{\mathbf{n}} \cdot \sigma, \tag{5.3.6}$$

where $\hat{\mathbf{n}}$ denotes the unit normal to the surface, we can express the surface force as

$$\oint_{\Gamma} \hat{\mathbf{n}} \cdot \sigma\, ds.$$

Using the divergence theorem, we can write

$$\oint_{\Gamma} \hat{\mathbf{n}} \cdot \sigma\, ds = \int_{\Omega} \nabla \cdot \sigma\, d\mathbf{x}. \tag{5.3.7}$$

The principle of conservation of linear momentum applied to a given mass of a medium $\mathcal{B}$, instantaneously occupying a region Ω with bounding surface Γ, and acted upon by external surface force $\mathbf{t}$ per unit area and body force $\mathbf{f}$ per unit mass, requires

$$\int_{\Omega} (\nabla \cdot \sigma + \rho \mathbf{f}) \, d\mathbf{x} = \frac{D}{Dt} \int_{\Omega} \rho \mathbf{v} \, d\mathbf{x}, \tag{5.3.8}$$

where $\mathbf{v}$ is the velocity vector. Using the Reynolds transport theorem, Eq. (5.2.28), we arrive at

$$0 = \int_{\Omega} \left[\nabla \cdot \sigma + \rho \mathbf{f} - \rho \frac{D\mathbf{v}}{Dt} \right] d\mathbf{x}, \tag{5.3.9}$$

which is the global form of the equation of motion. The local form is given by

$$\nabla \cdot \sigma + \rho \mathbf{f} = \rho \frac{D\mathbf{v}}{Dt} \tag{5.3.10}$$

or

$$\nabla \cdot \sigma + \rho \mathbf{f} = \rho \left(\frac{\partial \mathbf{v}}{\partial t} + \mathbf{v} \cdot \nabla \mathbf{v} \right). \tag{5.3.11}$$

In Cartesian rectangular system, we have

$$\frac{\partial \sigma_{ji}}{\partial x_j} + \rho f_i = \rho \left(\frac{\partial v_i}{\partial t} + v_j \frac{\partial v_i}{\partial x_j} \right). \tag{5.3.12}$$

In the case of steady-state conditions, Eq. (5.3.11) reduces to

$$\nabla \cdot \sigma + \rho \mathbf{f} = \rho \mathbf{v} \cdot \nabla \mathbf{v} \quad \text{or} \quad \frac{\partial \sigma_{ji}}{\partial x_j} + \rho f_i = \rho v_j \frac{\partial v_i}{\partial x_j}. \tag{5.3.13}$$

For kinematically infinitesimal deformation (i.e., when $\mathbf{S} \sim \sigma$) of solid bodies in static equilibrium, Eq. (5.3.10) reduces to the equations of equilibrium

$$\nabla \cdot \sigma + \rho \mathbf{f} = \mathbf{0} \quad \text{or} \quad \frac{\partial \sigma_{ji}}{\partial x_j} + \rho f_i = 0. \tag{5.3.14}$$

When the state of stress in the medium is of the form $\sigma = -p\mathbf{I}$ (i.e., hydrostatic state of stress), the equations of motion (5.3.10) reduce to

$$-\nabla p + \rho \mathbf{f} = \rho \frac{D\mathbf{v}}{Dt}. \tag{5.3.15}$$

Application of the stress equilibrium equations, Eq. (5.3.14), can be found in Examples 4.5.1 and 4.5.2.

5.3.2 Equations of Motion in Cylindrical and Spherical Coordinates

5.3.2.1 Cylindrical coordinates

To obtain the equations of motion in a cylindrical coordinate system, the operator ∇, velocity vector $\mathbf{v}$, body force vector $\mathbf{f}$, and stress tensor σ are written in cylindrical coordinates (r, θ, z) as

$$\nabla = \hat{\mathbf{e}}_r \frac{\partial}{\partial r} + \frac{1}{r}\hat{\mathbf{e}}_\theta \frac{\partial}{\partial \theta} + \hat{\mathbf{e}}_z \frac{\partial}{\partial z},$$

$$\mathbf{v} = \hat{\mathbf{e}}_r v_r + \hat{\mathbf{e}}_\theta v_\theta + \hat{\mathbf{e}}_z v_z,$$

$$\mathbf{f} = \hat{\mathbf{e}}_r f_r + \hat{\mathbf{e}}_\theta f_\theta + \hat{\mathbf{e}}_z f_z,$$

$$\boldsymbol{\sigma} = \sigma_{rr}\,\hat{\mathbf{e}}_r\hat{\mathbf{e}}_r + \sigma_{r\theta}\,\hat{\mathbf{e}}_r\hat{\mathbf{e}}_\theta + \sigma_{rz}\,\hat{\mathbf{e}}_r\hat{\mathbf{e}}_z$$

$$+ \sigma_{\theta r}\,\hat{\mathbf{e}}_\theta\hat{\mathbf{e}}_r + \sigma_{\theta\theta}\,\hat{\mathbf{e}}_\theta\hat{\mathbf{e}}_\theta + \sigma_{\theta z}\,\hat{\mathbf{e}}_\theta\hat{\mathbf{e}}_z$$

$$+ \sigma_{zr}\,\hat{\mathbf{e}}_z\hat{\mathbf{e}}_r + \sigma_{z\theta}\,\hat{\mathbf{e}}_z\hat{\mathbf{e}}_\theta + \sigma_{zz}\,\hat{\mathbf{e}}_z\hat{\mathbf{e}}_z. \tag{5.3.16}$$

Substituting these expressions into Eq. (5.3.11), we arrive at the following equations of motion in the cylindrical coordinate system:

$$\frac{\partial \sigma_{rr}}{\partial r} + \frac{1}{r}\frac{\partial \sigma_{\theta r}}{\partial \theta} + \frac{\partial \sigma_{zr}}{\partial z} + \frac{1}{r}(\sigma_{rr} - \sigma_{\theta\theta}) + \rho f_r$$

$$= \rho \left(\frac{\partial v_r}{\partial t} + v_r \frac{\partial v_r}{\partial r} + \frac{v_\theta}{r}\frac{\partial v_r}{\partial \theta} + v_z \frac{\partial v_r}{\partial z} - \frac{v_\theta^2}{r} \right),$$

$$\frac{\partial \sigma_{r\theta}}{\partial r} + \frac{1}{r}\frac{\partial \sigma_{\theta\theta}}{\partial \theta} + \frac{\partial \sigma_{z\theta}}{\partial z} + \frac{\sigma_{r\theta} + \sigma_{\theta r}}{r} + \rho f_\theta$$

$$= \rho \left(\frac{\partial v_\theta}{\partial t} + v_r \frac{\partial v_\theta}{\partial r} + \frac{v_\theta v_r}{r} + \frac{v_\theta}{r}\frac{\partial v_\theta}{\partial \theta} + v_z \frac{\partial v_\theta}{\partial z} \right),$$

$$\frac{\partial \sigma_{rz}}{\partial r} + \frac{1}{r}\frac{\partial \sigma_{\theta z}}{\partial \theta} + \frac{\partial \sigma_{zz}}{\partial z} + \frac{\sigma_{rz}}{r} + \rho f_z$$

$$= \rho \left(\frac{\partial v_z}{\partial t} + v_r \frac{\partial v_z}{\partial r} + \frac{v_\theta}{r}\frac{\partial v_z}{\partial \theta} + v_z \frac{\partial v_z}{\partial z} \right). \tag{5.3.17}$$

5.3.2.2 Spherical coordinates
In the spherical coordinate system (R, ϕ, θ), we write

$$\nabla = \hat{\mathbf{e}}_R \frac{\partial}{\partial R} + \frac{1}{R}\hat{\mathbf{e}}_\phi \frac{\partial}{\partial \phi} + \frac{1}{R\sin\phi}\hat{\mathbf{e}}_\theta \frac{\partial}{\partial \theta},$$

$$\mathbf{v} = \hat{\mathbf{e}}_R v_R + \hat{\mathbf{e}}_\phi v_\phi + \hat{\mathbf{e}}_\theta v_\theta,$$

$$\mathbf{f} = \hat{\mathbf{e}}_R f_R + \hat{\mathbf{e}}_\phi f_\phi + \hat{\mathbf{e}}_\theta f_\theta,$$

$$\boldsymbol{\sigma} = \sigma_{RR}\,\hat{\mathbf{e}}_R\hat{\mathbf{e}}_R + \sigma_{R\phi}\,\hat{\mathbf{e}}_R\hat{\mathbf{e}}_\phi + \sigma_{R\theta}\,\hat{\mathbf{e}}_R\hat{\mathbf{e}}_\theta$$

$$+ \sigma_{\phi R}\,\hat{\mathbf{e}}_\phi\hat{\mathbf{e}}_R + \sigma_{\phi\phi}\,\hat{\mathbf{e}}_\phi\hat{\mathbf{e}}_\phi + \sigma_{\phi\theta}\,\hat{\mathbf{e}}_\phi\hat{\mathbf{e}}_\theta$$

$$+ \sigma_{\theta R}\,\hat{\mathbf{e}}_\phi\hat{\mathbf{e}}_\theta + \sigma_{\theta\phi}\,\hat{\mathbf{e}}_\theta\hat{\mathbf{e}}_\phi + \sigma_{\theta\theta}\,\hat{\mathbf{e}}_\theta\hat{\mathbf{e}}_\theta. \tag{5.3.18}$$

Substituting these expressions into Eq. (5.3.11), we arrive at the following equations of motion in the spherical coordinate system (see Problem 5.11):

$$\frac{\partial \sigma_{RR}}{\partial R} + \frac{1}{R}\frac{\partial \sigma_{\phi R}}{\partial \phi} + \frac{1}{\sin\phi}\frac{\partial \sigma_{\theta R}}{\partial \theta} + \frac{1}{R}(2\sigma_{RR} - \sigma_{\phi\phi} - \sigma_{\theta\theta} + \sigma_{\phi R}\cot\phi) + \rho f_R$$

$$= \rho\left(\frac{\partial v_R}{\partial t} + v_R\frac{\partial u_R}{\partial R} + v_\phi\frac{\partial u_\phi}{\partial R} + v_\theta\frac{\partial v_\theta}{\partial R}\right),$$

$$\frac{\partial \sigma_{R\phi}}{\partial R} + \frac{1}{R}\frac{\partial \sigma_{\phi\phi}}{\partial \phi} + \frac{1}{R\sin\phi}\frac{\partial \sigma_{\theta\phi}}{\partial \theta} + \frac{1}{R}[(\sigma_{\phi\phi} - \sigma_{\theta\theta})\cot\phi + S_{\phi R} + 2\sigma_{R\phi}] + \rho f_\phi$$

$$= \frac{\rho}{R}\left[R\frac{\partial v_\phi}{\partial t} + v_R\left(\frac{\partial v_R}{\partial \phi} - v_\phi\right) + v_\phi\left(\frac{\partial u_\phi}{\partial \phi} + u_R\right) + v_\theta\frac{\partial u_\theta}{\partial \phi}\right],$$

$$\frac{\partial \sigma_{R\theta}}{\partial R} + \frac{1}{R}\frac{\partial \sigma_{\phi\theta}}{\partial \phi} + \frac{1}{R\sin\phi}\frac{\partial \sigma_{\theta\theta}}{\partial \theta} + \frac{1}{R}[(\sigma_{\theta\phi} + \sigma_{\phi\theta})\cot\phi + \sigma_{\theta R}] + \rho f_\theta$$

$$= \frac{\rho}{R\sin\phi}\left[R\sin\phi\frac{\partial v_\theta}{\partial t} + v_R\left(\frac{\partial v_R}{\partial \theta} - u_\theta\sin\phi\right)\right.$$

$$\left. + v_\phi\left(\frac{\partial v_\phi}{\partial \theta} - u_\theta\cos\phi\right) + v_\theta\left(\frac{\partial u_\theta}{\partial \theta} + v_R\sin\phi + u_\phi\cos\phi\right)\right]. \qquad (5.3.19)$$

5.3.3 Principle of Conservation of Angular Momentum

The principle of conservation of angular momentum states that *the time rate of change of the total moment of momentum for a continuum is equal to vector sum of the moments of external forces acting on the continuum.* The principle as applied to a control volume Ω with a control surface Γ can be expressed as

$$\mathbf{r} \times \mathbf{F} = \frac{\partial}{\partial t}\int_\Omega \rho\mathbf{r} \times \mathbf{v}\, dx + \int_\Gamma \rho\mathbf{r} \times \mathbf{v}\,(\mathbf{v}\cdot d\mathbf{s}). \qquad (5.3.20)$$

An application of the principle is presented next.

EXAMPLE 5.3.5: Consider the top view of a sprinkler as shown in Figure 5.3.5. The sprinkler discharges water outward in a horizontal plane (which is in the plane of the paper). The sprinkler exits are oriented at an angle of θ from the tangent line to the circle formed by rotating the sprinkler about its vertical centerline. The sprinkler has a constant cross-sectional flow area of A_0 and discharges a flow rate of Q when $\omega = 0$ at time $t = 0$. Hence, the radial velocity is equal to $v_r = Q/2A_0$. Determine ω (counterclockwise) as a function of time.

SOLUTION: Suppose that the moment of inertia of the empty sprinkler head is I_z and the resisting torque due to friction (from bearings and seals) is T (clockwise), we take the control volume to be the cylinder of unit height, formed by the rotating sprinkler head. The inflow, being along the axis, has no moment. Thus the time rate of moment of momentum of the sprinkler head plus the net

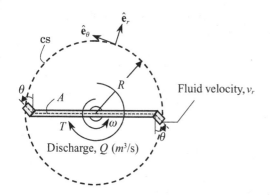

Figure 5.3.5. A rotating sprinkler system.

efflux of the moment of momentum from the control surface is equal to the torque T:

$$-T\hat{\mathbf{e}}_z = \left[2\frac{d}{dt}\int_0^R A_0\rho\omega r^2\,dr + I_z\frac{d\omega}{dt} + 2R\left(\rho\frac{Q}{2}\right)(\omega R - v_r\cos\theta)\right]\hat{\mathbf{e}}_z,$$

where the first term represents the time rate of change of the moment of momentum [moment arm times mass of a differential length dr times the velocity: $r \times (\rho A_0\,dr)(\omega r)$], the second term is the time rate of change of angular momentum, and the last term represents the efflux of the moment of momentum at the control surface (i.e., exit of the sprinkler nozzles). Simplifying the equation, we arrive at

$$\left(I_z + \frac{2}{3}\rho A_0 R^3\right)\frac{d\omega}{dt} + \rho Q R^2\omega = \rho Q R v_r\cos\theta - T.$$

The above equation indicates that for rotation to start $\rho Q R v_r\cos\theta - T > 0$. The final value of ω is obtained when the sprinkler motion reaches the steady state, i.e., $d\omega/dt = 0$. Thus, at steady state, we have

$$\omega_f = \frac{v_r}{R}\cos\theta - \frac{T}{\rho Q R^2}.$$

In the absence of body couples (i.e., volume-dependent couples $\mathbf{M}$)

$$\lim_{\Delta A\to 0}\frac{\Delta\mathbf{M}}{\Delta A} = \mathbf{0}, \tag{5.3.21}$$

the mathematical statement of the angular momentum principle as applied to a continuum is

$$\oint_\Gamma (\mathbf{x}\times\mathbf{t})ds + \int_v(\mathbf{x}\times\rho\mathbf{f})dv = \frac{D}{Dt}\int_v(\mathbf{x}\times\rho\mathbf{v})dv. \tag{5.3.22}$$

In index notation (kth component), Eq. (5.3.22) takes the form

$$\oint_\Gamma e_{ijk}x_i t_j\,ds + \int_v \rho e_{ijk}x_i f_j dv = \frac{D}{Dt}\int_v \rho e_{ijk}x_i v_j\,dv. \tag{5.3.23}$$

We use several steps to simplify the expression. First replace t_j with $t_j = n_p\sigma_{pj}$. Then transform the surface integral to a volume integral and use the Reynold's transport theorem for the material time derivative of a volume integral to obtain

$$\int_v e_{ijk} (x_i \sigma_{pj})_{,p} \, dv + \int_v \rho e_{ijk} x_i f_j dv = \int_v \rho e_{ijk} \frac{D}{Dt} (x_i v_j) \, dv.$$

Carrying out the indicated differentiations and noting $Dx_i / Dt = v_i$, we obtain

$$\int_v e_{ijk} (x_i \sigma_{pj,p} + \delta_{ip} \sigma_{pj} + \rho x_i f_j) \, dv = \int_v \rho e_{ijk} \left(v_i v_j + x_i \frac{D v_j}{Dt} \right) dv,$$

$$\int_v e_{ijk} \left[x_i \left(\sigma_{pj,p} + \rho f_j - \rho \frac{D v_j}{Dt} \right) + \sigma_{ij} \right] dv = 0,$$

or

$$e_{ijk} \sigma_{ij} = 0. \tag{5.3.24}$$

Equation (5.3.24) necessarily implies that $\sigma_{ij} = \sigma_{ji}$. To see this, expand the above expression for all values of the free index k:

$$k = 1: \quad \sigma_{23} - \sigma_{32} = 0,$$

$$k = 2: \quad \sigma_{31} - \sigma_{13} = 0, \tag{5.3.25}$$

$$k = 3: \quad \sigma_{21} - \sigma_{12} = 0.$$

These statements clearly show that $\sigma_{ij} = \sigma_{ji}$ or $\sigma = \sigma^T$. Thus, there are only six stress components that are independent, discussed in Section 4.4.

5.4 Thermodynamic Principles

5.4.1 Introduction

The first law of thermodynamics is commonly known as the *principle of conservation of energy*, and it can be regarded as a statement of the interconvertibility of heat and work. The law does not place any restrictions on the direction of the process. For instance, in the study of mechanics of particles and rigid bodies, the kinetic energy and potential energy can be fully transformed from one to the other in the absence of friction and other dissipative mechanisms. From our experience, we know that mechanical energy that is converted into heat cannot all be converted back into mechanical energy. For example, the motion (i.e., kinetic energy) of a flywheel can all be converted into heat (i.e., internal energy) by means of a friction brake; if the whole system is insulated, the internal energy causes the temperature of the system to rise. Although the first law does not restrict the reversal process, namely, the conversion of heat to internal energy and internal energy to motion (the flywheel), such a reversal cannot occur because the frictional dissipation is an *irreversible process*. The second law of thermodynamics provides the restriction on the interconvertibility of energies.

5.4.2 The First Law of Thermodynamics: Energy Equation

The first law of thermodynamics states that *the time-rate of change of the total energy is equal to the sum of the rate of work done by the external forces and the change of heat content per unit time.* The total energy is the sum of the kinetic energy and the internal energy. The principle of conservation of energy can be expressed as

$$\frac{D}{Dt}(K + U) = W + H. \tag{5.4.1}$$

Here, K denotes the kinetic energy, U is the internal energy, W is the power input, and H is the heat input to the system.

The kinetic energy of the system is given by

$$K = \frac{1}{2}\int_\Omega \rho\mathbf{v} \cdot \mathbf{v}\, d\mathbf{x}, \tag{5.4.2}$$

where $\mathbf{v}$ is the velocity vector. If e is the energy per unit mass (or *specific internal energy*), the total internal energy of the system is given by

$$U = \int_\Omega \rho e\, d\mathbf{x}. \tag{5.4.3}$$

The kinetic energy (K) of a system is the energy associated with the macroscopically observable velocity of the continuum. The kinetic energy associated with the (microscopic) motions of molecules of the continuum is a part of the internal energy; the elastic strain energy and other forms of energy are also parts of internal energy, U.

The power input, in the nonpolar case (i.e., without body couples), consists of the rate of work done by external surface tractions $\mathbf{t}$ per unit area and body forces $\mathbf{f}$ per unit volume of the region Ω bounded by Γ:

$$
\begin{aligned}
W &= \oint_\Gamma \mathbf{t} \cdot \mathbf{v}\, ds + \int_\Omega \rho\mathbf{f} \cdot \mathbf{v}\, d\mathbf{x} \\
&= \oint_\Gamma (\hat{\mathbf{n}} \cdot \boldsymbol{\sigma}) \cdot \mathbf{v}\, ds + \int_\Omega \rho\mathbf{f} \cdot \mathbf{v}\, d\mathbf{x} \\
&= \int_\Omega [\boldsymbol{\nabla} \cdot (\boldsymbol{\sigma} \cdot \mathbf{v}) + \rho\mathbf{f} \cdot \mathbf{v}]\, d\mathbf{x} \\
&= \int_\Omega [(\boldsymbol{\nabla} \cdot \boldsymbol{\sigma} + \rho\mathbf{f}) \cdot \mathbf{v} + \boldsymbol{\sigma} : \boldsymbol{\nabla}\mathbf{v}]\, d\mathbf{x} \\
&= \int_\Omega \left(\rho\frac{D\mathbf{v}}{Dt} \cdot \mathbf{v} + \boldsymbol{\sigma} : \boldsymbol{\nabla}\mathbf{v}\right) d\mathbf{x},
\end{aligned} \tag{5.4.4}
$$

where ":" denotes the double-dot product $\boldsymbol{\Phi} : \boldsymbol{\Psi} = \Phi_{ij}\Psi_{ji}$. The Cauchy formula, symmetry of the stress tensor, and the equation of motion (5.3.9) are used in arriving at the last line. Using the symmetry of σ, we can write $\sigma : \nabla \mathbf{v} = \sigma : \mathbf{D}$. Hence, we can write

$$
\begin{aligned}
W &= \frac{1}{2} \int_\Omega \rho \frac{D}{Dt} (\mathbf{v} \cdot \mathbf{v}) \, d\mathbf{x} + \int_\Omega \sigma : \mathbf{D} \, d\mathbf{x} \\
&= \frac{1}{2} \frac{D}{Dt} \int_\Omega \rho \, \mathbf{v} \cdot \mathbf{v} \, d\mathbf{x} + \int_\Omega \sigma : \mathbf{D} \, d\mathbf{x},
\end{aligned}
\tag{5.4.5}
$$

where $\mathbf{D}$ is the rate of deformation tensor [see Eq. (3.6.2)]

$$
\mathbf{D} = \frac{1}{2} \left[\nabla \mathbf{v} + (\nabla \mathbf{v})^T \right],
$$

and the Reynolds transport theorem (5.2.28) used to write the final expression.

The rate of heat input consists of conduction through the surface Γ and heat generation inside the region Ω (possibly from a radiation field or transmission of electric current). Let $\mathbf{q}$ be the heat flux vector and $\mathcal{E}$ be the internal heat generation per unit mass. Then the heat inflow across the surface element ds is $-\mathbf{q} \cdot \hat{\mathbf{n}} \, ds$, and internal heat generation in volume element $d\mathbf{x}$ is $\rho \mathcal{E} d\mathbf{x}$. Hence, the total heat input is

$$
H = - \oint_\Gamma \mathbf{q} \cdot \hat{\mathbf{n}} \, ds + \int_\Omega \rho \mathcal{E} \, d\mathbf{x} = \int_\Omega (-\nabla \cdot \mathbf{q} + \rho \mathcal{E}) \, d\mathbf{x}.
\tag{5.4.6}
$$

Substituting expressions for K, U, W, and H from Eqs. (5.4.2), (5.4.3), (5.4.5), and (5.4.6) into Eq. (5.4.1), we obtain

$$
\frac{D}{Dt} \int_\Omega \rho \left(\frac{1}{2} \mathbf{v} \cdot \mathbf{v} + e \right) d\mathbf{x} = \frac{1}{2} \frac{D}{Dt} \int_\Omega \rho \, \mathbf{v} \cdot \mathbf{v} \, d\mathbf{x} + \int_\Omega (\sigma : \mathbf{D} - \nabla \cdot \mathbf{q} + \rho \mathcal{E}) \, d\mathbf{x}
$$

or

$$
0 = \int_\Omega \left(\rho \frac{De}{Dt} - \sigma : \mathbf{D} + \nabla \cdot \mathbf{q} - \rho \mathcal{E} \right) d\mathbf{x},
\tag{5.4.7}
$$

which is the global form of the energy equation. The local form of the energy equation is given by

$$
\rho \frac{De}{Dt} = \sigma : \mathbf{D} - \nabla \cdot \mathbf{q} + \rho \mathcal{E},
\tag{5.4.8}
$$

which is known as the *thermodynamic form* of the energy equation for a continuum. The term $\sigma : \mathbf{D}$ is known as the *stress power*, which can be regarded as the internal production of energy. Special forms of this equation in various field problems will be discussed next.

5.4.3 Special Cases of Energy Equation

In the case of viscous fluids, the total stress tensor σ is decomposed into a viscous part and a pressure part:

$$
\sigma = \tau - p \mathbf{I},
\tag{5.4.9}
$$

where p is the hydrostatic pressure and τ is the viscous stress tensor. Then Eq. (5.4.8) can be written as (note that $\mathbf{I} : \mathbf{D} = \nabla \cdot \mathbf{v}$)

$$\rho \frac{De}{Dt} = \Phi - p\nabla \cdot \mathbf{v} - \nabla \cdot \mathbf{q} + \rho\mathcal{E}, \qquad (5.4.10)$$

where Φ is called the *viscous dissipation* function,

$$\Phi = \tau : \mathbf{D}. \qquad (5.4.11)$$

For incompressible materials (i.e., div $\mathbf{v} = 0$), Eq. (5.4.10) reduces to

$$\rho \frac{De}{Dt} = \Phi - \nabla \cdot \mathbf{q} + \rho\mathcal{E}. \qquad (5.4.12)$$

For heat transfer in a medium, the internal energy e is expressed as

$$e = h - \frac{P}{\rho} = h - Pv, \qquad (5.4.13)$$

where h is the *specific enthalpy,* P is the thermodynamic pressure, and $v = 1/\rho$ is the *specific volume.* Then we have

$$\frac{Dh}{Dt} = \frac{De}{Dt} + \frac{1}{\rho}\frac{DP}{Dt} - \frac{P}{\rho^2}\frac{D\rho}{Dt}. \qquad (5.4.14)$$

Substituting for De/Dt from Eq. (5.4.14) into Eq. (5.4.10), we arrive at the expression

$$\rho \frac{Dh}{Dt} = \Phi - \nabla \cdot \mathbf{q} + \rho\mathcal{E} + \frac{DP}{Dt} - \frac{P}{\rho}\left(\frac{D\rho}{Dt} + \rho\,\mathrm{div}\,\mathbf{v}\right), \qquad (5.4.15)$$

or, using the continuity of mass equation (5.3.14)

$$\rho \frac{Dh}{Dt} = \Phi - \nabla \cdot \mathbf{q} + \rho\mathcal{E} + \frac{DP}{Dt}. \qquad (5.4.16)$$

In general, the change in specific enthalpy, specific entropy and internal energy are expressed by the canonical relations

$$dh = \theta\,d\eta + v\,dP, \qquad (5.4.17)$$

$$de = \theta\,d\eta - P\,dv, \qquad (5.4.18)$$

where η is the *specific entropy* and θ is the absolute temperature. The *Gibb's energy* is defined to be

$$G = h - \theta\eta, \qquad (5.4.19)$$

which relates the enthalpy and entropy.

The concept of entropy is a difficult one to explain in simple terms; it has its roots in statistical physics and thermodynamics and is generally considered as a measure of the tendency of the atoms toward a disorder. For example, carbon has a lower entropy in the form of diamond, a hard crystal with atoms closely bound in a highly ordered array. Entropy is also considered as a variable conjugate to temperature θ (i.e., $\theta = \partial e/\partial \eta$).

5.4.3.1 Ideal Gas

An ideal fluid is inviscid (i.e., nonviscous) and incompressible. For an ideal gas, the specific internal energy, specific enthalpy, and specific entropy are given by

$$de = c_v \, d\theta, \quad dh = c_P \, d\theta, \quad d\eta = c_P \frac{d\theta}{\theta},$$

(5.4.20)

where c_v and c_P are specific heats at constant volume and constant pressure, respectively,

$$c_P = \left(\frac{\partial h}{\partial \theta} \right)_P, \quad c_v = \left(\frac{\partial e}{\partial \theta} \right)_v.$$

(5.4.21)

For this case, the energy equation (5.4.10) takes the form

$$\rho c_P \frac{D\theta}{Dt} = -\nabla \cdot \mathbf{q} + \rho \mathcal{E} + \frac{DP}{Dt}.$$

(5.4.22)

5.4.3.2 Incompressible Liquid

For an incompressible liquid, the specific internal energy, specific enthalpy, and specific entropy are given by

$$de = c \, d\theta, \quad dh = c \, d\theta + v \, dP, \quad d\eta = c \frac{d\theta}{\theta},$$

(5.4.23)

where c is the specific heat. The energy equation takes the form

$$\rho c \frac{D\theta}{Dt} = \Phi - \nabla \cdot \mathbf{q} + \rho \mathcal{E}.$$

(5.4.24)

5.4.3.3 Pure Substance

In general, the specific internal energy, specific enthalpy, and specific entropy are given by

$$de = c_v \, d\theta + \left[\theta \left(\frac{\partial P}{\partial \theta} \right)_v - P \right] dv, \quad dh = c \, d\theta + \left[-\theta \left(\frac{\partial v}{\partial \theta} \right)_P + v \right] dP,$$

$$d\eta = c_P \frac{d\theta}{\theta} - \left(\frac{\partial v}{\partial \theta} \right)_P dP = c_v \frac{d\theta}{\theta} + \left(\frac{\partial P}{\partial \theta} \right)_v dv.$$

(5.4.25)

The energy equation takes the form

$$\rho c_P \frac{D\theta}{Dt} = \Phi - \nabla \cdot \mathbf{q} + \rho \mathcal{E} + \beta T \frac{DP}{Dt},$$

(5.4.26)

where β is the thermal coefficient of thermal expansion

$$\beta = -\frac{1}{\rho} \left(\frac{\partial \rho}{\partial \theta} \right)_P.$$

(5.4.27)

5.4.4 Energy Equation for One-Dimensional Flows

Various forms of energy equation derived in the preceding sections are valid for any continuum. For simple, one-dimensional flow problems (i.e., problems with one stream of fluid particles), the equations derived are too complicated to be of use.

In this section, a simple form of the energy equation is derived for use with one-dimensional fluid flow problems.

The first law of thermodynamics for a system occupying the domain (control volume) Ω can be written as

$$\frac{D}{Dt} \int_\Omega \rho \epsilon \, dV = W_{\text{net}} + H_{\text{net}}, \qquad (5.4.28)$$

where ϵ is the total energy stored per unit mass, W_{net} is the net rate of work transferred into the system, and H_{net} is the net rate of heat transfer into the system. The total stored energy per unit mass ϵ consists of the internal energy per unit mass e, the kinetic energy per unit mass $v^2/2$, and the potential energy per unit mass gz (g is the gravitational acceleration and z is the vertical distance above a reference value)

$$\epsilon = e + \frac{v^2}{2} + gz. \qquad (5.4.29)$$

The rate of work done in the absence of body forces is given by ($\sigma = \tau - P\mathbf{I}$)

$$W_{\text{net}} = W_{\text{shaft}} - \oint_\Gamma P\mathbf{v} \cdot \hat{\mathbf{n}} \, ds, \qquad (5.4.30)$$

where P is the pressure (normal stress) and W_{shaft} is the rate of work done by the tangential force (due to shear stress, in rotary devices such as fans, propellers, and turbines).

Using the Reynolds transport theorem (5.2.26) and Eqs. (5.4.29) and (5.4.30), we can write (5.4.28) as

$$\frac{\partial}{\partial t} \int_\Omega \rho \epsilon \, dV + \oint_\Gamma \left(e + \frac{P}{\rho} + \frac{v^2}{2} + gz \right) \rho \mathbf{v} \cdot \hat{\mathbf{n}} \, ds = W_{\text{shaft}} + H_{\text{net}}. \qquad (5.4.31)$$

If only one stream of fluid (compressible or incompressible) enters the control volume, the integral over the control surface in Eq. (5.4.31) can be written as

$$\left(e + \frac{P}{\rho} + \frac{v^2}{2} + gz \right)_{\text{out}} (\rho Q)_{\text{out}} - \left(e + \frac{P}{\rho} + \frac{v^2}{2} + gz \right)_{\text{in}} (\rho Q)_{\text{in}}, \qquad (5.4.32)$$

where ρQ denotes the mass flow rate. Finally, if the flow is steady, Eq. (5.4.31) can be written as

$$\left(e + \frac{P}{\rho} + \frac{v^2}{2} + gz \right)_{\text{out}} (\rho Q)_{\text{out}} - \left(e + \frac{P}{\rho} + \frac{v^2}{2} + gz \right)_{\text{in}} (\rho Q)_{\text{in}} = W_{\text{shaft}} + H_{\text{net}}.$$
$$(5.4.33)$$

In writing the above equation, it is assumed that the flow is one-dimensional and the velocity field is uniform. If the velocity profile at sections crossing the control surface is not uniform, correction must be made to Eq. (5.4.33). In particular, when the velocity profile is not uniform, the integral

$$\oint_\Gamma \frac{v^2}{2} \rho \mathbf{v} \cdot \hat{\mathbf{n}} \, ds$$

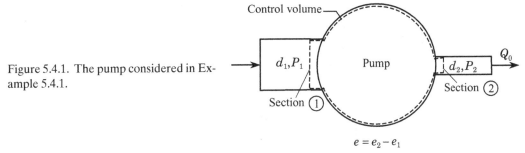

Figure 5.4.1. The pump considered in Example 5.4.1.

$$e = e_2 - e_1$$

cannot be replaced with $(v^2/2)(\rho Q) = \rho A v^3/2$, where A is the cross-section area of the flow because integral of v^3 is different when v is uniform or varies across the section. If we define the ratio, called the *kinetic energy coefficient*

$$\alpha = \frac{\oint_\Gamma \frac{v^2}{2} \rho \mathbf{v} \cdot \hat{\mathbf{n}} \, ds}{(\rho Q v^2/2)}, \tag{5.4.34}$$

Eq. (5.4.33) can be expressed as

$$\left(e + \frac{P}{\rho} + \frac{\alpha v^2}{2} + gz \right)_{\text{out}} (\rho Q)_{\text{out}} - \left(e + \frac{P}{\rho} + \frac{\alpha v^2}{2} + gz \right)_{\text{in}} (\rho Q)_{\text{in}} = W_{\text{shaft}} + H_{\text{net}}. \tag{5.4.35}$$

An example of the application of energy equation (5.4.35) is presented next.

EXAMPLE 5.4.1: A pump delivers water at a steady rate of Q_0 (gal/min), as shown in Figure 5.4.1. If the left-side pipe is of diameter d_1 (in.) and the right-side pipe is of diameter d_2 (in.), and the pressures in the two pipes are p_1 and p_2 (psi), respectively, determine the horsepower (hp) required by the pump if the rise in the internal energy across the pump is e. Assume that there is no change of elevation in water level across the pump, and the pumping process is adiabatic (i.e., the heat transfer rate is zero). Use the following data ($\alpha = 1$):

$$\rho = 1.94 \text{ slugs/ft}^3, \quad d_1 = 4 \text{ in.}, d_2 = 1 \text{ in.},$$

$$P_1 = 20 \text{ psi}, \quad P_2 = 50 \text{ psi}, \quad Q_0 = 350 \text{ gal/min}, \quad e = 3300 \text{ lb-ft/slug}.$$

SOLUTION: We take the control volume between the entrance and exit sections of the pump, as shown in dotted lines in Figure 5.4.1. The mass flow rate entering and exiting the pump is the same (conservation of mass) and equal to

$$\rho Q_0 = \frac{1.94 \times 350}{7.48 \times 60} = 1.513 \text{ slugs/s}.$$

The velocities at Sections 1 and 2 are (converting all quantities to proper units) are

$$v_1 = \frac{Q_0}{A_1} = \frac{350}{7.48 \times 60} \frac{4 \times 144}{16\pi} = 8.94 \text{ ft/s},$$

$$v_2 = \frac{Q_0}{A_2} = \frac{350}{7.48 \times 60} \frac{4 \times 144}{\pi} = 143 \text{ ft/s}.$$

For adiabatic flow $H_{net} = 0$, the potential energy term is zero on account of no elevation difference between the entrance and exits, and $e = e_2 - e_1 = 3300$ ft-lb/slug. Thus, we have

$$W_{shaft} = \rho Q_0 \left[\left(e + \frac{P}{\rho} + \frac{v^2}{2} \right)_2 - \left(e + \frac{P}{\rho} + \frac{v^2}{2} \right)_1 \right]$$

$$= (1.513) \left[3300 + \frac{(50 - 20) \times 144}{1.94} + \frac{(143)^2 - (8.94)^2}{2} \right] \frac{1}{550} = 43.22 \text{ hp.}$$

5.4.5 The Second Law of Thermodynamics

For the sake of completeness, we briefly review the second law of thermodynamics. The second law of thermodynamics for a reversible process states that there exists a function $\eta = \eta(\varepsilon, \theta)$, called the *specific entropy* (or entropy per unit mass), such that

$$d\eta = \left[\frac{\rho\varepsilon - \nabla \cdot \mathbf{q}}{\rho\theta} \right] dt \qquad (5.4.35)$$

is a perfect differential. Here θ denotes the temperature. The product $-\theta\eta$ is the irreversible heat energy due to entropy as related to temperature. Equation (5.4.35) is called the *entropy equation of state*. Using the energy equation (5.4.8), Eq. (5.4.35) can be expressed as

$$\rho \frac{De}{Dt} = \sigma : \mathbf{D} + \rho\theta\dot{\eta}, \qquad (5.4.36)$$

where the superposed dot indicates the time derivative.

For an irreversible process, the second law of thermodynamics requires that the sum of viscous and thermal dissipation rates (i.e., entropy production) must be positive. The entropy production is

$$\int_\Omega \rho\eta \, d\mathbf{x}, \qquad (5.4.37)$$

where η is the entropy per unit mass. The entropy input rate is

$$\int_\Omega \left(\frac{\rho\varepsilon}{\theta} \right) d\mathbf{x} - \oint_\Gamma \left(\frac{\mathbf{q}}{\theta} \right) \cdot \hat{\mathbf{n}} \, ds. \qquad (5.4.38)$$

The second law of thermodynamics places the restriction that the rate of entropy increase must be greater than the entropy input rate

$$\frac{D}{Dt} \int_\Omega \rho\eta \, d\mathbf{x} \geq \int_\Omega \left[\left(\frac{\rho\varepsilon}{\theta} \right) - \nabla \cdot \left(\frac{\mathbf{q}}{\theta} \right) - \right] d\mathbf{x}. \qquad (5.4.39)$$

The local form of the second law of thermodynamics, known as the *Clausius–Duhem inequality*, or *entropy inequality*, is

$$\frac{D\eta}{Dt} \geq \frac{\varepsilon}{\theta} - \frac{1}{\rho} \nabla \cdot \left(\frac{\mathbf{q}}{\theta} \right). \qquad (5.4.40)$$

The quantity $\mathbf{q}/\theta$ is known as the *entropy flux* and ε/θ as the *entropy supply density*.

The sum of internal energy (e) and irreversible heat energy ($-\theta\eta$) is known as *Helmhotz free energy*

$$\Psi = e - \theta\eta. \tag{5.4.41}$$

Substituting Eq. (5.4.41) into Eq. (5.4.8), we obtain

$$\rho\frac{D\Psi}{Dt} = \boldsymbol{\sigma} : \mathbf{D} - \rho\frac{D\theta}{Dt}\eta - \mathcal{D}, \tag{5.4.42}$$

where $\mathcal{D}$ is the *internal dissipation*

$$\mathcal{D} = \rho\theta\frac{D\eta}{Dt} + \nabla \cdot \mathbf{q} - \rho\mathcal{E}. \tag{5.4.43}$$

In view of Eq. (5.4.40), we can write

$$\mathcal{D} - \frac{1}{\theta}\mathbf{q} \cdot \nabla\theta \geq 0. \tag{5.4.44}$$

We have $\mathcal{D} > 0$ for an irreversible process, and $\mathcal{D} = 0$ for a reversible process because it is always true that

$$-\frac{1}{\theta}\mathbf{q} \cdot \nabla\theta \geq 0. \tag{5.4.45}$$

5.5 Summary

This chapter was devoted to the derivation of the field equations governing a continuous medium using the principles of conservation of mass, momenta, and energy and therefore constitutes the heart of the book. The equations are derived in invariant (i.e., vector and tensor) form so that they can be expressed in any chosen coordinate system (e.g., rectangular, cylindrical, spherical, or even a curvilinear system). The principle of conservation of mass results in the continuity equation; the principle of conservation of linear momentum, which is equivalent to Newton's second law of motion, leads to the equations of motion in terms of the Cauchy stress tensor; the principle of conservation of angular momentum yields, in the absence of body couples, in the symmetry of Cauchy stress tensor; and the principles of thermodynamics – the first and second laws of thermodynamics – give rise to the energy equation and Clausius–Duhem inequality. Examples to illustrate the conservation principles are also presented.

In closing this chapter, we summarize the invariant form of the equations resulting from the application of conservation principles to a continuum. The variables appearing in the equations were already defined and will not be repeated here.

Conservation of mass

$$\frac{\partial\rho}{\partial t} + \text{div}(\rho\mathbf{v}) = 0. \tag{5.5.1}$$

Conservation of linear momentum

$$\nabla \cdot \boldsymbol{\sigma} + \rho\mathbf{f} = \rho\left(\frac{\partial\mathbf{v}}{\partial t} + \mathbf{v} \cdot \nabla\mathbf{v}\right). \tag{5.5.2}$$

Conservation of angular momentum

$$\sigma^T = \sigma. \tag{5.5.3}$$

Conservation of energy

$$\rho \frac{De}{Dt} = \sigma : \mathbf{D} - \nabla \cdot \mathbf{q} + \rho \mathcal{E}. \tag{5.5.5}$$

Entropy inequality

$$\rho \theta \frac{D\eta}{Dt} - \rho \mathcal{E} + \nabla \cdot \mathbf{q} - \frac{1}{\theta} \mathbf{q} \cdot \nabla \theta \geq 0. \tag{5.5.6}$$

The subject of continuum mechanics is primarily concerned with the determination of the behavior (e.g., ρ, $\mathbf{v}$, θ) of a body under externally applied causes (e.g., $\mathbf{f}$, $\mathcal{E}$). After introducing suitable constitutive relations for σ, e, and $\mathbf{q}$ (to be discussed in the next chapter), this task involves solving initial-boundary-value problem described by partial differential equations (5.5.1)–(5.5.5) under specified initial and boundary conditions. The role of the entropy inequality in this exercise is to make sure that the behavior of a body is consistent with the inequality (5.5.6). Often, the constitutive relations developed are required to be consistent with the second law of thermodynamics (i.e., satisfy the entropy inequality). The *entropy principle* states that constitutive relations be such that the entropy inequality is satisfied identically for any thermodynamic process.

To complete the mathematical description of the behavior of a continuous medium, the conservation equations derived in this chapter must be supplemented with the constitutive equations that relate σ, e, and $\mathbf{q}$ to $\mathbf{v}$, ρ, and θ. Chapter 6 is dedicated to the discussion of the constitutive relations. Applications of the governing equations of a continuum to linearized elasticity problems, and fluid mechanics and heat transfer problems will be discussed in Chapters 7 and 8, respectively.

PROBLEMS

5.1 The acceleration of a material element in a continuum is described by

$$\frac{D\mathbf{v}}{Dt} \equiv \frac{\partial \mathbf{v}}{\partial t} + \mathbf{v} \cdot \text{grad } \mathbf{v}, \tag{a}$$

where $\mathbf{v}$ is the velocity vector. Show by means of vector identities that the acceleration can also be written as

$$\frac{D\mathbf{v}}{Dt} \equiv \frac{\partial \mathbf{v}}{\partial t} + \text{grad}\left(\frac{v^2}{2}\right) - \mathbf{v} \times \text{curl}\mathbf{v}. \tag{b}$$

This form displays the role of the *vorticity vector*, $\boldsymbol{\omega} = \text{curl } \mathbf{v}$.

5.2 Show that the local form of the principle of conservation of mass, Eq. (5.2.10), can be expressed as

$$\frac{D}{Dt}(\rho J) = 0.$$

5.3 Derive the continuity equation in the cylindrical coordinate system by considering a differential volume element shown in Figure P5.3.

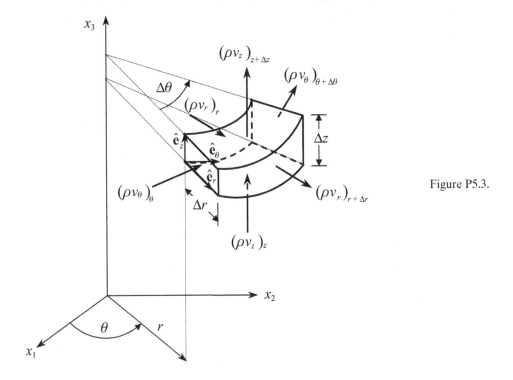

Figure P5.3.

5.4 Express the continuity equation (5.2.12) in the cylindrical coordinate system (see Table 2.4.2 for various operators). The result should match the one in Eq. (5.2.18).

5.5 Express the continuity equation (5.2.12) in the spherical coordinate system (see Table 2.4.2 for various operators). The result should match the one in Eq. (5.2.19).

5.6 Determine whether the following velocity fields for an incompressible flow satisfies the continuity equation:

(a) $v_2(x_1, x_2) = -\frac{x_1}{r^2}$, $v_2(x_1, x_2) = -\frac{x_2}{r^2}$, where $r^2 = x_1^2 + x_2^2$.

(b) $v_r = 0$, $v_\theta = 0$, $v_z = c\left(1 - \frac{r^2}{R^2}\right)$, where c and R are constants.

5.7 The velocity distribution between two parallel plates separated by distance b is

$$v_x(y) = \frac{y}{b}v_0 - c\frac{y}{b}\left(1 - \frac{y}{b}\right), \quad v_y = 0, \quad v_z = 0, \quad 0 < y < b,$$

where y is measured from and normal to the bottom plate, x is taken along the plates, v_x is the velocity component parallel to the plates, v_0 is the velocity of the top plate in the x direction, and c is a constant. Determine whether the velocity field satisfies the continuity equation and find the volume rate of flow and the average velocity.

5.8 A jet of air ($\rho = 1.206$ kg/m^3) impinges on a smooth vane with a velocity $v = 50$ m/s at the rate of $Q = 0.4$ m^3/s. Determine the force required to hold the plate in position for the two different vane configurations shown in Figure P5.8. Assume that the vane splits the jet into two equal streams, and neglect any energy loss in the streams.

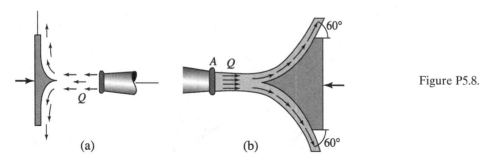

Figure P5.8.

(a)   (b)

5.9 In Part 1 of Example 5.3.3, determine (a) the velocity and accelerations as functions of x and (b) the velocity as the chain leaves the table.

5.10 Using the definition of ∇, vector forms of the velocity vector, body force vector, and the dyadic form of σ [see Eq. (5.3.16)], express the equations of motion (5.3.11) in the cylindrical coordinate system as given in Eq. (5.3.17).

5.11 Using the definition of ∇, vectors forms of the velocity vector, body force vector, and the dyadic form of σ [see Eq. (5.2.18)], express the equations of motion (5.3.11) in the spherical coordinate system as given in Eq. (5.3.19).

5.12 Use the continuity (i.e., conservation of mass) equation and the equation of motion to obtain the so-called conservation form of the momentum equation

$$\frac{\partial}{\partial t}(\rho \mathbf{v}) + \text{div}\,(\rho \mathbf{v}\mathbf{v} - \boldsymbol{\sigma}) = \rho \mathbf{f}.$$

5.13 Show that

$$\rho \frac{D}{Dt}\left(\frac{v^2}{2}\right) = \mathbf{v} \cdot \text{div}\,\boldsymbol{\sigma} + \rho \mathbf{v} \cdot \mathbf{f} \qquad (v = |\mathbf{v}|).$$

5.14 Deduce that

$$\text{curl}\left(\frac{D\mathbf{v}}{Dt}\right) \equiv \frac{D\boldsymbol{\omega}}{Dt} + \boldsymbol{\omega}\,\text{div}\,\mathbf{v} - \boldsymbol{\omega} \cdot \nabla\mathbf{v}, \qquad (a)$$

where $\boldsymbol{\omega} \equiv \text{curl}\,\mathbf{v}$ is the vorticity vector. *Hint:* Use the result of Problem 5.1 and the identity (you need to prove it)

$$\nabla \times (\mathbf{A} \times \mathbf{B}) = \mathbf{B} \cdot \nabla\mathbf{A} - \mathbf{A} \cdot \nabla\mathbf{B} + \mathbf{A}\nabla \cdot \mathbf{B} - \mathbf{B}\nabla \cdot \mathbf{A}.$$

5.15 Derive the following vorticity equation for a fluid of constant density and viscosity

$$\frac{\partial\boldsymbol{\omega}}{\partial t} + (\mathbf{v} \cdot \nabla)\boldsymbol{\omega} = (\boldsymbol{\omega} \cdot \nabla)\mathbf{v} + \nu\nabla^2\boldsymbol{\omega},$$

where $\boldsymbol{\omega} = \nabla \times \mathbf{v}$ and $\nu = \mu/\rho$.

5.16 *Bernoulli's Equations.* Consider a flow with hydrostatic pressure $\sigma = -P\mathbf{I}$ and conservative body force $\mathbf{f} = -\operatorname{grad}\phi$.

(a) For steady flow, show that

$$\mathbf{v} \cdot \operatorname{grad}\left(\frac{v^2}{2} + \phi\right) + \frac{1}{\rho}\mathbf{v} \cdot \operatorname{grad} P = 0.$$

(b) For steady and irrotational (i.e., $\operatorname{curl}\mathbf{v} = \mathbf{0}$) flow, show that

$$\operatorname{grad}\left(\frac{v^2}{2} + \phi\right) + \frac{1}{\rho}\operatorname{grad} P = 0.$$

5.17 Use the Bernoulli's equation (which is valid for *steady, frictionless, incompressible* flow) derived in Problem 5.16 to determine the velocity and discharge of the fluid at the exit of the nozzle in the wall of the reservoir shown in Fig. P5.17.

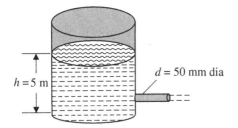

Figure P5.17.

$d = 50$ mm dia

$h = 5$ m

5.18 If the stress field in a body has the following components in a rectangular Cartesian coordinate system

$$[\sigma] = a \begin{bmatrix} x_1^2 x_2 & (b^2 - x_2^2)x_1 & 0 \\ (b^2 - x_2^2)x_1 & \frac{1}{3}(x_2^2 - 3b^2)x_2 & 0 \\ 0 & 0 & 2bx_3^2 \end{bmatrix},$$

where a and b are constants, determine the body force components necessary for the body to be in equilibrium.

5.19 A two-dimensional state of stress exists in a body with no body forces. The following components of stress are given:

$$\sigma_{11} = c_1 x_2^3 + c_2 x_1^2 x_2 - c_3 x_1, \quad \sigma_{22} = c_4 x_2^3 - c_5, \quad \sigma_{12} = c_6 x_1 x_2^2 + c_7 x_1^2 x_2 - c_8,$$

where c_i are constants. Determine the conditions on the constants so that the stress field is in equilibrium.

5.20 For a cantilevered beam bent by a point load at the free end, the bending moment M_3 about the x_3-axis is given by $M_3 = -Px_1$ (see Figure 3.8.1). The bending stress σ_{11} is given by

$$\sigma_{11} = \frac{M_3 x_2}{I_3} = -\frac{Px_1 x_2}{I_3},$$

where I_3 is the moment of inertia of the cross section about the x_3-axis. Starting with this equation, use the two-dimensional equilibrium equations to determine stresses σ_{22} and σ_{12} as functions of x_1 and x_2.

5.21 A sprinkler with four nozzles, each nozzle having an exit area of $A = 0.25$ cm^2, rotates at a constant angular velocity of $\omega = 20$ rad/s and distributes water ($\rho = 10^3$ kg/m^3) at the rate of $Q = 0.5$ L/s (see Figure P5.21). Determine (a) the torque T required on the shaft of the sprinkler to maintain the given motion and (b) the angular velocity ω_0 at which the sprinkler rotates when no external torque is applied.

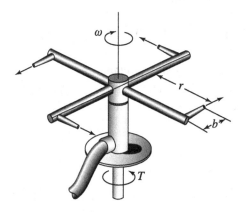

Figure P5.21.

5.22 Consider an unsymmetric sprinkler head shown in Figure P5.22. If the discharge is $Q = 0.5$ L/s through each nozzle, determine the angular velocity of the sprinkler. Assume that no external torque is exerted on the system.

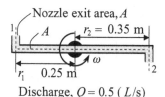

Figure P5.22.

5.23 Establish the following alternative form of the energy equation

$$\rho \frac{D}{Dt} \left(e + \frac{v^2}{2} \right) = \text{div} \left(\sigma \cdot \mathbf{v} \right) + \rho \mathbf{f} \cdot \mathbf{v} + \rho \mathcal{E} - \nabla \cdot \mathbf{q}.$$

5.24 The fan shown in Figure P5.24 moves air ($\rho = 1.23$ kg/m^3) at a mass flow rate of 0.1 kg/min. The upstream side of the fan is connected to a pipe of diameter $d_1 = 50$ mm, the flow is laminar, the velocity distribution is parabolic, and the kinetic energy coefficient is $\alpha = 2$. The downstream of the fan is connected to a pipe of diameter $d_2 = 25$ mm, the flow is turbulent, the velocity profile is uniform, and the kinetic energy coefficient is $\alpha = 1$. If the rise in static pressure between upstream and downstream is 100 Pa and the fan motor draws 0.15 W, determine the loss ($-H_{\text{net}}$).

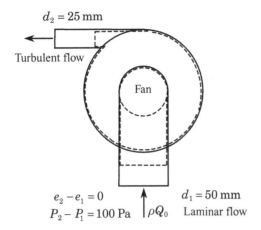

$d_2 = 25\,\text{mm}$

Turbulent flow

Fan

Figure P5.24.

$e_2 - e_1 = 0$

$P_2 - P_1 = 100\,\text{Pa}$ ρQ_0 Laminar flow

$d_1 = 50\,\text{mm}$

5.25 The rate of internal work done (power) in a continuous medium in the current configuration can be expressed as

$$W = \frac{1}{2} \int_v \boldsymbol{\sigma} : \mathbf{D}\, dv, \tag{1}$$

where $\boldsymbol{\sigma}$ is the Cauchy stress tensor and $\mathbf{D}$ is the rate of deformation tensor (i.e., symmetric part of the velocity gradient tensor)

$$\mathbf{D} = \frac{1}{2}\left[(\nabla\mathbf{v})^{\mathrm{T}} + \nabla\mathbf{v}\right], \quad \mathbf{v} = \frac{d\mathbf{x}}{dt}. \tag{2}$$

The pair $(\boldsymbol{\sigma}, \mathbf{D})$ is said to be *energetically conjugate* since it produces the (strain) energy stored in a deformable medium. Show that (a) the first Piola–Kirchhoff stress tensor $\mathbf{P}$ is energetically conjugate to the rate of deformation gradient tensor $\dot{\mathbf{F}}$ and (b) the second Piola–Kirchhoff stress tensor $\mathbf{S}$ is energetically conjugate to the rate of Green strain tensor $\dot{\mathbf{E}}$. *Hints:* Note the following identities:

$$d\mathbf{x} = J\, d\mathbf{X}, \quad \mathbf{L} \equiv \nabla\mathbf{v} = \dot{\mathbf{F}} \cdot \mathbf{F}^{-1}, \quad \mathbf{P} = J\mathbf{F}^{-1} \cdot \boldsymbol{\sigma}, \quad \boldsymbol{\sigma} = \frac{1}{J}\mathbf{F} \cdot \mathbf{S} \cdot \mathbf{F}^{\mathrm{T}}.$$

Constitutive Equations

What we need is imagination. We have to find a new view of the world.

Richard Feynman

The farther the experiment is from theory, the closer it is to the Nobel Prize.

Joliet-Curie

6.1 Introduction

The kinematic relations developed in Chapter 3, and the principles of conservation of mass and momenta and thermodynamic principles discussed in Chapter 5, are applicable to any continuum irrespective of its physical constitution. The kinematic variables such as the strains and temperature gradient, and kinetic variables such as the stresses and heat flux were introduced independently of each other. *Constitutive equations* are those relations that connect the *primary* field variables (e.g., ρ, T, $\mathbf{x}$, and $\mathbf{u}$ or $\mathbf{v}$) to the *secondary* field variables (e.g., e, $\mathbf{q}$, and $\boldsymbol{\sigma}$). Constitutive equations are *not* derived from any physical principles, although they are subject to obeying certain rules and the entropy inequality. In essence, constitutive equations are mathematical models of the behavior of materials that are validated against experimental results. The differences between theoretical predictions and experimental findings are often attributed to inaccurate representation of the constitutive behavior.

First, we review certain terminologies that were already introduced in beginning courses on mechanics of materials. A material body is said to be *homogeneous* if the material properties are the same throughout the body (i.e., independent of position). In a *heterogeneous* body, the material properties are a function of position. An *anisotropic* body is one that has different values of a material property in different directions at a point, i.e., material properties are direction dependent. An *isotropic* material is one for which every material property is the same in all directions at a point. An isotropic or anisotropic material can be nonhomogeneous or homogeneous.

Materials for which the constitutive behavior is only a function of the current state of deformation are known as *elastic*. If the constitutive behavior is only a function of the current state of rate of deformation, such materials are termed *viscous*. In this study, we shall be concerned with (a) elastic materials for which the stresses are functions of the current deformation and temperature and (b) viscous fluids for which the stresses are functions of density, temperature, and rate of deformation. Special cases of these materials are the Hookean solids and Newtonian fluids. A study of these "theoretical" materials is important because these materials provide good mathematical models for the behavior of "real" materials. There exist other materials, for example, polymers and elastomers, whose constitutive relations cannot be adequately described by those of a Hookean solid or Newtonian fluid.

Constitutive equations are often postulated directly from experimental observations. While experiments are necessary in the determination of various parameters (e.g., elastic constants, thermal conductivity, thermal coefficient of expansion, and coefficients of viscosity) appearing in the constitutive equations, the formulation of the constitutive equations for a given material is guided by certain rules. The approach typically involves assuming the form of the constitutive equation and then restricting the form to a specific one by appealing to certain physical requirements, including invariance of the equations and material frame indifference discussed in Section 3.9 (see Problem 6.8 for the axioms of constitutive theory).

This chapter is primarily focused on Hookean solids and Newtonian fluids. The constitutive equations presented in Section 6.2 for elastic solids are based on small strain assumption. Thus, we make no distinction between the material coordinates $\mathbf{X}$ and spatial coordinates $\mathbf{x}$ and between the Cauchy stress tensor $\boldsymbol{\sigma}$ and second Piola–Kirchhoff stress tensor $\mathbf{S}$. A brief discussion of some well-known nonlinear constitutive models (e.g., Mooney–Rivlin solids and non-Newtonian fluids) will be presented in Sections 6.2.11 and 6.3.4.

6.2 Elastic Solids

6.2.1 Introduction

A material is said to be (ideally or simple) *elastic* or *Cauchy elastic* when, under isothermal conditions, the body recovers its original form completely upon removal of the forces causing deformation, and there is a one-to-one relationship between the state of stress and the state of strain in the current configuration. The work done by the stress is, in general, dependent on the deformation path. For Cauchy elastic materials, the Cauchy stress $\boldsymbol{\sigma}$ does not depend on the path of deformation, and the state of stress in the current configuration is determined solely by the state of deformation

$$\boldsymbol{\sigma} = \boldsymbol{\sigma}(\mathbf{F}), \qquad (6.2.1)$$

where $\mathbf{F}$ is the deformation gradient tensor with respect to an arbitrary choice of reference configuration κ_0. For Cauchy elastic material, in contrast to the Green elastic material (see below), the stress is not derivable from a scalar potential function.

The constitutive equations to be developed here for stress tensor σ do not include creep at constant stress and stress relaxation at constant strain. Thus, the material coefficients that specify the constitutive relationship between the stress and strain components are assumed to be constant during the deformation. This does not automatically imply that we neglect temperature effects on deformation. We account for the thermal expansion of the material, which can produce strains or stresses as large as those produced by the applied mechanical forces.

A material is said to be *hyperelastic* or *Green elastic* if there exists a *strain energy density* function $U_0(\varepsilon)$ such that

$$\sigma = \frac{\partial U_0}{\partial \varepsilon}, \quad \left(\sigma_{ij} = \frac{\partial U_0}{\partial \varepsilon_{ij}}\right). \tag{6.2.2}$$

For an incompressible elastic material (i.e., material for which the volume is preserved and hence $J = 1$ or div $\mathbf{u} = 0$), the above relation is written as

$$\sigma = -p\mathbf{I} + \frac{\partial U_0}{\partial \varepsilon}, \quad \left(\sigma_{ij} = -p\delta_{ij} + \frac{\partial U_0}{\partial \varepsilon_{ij}}\right), \tag{6.2.3}$$

where p is the hydrostatic pressure. In developing a mathematical model of the constitutive behavior of an hyperelastic material, U_0 is expanded in Taylor's series about $\varepsilon = 0$:

$$U_0 = C_0 + C_{ij}\varepsilon_{ij} + \frac{1}{2!}\hat{C}_{ijk\ell}\varepsilon_{ij}\varepsilon_{k\ell} + \frac{1}{3!}\hat{C}_{ijk\ell mn}\varepsilon_{ij}\varepsilon_{k\ell}\varepsilon_{mn} + \dots, \tag{6.2.4}$$

where C_0, C_{ij}, $\hat{C}$, and so on are material stiffnesses. For nonlinear elastic materials, U_0 is a cubic and higher-order function of the strains. For linear elastic materials, U_0 is a quadratic function of strains.

In Sections 6.2.2–6.2.10 we discuss the constitutive equations of Hookean solids (i.e., relations between stress and strain are linear) for the case of infinitesimal deformation (i.e., $|\nabla\mathbf{u}| << 1$). Hence, we will not distinguish between various measures of stress and strain, and use $\mathbf{S} \approx \sigma$ for the stress tensor and $\mathbf{E} \approx \varepsilon$ for strain tensor in the material description used in solid mechanics. The linear constitutive model for infinitesimal deformations is referred to as the *generalized Hooke's law*.

6.2.2 Generalized Hooke's Law

To derive the stress–strain relations for a linear elastic solid, begin with the quadratic form of U_0

$$U_0 = C_0 + C_{ij}\varepsilon_{ij} + \frac{1}{2!}\hat{C}_{ijk\ell}\varepsilon_{ij}\varepsilon_{k\ell}, \tag{6.2.5}$$

where C_0 is a reference value of U_0 from which the strain energy density function is measured. From Eq. (6.2.2), we have

$$\sigma_{mn} = \frac{\partial U_0}{\partial \varepsilon_{mn}} = C_{ij}\delta_{mi}\delta_{nj} + \frac{1}{2}\hat{C}_{ijk\ell}\left(\varepsilon_{k\ell}\delta_{im}\delta_{jn} + \varepsilon_{ij}\delta_{km}\delta_{\ell n}\right)$$

$$= C_{mn} + \frac{1}{2}\hat{C}_{mnk\ell}\varepsilon_{k\ell} + \frac{1}{2}\hat{C}_{ijmn}\varepsilon_{ij}$$

$$= C_{mn} + \frac{1}{2}\left(\hat{C}_{mnij} + \hat{C}_{ijmn}\right)\varepsilon_{ij}$$

$$= C_{mn} + C_{mnij}\varepsilon_{ij}, \tag{6.2.6}$$

where

$$C_{mnij} = \frac{1}{2}\left(\hat{C}_{mnij} + \hat{C}_{ijmn}\right) = \frac{\partial^2 U_0}{\partial \varepsilon_{ij}\partial \varepsilon_{mn}} = \frac{\partial^2 U_0}{\partial \varepsilon_{mn}\partial \varepsilon_{ij}} = C_{ijmn}. \tag{6.2.7}$$

Clearly, C_{mn} have the same units as σ_{mn}, and they represent the *residual stress* components of a solid. We shall assume, without loss of generality, that the body is free of stress prior to the load application so that we may write

$$\sigma_{ij} = C_{ijk\ell}\varepsilon_{k\ell}. \tag{6.2.8}$$

The coefficients $C_{ijk\ell}$ are called elastic *stiffness* coefficients. In general, there are $81(=3^4)$ scalar components of the fourth-order tensor $\mathbf{C}$. However, the components $C_{ijk\ell}$ satisfy the following symmetry conditions by virtue of definition (6.2.7) and the symmetry of stress and strain tensor components:

$$C_{ijk\ell} = C_{k\ell ij}, \quad C_{ijk\ell} = C_{jik\ell}, \quad C_{ijk\ell} = C_{ji\ell k}. \tag{6.2.9}$$

Thus, the number of independent coefficients in C_{ijmn} is reduced to 21:

$$\begin{bmatrix} C_{1111} & C_{1122} & C_{1133} & C_{1123} & C_{1113} & C_{1112} \\ & C_{2222} & C_{2233} & C_{2223} & C_{2213} & C_{2212} \\ & & C_{3333} & C_{3323} & C_{3313} & C_{3312} \\ & & & C_{2323} & C_{2313} & C_{2312} \\ & & & & C_{1313} & C_{1312} \\ & & & & & C_{1212} \end{bmatrix}. \tag{6.2.10}$$

We express Eq. (6.2.8) in an alternate form using single subscript notation for stresses and strains and two subscript notation for the material stiffness coefficients:

$$\sigma_1 = \sigma_{11}, \quad \sigma_2 = \sigma_{22}, \quad \sigma_3 = \sigma_{33}, \quad \sigma_4 = \sigma_{23}, \quad \sigma_5 = \sigma_{13}, \quad \sigma_6 = \sigma_{12},$$

$$\varepsilon_1 = \varepsilon_{11}, \quad \varepsilon_2 = \varepsilon_{22}, \quad \varepsilon_3 = \varepsilon_{33}, \quad \varepsilon_4 = 2\varepsilon_{23}, \quad \varepsilon_5 = 2\varepsilon_{13}, \quad \varepsilon_6 = 2\varepsilon_{12}. \tag{6.2.11}$$

$$11 \rightarrow 1 \quad 22 \rightarrow 2 \quad 33 \rightarrow 3 \quad 23 \rightarrow 4 \quad 13 \rightarrow 5 \quad 12 \rightarrow 6. \tag{6.2.12}$$

It should be cautioned that the single subscript notation used for stresses and strains and the two-subscript components C_{ij} render them nontensor components (i.e., σ_i, ε_i, and C_{ij} do not transform like the components of a tensor). The single subscript notation for stresses and strains is called the *engineering notation* or the *Voigt-Kelvin notation*. Equation (6.2.8) now takes the form

$$\sigma_i = C_{ij}\varepsilon_j, \tag{6.2.13}$$

where summation on repeated subscripts is implied (now from 1 to 6). The coefficients C_{ij} are symmetric ($C_{ij} = C_{ji}$), and we have $21(= 6+5+4+3+2+1)$ independent stiffness coefficients for the most general elastic material. In matrix notation Eq. (6.2.13) can be written as

$$\begin{Bmatrix} \sigma_1 \\ \sigma_2 \\ \sigma_3 \\ \sigma_4 \\ \sigma_5 \\ \sigma_6 \end{Bmatrix} = \begin{bmatrix} C_{11} & C_{12} & C_{13} & C_{14} & C_{15} & C_{16} \\ C_{21} & C_{22} & C_{23} & C_{24} & C_{25} & C_{26} \\ C_{31} & C_{32} & C_{33} & C_{34} & C_{35} & C_{36} \\ C_{41} & C_{42} & C_{43} & C_{44} & C_{45} & C_{46} \\ C_{51} & C_{52} & C_{53} & C_{54} & C_{55} & C_{56} \\ C_{61} & C_{62} & C_{63} & C_{64} & C_{65} & C_{66} \end{bmatrix} \begin{Bmatrix} \varepsilon_1 \\ \varepsilon_2 \\ \varepsilon_3 \\ \varepsilon_4 \\ \varepsilon_5 \\ \varepsilon_6 \end{Bmatrix}. \tag{6.2.14}$$

We assume that the stress–strain relations (6.2.14) are invertible. Thus, the components of strain are related to the components of stress by

$$\varepsilon_i = S_{ij}\sigma_j, \tag{6.2.15}$$

where S_{ij} are the material *compliance* coefficients with $[S] = [C]^{-1}$ (i.e., the compliance tensor is the inverse of the stiffness tensor: $\mathbf{S} = \mathbf{C}^{-1}$). In matrix notation, Eq. (6.2.15) has the form

$$\begin{Bmatrix} \varepsilon_1 \\ \varepsilon_2 \\ \varepsilon_3 \\ \varepsilon_4 \\ \varepsilon_5 \\ \varepsilon_6 \end{Bmatrix} = \begin{bmatrix} S_{11} & S_{12} & S_{13} & S_{14} & S_{15} & S_{16} \\ S_{21} & S_{22} & S_{23} & S_{24} & S_{25} & S_{26} \\ S_{31} & S_{32} & S_{33} & S_{34} & S_{35} & S_{36} \\ S_{41} & S_{42} & S_{43} & S_{44} & S_{45} & S_{46} \\ S_{51} & S_{52} & S_{53} & S_{54} & S_{55} & S_{56} \\ S_{61} & S_{62} & S_{63} & S_{64} & S_{65} & S_{66} \end{bmatrix} \begin{Bmatrix} \sigma_1 \\ \sigma_2 \\ \sigma_3 \\ \sigma_4 \\ \sigma_5 \\ \sigma_6 \end{Bmatrix}. \tag{6.2.16}$$

6.2.3 Material Symmetry

Further reduction in the number of independent stiffness (or compliance) parameters comes from the so-called material symmetry. Suppose that (x_1, x_2, x_3) denote the coordinate system with respect to which Eqs. (6.2.5)–(6.2.16) are defined. We shall call it a *material coordinate system*. The coordinate system (x, y, z) used to write the equations of motion and strain-displacement equations will be called the *problem coordinates* to distinguish them from the material coordinate system. The phrase "material coordinates" used in connection with the material description should not be confused with the present term. In the remaining discussion, we will use the material description for everything, but we may use one material coordinate system, say (x, y, z), to describe the kinematics as well as stress state in the body and another material coordinate system (x_1, x_2, x_3) to describe the stress–strain behavior. Both are fixed in the body, and the two systems are oriented with respect to each other. When elastic material parameters at a point have *the same values* for every pair of coordinate systems that are mirror images of each other in a certain plane, that plane is called a *material plane of symmetry* (e.g., symmetry of internal structure due to crystallographic form, *regular* arrangement of fibers or molecules). We note that the symmetry under discussion is a directional property and not a positional property. Thus, a material may have certain elastic symmetry at every point of a material body and the properties may vary from point to point. Positional dependence of material properties is what we called the *inhomogeneity of the material*.

In the following, we discuss various planes of symmetry and forms of associated stress–strain relations. Use of the tensor components of stress and strain is necessary because the transformations are valid only for the tensor components. The second-order tensor components σ_{ij} and ε_{ij} and the fourth-order tensor components C_{ijkl} transform according to the formulae

$$\bar{\sigma}_{ij} = \ell_{ip}\,\ell_{jq}\,\sigma_{pq}, \quad \bar{\varepsilon}_{ij} = \ell_{ip}\,\ell_{jq}\,\varepsilon_{pq}, \quad \bar{C}_{ijkl} = \ell_{ip}\,\ell_{jq}\,\ell_{kr}\,\ell_{ls}\,C_{pqrs}, \qquad (6.2.17)$$

where ℓ_{ij} are the direction cosines associated with the coordinate systems $(\bar{x}_1, \bar{x}_2, \bar{x}_3)$ and (x_1, x_2, x_3), and $\bar{C}_{ijkl}$ and C_{pqrs} are the components of the fourth-order tensor **C** in the barred and unbarred coordinates systems, respectively. A trivial symmetry transformation is one in which the barred coordinate system is obtained from the unbarred coordinate system by simply reversing their directions: $\bar{x}_1 = -x_1$, $\bar{x}_2 = -x_2$, and $\bar{x}_3 = -x_3$. This transformation is satisfied by all materials, and they are called *triclinic* materials. The associated transformation matrix is given by

$$[L] = \begin{bmatrix} -1 & 0 & 0 \\ 0 & -1 & 0 \\ 0 & 0 & -1 \end{bmatrix}. \qquad (6.2.18)$$

For this transformation, one can show that Eq. (6.2.17) gives the trivial result $\bar{C}_{ijkl} = C_{ijkl}$. Next, we consider some commonly known nontrivial symmetry transformations.

6.2.4 Monoclinic Materials

When the elastic coefficients at a point have the same value for every pair of coordinate systems which are the mirror images of each other with respect to a plane, the material is called a *monoclinic material*. For example, let (x_1, x_2, x_3) and $(\bar{x}_1, \bar{x}_2, \bar{x}_3)$ be two coordinates systems, with the x_1, x_2-plane parallel to the plane of symmetry. Choosing the $\bar{x}_3$-axis such that $\bar{x}_3 = -x_3$ (never mind about the left-handed coordinate system as it does not affect the discussion) so that one system is the mirror image of the other. This symmetry transformation can be expressed by the transformation matrix $(\bar{x}_1 = x_1, \bar{x}_2 = x_2, \bar{x}_3 = -x_3)$.

$$[L] = \begin{bmatrix} 1 & 0 & 0 \\ 0 & 1 & 0 \\ 0 & 0 & -1 \end{bmatrix}. \qquad (6.2.19)$$

The requirement that $\bar{C}_{ijkl}$ be the same as C_{ijkl} under the transformation (6.2.19) yields (because $\bar{\sigma}_{23} = -\sigma_{23}, \bar{\sigma}_{31} = -\sigma_{31}, \bar{\varepsilon}_{23} = -\varepsilon_{23}, \bar{\varepsilon}_{31} = -\varepsilon_{31}$ under the same transformation):

$$C_{1113} = \bar{C}_{1113} = -C_{1113}, \quad C_{1123} = \bar{C}_{1123} = -C_{1123},$$

$$C_{1213} = \bar{C}_{1213} = -C_{1213}, \quad C_{2213} = \bar{C}_{2213} = -C_{2213},$$

$$C_{2223} = \bar{C}_{2223} = -C_{2223}, \quad C_{2333} = \bar{C}_{2333} = -C_{2333},$$

$$C_{3323} = \bar{C}_{3323} = -C_{3323}, \quad C_{3313} = \bar{C}_{3313} = -C_{3313}.$$

Thus, all Cs with an odd number of index "3" are zero.

Alternatively, if we use Eq. (6.2.13) and $\bar{\sigma}_4 = -\sigma_4, \bar{\sigma}_5 = -\sigma_5, \bar{\varepsilon}_4 = -\varepsilon_4, \bar{\varepsilon}_5 = -\varepsilon_5$, we obtain

$$\bar{\sigma}_1 = C_{11}\bar{\varepsilon}_1 + C_{12}\bar{\varepsilon}_2 + C_{13}\bar{\varepsilon}_3 + C_{14}\bar{\varepsilon}_4 + C_{15}\bar{\varepsilon}_5 + C_{16}\bar{\varepsilon}_6,$$

$$\sigma_1 = C_{11}\varepsilon_1 + C_{12}\varepsilon_2 + C_{13}\varepsilon_3 - C_{14}\varepsilon_4 - C_{15}\varepsilon_5 + C_{16}\varepsilon_6.$$

But we also have

$$\sigma_1 = C_{11}\varepsilon_1 + C_{12}\varepsilon_2 + C_{13}\varepsilon_3 + C_{14}\varepsilon_4 + C_{15}\varepsilon_5 + C_{16}\varepsilon_6.$$

The elastic parameters C_{ij} are the same for the two coordinate systems because they are the mirror images in the plane of symmetry. From the above two equations we arrive at

$$C_{14}\varepsilon_4 + C_{15}\varepsilon_5 = 0 \quad \text{for all values of } \varepsilon_4 \text{ and } \varepsilon_5.$$

The above equation holds only if $C_{14} = 0$ and $C_{15} = 0$. Similar discussion with the two alternative expressions of the remaining stress components yield $C_{24} = 0$ and $C_{25} = 0$; $C_{34} = 0$ and $C_{35} = 0$; and $C_{46} = 0$ and $C_{56} = 0$. Thus, of 21 material parameters, we have only $21 - 8 = 13$ independent parameters, as indicated below

$$[C] = \begin{bmatrix} C_{11} & C_{12} & C_{13} & 0 & 0 & C_{16} \\ C_{12} & C_{22} & C_{23} & 0 & 0 & C_{26} \\ C_{13} & C_{23} & C_{33} & 0 & 0 & C_{36} \\ 0 & 0 & 0 & C_{44} & C_{45} & 0 \\ 0 & 0 & 0 & C_{45} & C_{55} & 0 \\ C_{16} & C_{26} & C_{36} & 0 & 0 & C_{66} \end{bmatrix}. \tag{6.2.20}$$

Monoclinic materials exhibit shear-extensional coupling, that is, a shear strain can produce a normal stress.

6.2.5 Orthotropic Materials

When three mutually orthogonal planes of material symmetry exist, the number of elastic coefficients is reduced to 9 using arguments similar to those given for single material symmetry plane, and such materials are called *orthotropic*. The transformation matrices associated with the planes of symmetry are

$$[L^{(1)}] = \begin{bmatrix} 1 & 0 & 0 \\ 0 & 1 & 0 \\ 0 & 0 & -1 \end{bmatrix}, \; [L^{(2)}] = \begin{bmatrix} -1 & 0 & 0 \\ 0 & 1 & 0 \\ 0 & 0 & 1 \end{bmatrix}, \; [L^{(3)}] = \begin{bmatrix} 1 & 0 & 0 \\ 0 & -1 & 0 \\ 0 & 0 & 1 \end{bmatrix}. \quad (6.2.21)$$

Under these transformations, we obtain $C_{1112} = C_{16} = 0$, $C_{2212} = C_{26} = 0$, $C_{3312} = C_{36} = 0$, and $C_{2313} = C_{45} = 0$. The stress–strain relations for an orthotropic material take the form

$$\begin{Bmatrix} \sigma_1 \\ \sigma_2 \\ \sigma_3 \\ \sigma_4 \\ \sigma_5 \\ \sigma_6 \end{Bmatrix} = \begin{bmatrix} C_{11} & C_{12} & C_{13} & 0 & 0 & 0 \\ C_{12} & C_{22} & C_{23} & 0 & 0 & 0 \\ C_{13} & C_{23} & C_{33} & 0 & 0 & 0 \\ 0 & 0 & 0 & C_{44} & 0 & 0 \\ 0 & 0 & 0 & 0 & C_{55} & 0 \\ 0 & 0 & 0 & 0 & 0 & C_{66} \end{bmatrix} \begin{Bmatrix} \varepsilon_1 \\ \varepsilon_2 \\ \varepsilon_3 \\ \varepsilon_4 \\ \varepsilon_5 \\ \varepsilon_6 \end{Bmatrix}. \quad (6.2.22)$$

Most simple characterization tests are performed with a known load or stress. Hence, it is convenient to write the inverse of relations in Eq. (6.2.22)

$$\begin{Bmatrix} \varepsilon_1 \\ \varepsilon_2 \\ \varepsilon_3 \\ \varepsilon_4 \\ \varepsilon_5 \\ \varepsilon_6 \end{Bmatrix} = \begin{bmatrix} S_{11} & S_{12} & S_{13} & 0 & 0 & 0 \\ S_{12} & S_{22} & S_{23} & 0 & 0 & 0 \\ S_{13} & S_{23} & S_{33} & 0 & 0 & 0 \\ 0 & 0 & 0 & S_{44} & 0 & 0 \\ 0 & 0 & 0 & 0 & S_{55} & 0 \\ 0 & 0 & 0 & 0 & 0 & S_{66} \end{bmatrix} \begin{Bmatrix} \sigma_1 \\ \sigma_2 \\ \sigma_3 \\ \sigma_4 \\ \sigma_5 \\ \sigma_6 \end{Bmatrix}, \quad (6.2.23)$$

where $[S]$ is the matrix of compliance coefficients S_{ij}, and $[S] = [C]^{-1}$.

Most often, the material properties are determined in a laboratory in terms of the engineering constants such as Young's modulus, shear modulus, and so on. These constants are measured using simple tests like uniaxial tension test or pure shear test. Because of their direct and obvious physical meaning, engineering constants are used in place of the more abstract stiffness coefficients C_{ij} and compliance coefficients S_{ij}. Next we discuss how to obtain the strain–stress relations (6.2.23) and relate S_{ij} to the engineering constants.

One of the consequences of linearity (both kinematic and material linearizations) is that the principle of superposition applies. That is, if the applied loads and geometric constraints are independent of deformation, the sum of the displacements (and hence strains) produced by two sets of loads is equal to the displacements (and strains) produced by the sum of the two sets of loads. In particular, the strains of the same type produced by the application of individual stress components can be superposed. For example, the extensional strain $\varepsilon_{11}^{(1)}$ in the material coordinate direction x_1 due to the stress σ_{11} in the same direction is σ_{11}/E_1, where E_1 denotes Young's modulus of the material in the x_1 direction, as shown in Figure 6.2.1. The extensional strain $\varepsilon_{11}^{(2)}$ due to the stress σ_{22} applied in the x_2-direction is (a result of the Poisson

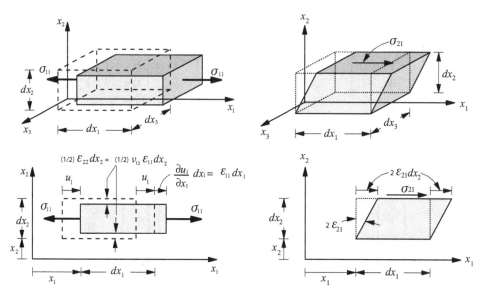

Figure 6.2.1. Strains produced by stresses in a cube of material.

effect) $-\nu_{21}(\sigma_{22}/E_2)$, where ν_{21} is Poisson's ratio (note that the first subscript in ν_{ij}, $i \neq j$, corresponds to the load direction and the second subscript refers to the directions of the strain)

$$\nu_{21} = -\frac{\varepsilon_{11}}{\varepsilon_{22}}$$

and E_2 is Young's modulus of the material in the x_2-direction. Similarly, σ_{33} produces a strain $\varepsilon_{11}^{(3)}$ equal to $-\nu_{31}(\sigma_{33}/E_3)$. Hence, the total strain ε_{11} due to the simultaneous application of all three normal stress components is

$$\varepsilon_{11} = \varepsilon_{11}^{(1)} + \varepsilon_{11}^{(2)} + \varepsilon_{11}^{(3)} = \frac{\sigma_{11}}{E_1} - \nu_{21}\frac{\sigma_{22}}{E_2} - \nu_{31}\frac{\sigma_{33}}{E_3}, \qquad (6.2.24)$$

where the direction of loading is denoted by the superscript. Similarly, we can write

$$\varepsilon_{22} = -\nu_{12}\frac{\sigma_{11}}{E_1} + \frac{\sigma_{22}}{E_2} - \nu_{32}\frac{\sigma_{33}}{E_3},$$

$$\varepsilon_{33} = -\nu_{13}\frac{\sigma_{11}}{E_1} - \nu_{23}\frac{\sigma_{22}}{E_2} + \frac{\sigma_{33}}{E_3}. \qquad (6.2.25)$$

The simple shear tests with an orthotropic material give the results

$$2\varepsilon_{12} = \frac{\sigma_{12}}{G_{12}}, \quad 2\varepsilon_{13} = \frac{\sigma_{13}}{G_{13}}, \quad 2\varepsilon_{23} = \frac{\sigma_{23}}{G_{23}}. \qquad (6.2.26)$$

Recall from Section 3.5.2 that $2\varepsilon_{ij}$ $(i \neq j)$ is the change in the right angle between two lines parallel to the x_1- and x_2-directions at a point, σ_{ij} $(i \neq j)$ denotes the corresponding shear stress in the x_i-x_j plane, and G_{ij} $(i \neq j)$ are the shear moduli in the x_i-x_j plane.

Writing Eqs. (6.2.24)–(6.2.26) in matrix form, we obtain

$$
\begin{Bmatrix} \varepsilon_1 \\ \varepsilon_2 \\ \varepsilon_3 \\ \varepsilon_4 \\ \varepsilon_5 \\ \varepsilon_6 \end{Bmatrix} = \begin{bmatrix} \frac{1}{E_1} & -\frac{\nu_{21}}{E_2} & -\frac{\nu_{31}}{E_3} & 0 & 0 & 0 \\ -\frac{\nu_{12}}{E_1} & \frac{1}{E_2} & -\frac{\nu_{32}}{E_3} & 0 & 0 & 0 \\ -\frac{\nu_{13}}{E_1} & -\frac{\nu_{23}}{E_2} & \frac{1}{E_3} & 0 & 0 & 0 \\ 0 & 0 & 0 & \frac{1}{G_{23}} & 0 & 0 \\ 0 & 0 & 0 & 0 & \frac{1}{G_{13}} & 0 \\ 0 & 0 & 0 & 0 & 0 & \frac{1}{G_{12}} \end{bmatrix} \begin{Bmatrix} \sigma_1 \\ \sigma_2 \\ \sigma_3 \\ \sigma_4 \\ \sigma_5 \\ \sigma_6 \end{Bmatrix}, \tag{6.2.27}
$$

where E_1, E_2, E_3 are Young's moduli in 1, 2, and 3 material directions, respectively, ν_{ij} is Poisson's ratio, defined as the ratio of transverse strain in the jth direction to the axial strain in the ith direction when stressed in the i-direction, and G_{23}, G_{13}, G_{12} are shear moduli in the 2-3, 1-3, and 1-2 planes, respectively. Since $[S]$ is the inverse of $[C]$ and the $[C]$ is symmetric, then $[S]$ is also a symmetric matrix. This in turn implies that the following reciprocal relations hold [i.e., compare the off-diagonal terms in Eq. (6.2.27)]:

$$
\frac{\nu_{21}}{E_2} = \frac{\nu_{12}}{E_1}; \quad \frac{\nu_{31}}{E_3} = \frac{\nu_{13}}{E_1}; \quad \frac{\nu_{32}}{E_3} = \frac{\nu_{23}}{E_2} \quad \rightarrow \quad \frac{\nu_{ij}}{E_i} = \frac{\nu_{ji}}{E_j} \tag{6.2.28}
$$

for $i, j = 1, 2, 3$. The nine independent material coefficients for an orthotropic material are

$$
E_1, \ E_2, \ E_3, \ G_{23}, \ G_{13}, \ G_{12}, \ \nu_{12}, \ \nu_{13}, \ \nu_{23}.
$$

6.2.6 Isotropic Materials

Isotropic materials are those for which the material properties are independent of the direction, and we have

$$
E_1 = E_2 = E_3 = E, \quad G_{12} = G_{13} = G_{23} = G, \quad \nu_{12} = \nu_{23} = \nu_{13} = \nu.
$$

The stress–strain relations take the form

$$
\sigma_{ij} = \frac{E}{1+\nu}\varepsilon_{ij} + \frac{\nu E}{(1+\nu)(1-2\nu)}\varepsilon_{kk}\delta_{ij}, \tag{6.2.29}
$$

where summation on repeated indices is implied. The inverse relations are ($G = E/[2(1+\nu)]$)

$$
\varepsilon_{ij} = \frac{1+\nu}{E}\sigma_{ij} - \frac{\nu}{E}\sigma_{kk}\delta_{ij}, \tag{6.2.30}
$$

Application of a normal stress to a rectangular block of isotropic or orthotropic material leads to only extension in the direction of the applied stress and contraction perpendicular to it, whereas an anisotropic material experiences extension in the direction of the applied normal stress, contraction perpendicular to it, and shearing strain, as shown in Figure 6.2.2. Conversely, the application of a shearing stress to an anisotropic material causes shearing strain as well as normal strains. Normal stress applied to an orthotropic material at an angle to its principal material directions causes it to behave like an anisotropic material.

Normal Stress   Shear Stress

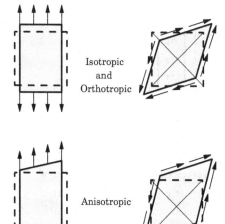

Isotropic
and
Orthotropic

Figure 6.2.2. Deformation of orthotropic and anisotropic rectangular block under uniaxial tension.

Anisotropic

6.2.7 Transformation of Stress and Strain Components

The constitutive relations (6.2.22) and (6.2.23) for an orthotropic material were written in terms of the stress and strain components that are referred to the material
coordinate system. The coordinate system used in the problem formulation, in general, does not coincide with the material coordinate system. Thus, there is a need to
express the constitutive equations of an orthotropic material in terms of the stress
and strain components referred to the problem coordinate system. We can use the
transformation equations (6.2.17) of a second-order tensor to write the stress and
strain components $(\sigma_i, \varepsilon_i)$ referred to the material coordinate system in terms of
those referred to the problem coordinates.

Let (x, y, z) denote the coordinate system used to write the governing equations
of a problem, and let (x_1, x_2, x_3) be the principal material coordinates such that x_3-
axis is parallel to the z-axis (i.e., the x_1x_2-plane and the xy-plane are parallel) and
the x_1-axis is oriented at an angle of $+\theta$ counterclockwise (when looking down) from
the x-axis, as shown in Figure 6.2.3. The coordinates of a material point in the two
coordinate systems are related as follows $(z = x_3)$:

$$\begin{Bmatrix} x_1 \\ x_2 \\ x_3 \end{Bmatrix} = \begin{bmatrix} \cos\theta & \sin\theta & 0 \\ -\sin\theta & \cos\theta & 0 \\ 0 & 0 & 1 \end{bmatrix} \begin{Bmatrix} x \\ y \\ z \end{Bmatrix} = [L] \begin{Bmatrix} x \\ y \\ z \end{Bmatrix}. \tag{6.2.31}$$

The inverse of Eq. (6.2.31) is

$$\begin{Bmatrix} x \\ y \\ z \end{Bmatrix} = \begin{bmatrix} \cos\theta & -\sin\theta & 0 \\ \sin\theta & \cos\theta & 0 \\ 0 & 0 & 1 \end{bmatrix} \begin{Bmatrix} x_1 \\ x_2 \\ x_3 \end{Bmatrix} = [L]^T \begin{Bmatrix} x_1 \\ x_2 \\ x_3 \end{Bmatrix}. \tag{6.2.32}$$

The inverse of $[L]$ is equal to its transpose: $[L]^{-1} = [L]^T$.

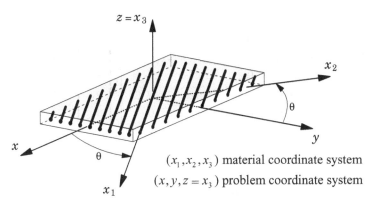

Figure 6.2.3. A material with material and problem coordinate systems.

6.2.7.1 Transformation of Stress Components

Let σ denote the stress tensor, which has components $\sigma_{11}, \sigma_{12}, \ldots, \sigma_{33}$ in the material (m) coordinates (x_1, x_2, x_3) and components $\sigma_{xx}, \sigma_{xy}, \ldots, \sigma_{zz}$ in the problem (p) coordinates (x, y, z). Since stress tensor is a second-order tensor, it transforms according to the formula

$$(\sigma_{kq})_m = \ell_{ki}\ell_{qj}(\sigma_{ij})_p \quad \text{or} \quad [\sigma]_m = [L][\sigma]_p[L]^{\mathrm{T}},$$
$$(\sigma_{kq})_p = \ell_{ik}\ell_{jq}(\sigma_{ij})_m \quad \text{or} \quad [\sigma]_p = [L]^{\mathrm{T}}[\sigma]_m[L]. \tag{6.2.33}$$

where $(\sigma_{ij})_m$ are the components of the stress tensor σ in the material coordinates (x_1, x_2, x_3), whereas $(\sigma_{ij})_p$ are the components of the same stress tensor σ in the problem coordinates (x, y, z),

$$[\sigma]_p = \begin{bmatrix} \sigma_{xx} & \sigma_{xy} & \sigma_{xz} \\ \sigma_{xy} & \sigma_{yy} & \sigma_{yz} \\ \sigma_{xz} & \sigma_{yz} & \sigma_{zz} \end{bmatrix}, \quad [\sigma]_m = \begin{bmatrix} \sigma_{11} & \sigma_{12} & \sigma_{13} \\ \sigma_{12} & \sigma_{22} & \sigma_{23} \\ \sigma_{13} & \sigma_{23} & \sigma_{33} \end{bmatrix}, \tag{6.2.34}$$

and ℓ_{ij} are the direction cosines defined by

$$\ell_{ij} = (\hat{\mathbf{e}}_i)_m \cdot (\hat{\mathbf{e}}_j)_p, \tag{6.2.35}$$

$(\hat{\mathbf{e}}_i)_m$ and $(\hat{\mathbf{e}}_i)_p$ being the orthonormal basis vectors in the material and problem coordinate systems, respectively.

Carrying out the matrix multiplications in Eq. (6.2.33), with $[L]$ defined by Eq. (6.2.31), and rearranging the equations in terms of the single-subscript stress components in (x, y, z) and (x_1, x_2, x_3) coordinate systems, we obtain

$$\begin{Bmatrix} \sigma_{xx} \\ \sigma_{yy} \\ \sigma_{zz} \\ \sigma_{yz} \\ \sigma_{xz} \\ \sigma_{xy} \end{Bmatrix} = \begin{bmatrix} \cos^2\theta & \sin^2\theta & 0 & 0 & 0 & -\sin 2\theta \\ \sin^2\theta & \cos^2\theta & 0 & 0 & 0 & \sin 2\theta \\ 0 & 0 & 1 & 0 & 0 & 0 \\ 0 & 0 & 0 & \cos\theta & \sin\theta & 0 \\ 0 & 0 & 0 & -\sin\theta & \cos\theta & 0 \\ \frac{1}{2}\sin 2\theta & -\frac{1}{2}\sin 2\theta & 0 & 0 & 0 & \cos 2\theta \end{bmatrix} \begin{Bmatrix} \sigma_1 \\ \sigma_2 \\ \sigma_3 \\ \sigma_4 \\ \sigma_5 \\ \sigma_6 \end{Bmatrix}. \tag{6.2.36}$$

The inverse relationship between $\{\sigma\}_m$ and $\{\sigma\}_p$ is given by

$$
\begin{Bmatrix} \sigma_1 \\ \sigma_2 \\ \sigma_3 \\ \sigma_4 \\ \sigma_5 \\ \sigma_6 \end{Bmatrix} =
\begin{bmatrix}
\cos^2\theta & \sin^2\theta & 0 & 0 & 0 & \sin 2\theta \\
\sin^2\theta & \cos^2\theta & 0 & 0 & 0 & -\sin 2\theta \\
0 & 0 & 1 & 0 & 0 & 0 \\
0 & 0 & 0 & \cos\theta & -\sin\theta & 0 \\
0 & 0 & 0 & \sin\theta & \cos\theta & 0 \\
-\frac{1}{2}\sin 2\theta & \frac{1}{2}\sin 2\theta & 0 & 0 & 0 & \cos 2\theta
\end{bmatrix}
\begin{Bmatrix} \sigma_{xx} \\ \sigma_{yy} \\ \sigma_{zz} \\ \sigma_{yz} \\ \sigma_{xz} \\ \sigma_{xy} \end{Bmatrix}. \qquad (6.2.37)
$$

6.2.7.2 Transformation of Strain Components

Transformation equations derived for stresses are also valid for *tensor* components of strains

$$
[\varepsilon]_m = [L][\varepsilon]_p[L]^T; \quad [\varepsilon]_p = [L]^T[\varepsilon]_m[L]. \qquad (6.2.38)
$$

However, the single-column formats in Eqs. (6.2.36) and (6.2.37) for stresses are not valid for single-column formats of strains because of the definition:

$$
2\varepsilon_{12} = \varepsilon_6, \quad 2\varepsilon_{13} = \varepsilon_5, \quad 2\varepsilon_{23} = \varepsilon_4.
$$

Slight modification of the results in Eqs. (6.2.36) and (6.2.37) will yield the proper relations for the engineering components of strains. We have

$$
\begin{Bmatrix} \varepsilon_{xx} \\ \varepsilon_{yy} \\ \varepsilon_{zz} \\ 2\varepsilon_{yz} \\ 2\varepsilon_{xz} \\ 2\varepsilon_{xy} \end{Bmatrix} =
\begin{bmatrix}
\cos^2\theta & \sin^2\theta & 0 & 0 & 0 & -\sin\theta\cos\theta \\
\sin^2\theta & \cos^2\theta & 0 & 0 & 0 & \sin\theta\cos\theta \\
0 & 0 & 1 & 0 & 0 & 0 \\
0 & 0 & 0 & \cos\theta & \sin\theta & 0 \\
0 & 0 & 0 & -\sin\theta & \cos\theta & 0 \\
\sin 2\theta & -\sin 2\theta & 0 & 0 & 0 & \cos^2\theta - \sin^2\theta
\end{bmatrix}
\begin{Bmatrix} \varepsilon_1 \\ \varepsilon_2 \\ \varepsilon_3 \\ \varepsilon_4 \\ \varepsilon_5 \\ \varepsilon_6 \end{Bmatrix}.
$$

$$(6.2.39)$$

The inverse relation is given by

$$
\begin{Bmatrix} \varepsilon_1 \\ \varepsilon_2 \\ \varepsilon_3 \\ \varepsilon_4 \\ \varepsilon_5 \\ \varepsilon_6 \end{Bmatrix} =
\begin{bmatrix}
\cos^2\theta & \sin^2\theta & 0 & 0 & 0 & \sin\theta\cos\theta \\
\sin^2\theta & \cos^2\theta & 0 & 0 & 0 & -\sin\theta\cos\theta \\
0 & 0 & 1 & 0 & 0 & 0 \\
0 & 0 & 0 & \cos\theta & -\sin\theta & 0 \\
0 & 0 & 0 & \sin\theta & \cos\theta & 0 \\
-\sin 2\theta & \sin 2\theta & 0 & 0 & 0 & \cos^2\theta - \sin^2\theta
\end{bmatrix}
\begin{Bmatrix} \varepsilon_{xx} \\ \varepsilon_{yy} \\ \varepsilon_{zz} \\ 2\varepsilon_{yz} \\ 2\varepsilon_{xz} \\ 2\varepsilon_{xy} \end{Bmatrix}.
$$

$$(6.2.40)$$

Although not discussed here [see Reddy (2004)], the elasticity tensor **C** must also be transformed from the material coordinate system to the problem coordinates.

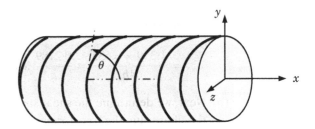

Figure 6.2.4. A filament-wound cylindrical
pressure vessel.

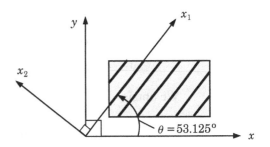

Next, we consider an example of application of the transformation equations
(see Reddy, 2004; also, see Example 4.3.2).

EXAMPLE 6.2.1: Consider a thin, filament-wound, closed cylindrical pressure vessel shown in Figure 6.2.4. The vessel is of 63.5 cm (25 in.) internal diameter,
2 cm thickness (0.7874 in.), and pressurized to 1.379 MPa (200 psi); Note that
MPa means mega (10^6) Pascal (Pa), Pa $= N/m^2$, and 1 psi $= 6{,}894.76$ Pa. A giga
Pascal (GPa) is 1,000 MPa. Determine

(a) stresses σ_{xx}, σ_{yy}, and σ_{xy} in the vessel,
(b) stresses σ_{11}, σ_{22}, and σ_{12} in the material coordinates (x_1, x_2, x_3) with x_1 being
   along the filament direction,
(c) strains ε_{11}, ε_{22}, and $2\varepsilon_{12}$ in the material coordinates, and
(d) strains ε_{xx}, ε_{yy}, and γ_{xy} in the vessel. Assume a filament winding angle
   of $\theta = 53.125°$ from the longitudinal axis of the pressure vessel, and use
   the following material properties, typical of graphite-epoxy material: $E_1 =$
   140 GPa (20.3 Msi), $E_2 = 10$ GPa (1.45 Msi), $G_{12} = 7$ GPa (1.02 Msi), and
   $\nu_{12} = 0.3$.

SOLUTION:
(a) The equations of equilibrium of forces in a structure do not depend on the
   material properties. Hence, equations derived for the longitudinal (σ_{xx}) and
   circumferential (σ_{yy}) stresses in a thin-walled cylindrical pressure vessel are
   valid here:

$$\sigma_{xx} = \frac{pD_i}{4h}, \quad \sigma_{yy} = \frac{pD_i}{2h},$$

where p is internal pressure, D_i is internal diameter, and h is thickness of the pressure vessel. We obtain (the shear stress σ_{xy} is zero)

$$\sigma_{xx} = \frac{1.379 \times 0.635}{4h} = \frac{0.2189}{h} \text{ MPa}, \quad \sigma_{yy} = \frac{1.379 \times 0.635}{2h} = \frac{0.4378}{h} \text{ MPa}.$$

(b) Next, we determine the shear stress along the fiber and the normal stress in the fiber using the transformation equations

$$\sigma_{11} = \sigma_{xx} \cos^2\theta + \sigma_{yy} \sin^2\theta + 2\sigma_{xy}\cos\theta\sin\theta,$$

$$\sigma_{22} = \sigma_{xx} \sin^2\theta + \sigma_{yy} \cos^2\theta - 2\sigma_{xy}\cos\theta\sin\theta,$$

$$\sigma_{12} = -\sigma_{xx}\sin\theta\cos\theta + \sigma_{yy}\cos\theta\sin\theta + \sigma_{xy}(\cos^2\theta - \sin^2\theta).$$

We obtain

$$\sigma_{11} = \frac{0.2189}{h}(0.6)^2 + \frac{0.4378}{h}(0.8)^2 = \frac{0.3590}{h} \text{ MPa},$$

$$\sigma_{22} = \frac{0.2189}{h}(0.8)^2 + \frac{0.4378}{h}(0.6)^2 = \frac{0.2977}{h} \text{ MPa},$$

$$\sigma_{12} = \left(\frac{0.4378}{h} - \frac{0.2189}{h}\right) \times 0.6 \times 0.8 = \frac{0.1051}{h} \text{ MPa}.$$

Thus, the normal and shear forces per unit length along the fiber-matrix interface are $F_{22} = 0.2977$ MN and $F_{12} = 0.1051$ MN, whereas the force per unit length in the fiber direction is $F_{11} = 0.359$ MN. For $h = 2$ cm, the stress field in the material coordinates becomes

$$\sigma_{11} = 17.95 \text{ MPa}, \quad \sigma_{22} = 14.885 \text{ MPa}, \quad \sigma_{12} = 5.255 \text{ MPa}.$$

(c) The strains in the material coordinates can be calculated using the strain–stress relations (6.2.27). We have $(\nu_{21}/E_2 = \nu_{12}/E_1, \sigma_{33} = 0)$

$$\varepsilon_{11} = \frac{\sigma_{11}}{E_1} - \frac{\sigma_{22}\nu_{12}}{E_1} = \frac{17.95}{140 \times 10^3} - \frac{14.885 \times 0.3}{140 \times 10^3} = 0.0963 \times 10^{-3} \text{ m/m},$$

$$\varepsilon_{22} = -\frac{\sigma_{11}\nu_{12}}{E_1} + \frac{\sigma_{22}}{E_2} = -\frac{17.95 \times 0.3}{140 \times 10^3} + \frac{14.885}{10 \times 10^3} = 1.45 \times 10^{-3} \text{ m/m},$$

$$\varepsilon_{12} = \frac{\sigma_{12}}{2G_{12}} = \frac{5.255}{2 \times 7} = 0.3757 \times 10^{-3}.$$

(d) The strains in the (x, y) coordinates can be computed using

$$\varepsilon_{xx} = \varepsilon_{11}\cos^2\theta + \varepsilon_{22}\sin^2\theta - 2\varepsilon_{12}\cos\theta\sin\theta,$$

$$\varepsilon_{yy} = \varepsilon_{11}\sin^2\theta + \varepsilon_{22}\cos^2\theta + 2\varepsilon_{12}\cos\theta\sin\theta,$$

$$\varepsilon_{xy} = (\varepsilon_{11} - \varepsilon_{22})\cos\theta\sin\theta + \varepsilon_{12}(\cos^2\theta - \sin^2\theta),$$

or

$$\varepsilon_{xx} = 10^{-3}[0.0963 \times (0.6)^2 + 1.45 \times (0.8)^2 - 0.3757 \times 0.6 \times 0.8]$$

$$= 0.782 \times 10^{-3} \text{ m/m},$$

$$\varepsilon_{yy} = 10^{-3}[0.0963 \times (0.8)^2 + 1.45 \times (0.6)^2 + 0.3757 \times 0.6 \times 0.8]$$

$$= 0.764 \times 10^{-3} \text{ m/m},$$

$$\varepsilon_{xy} = 10^{-3}\{2(0.0963 - 1.45) \times (0.6) \times 0.8 + 0.3757[(0.6)^2 - (0.8)^2]\}$$

$$= -1.405 \times 10^{-3}.$$

6.2.8 Nonlinear Elastic Constitutive Relations

Most materials exhibit nonlinear elastic behavior for certain strain threshold, that is, the stress-strain relation is no longer linear but recovers all its deformation upon the removal of the loads, and Hooke's law is no longer valid. Past certain nonlinear elastic range, permanent deformation ensues, and the material is said be inelastic or plastic, as shown in Figure 6.2.5. Here, we briefly review constitutive relations for two well-known nonlinear elastic materials, namely, the Mooney–Rivlin and neo-Hookean materials. More discussion can be found in Truesdell and Noll (1965) and Liu (2002).

Recall from Eq. (6.2.1) that for elastic materials under isothermal conditions the constitutive equation can be expressed as $\sigma = \sigma(\mathbf{F})$, where $\mathbf{F}$ is deformation gradient tensor with respect to some reference configuration κ_0 (for which det $\mathbf{F} = J > 0$). For a hyperelastic material, there exists a free energy function $\psi = \psi(\mathbf{F})$ such that

$$\sigma(\mathbf{F}) = \rho \frac{\partial \psi}{\partial \mathbf{F}} \cdot \mathbf{F}^{\mathrm{T}} \tag{6.2.41}$$

for compressible elastic materials, where ρ is the material density.

Some materials (e.g., rubber-like materials) undergo large deformations without appreciable change in volume (i.e., $J \approx 1$). Such materials are called *incompressible* materials. For incompressible elastic materials, the stress tensor is not

Figure 6.2.5. A typical stress–strain curve.

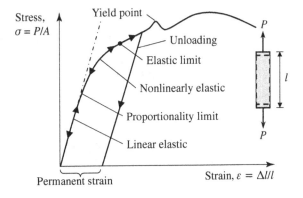

completely determined by deformation. The hydrostatic pressure affects the stress. For incompressible elastic materials, Eq. (6.2.41) takes the form

$$\sigma(\mathbf{F}) = -p\mathbf{I} + \rho \frac{\partial \psi}{\partial \mathbf{F}} \cdot \mathbf{F}^{\mathrm{T}}, \tag{6.2.42}$$

where p is the thermodynamic pressure.

For an hyperelastic elastic material, Eq. (6.2.41) can also be expressed as

$$\sigma(\mathbf{B}) = 2\rho \frac{\partial \psi}{\partial \mathbf{B}} \cdot \mathbf{B}, \tag{6.2.43}$$

where the free-energy function ψ is written as $\psi = \psi(\mathbf{B})$ and $\mathbf{B}$ is the left Cauchy–Green tensor $\mathbf{B} = \mathbf{F} \cdot \mathbf{F}^{\mathrm{T}}$ [see Eq. (3.4.22)]. Equations (6.2.41)–(6.2.43), in general, are nonlinear. The free energy function ψ takes different forms for different materials. It is often expressed as a linear combination of unknown parameters and principal invariants of Green strain tensor $\mathbf{E}$, deformation gradient tensor $\mathbf{F}$, or left Cauchy–Green strain tensor $\mathbf{B}$. The parameters characterize the material and they are determined through suitable experiments.

For incompressible materials, the free energy function ψ is taken as a linear function of the principal invariants of $\mathbf{B}$

$$\psi = C_1(I_B - 3) + C_2(II_B - 3), \tag{6.2.44}$$

where C_1 and C_2 are constants and I_B and II_B are the two principal invariants of $\mathbf{B}$ (the third invariant III_B is equal to unity for incompressible materials). Materials for which the strain energy functional is given by Eq. (6.2.44) are known as the *Mooney–Rivlin materials*. The stress tensor in this case has the form

$$\sigma = -p\mathbf{I} + \alpha\mathbf{B} + \beta\mathbf{B}^{-1}, \tag{6.2.45}$$

where α and β are given by

$$\alpha = 2\rho \frac{\partial \psi}{\partial I_B} = 2\rho C_1, \quad \beta = -2\rho \frac{\partial \psi}{\partial II_B} = -2\rho C_2. \tag{6.2.46}$$

The Mooney–Rivlin incompressible material model is most commonly used to represent the stress-strain behavior of rubber-like solid materials.

If the free energy function is of the form $\psi = C_1(I_B - 3)$, that is, $C_2 = 0$, the constitutive equation in Eq. (6.2.45) takes the form

$$\sigma = -p\mathbf{I} + 2\rho C_1 \mathbf{B}. \tag{6.2.47}$$

Materials whose constitutive behavior is described by Eq. (6.2.47) are called the *neo-Hookean materials*. The neo-Hookean model provides a reasonable prediction of the constitutive behavior of natural rubber for moderate strains.

6.3 Constitutive Equations for Fluids

6.3.1 Introduction

All bulk matter in nature exists in one of two forms: solid or fluid. A solid body is characterized by relative immobility of its molecules, whereas a fluid state is characterized by their relative mobility. Fluids can exist either as gases or liquids.

The stress in a fluid is proportional to the time rate of strain (i.e., time rate of deformation). The proportionality parameter is known as the *viscosity*. It is a measure of the intermolecular forces exerted as layers of fluid attempt to slide past one another. The viscosity of a fluid, in general, is a function of the thermodynamic state of the fluid and in some cases the strain rate. A *Newtonian fluid* is one for which the stresses are linearly proportional to the velocity gradients. If the constitutive equation for stress tensor is nonlinear, the fluid is said to *non-Newtonian*. A non-Newtonian constitutive relation can be of algebraic (e.g., power-law), differential, or integral type. A number of non-Newtonian models are presented in Section 6.3.4.

6.3.2 Ideal Fluids

A fluid is said to be *incompressible* if the volume change is zero:

$$\nabla \cdot \mathbf{v} = 0, \tag{6.3.1}$$

where $\mathbf{v}$ is the velocity vector. A fluid is termed *inviscid* if the viscosity is zero, $\mu = 0$. An *ideal fluid* is one that has zero viscosity and is incompressible.

The simplest constitutive equations are those for an ideal fluid. The most general constitutive equations for an ideal fluid are of the form

$$\sigma = -p(\rho, \theta)\mathbf{I}, \tag{6.3.2}$$

where p is the pressure and θ is the absolute temperature. The dependence of p on ρ and θ has been experimentally verified many times during several centuries. The thermomechanical properties of an ideal fluid are the same in all directions, that is, the material is isotropic. It can be verified that Eq. (6.3.2) satisfies the frame indifference requirement (see Section 3.9) because $\sigma* = \mathbf{Q} \cdot \sigma \cdot \mathbf{Q}^{\mathrm{T}} = -p\mathbf{Q} \cdot \mathbf{I} \cdot \mathbf{Q}^{\mathrm{T}} = -p\mathbf{I}$.

An explicit functional form of $p(\rho, \theta)$ valid for gases over a wide range of temperature and density is

$$p = R\rho\theta/m, \tag{6.3.3}$$

where R is the universal gas constant, m is the mean molecular weight of the gas, and θ is the absolute temperature. Equation (6.3.3) is known to define a "perfect" gas. When p is only a function of the density, the fluid is said to be "barotropic," and the barotropic constitutive model is applicable under isothermal conditions. If p is independent of both ρ and θ ($\rho = \rho_0 = $ constant), p is determined from the equations of motion.

6.3.3 Viscous Incompressible Fluids

The constitutive equation for stress tensor in a fluid motion is assumed to be of the general form[1]

$$\sigma = \mathbf{F}(\mathbf{D}) - p\,\mathbf{I}, \tag{6.3.4}$$

where $\mathbf{F}$ is a tensor-valued function of the rate of deformation $\mathbf{D}$ and p is the thermodynamic pressure. The viscous stress τ is equal to the total stress σ minus the equilibrium stress $-p\mathbf{I}$

$$\sigma = \tau - p\,\mathbf{I}, \quad \tau = \mathbf{F}(\mathbf{D}). \tag{6.3.5}$$

For a Newtonian fluid, $\mathbf{F}$ is assumed to be a linear function of $\mathbf{D}$,

$$\tau = \mathbf{C} : \mathbf{D} \quad \text{or} \quad \tau_{ij} = C_{ijkl}\,D_{kl}, \tag{6.3.6}$$

where $\mathbf{C}$ is the fourth-order tensor of viscosities of the fluid. For an isotropic viscous fluid, Eq. (6.3.6) reduces to [analogous to Eq. (6.2.29) for a Hookean solid]

$$\tau = 2\mu\mathbf{D} + \lambda(\text{tr}\,\mathbf{D})\mathbf{I} \quad \text{or} \quad \tau_{ij} = 2\mu D_{ij} + \lambda D_{kk}\delta_{ij}, \tag{6.3.7}$$

where μ and λ are the Lamé constants. Equation (6.3.5) takes the form

$$\sigma = 2\mu\mathbf{D} + \lambda(\text{tr}\,\mathbf{D})\mathbf{I} - p\,\mathbf{I}, \quad \sigma_{ij} = 2\mu D_{ij} + (\lambda D_{kk} - p)\,\delta_{ij}. \tag{6.3.8}$$

In terms of the deviatoric components of stress and rate of deformation tensors,

$$\sigma' = \sigma - \tilde{\sigma}\mathbf{I}, \quad \mathbf{D}' = \mathbf{D} - \frac{1}{3}(\text{tr}\,\mathbf{D})\,\mathbf{I}, \quad \tilde{\sigma} = \frac{1}{3}\text{tr}\,\sigma, \tag{6.3.9}$$

the Newtonian constitutive equation (6.3.8) takes the form

$$\sigma' = 2\mu\mathbf{D}' + \left(\frac{2}{3}\mu + \lambda\right)(\text{tr}\,\mathbf{D})\mathbf{I} - (\tilde{\sigma} + p)\,\mathbf{I},$$

$$\sigma'_{ij} = 2\mu D'_{ij} + \left(\frac{2}{3}\mu + \lambda\right)D_{kk}\delta_{ij} - (\tilde{\sigma} + p)\,\delta_{ij}. \tag{6.3.10}$$

Since

$$\sigma'_{ii} = 2\mu D'_{ii} + (2\mu + 3\lambda)\,D_{kk} - 3\,(\tilde{\sigma} + p) = 0, \tag{6.3.11}$$

the last two terms in Eq. (6.3.10) vanish, and we obtain

$$\sigma' = 2\mu\mathbf{D}', \quad \sigma'_{ij} = 2\mu D'_{ij}. \tag{6.3.12}$$

The mean stress $\tilde{\sigma}$ is equal to the thermodynamic pressure $-p$ if and only if one of the following two conditions are satisfied:

$$\text{Fluid is incompressible:} \quad \nabla \cdot D = 0, \tag{6.3.13}$$

$$\text{Stokes condition:} \quad K = \frac{2}{3}\mu + \lambda = 0. \tag{6.3.14}$$

---

[1] The dependence of $\mathbf{F}$ on the rotation tensor ω is eliminated to satisfy the frame indifference requirement.

In general, the Stokes condition does not hold. For Newtonian fluids, incompressibility does not necessarily imply that $\tilde{\sigma} = -p$.

Thus, the constitutive equation for a viscous, isotropic, incompressible fluid reduces to

$$\boldsymbol{\sigma} = -p\mathbf{I} + 2\mu\mathbf{D}, \quad (\sigma_{ij} = -p\,\delta_{ij} + 2\mu\,D_{ij}). \tag{6.3.15}$$

For inviscid fluids, the constitutive equation for the stress tensor has the form

$$\boldsymbol{\sigma} = -p\mathbf{I} \quad (\sigma_{ij} = -p\,\delta_{ij}), \tag{6.3.16}$$

and p in this case represents the mean normal stress or *hydrostatic pressure.*

6.3.4 Non-Newtonian Fluids

Non-Newtonian fluids are those for which the constitutive behavior is nonlinear. Non-Newtonian fluids include motor oils; high molecular weight liquids such as polymers, slurries, pastes; and other complex mixtures. The processing and transport of such fluids are central problems in the chemical, food, plastics, petroleum, and polymer industries. The non-Newtonian constitutive models presented in this section for viscous fluids are only a few of the many available in literature [see Reddy and Gartling (2001)].

Most non-Newtonian fluids exhibit a shear rate dependent viscosity, with "shear thinning" characteristic (i.e., decreasing viscosity with increasing shear rate). Other characteristics associated with non-Newtonian fluids are elasticity, memory effects, the Weissenberg effect, and the curvature of the free surface in an open-channel flow. A discussion of these and other non-Newtonian effects is presented in the book by Bird et al. (1971).

Non-Newtonian fluids can be classified into two groups: (1) inelastic fluids or fluids without memory and (2) viscoelastic fluids, in which memory effects are significant. For inelastic fluids, the viscosity depends on the rate of deformation of the fluid, much like nonlinear elastic solids. Viscoelastic fluids exhibit time-dependent "memory"; that is, the motion of a material point depends not only on the present stress state but also on the deformation history of the material element. This history dependence leads to very complex constitutive equations.

The constitutive equation for the stress tensor for a non-Newtonian fluid can be expressed as

$$\boldsymbol{\sigma} = -p\mathbf{I} + \boldsymbol{\tau} \quad (\sigma_{ij} = -p\,\delta_{ij} + \tau_{ij}), \tag{6.3.17}$$

where $\boldsymbol{\tau}$ is known as the viscous or extra stress tensor.

6.3.4.1 Inelastic Fluids

The viscosity for inelastic fluids is found to depend on the rate of deformation tensor **D**. Often the viscosity is expressed as a function of the principal invariants of the deformation tensor **D**

$$\mu = \mu(I_1, I_2, I_3), \tag{6.3.18}$$

where the I_1, I_2, and I_3 are the principal invariants of $\mathbf{D}$,

$$I_1 = tr(\mathbf{D}) = D_{ii},$$

$$I_2 = \frac{1}{2}tr(\mathbf{D}^2) = \frac{1}{2}D_{ij}D_{ji}, \tag{6.3.19}$$

$$I_3 = \frac{1}{3}tr(\mathbf{D}^3) = \frac{1}{3}D_{ij}D_{jk}D_{ki},$$

where tr denotes the trace.

For an incompressible fluid, $I_1 = \nabla \cdot \mathbf{v} = 0$. Also, there is no theoretical or experimental evidence to suggest that the viscosity depends on I_3; thus, the dependence on the third invariant is eliminated. Equation (6.3.18) reduces to

$$\mu = \mu(I_2). \tag{6.3.20}$$

The viscosity can also depend on the thermodynamic state of the fluid, which for incompressible fluids usually implies a dependence only on the temperature. Equation (6.3.20) gives the general functional form for the viscosity function, and experimental observations and a limited theoretical base are used to provide specific forms of Eq. (6.3.20) for non-Newtonian viscosities. A variety of inelastic models have been proposed and correlated with experimental data, as discussed by Bird et al. (1971). Several of the most useful and popular models are presented next [see Reddy and Gartling (2001)].

POWER-LAW MODEL. The simplest and most familiar non-Newtonian viscosity model is the power-law model, which has the form

$$\mu = K I_2^{(n-1)/2}, \tag{6.3.21}$$

where n and K are parameters, which are, in general, functions of temperature; n is termed the *power-law index* and K is called *consistency*. Fluids, with an index $n < 1$, are termed *shear thinning* or *pseudoplastic*. A few materials are *shear thickening* or *dilatant* and have an index $n > 1$. The Newtonian viscosity is obtained with $n = 1$. The admissible range of the index n is bounded below by zero due to stability considerations.

When considering nonisothermal flows, the following empirical relations for n and K are used:

$$n = n_0 + B\left(\frac{T - T_0}{T_0}\right), \tag{6.3.22}$$

$$K = K_0 exp\left(-A[T - T_0]/T_0\right). \tag{6.3.23}$$

where subscript '0' indicates a reference value and A and B are material constants.

CARREAU MODEL. A major deficiency in the power-law model is that it fails to predict upper and lower limiting viscosities for extreme values of the deformation rate. This problem is alleviated in the Carreau model

$$\mu = \mu_\infty + (\mu_0 - \mu_\infty)\left(1 + [\lambda I_2]^2\right)^{(n-1)/2}, \tag{6.3.24}$$

wherein μ_0 and μ_∞ are the initial and infinite shear rate viscosities, respectively, and λ is a time constant.

BINGHAM MODEL. The *Bingham fluid* differs from most other fluids in that it can sustain an applied stress without fluid motion occurring. The fluid possesses a yield stress, τ_0, such that when the applied stresses are below τ_0 no motion occurs; when the applied stresses exceed τ_0 the material flows, with the viscous stresses being proportional to the excess of the stress over the yield condition. Typically, the constitutive equation after yield is taken to be Newtonian (Bingham model), though other forms such as a power-law equation are possible. In a general form, the Bingham model can be expressed as

$$\tau = \left(\frac{\tau_0}{\sqrt{I_2}} + 2\mu \right) \mathbf{D} \quad \text{when } \frac{1}{2} tr(\tau^2) \geq \tau_0^2, \tag{6.3.25}$$

$$\tau = 0 \quad \text{when } \frac{1}{2} tr(\tau^2) < \tau_0^2. \tag{6.3.26}$$

From Eq. (6.3.25) the apparent viscosity of the material beyond the yield point is $(\tau_0/\sqrt{I_2} + 2\mu)$. For a Herschel–Buckley fluid, the μ in Eq. (6.3.25) is given by Eq. (6.3.21). The inequalities in Eqs. (6.3.25) and (6.3.26) describe a von Mises yield criterion.

6.3.4.2 Viscoelastic Constitutive Models

For a viscoelastic fluid, the choice of the constitutive equation for the extra-stress τ in Eq. (6.3.17) is time-dependent. Such a relationship is often expressed in abstract form where the current extra-stress is related to the history of deformation in the fluid as

$$\tau = \mathcal{F}[\mathbf{G}(s),\ 0 < s < \infty], \tag{6.3.27}$$

where $\mathcal{F}$ is a tensor-valued functional, $\mathbf{G}$ is a finite deformation tensor (related to the Cauchy–Green tensor) and $s = t - t'$ is the time lapse from time t' to the present time, t. Fluids that obey constitutive equation of the form in Eq. (6.3.27) are called *simple fluids*. The functional form in Eq. (6.3.27) is not useful for general flow problems, and therefore numerous approximations of (6.3.27) have been proposed in several different forms. Several of them are reviewed here.

The two major categories of approximate constitutive relations include the integral and differential models. The integral model represents the extra-stress in terms of an integral over past time of the fluid deformation history. For a differential model the extra-stress is determined from a differential equation that relates the stress and stress rate to the flow kinematics. In general, the specific choice is dictated by the ability of a given model to predict the non-Newtonian effects expected in a particular application.

DIFFERENTIAL MODELS. The well-known differential constitutive equations are generally associated with the names of Oldroyd, Maxwell, and Jeffrey. First, we define various types of material time derivatives used in these models. For an Eulerian reference frame, the material time derivative of a symmetric second-order tensor can be defined in several ways, all of which are frame invariant. Let **S** denote a second-order tensor. Then, the *upper-convected* (or co deformational) derivative is defined by

$$\overset{\triangledown}{\mathbf{S}} = \frac{\partial \mathbf{S}}{\partial t} + \mathbf{v} \cdot \nabla \mathbf{S} - \mathbf{L} \cdot \mathbf{S} - (\mathbf{L} \cdot \mathbf{S})^{\mathrm{T}}, \tag{6.3.28}$$

and the *lower-convected* derivative is defined as

$$\overset{\triangle}{\mathbf{S}} = \frac{\partial \mathbf{S}}{\partial t} + \mathbf{v} \cdot \nabla \mathbf{S} + \mathbf{L}^{\mathrm{T}} \cdot \mathbf{S} + \mathbf{S}^{\mathrm{T}} \cdot \mathbf{L}, \tag{6.3.29}$$

where **v** is the velocity vector and **L** is the velocity gradient tensor

$$\mathbf{L} = \nabla \mathbf{v} \quad \left(L_{ij} = \frac{\partial v_j}{\partial x_i} \right). \tag{6.3.30}$$

Since both Eqs. (6.3.28) and (6.3.29) are admissible convected derivatives, their linear combination is also admissible:

$$\overset{\circ}{\mathbf{S}} = (1 - \alpha) \overset{\triangledown}{\mathbf{S}} + \alpha \overset{\triangle}{\mathbf{S}}. \tag{6.3.31}$$

Equation (6.3.31) is a general convected derivative, which reduces to (6.3.28) for $\alpha = 0$ and (6.3.29) for $\alpha = 1$. When $\alpha = 0.5$ [average of Eqs. (6.3.28) and (6.3.29)], the convected derivative in Eq. (6.3.31) is termed a *corotational* or *Jaumann derivative*. All of these derivatives have been used in various differential constitutive equations. The selection of one type of derivative over other is usually based on the physical plausibility of the resulting constitutive equation and the matching of experimental data to the model for simple (viscometric) flows.

The simplest differential constitutive models are the upper- and lower-convected Maxwell fluids, which are defined by the following equations:

Upper-convected Maxwell fluid: $\boldsymbol{\tau} + \lambda \overset{\triangledown}{\boldsymbol{\tau}} = 2\mu^p \mathbf{D}$ \hfill (6.3.32)

Lower-convected Maxwell fluid: $\boldsymbol{\tau} + \lambda \overset{\triangle}{\boldsymbol{\tau}} = 2\mu^p \mathbf{D},$ \hfill (6.3.33)

where λ is a relaxation time for the fluid, μ^p is a viscosity, and **D** are the components of the rate of deformation tensor. The upper-convected Maxwell model in Eq. (6.3.32) has been used extensively in testing numerical algorithms; the lower-convected and corotational forms of the Maxwell fluid predict physically unrealistic behavior and are not generally used.

JOHNSON–SEGALMAN MODEL. By employing the general convected derivative (6.3.31) in a Maxwell-like model the Johnson–Segalman model is produced

$$\boldsymbol{\tau} + \lambda \overset{\circ}{\boldsymbol{\tau}} = 2\mu^p \mathbf{D}. \tag{6.3.34}$$

PHAN THIEN–TANNER MODEL. By slightly modifying Eq. (6.3.34) to include a variable coefficient for τ, the Phan Thien–Tanner model is obtained.

$$Y(\tau)\tau + \lambda\overset{\circ}{\tau} = 2\mu^p \mathbf{D}, \tag{6.3.35}$$

where

$$Y(\tau) = 1 + \frac{\epsilon\lambda}{\mu^p} \, \mathrm{tr}(\tau) \tag{6.3.36}$$

and ϵ is a constant. This equation is somewhat better than Eq. (6.3.34) in representing actual material behavior.

OLDROYD MODEL. The Johnson–Segalman and Phan Thien–Tanner models suffer from a common defect. For a monotonically increasing shear rate, there is a region where the shear stress decreases, which is a physically unrealistic behavior. To correct this anomaly, the constitutive equations are altered using the following procedure. First, the extra-stress is decomposed into two partial stresses, τ^s and τ^p such that

$$\tau = \tau^s + \tau^p, \tag{6.3.37}$$

where τ^s is a purely viscous and τ^p is a viscoelastic stress component. Then, τ^s and τ^p are expressed in terms of the deformation gradient tensor $\mathbf{D}$, using the Johnson–Segalman fluid as an example, as

$$\tau^s = 2\mu^s \, \mathbf{D}, \quad \tau^p + \lambda\overset{\circ}{\tau}^p = 2\mu^p \, \mathbf{D}. \tag{6.3.38}$$

Finally, the partial stresses in Eqs. (6.3.37) and (6.3.38) are eliminated to produce a new constitutive relation

$$\tau + \lambda\overset{\circ}{\tau} = 2\bar{\mu}(\mathbf{D} + \lambda'\overset{\circ}{\mathbf{D}}), \tag{6.3.39}$$

where $\bar{\mu} = (\mu^s + \mu^p)$ and $\lambda' = \lambda\mu^s/\bar{\mu}$; and λ' is a retardation time. The constitutive equation in (6.3.39) is known as a type of Oldroyd fluid. For particular choices of the convected derivative in Eq. (6.3.39), specific models can be generated. When $\alpha = 0$ ($\overset{\circ}{\tau}$ becomes $\overset{\triangledown}{\tau}$), then Eq. (6.3.39) becomes the Oldroyd B fluid; the case $\alpha = 1$ ($\overset{\circ}{\tau}$ becomes $\overset{\triangledown}{\tau}$) produces the Oldroyd A fluid. In order to ensure a monotonically increasing shear stress, the inequality $\mu^s \geq \mu^p/8$ must be satisfied. The stress decomposition employed above can also be used with the Phan Thien–Tanner model to produce a correct shear stress behavior.

WHITE–METZNER MODEL. In all of the constitutive equations the material parameters, λ and μ^p, were assumed to be constants. For some constitutive equations, the constancy of these parameters leads to material (or viscometric) functions that do not accurately represent the behavior of real elastic fluids. For example, the shear viscosity predicted by a Maxwell fluid is a constant, when infact viscoelastic fluids normally exhibit a shear thinning behavior. This situation can be remedied to some degree by allowing the parameters λ and μ^p to be functions of the invariants of the rate of deformation tensor **D**. Using the upper-convected Maxwell fluid as an example, then

$$\boldsymbol{\tau} + \lambda(I_2)\overset{\triangledown}{\boldsymbol{\tau}} = 2\mu^p(I_2)\mathbf{D}, \tag{6.3.40}$$

where I_2 is the second invariant of the deformation tensor **D**, $I_2 = 1/2(\mathbf{D}:\mathbf{D})$. The constitutive equation in Eq. (6.3.40) is termed a *White–Metzner model*. White–Metzner forms of other differential models, such as the Oldroyd fluids, have also been developed and used in various situations.

INTEGRAL MODELS. An approximate integral model for a viscoelastic fluid represents the extra-stress in terms of an integral over the past history of the fluid deformation. A general form for a single integral model can be expressed as

$$\boldsymbol{\tau} = \int_{-\infty}^{t} 2m(t - t')\mathbf{H}(t, t')dt', \tag{6.3.41}$$

where t is the current time, m is a scalar memory function (or relaxation kernel), and **H** is a nonlinear deformation measure (tensor) between the past time t' and current time t.

There are many possible forms for both the memory function m and the deformation measure **H**. Normally the memory function is a decreasing function of the time lapse $s = t - t'$. Typical of such a function is the exponential given by

$$m(t - t') = m(s) = \frac{\mu_0}{\lambda^2} e^{-s/\lambda}, \tag{6.3.42}$$

where the parameters μ_0, λ, and s were defined previously. Like the choice of a convected derivative in a differential model, the selection of a deformation measure for use in Eq. (6.3.41) is somewhat arbitrary. One particular form that has received some attention is given by

$$\mathbf{H} = \phi_1(I_B, \tilde{I}_B)\mathbf{B} + \phi_2(I_B, \tilde{I}_B)\tilde{\mathbf{B}}. \tag{6.3.43}$$

In Eq. (6.3.43), $\tilde{\mathbf{B}}$ is the Cauchy–Green deformation tensor, **B** is its inverse, called the *Finger tensor* [see Eq. (3.4.14)], and the ϕ_1 and ϕ_2 are scalar functions of the invariants of the deformation tensors, $I_B = \text{tr}(\mathbf{B})$ and $\tilde{I}_B = \text{tr}(\tilde{\mathbf{B}})$. The form of the deformation measure in Eq. (6.3.43) is still quite general, though specific choices for the functions ϕ_i and the memory function m lead to several well-known constitutive models. Among these are the Kaye–BKZ fluid and the Lodge rubber-like liquid.

As a specific example of an integral model, we consider the Maxwell fluid. Setting $\phi_1 = 1$ and $\phi_2 = 0$ in Eq. (6.3.43) and using the memory function of Eq. (6.3.42), we obtain a constitutive equation of the form

$$\boldsymbol{\tau} = \frac{\mu_0}{\lambda^2} \int_{-\infty}^{t} exp\left[-(t - t')/\lambda\right] \left[\mathbf{B}(t') - \mathbf{I}\right] dt'. \tag{6.3.44}$$

The constitutive equation, Eq. (6.3.44), is an integral equivalent to the upper-convected Maxwell model shown in differential form in Eq. (6.3.32). In this case, the extra-stress is given in an explicit form, though its evaluation requires that the strain history be known for each fluid particle. Although the Maxwell fluid has both differential and integral form, this is not generally true for other constitutive equations.

6.4 Heat Transfer

6.4.1 General Introduction

Heat transfer is a branch of engineering that deals with the transfer of thermal energy within a medium or from one medium to another due to a temperature difference. Heat transfer may take place in one or more of the three basic forms: *conduction*, *convection*, and *radiation* (see Reddy and Gartling, 2001). The transfer of heat within a medium due to diffusion process is called *conduction heat transfer*. *Fourier's law* states that the heat flow is proportional to the temperature gradient. The constant of proportionality depends, among other things, on a material parameter known as the *thermal conductivity* of the material. For heat conduction to occur, there must be temperature differences between neighboring points.

Convection heat transfer is the energy transport effected by the motion of a fluid. The convection heat transfer between two dissimilar media is governed by *Newton's law* of cooling. It states that the heat flow is proportional to the difference of the temperatures of the two media. The proportionality constant is called the *convection heat transfer coefficient* or *film conductance*. For heat convection to occur, there must be a fluid that is free to move and transport energy with it.

Radiation is a mechanism that is different from the three transport processes we discussed so far: (1) momentum transport in Newtonian fluids that is proportional to the velocity gradient, (2) energy transport by conduction that is proportional to the negative of the temperature gradient, and (3) energy transport by convection that is proportional to the difference in temperatures of the body and the moving fluid in contact with the body. Thermal radiation is an electromagnetic mechanism, which allows energy transport with the speed of light through regions of space that are devoid of any matter. Radiant energy exchange between surfaces or between a region and its surroundings is described by the *Stefan–Boltzmann law*, which states that the radiant energy transmitted is proportional to the difference of the fourth power of the temperatures of the surfaces. The proportionality parameter is known as the *Stefan–Boltzmann constant*.

6.4.2 Fourier's Heat Conduction Law

The Fourier heat conduction law states that the heat flow $\mathbf{q}$ is related to the temperature gradient by

$$\mathbf{q} = -\mathbf{k} \cdot \nabla\theta, \tag{6.4.1}$$

where $\mathbf{k}$ is the thermal conductivity tensor of order two. The negative sign in Eq. (6.4.1) indicates that heat flows downhill on the temperature scale. The balance of energy [Eq. (5.4.12)] requires that

$$\rho c \frac{D\theta}{Dt} = \Phi - \nabla \cdot \mathbf{q} + \rho \mathcal{E}, \quad \Phi = \boldsymbol{\tau} : \mathbf{D}, \tag{6.4.2}$$

which, in view of Eq. (6.4.1), becomes

$$\rho c \frac{D\theta}{Dt} = \Phi + \nabla \cdot (\mathbf{k} \cdot \nabla \theta) + \rho \mathcal{E}, \tag{6.4.3}$$

where $\rho \mathcal{E}$ is the heat energy generated per unit volume, ρ is the density, and c is the specific heat of the material.

For heat transfer in a solid medium, Eq. (6.4.3) reduces to

$$\rho c \frac{\partial \theta}{\partial t} = \nabla \cdot (\mathbf{k} \cdot \nabla \theta) + \rho \mathcal{E}, \tag{6.4.4}$$

which forms the subject of the field of conduction heat transfer. For a fluid medium, Eq. (6.4.3) becomes

$$\rho c \left(\frac{\partial \theta}{\partial t} + \mathbf{v} \cdot \nabla \theta \right) = \Phi + \nabla \cdot (\mathbf{k} \cdot \nabla \theta) + \rho \mathcal{E}, \tag{6.4.5}$$

where $\mathbf{v}$ is the velocity field and Φ is the viscous dissipation function.

6.4.3 Newton's Law of Cooling

At a solid–fluid interface the heat flux is related to the difference between the temperature at the interface and that in the fluid

$$q_n \equiv \hat{\mathbf{n}} \cdot \mathbf{q} = h \left(\theta - \theta_{\text{fluid}} \right), \tag{6.4.6}$$

where $\hat{\mathbf{n}}$ is the unit normal to the surface of the body and h is known as the *heat transfer coefficient* or *film conductance*. This relation is known as Newton's law of cooling, which also defines h. Clearly, Eq. (6.4.6) defines a boundary condition on the bounding surface of a conducting medium.

6.4.4 Stefan–Boltzmann Law

The heat flow from surface 1 to surface 2 by radiation is governed by the Stefan–Boltzmann law

$$q_n = \sigma \left(\theta_1^4 - \theta_2^4 \right), \tag{6.4.7}$$

where θ_1 and θ_2 are the temperatures of surfaces 1 and 2, respectively, and σ is the Stefan–Boltzmann constant. Again, Eq. (6.4.7) defines a boundary condition on the surface 1 of a body.

6.5 Electromagnetics

6.5.1 Introduction

Problems involving the coupling of electromagnetic fields with fluid and thermal transport have a broad spectrum of applications ranging from astrophysics to manufacturing and to electromechanical devices and sensors. A good introduction to coupled fluid-electromagnetic problems is available in Hughes and Young (1966); general electromagnetic field theory is available in such texts as Jackson (1975). Here, we present a brief discussion of pertinent equations for the sake of completeness. No attempt is made in this book to make use of these constitutive equations.

6.5.2 Maxwell's Equations

The appropriate mathematical description of electromagnetic phenomena in a conducting material region, Ω_C, is given by the following Maxwell's equations [see Reddy and Gartling (2001), Hughes and Young (1966), and Jackson (1975); *caution:* the notation used here for various fields is standard in the literature; unfortunately, some of the symbols used here were already used previously for other variables]:

$$\nabla \times \mathbf{E} = -\frac{\partial \mathbf{B}}{\partial t}, \tag{6.5.1}$$

$$\nabla \times \mathbf{H} = \mathbf{J} + \frac{\partial \mathbf{D}}{\partial t}, \tag{6.5.2}$$

$$\nabla \cdot \mathbf{B} = 0, \tag{6.5.3}$$

$$\nabla \cdot \mathbf{D} = \rho, \tag{6.5.4}$$

where $\mathbf{E}$ is the electric field intensity, $\mathbf{H}$ the magnetic field intensity, $\mathbf{B}$ the magnetic flux density, $\mathbf{D}$ the electric flux (displacement) density, $\mathbf{J}$ the conduction current density, and ρ is the source charge density. Equation (6.5.1) is referred to as Faraday's law, Eq. (6.5.2) as Ampere's law (as modified by Maxwell), and Eq. (6.5.4) as Gauss' law. A continuity condition on the current density is also defined by

$$\nabla \cdot \mathbf{J} = \frac{\partial \rho}{\partial t}. \tag{6.5.5}$$

Only three of the previous five equations are independent; either Eqs. (6.5.1), (6.5.2), and (6.5.4) or Eqs. (6.5.1), (6.5.2), and (6.5.5) form valid sets of equations for the fields.

6.5.3 Constitutive Relations

To complete the formulation, the constitutive relations for the material are required. The fluxes are functionally related to the field variables by

$$\mathbf{D} = f_D(\mathbf{E}, \mathbf{B}), \tag{6.5.6}$$

$$\mathbf{H} = f_H(\mathbf{E}, \mathbf{B}), \tag{6.5.7}$$

$$\mathbf{J} = f_J(\mathbf{E}, \mathbf{B}), \tag{6.5.8}$$

where the functions (f_D, f_H, f_J) may also depend on external variables such as temperature or mechanical stress. The form of the material response to applied $\mathbf{E}$ or $\mathbf{B}$ fields can vary strongly depending on the state of the material, its microstructure and the strength, and time-dependent behavior of the applied field.

6.5.3.1 Conductive and Dielectric Materials

For conducting materials, the standard f_J relation is Ohm's law, which relates the current density $\mathbf{J}$ to the electric field intensity $\mathbf{E}$

$$\mathbf{J} = \mathbf{k}_\sigma \cdot \mathbf{E}, \tag{6.5.9}$$

where $\mathbf{k}_\sigma$ is the conductivity tensor. For isotropic materials, we have $\mathbf{k}_\sigma = k_\sigma \mathbf{I}$, where k_σ is a scalar. In general, the conductivity may be a function of $\mathbf{E}$ or an external variable such as temperature. This form of Ohm's law applies to stationary conductors. If the conductive material is moving in a magnetic field, then Eq. (6.5.9) is modified to read

$$\mathbf{J} = \mathbf{k}_\sigma \cdot \mathbf{E} + \mathbf{k}_\sigma \cdot (\mathbf{v} \times \mathbf{B}), \tag{6.5.10}$$

where $\mathbf{v}$ is the velocity vector describing the motion of the conductor and $\mathbf{B}$ is the magnetic flux vector.

For dielectric materials, the standard f_D function relates the electric flux density $\mathbf{D}$ to the electric field $\mathbf{E}$ and polarization vector $\mathbf{P}$:

$$\mathbf{D} = \epsilon_0 \cdot \mathbf{E} + \mathbf{P}, \tag{6.5.11}$$

where ϵ_0 is the permittivity of free space. The polarization is generally related to the electric field through

$$\mathbf{P} = \epsilon_0 \mathbf{S}_e \cdot \mathbf{E} + \mathbf{P}_0, \tag{6.5.12}$$

where $\mathbf{S}_e$ is the electric susceptibility tensor that accounts for the different types of polarization and $\mathbf{P}_0$ is the remnant polarization that may be present in some materials.

6.5.3.2 Magnetic Materials

For magnetic materials, the standard f_H function relates the magnetic field intensity $\mathbf{H}$ to the magnetic flux $\mathbf{B}$

$$\mathbf{H} = \frac{1}{\mu_0}\mathbf{B} - \mathbf{M}, \tag{6.5.13}$$

where μ_0 is the permeability of free space and $\mathbf{M}$ is the magnetization vector. The magnetization vector $\mathbf{M}$ can be related to either the magnetic flux $\mathbf{B}$ or magnetic field intensity $\mathbf{H}$ by

$$\mathbf{M} = \frac{1}{\mu_0}\frac{\mathbf{S}_m}{(\mathbf{I}+\mathbf{S}_m)} \cdot \mathbf{B} + \mathbf{M}_0, \tag{6.5.14}$$

$$\mathbf{M} = \mathbf{S}_m \cdot \mathbf{H} + (\mathbf{I}+\mathbf{S}_m) \cdot \mathbf{M}_0, \tag{6.5.15}$$

where $\mathbf{S}_m$ is the magnetic susceptibility for the material and $\mathbf{M}_0$ is the remnant magnetization. If the susceptibility is negative, the material is diamagnetic; while a positive susceptibility defines a paramagnetic material. Generally, these susceptibilities are quite small and are often neglected. Ferromagnetic materials have large positive susceptibilities and produce a nonlinear (hysteretic) relationship between $\mathbf{B}$ and $\mathbf{H}$. These materials may also exhibit spontaneous and remnant magnetization.

6.5.3.3 Electromagnetic Forces and Volume Heating

The coupling of electromagnetic fields with a fluid or thermal problem occurs through the dependence of material properties on electromagnetic field quantities and the production of electromagnetic-induced body forces and volumetric energy production. The Lorentz body force in a conductor due to the presence of electric currents and magnetic fields is given by

$$\mathbf{F}_B = \rho\mathbf{E} + \mathbf{J} \times \mathbf{B}, \tag{6.5.16}$$

where, in the general case, the current is defined by Eq. (6.5.10). The first term on the right-hand side of Eq. (6.5.16) is the electric field contribution to the Lorentz force; the magnetic term $\mathbf{J} \times \mathbf{B}$ is usually of more interest in applied mechanics problems. The energy generation or Joule heating in a conductor is described by

$$Q_J = \mathbf{J} \cdot \mathbf{E}, \tag{6.5.17}$$

which takes on a more familiar form if the simplified ($\mathbf{v} = \mathbf{0}$) form of Eq. (6.5.10) is used to produce

$$Q_J = \sigma^{-1}(\mathbf{J} \cdot \mathbf{J}). \tag{6.5.18}$$

The above forces and heat source occur in the fluid momentum and energy equations, respectively.

6.6 Summary

This chapter was dedicated to a discussion of the constitutive equations, that is, relations between the primary variables such as the displacements, velocities, and temperature to the secondary variables such as the stresses, pressure, and heat flux of continua. Although there are no physical principles to derive these mathematical relations, there are rules or guidelines that help to develop mathematical models of the constitutive behavior which must be, ultimately, validated against actual response characteristics observed in physical experiments. The constitutive relations, in general, can be algebraic, differential, or integral relations, depending on the nature of the material behavior being modeled.

In this chapter, the generalized Hooke's law governs linear elastic solids, Newtonian relations for viscous fluids, and the Fourier heat conduction equation for heat transfer in solids are presented. These equations are used in Chapters 7 and 8 to analyze problems of solid mechanics, fluid mechanics, and heat transfer. Constitutive relations of nonlinear elastic solids, non-Newtonian fluids, and electromagnetics are also presented for the sake of completeness. Constitutive relations of linear viscoelastic materials are discussed in Chapter 9.

PROBLEMS

6.1 Establish the following relations between the Lamé constants μ and λ and engineering constants E, v, and K:

$$\lambda = \frac{vE}{(1+v)(1-2v)}, \quad \mu = G = \frac{E}{2(1+v)}, \quad K = \frac{E}{3(1-2v)}.$$

6.2 Determine the stress tensor components at a point in 7075-T6 aluminum alloy body ($E = 72$ GPa and $G = 27$ GPa) if the strain tensor at the point has the following components with respect to the Cartesian basis vectors $\hat{e}_i$:

$$[\varepsilon] = \begin{bmatrix} 200 & 100 & 0 \\ 100 & 300 & 400 \\ 0 & 400 & 0 \end{bmatrix} \times 10^{-6}.$$

6.3 For the state of stress and strain given in Problem 6.2, determine the stress and strain invariants.

6.4 If the components of strain at a point in a body made of structural steel are

$$[\varepsilon] = \begin{bmatrix} 36 & 12 & 30 \\ 12 & 40 & 0 \\ 30 & 0 & 25 \end{bmatrix} \times 10^{-6}.$$

Assuming that the Lamé constants for the structural steel are $\lambda = 207$ GPa (30×10^6 psi) and $\mu = 79.6$ GPa (11.54×10^6 psi), determine the stress invariants.

6.5 If the components of stress at a point in a body are

$$[\sigma] = \begin{bmatrix} 42 & 12 & 30 \\ 12 & 15 & 0 \\ 30 & 0 & -5 \end{bmatrix} \text{ MPa.}$$

Assuming that the Lamé constants for are $\lambda = 207$ GPa (30×10^6 psi) and $\mu = 79.6$ GPa (11.54×10^6 psi), determine the strain invariants.

6.6 Given the following motion of an isotropic continuum,

$$\chi(\mathbf{X}) = \left(X_1 + kt^2 X_2^2 \right) \hat{\mathbf{e}}_1 + (X_2 + kt\, X_2)\, \hat{\mathbf{e}}_2 + X_3\, \hat{\mathbf{e}}_3,$$

determine the components of the viscous stress tensor as a function of position and time.

6.7 Express the upper and lower convective derivatives of Eqs. (6.3.28) and (6.3.29) in Cartesian component form.

6.8 Most advanced books on continuum mechanics discuss the general axioms of constitutive theory. This exercise has the objective of making the reader to get familiar with the axioms of the constitutive theory. List the axioms of the constitutive theory and explain briefly what the axioms mean.

7 Linearized Elasticity Problems

You cannot depend on your eyes when your imagination is out of focus.

Mark Twain

Research is to see what everybody else has seen, and to think what nobody else has thought.

Albert Szent-Gyoergi

7.1 Introduction

This chapter is dedicated to the study of deformation and stress in solid bodies under a prescribed set of forces and kinematic constraints. We assume that stresses and strains are small so that linear strain–displacement relations and Hooke's law are valid, and we use appropriate governing equations, called field equations, derived in the previous chapters. Mathematically, we seek solutions to coupled partial differential equations over an elastic domain occupied by the reference (or undeformed) configuration of the body, subject to specified boundary conditions on displacements and forces. Such problems are called *boundary value problems of elasticity*.

Most practical problems of even linearized elasticity involve geometries that are complicated and analytical solutions to such problems cannot be obtained. Therefore, the objective here is to familiarize the reader with the certain solution methods as applied to simple boundary value problems. Problems discussed in most elasticity books are about the same and they illustrate the methodologies used in the analytical solution of problems of elasticity. Since this is a book on continuum mechanics, the coverage is some what limited. Most problems discussed here can be found in elasticity books, for example, by Timoshenko and Goodier (1970) and Slaughter (2002). While the methods discussed here may not be useful in solving practical engineering problems, the discussion provides certain insights into the formulation of boundary value problems. These insights are useful irrespective of the specific methods of solution.

7.2 Governing Equations

It is useful to summarize the equations of linearized elasticity for use in the remainder of the chapter. For the moment, we consider isothermal elasticity and study only equilibrium (i.e., static) problems. The governing equations of a three-dimensional elastic body involve: (1) six strain-displacement relations among nine variables, six strain components, and three displacements; (2) three equilibrium equations among six components of stress, assuming symmetry of the stress tensor; and (3) six stress-strain equations among the six stress and six strain components that are already counted. Thus, there are a total of 15 coupled equations among 15 scalar fields. These equations are listed here in vector and Cartesian component forms for an isotropic body occupying a domain Ω with closed boundary Γ in the reference configuration.

Strain–displacement equations

$$\varepsilon = \frac{1}{2}\left[\nabla \mathbf{u} + (\nabla \mathbf{u})^{\mathrm{T}}\right], \quad \varepsilon_{ij} = \frac{1}{2}(u_{i,j} + u_{j,i}). \tag{7.2.1}$$

Equilibrium equations

$$\nabla \cdot \boldsymbol{\sigma} + \mathbf{f} = \mathbf{0} \quad (\boldsymbol{\sigma}^{\mathrm{T}} = \boldsymbol{\sigma}), \quad \sigma_{ji,j} + f_i = 0, \quad (\sigma_{ji} = \sigma_{ij}), \tag{7.2.2}$$

where $\mathbf{f}$ is the body force measured per unit volume.

Constitutive equations

$$\boldsymbol{\sigma} = 2\mu\varepsilon + \lambda\,(\mathrm{tr}\,\varepsilon)\,\mathbf{I}, \quad \sigma_{ij} = 2\mu\varepsilon_{ij} + \lambda\varepsilon_{kk}\delta_{ij}. \tag{7.2.3}$$

These equations are valid for all problems of linearized elasticity; different problems differ from each other only in (a) geometry of the domain, (b) boundary conditions, and (c) material constitution. The general form of the boundary condition is given below.

Boundary conditions

$$\mathbf{t} \equiv \hat{\mathbf{n}} \cdot \boldsymbol{\sigma} = \hat{\mathbf{t}}, \quad t_i \equiv n_j\sigma_{ji} = \hat{t}_i \text{ on } \Gamma_\sigma \tag{7.2.4}$$

and

$$\mathbf{u} = \hat{\mathbf{u}}, \quad u_i = \hat{u}_i \text{ on } \Gamma_u, \tag{7.2.5}$$

where Γ_σ and Γ_u are disjoint portions (except for a point) of the boundary whose union is equal to the total boundary Γ. Only one element of the pair (t_i, u_i), for any $i = 1, 2, 3$, may be specified at a point on the boundary.

In addition to the 15 equations listed in (7.2.1)–(7.2.3), there are 6 *compatibility conditions* among 6 components of strain:

$$\nabla \times (\nabla \times \varepsilon)^{\mathrm{T}} = \mathbf{0}, \quad e_{ikr}e_{j\ell s}\varepsilon_{ij,k\ell} = 0. \tag{7.2.6}$$

Recall that the compatibility equations are necessary and sufficient conditions on the strain field to ensure the existence of a corresponding displacement field.

Associated with each displacement field, there is a unique strain field as given by Eq. (7.2.1), and there is no need to use the compatibility conditions. The compatibility conditions are required only when the strain field is given and displacement field is to be determined.

In most formulations of boundary value problems of elasticity, one does not use the 15 equations in 15 unknowns. Most often, the 15 equations are reduced to either 3 equations in terms of displacement field or 6 equations in terms of stress field. The two sets of equations are presented next.

7.3 The Navier Equations

The 15 equations can be combined into 3 equations by substituting strain–displacement equations into the stress–strain relations and the result into the equations of equilibrium. We shall carry out this process using the Cartesian component form and then express the final result in vector as well as Cartesian component forms.

From Eqs. (7.2.1) and (7.2.3), we obtain

$$\sigma_{ij} = \mu \left(u_{i,j} + u_{j,i} \right) + \lambda u_{k,k} \delta_{ij}. \tag{7.3.1}$$

Substituting into Eq. (7.2.2), we arrive at the equations

$$0 = \sigma_{ji,j} + f_i$$

$$= \mu \left(u_{i,jj} + u_{j,ij} \right) + \lambda u_{k,ki} + f_i$$

$$= \mu u_{i,jj} + (\mu + \lambda) u_{j,ji} + f_i. \tag{7.3.2}$$

Thus, we have

$$\mu \nabla^2 \mathbf{u} + (\mu + \lambda) \nabla \left(\nabla \cdot \mathbf{u} \right) + \mathbf{f} = \mathbf{0},$$

$$\mu u_{i,jj} + (\mu + \lambda) u_{j,ji} + f_i = 0. \tag{7.3.3}$$

These are called *Lamé–Navier equations* of elasticity, and they represent the equilibrium equations expressed in terms of the displacement field. The boundary conditions (7.2.4) and (7.2.5) can be expressed in terms of the displacement field as

$$[n_j \mu \left(u_{i,j} + u_{j,i} \right) + n_i \lambda u_{k,k}] = \hat{t}_i \text{ on } \Gamma_\sigma, \quad u_i = \hat{u}_i \text{ on } \Gamma_u. \tag{7.3.4}$$

Equations (7.3.3) and (7.3.4) together describe the boundary value problem of linearized elasticity.

7.4 The Beltrami–Michell Equations

Alternative to the formulation of Section 7.3, the 12 equations from (7.2.2) and (7.2.3) and 6 equations from (7.2.6) can be combined into 6 equations in terms of the stress field. Substitution of the constitutive (strain–stress) equations

$$\varepsilon_{ij} = \frac{1}{E} [(1 + \nu)\sigma_{ij} - \nu \sigma_{mm} \delta_{ij}] \tag{7.4.1}$$

into the compatibility equations (7.2.6) yields

$$\begin{aligned}
0 &= e_{ikr} e_{jls} \, \varepsilon_{ij,kl} \\
&= e_{ikr} e_{jls} \left[(1+v)\sigma_{ij,kl} - v\sigma_{mm,kl}\delta_{ij} \right] \\
&= (1+v)e_{ikr} e_{jls}\sigma_{ij,kl} - v e_{ikr} e_{ils}\sigma_{mm,kl} \\
&= (1+v)e_{ikr} e_{jls}\sigma_{ij,kl} - v\left(\delta_{kl}\delta_{rs} - \delta_{ks}\delta_{lr} \right)\sigma_{mm,kl} \\
&= (1+v)e_{ikr} e_{jls}\sigma_{ij,kl} - v\left(\delta_{rs}\sigma_{mm,kk} - \sigma_{mm,rs} \right).
\end{aligned} \tag{7.4.2}$$

Since [see Problem 2.5(f)]

$$e_{ikr} e_{jls} = \begin{vmatrix} \delta_{ij} & \delta_{il} & \delta_{is} \\ \delta_{kj} & \delta_{kl} & \delta_{ks} \\ \delta_{rj} & \delta_{rl} & \delta_{rs} \end{vmatrix}$$

$$= \delta_{ij}\delta_{kl}\delta_{rs} - \delta_{ij}\delta_{ks}\delta_{rl} - \delta_{kj}\delta_{il}\delta_{rs} + \delta_{kj}\delta_{rl}\delta_{is} + \delta_{rj}\delta_{il}\delta_{ks} - \delta_{rj}\delta_{kl}\delta_{is},$$

$$\tag{7.4.3}$$

Eq. (7.4.2) simplifies to

$$\delta_{rs}\sigma_{ii,jj} - \sigma_{ii,rs} - (1+v)\left(\delta_{rs}\sigma_{ij,ij} + \sigma_{rs,ii} - \sigma_{is,ir} - \sigma_{ir,is} \right) = 0. \tag{7.4.4}$$

Contracting the indices r and s $(s \to r)$ gives

$$2\sigma_{ii,jj} - (1+v)\left(\sigma_{ij,ij} + \sigma_{jj,ii} \right) = 0.$$

Simplifying the above result, we obtain

$$\sigma_{ii,jj} = \frac{(1+v)}{(1-v)}\sigma_{ij,ij}. \tag{7.4.5}$$

Substituting this result back into Eq. (7.4.4) leads to

$$\sigma_{ij,kk} + \frac{1}{1+v}\sigma_{kk,ij} = \frac{v}{1-v}\sigma_{rs,rs}\delta_{ij} + \sigma_{kj,ki} + \sigma_{ki,kj}. \tag{7.4.6}$$

Next, we use the equilibrium equations to compute the second derivative of the stress components, $\sigma_{rs,rk} = -f_{s,k}$. We have

$$\sigma_{ij,kk} + \frac{1}{1+v}\sigma_{kk,ij} = -\frac{v}{1-v}f_{k,k}\delta_{ij} - \left(f_{j,i} + f_{i,j} \right). \tag{7.4.7}$$

or in vector form

$$\nabla^2 \boldsymbol{\sigma} + \frac{1}{1+v}\nabla[\nabla(\mathrm{tr}\,\boldsymbol{\sigma})] = -\frac{v}{1-v}(\nabla \cdot \mathbf{f})\mathbf{I} - \left[\nabla\mathbf{f} + (\nabla\mathbf{f})^{\mathrm{T}}\right]. \tag{7.4.8}$$

The six equations in (7.4.7) or (7.4.8), called *Michell's equations*, provide the necessary and sufficient conditions for an equilibrated stress field to be compatible with the displacement field in the body. The traction boundary conditions in Eq. (7.3.4) are valid for this formulation.

When the body force is uniform, we have $\nabla \cdot \mathbf{f} = 0$ and $\nabla \mathbf{f} = \mathbf{0}$, and Michell's equations (7.4.8) reduce to *Beltrami's equations*

$$\nabla^2 \sigma + \frac{1}{1+\nu}\nabla[\nabla (\text{tr } \sigma)] = \mathbf{0}, \quad \sigma_{ij,kk} + \frac{1}{1+\nu}\sigma_{kk,ij} = 0. \tag{7.4.9}$$

7.5 Type of Boundary Value Problems and Superposition Principle

The boundary value problems of elasticity can be classified into three types on the basis of the nature of specified boundary conditions. They are discussed next.

TYPE I. Boundary value problems in which if all specified boundary conditions are of the displacement type

$$\mathbf{u} = \hat{\mathbf{u}} \text{ on } \Gamma \tag{7.5.1}$$

are called *boundary value problems of Type I* or *displacement boundary value problems*.

TYPE II. Boundary value problems in which if all specified boundary conditions are of the traction type

$$\mathbf{t} = \hat{\mathbf{t}} \text{ on } \Gamma \tag{7.5.2}$$

are called *boundary value problems of Type II* or *stress boundary value problems*.

TYPE III. Boundary value problems in which if all specified boundary conditions are of the mixed type,

$$\mathbf{u} = \hat{\mathbf{u}} \text{ on } \Gamma_u \quad \text{and} \quad \mathbf{t} = \hat{\mathbf{t}} \text{ on } \Gamma_\sigma, \tag{7.5.3}$$

are called *boundary value problems of Type III* or *mixed boundary value problems*.

Most practical problems fall into the category of boundary value problems of Type III.

While existence of solutions is a difficult question to answer, uniqueness of solutions is rather easy to prove for linear boundary value problems of elasticity. Another advantage of linear boundary value problems is that the principle of superposition holds. The principle of superposition is said to hold for a solid body if the displacements obtained under two sets of boundary conditions and forces is equal to the sum of the displacements that would be obtained by applying each set of boundary conditions and forces separately.

To be more specific, consider the following two sets of boundary conditions and forces

$$\textbf{Set1}: \quad \mathbf{u} = \mathbf{u}^{(1)} \text{ on } \Gamma_u; \quad \mathbf{t} = \mathbf{t}^{(1)} \text{ on } \Gamma_\sigma; \quad \mathbf{f} = \mathbf{f}^{(1)} \text{ in } \Omega \tag{7.5.4}$$

$$\textbf{Set2}: \quad \mathbf{u} = \mathbf{u}^{(2)} \text{ on } \Gamma_u; \quad \mathbf{t} = \mathbf{t}^{(2)} \text{ on } \Gamma_\sigma; \quad \mathbf{f} = \mathbf{f}^{(2)} \text{ in } \Omega \tag{7.5.5}$$

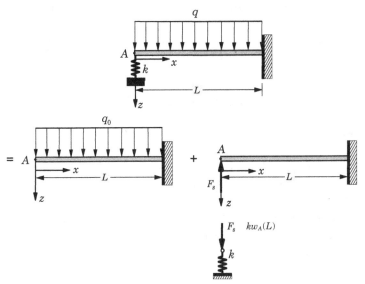

Figure 7.5.1. Representation of an indeterminate beam as a superposition of two determinate beams.

where the specified data $(\mathbf{u}^{(1)}, \mathbf{t}^{(1)}, \mathbf{f}^{(1)})$ and $(\mathbf{u}^{(2)}, \mathbf{t}^{(2)}, \mathbf{f}^{(2)})$ is independent of the deformation. Suppose that the solution to the two problems be $\mathbf{u}(\mathbf{x})^{(1)}$ and $\mathbf{u}(\mathbf{x})^{(2)}$, respectively. The superposition of the two sets of boundary conditions is

$$\mathbf{u} = \mathbf{u}^{(1)} + \mathbf{u}^{(2)} \text{ on } \Gamma_u; \quad \mathbf{t} = \mathbf{t}^{(1)} + \mathbf{t}^{(1)} \text{ on } \Gamma_\sigma; \quad \mathbf{f} = \mathbf{f}^{(1)} + \mathbf{f}^{(2)} \text{ in } \Omega. \quad (7.5.6)$$

Because of the linearity of the elasticity equations, the solution of the boundary value problem with the superposed data is $\mathbf{u}(\mathbf{x}) = \mathbf{u}^{(1)}(\mathbf{x}) + \mathbf{u}^{(2)}(\mathbf{x})$ in Ω. This is known as the *superposition principle.*

The principle of superposition can be used to represent a linear problem with complicated boundary conditions or loads as a combination of linear problems that are equivalent to the original problem. The next example illustrates this point (see Reddy, 2002).

EXAMPLE 7.5.1: Consider the indeterminate beam shown in Figure 7.5.1. Determine the deflection of point A using the principle of superposition.

SOLUTION: The problem can be viewed as one equivalent to the two beam problems shown there. The sum of the deflections from each problem is the solution of the original problem. Within the restrictions of the linear Euler–Bernoulli beam theory, the deflections are linear functions of the loads. Therefore, the principle of superposition is valid. In particular, the deflection w_A at point A is equal to the sum of w_A^q and w_A^s due to the distributed load q_0 and spring force F_s, respectively, at point A:

$$w_A = w_A^q + w_A^s = \frac{q_0 L^4}{8EI} - \frac{F_s L^3}{3EI}.$$

Because the spring force F_s is equal to kw_A, we can calculate w_A from

$$w_A = \frac{q_0 L^4}{8EI\left(1 + \frac{kL^3}{3EI}\right)}.$$

7.6 Clapeyron's Theorem and Reciprocity Relations

7.6.1 Clapeyron's Theorem

The principle of superposition is not valid for energies because they are quadratic functions of displacements and forces. In other words, when a linear elastic body $\mathcal{B}$ is subjected to more than one external force, the total work done due to external forces is *not* equal to the sum of the works that are obtained by applying the single forces separately. However, there exist theorems that relate the work done by two different forces applied in different orders. We will consider them in this section.

Recall from Chapter 6 that the strain energy density due to linear elastic deformation is given by

$$U_0 = \frac{1}{2}\,\boldsymbol{\sigma} : \boldsymbol{\varepsilon} = \frac{1}{2}\,\sigma_{ij}\varepsilon_{ij}. \tag{7.6.1}$$

The total strain energy stored in the body $\mathcal{B}$ occupying the region Ω with surface Γ is equal to

$$U = \int_\Omega U_0\,d\mathbf{x} = \frac{1}{2}\int_\Omega \boldsymbol{\sigma} : \boldsymbol{\varepsilon}\,d\mathbf{x} = \frac{1}{2}\int_\Omega \sigma_{ij}\varepsilon_{ij}\,d\mathbf{x}, \tag{7.6.2}$$

where $d\mathbf{x}$ denotes the line element dx_1, the area element $dx_1\,dx_2$, or the volume element $dx_1\,dx_2\,dx_3$, depending on the dimension of the domain Ω. The work done by externally applied body force $\mathbf{f}$ and surface tractions $\mathbf{t}$ in moving through the displacement vector $\mathbf{u}$ is given by

$$W_E = \int_\Omega \mathbf{f} \cdot \mathbf{u}\,d\mathbf{x} + \oint_\Gamma \mathbf{t} \cdot \mathbf{u}\,ds. \tag{7.6.3}$$

Because of the symmetry of the stress tensor, $\sigma_{ij} = \sigma_{ji}$, we can write $\sigma_{ij}\varepsilon_{ij} = \sigma_{ij}u_{i,j}$. Consequently, the strain energy U can be expressed as

$$\begin{aligned}
U &= \frac{1}{2}\int_\Omega \sigma_{ij}u_{i,j}\,d\mathbf{x} \\
&= \frac{1}{2}\left[-\int_\Omega \sigma_{ij,j}u_i\,d\mathbf{x} + \oint_\Gamma n_j\sigma_{ij}u_i\,ds\right] \\
&= \frac{1}{2}\left[\int_\Omega f_i u_i\,d\mathbf{x} + \oint_\Gamma t_i u_i\,ds\right] \\
&= \frac{1}{2}\left[\int_\Omega \mathbf{f} \cdot \mathbf{u}\,d\mathbf{x} + \oint_\Gamma \mathbf{t} \cdot \mathbf{u}\,ds\right],
\end{aligned}$$

where, in arriving at the last line, we have used the equilibrium equation $\sigma_{ji,j} + f_i = 0$, the Cauchy's formula $t_i = n_j\sigma_{ji}$, and the divergence theorem (2.4.34). Thus, the

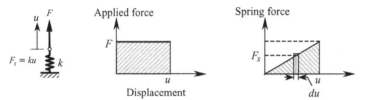

Figure 7.6.1. Strain energy stored in a linear elastic spring.

total strain energy in a body undergoing linear elastic deformation is

$$U = \frac{1}{2} \left[\int_\Omega \mathbf{f} \cdot \mathbf{u} \, dx + \oint_\Gamma \mathbf{t} \cdot \mathbf{u} \, ds \right]. \tag{7.6.4}$$

The first term in the square brackets on the right-hand side represents the work done by body force $\mathbf{f}$ in moving through the displacement $\mathbf{u}$ while the second term represents the work done by surface forces $\mathbf{t}$ in moving through the displacements $\mathbf{u}$ during linear elastic deformation. Equation (7.6.4) is a statement of *Clapeyron's theorem*, which states that the total strain energy stored in a body during linear elastic deformation is equal to the half of the work done by external forces acting on the body.

EXAMPLE 7.6.1:

1. Consider a linear elastic spring with spring constant k. Let F be the external force applied on the spring to elongate it and u be the resulting elongation of the spring (see Figure 7.6.1). Verify Clapeyron's theorem.

SOLUTION: The internal force developed in the spring is $F_s = ku$. The work done by F_s in moving through an increment of displacement du is $F_s \cdot du$. The total strain energy stored in the spring is

$$U = \int_0^u F_s \, du = \int_0^u ku \, du = \frac{1}{2} ku^2. \tag{7.6.5}$$

The work done by external force F is equal to Fu. But by equilibrium, $F = F_s = ku$. Hence,

$$U = \frac{1}{2} ku^2 = \frac{1}{2} Fu,$$

which proves Clapeyron's theorem.

2. Consider a uniform elastic bar of length L, cross-sectional area A, and modulus of elasticity E. The bar is fixed at $x = 0$ and subjected to a tensile force of P at $x = L$, as shown in Figure 7.6.2. Determine the deflection $w(L)$ using Clapeyron's theorem.

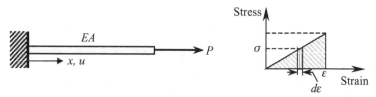

Figure 7.6.2. A bar subjected to an end load.

SOLUTION: If the axial displacement in the bar is equal to $u(x)$, then the work done by external point force P is equal to $W = Pu(L)$. The strain energy in the bar is given by

$$U = \frac{1}{2} \int_A \int_0^L \sigma_{xx}\varepsilon_{xx}\, dx\, dA = \frac{EA}{2} \int_0^L \varepsilon_{xx}^2\, dx = \frac{EA}{2} \int_0^L \left(\frac{du}{dx}\right)^2 dx.$$
(7.6.6)

Hence, by Clapeyron's theorem, we have

$$\frac{Pu(L)}{2} = \frac{EA}{2} \int_0^L \left(\frac{du}{dx}\right)^2 dx.$$

To make use of the above equation to determine $u(x)$, let us assume that $u(x) = u(L)x/L$, which certainly satisfies the geometric boundary condition, $u(0) = 0$. Then we have

$$u(L) = \frac{EA}{P} \int_0^L \left(\frac{du}{dx}\right)^2 dx = \frac{EA}{PL}[u(L)]^2,$$

or $u(L) = PL/AE$ and the solution is $u(x) = Px/AE$, which happens to coincide with the exact solution to the problem.

3. Consider a cantilever beam of length L and flexural rigidity EI and bent by a point load F at the free end (see Figure 7.6.3). Determine $w(0)$ using Clapeyron's theorem.

SOLUTION: By Clapeyron's theorem we have

$$\frac{1}{2} Fw(0) = \frac{1}{2} \int_A \int_0^L \sigma_{xx}\varepsilon_{xx}\, dx\, dA.$$

But according to the Euler–Bernoulli beam theory the strain in the beam is given by

$$\varepsilon_{xx} = -z\frac{d^2w}{dx^2},$$
(7.6.7)

where w is the transverse deflection. Then we have

$$\frac{1}{2} Fw(0) = \frac{1}{2} \int_A \int_0^L E\varepsilon_{xx}^2\, dx\, dA = \frac{1}{2} \int_A \int_0^L Ez^2 \left(\frac{d^2w}{dx^2}\right)^2 dA\, dx$$

$$= \frac{1}{2} \int_0^L EI \left(\frac{d^2w}{dx^2}\right)^2 dx = \frac{1}{2} \int_0^L \frac{M^2}{EI}\, dx,$$
(7.6.8)

Figure 7.6.3. A beam subjected to an end load.

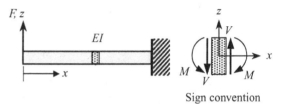

Sign convention

where $M(x)$ is the bending moment at x

$$M(x) = \int_A z\sigma_{xx}\, dA = -E \int_A z^2 \frac{d^2w}{dx^2}\, dA = -EI \frac{d^2w}{dx^2}. \qquad (7.6.9)$$

Equation (7.6.8) can be used to determine the deflection $w(0)$. The bending moment at any point x is $M(x) = -Fx$. Hence, we have

$$Fw(0) = \frac{1}{EI} \int_0^L F^2 x^2\, dx = \frac{F^2 L^3}{3EI} \quad \text{or} \quad w(0) = \frac{FL^3}{3EI}. \qquad (7.6.10)$$

7.6.2 Betti's Reciprocity Relations

Consider the equilibrium state of a linear elastic solid under the action of two different external forces, F_1 and F_2, as shown in Figure 7.6.4 [see Reddy, (2002)]. Since the order of application of the forces is arbitrary for linearized elasticity, we suppose that force F_1 is applied first. Let W_1 be the work produced by F_1. Then, we apply force F_2, which produces work W_2. This work is the same as that produced by force F_2, if it alone were acting on the body. When force F_2 is applied, force F_1 (which is already acting on the body) does additional work because its point of application is displaced due to the deformation caused by force F_2. Let us denote this work by W_{12}. Thus the total work done by the application of forces F_1 and F_2, F_1 first and F_2 next, is

$$W = W_1 + W_2 + W_{12}. \qquad (7.6.11)$$

Work W_{12}, which can be positive or negative, is zero if and only if the displacement of the point of application of force F_1 produced by force F_2 is zero or perpendicular to the direction of F_1.

Figure 7.6.4. Configurations of an elastic body due to the application of loads F_1 and F_2. —— Undeformed configuration.
- - - - Deformed configuration after the application of F_1.
...... Deformed configuration after the application of F_2.

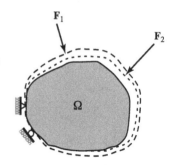

Now suppose that we change the order of application. Then the total work done is equal to

$$W = W_1 + W_2 + W_{21}, \tag{7.6.12}$$

where W_{21} is the work done by force F_2 due to the application of force F_1. The work done in both cases should be the same because at the end elastic body is loaded by the same pair of external forces. Thus, we have $W = \overline{W}$, or

$$W_{12} = W_{21}. \tag{7.6.13}$$

Equation (7.6.13) is a mathematical statement of Betti's (1823–1892) reciprocity theorem: *if a linear elastic body is subjected to two different sets of forces, the work done by the first system of forces in moving through the displacements produced by the second system of forces is equal to the work done by the second system of forces in moving through the displacements produced by the first system of forces.* Applied to a three-dimensional elastic body Ω with closed surface s, Eq. (7.6.13) takes the form

$$\int_\Omega \mathbf{f}^{(1)} \cdot \mathbf{u}^{(2)} \, d\mathbf{x} + \oint_s \mathbf{t}^{(1)} \cdot \mathbf{u}^{(2)} \, ds = \int_\Omega \mathbf{f}^{(2)} \cdot \mathbf{u}^{(1)} \, d\mathbf{x} + \oint_s \mathbf{t}^{(2)} \cdot \mathbf{u}^{(1)} \, ds, \tag{7.6.14}$$

where $\mathbf{u}^{(i)}$ are the displacements produced by body forces $\mathbf{f}^{(i)}$ and surface forces $\mathbf{t}^{(i)}$.

The proof of Betti's reciprocity theorem is straightforward. Let W_{12} denote the work done by forces $(\mathbf{f}^{(1)}, \mathbf{t}^{(1)})$ acting through the displacement $\mathbf{u}^{(2)}$. Then

$$\begin{aligned}
W_{12} &= \int_\Omega \mathbf{f}^{(1)} \cdot \mathbf{u}^{(2)} \, d\mathbf{x} + \oint_s \mathbf{t}^{(1)} \cdot \mathbf{u}^{(2)} \, ds \\
&= \int_\Omega f_i^{(1)} u_i^{(2)} \, d\mathbf{x} + \oint_s t_i^{(1)} u_i^{(2)} \, ds \\
&= \int_\Omega f_i^{(1)} u_i^{(2)} \, d\mathbf{x} + \oint_s n_j \sigma_{ji}^{(1)} u_i^{(2)} \, ds \\
&= \int_\Omega f_i^{(1)} u_i^{(2)} \, d\mathbf{x} + \int_\Omega \left(\sigma_{ji}^{(1)} u_i^{(2)} \right)_{,j} \, d\mathbf{x} \\
&= \int_\Omega \left(\sigma_{ij,j}^{(1)} + f_i^{(1)} \right) u_i^{(2)} \, d\mathbf{x} + \int_\Omega \sigma_{ij}^{(1)} u_{i,j}^{(2)} \, d\mathbf{x} \\
&= \int_\Omega \sigma_{ij}^{(1)} u_{i,j}^{(2)} \, d\mathbf{x} = \int_\Omega \sigma_{ij}^{(1)} \varepsilon_{ij}^{(2)} \, d\mathbf{x}. \tag{7.6.15}
\end{aligned}$$

Since $\sigma_{ij}^{(1)} = C_{ijk\ell} \varepsilon_{k\ell}^{(1)}$, we obtain

$$W_{12} = \int_\Omega C_{ijk\ell} \varepsilon_{k\ell}^{(1)} \varepsilon_{ij}^{(2)} \, d\mathbf{x}. \tag{7.6.16}$$

Since $C_{ijk\ell} = C_{k\ell ij}$, it follows that

$$\begin{aligned}
W_{12} &= \int_\Omega C_{ijk\ell} \varepsilon_{k\ell}^{(1)} \varepsilon_{ij}^{(2)} \, d\mathbf{x} \\
&= \int_\Omega C_{k\ell ij} \varepsilon_{ij}^{(2)} \varepsilon_{k\ell}^{(1)} \, d\mathbf{x} = W_{21}. \tag{7.6.17}
\end{aligned}$$

Thus, we have established the equality in Eq. (7.6.14).

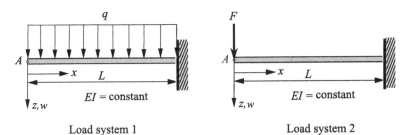

Figure 7.6.5. A cantilever beam subjected to two different types of loads.

From Eq. (7.6.17), we also have

$$\int_\Omega \sigma_{ij}^{(1)} \varepsilon_{ij}^{(2)} \, d\mathbf{x} = \int_\Omega \sigma_{ij}^{(2)} \varepsilon_{ij}^{(1)} \, d\mathbf{x},$$

$$\int_\Omega \sigma^{(1)} : \varepsilon^{(2)} \, d\mathbf{x} = \int_\Omega \sigma^{(2)} : \varepsilon^{(1)} \, d\mathbf{x}.$$

(7.6.18)

EXAMPLE 7.6.2: Consider a cantilever beam of length L subjected to two different types of loads: a concentrated load F at the free end and to a uniformly distributed load of intensity q (see Figure 7.6.5). Verify that the work done by the point load F in moving through the displacement w^q produced by q is equal to the work done by the distributed force q in moving through the displacement w^F produced by the point load F, $W_{12} = W_{21}$.

SOLUTION: The deflection $w^F(x)$ due to the concentrated load alone is

$$w^F(x) = \frac{F}{6EI}(x^3 - 3L^2x + 2L^3),$$

and the deflection equation due to the distributed load alone is

$$w^q(x) = \frac{q}{24EI}(x^4 - 4L^3x + 3L^4).$$

The work done by the load F in moving through the displacement due to the application of the uniformly distributed load q is

$$W_{12} = Fw^q(0) = \frac{FqL^4}{8EI} \,,$$

The work done by the uniformly distributed q in moving through the displacement field due to the application of point load F is

$$W_{21} = \int_0^L \frac{F}{6EI}(x^3 - 3L^2x + 2L^3)q \, dx = \frac{FqL^4}{8EI} \,,$$

which is in agreement with W_{12}.

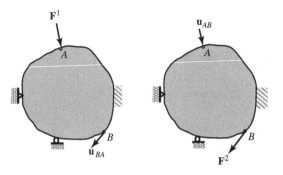

Figure 7.6.6. Configurations of the body discussed in Maxwell's theorem.

7.6.3 Maxwell's Reciprocity Relation

An important special case of Betti's reciprocity theorem is given by Maxwell's (1831–1879) reciprocity theorem. Maxwell's theorem was given in 1864, whereas Betti's theorem was given in 1872. Therefore, it may be considered that Betti generalized the work of Maxwell. We derive Maxwell's reciprocity theorem from Betti's reciprocity theorem.

Consider a linear elastic solid subjected to force $\mathbf{F}^1$ of unit magnitude acting at point A, and force $\mathbf{F}^2$ of unit magnitude acting at a different point B of the body. Let $\mathbf{u}_{AB}$ be the displacement of point A in the direction of force $\mathbf{F}^1$ produced by unit force $\mathbf{F}^2$, and $\mathbf{u}_{BA}$ by the displacement of point B in the direction of force $\mathbf{F}^2$ produced by unit force $\mathbf{F}^1$ (see Figure 7.6.6). From Betti's theorem, it follows that

$$\mathbf{F}^1 \cdot \mathbf{u}_{AB} = \mathbf{F}^2 \cdot \mathbf{u}_{BA} \quad \text{or} \quad u_{AB} = u_{BA}. \tag{7.6.19}$$

Equation (7.6.19) is a statement of Maxwell's theorem. If $\hat{\mathbf{e}}_1$ and $\hat{\mathbf{e}}_2$ denote the unit vectors along forces $\mathbf{F}^1$ and $\mathbf{F}^2$, respectively, Maxwell's theorem states that the displacement of point A in the $\hat{\mathbf{e}}_1$ direction produced by unit force acting at point B in the $\hat{\mathbf{e}}_2$ direction is equal to the displacement of point B in the $\hat{\mathbf{e}}_2$-direction produced by unit force acting at point A in the $\hat{\mathbf{e}}_1$ direction.

We close this section with the following example that illustrates the usefulness of Maxwell's theorem.

EXAMPLE 7.6.3:

1. Consider a cantilever beam ($E = 24 \times 10^6$ psi, $I = 120$ in^4) of length 12 ft subjected to a point load 4,000 lb at the free end. Find the deflection at a point 3 ft from the free end (see Figure 7.6.7) using Maxwell's theorem.

SOLUTION: By Maxwell's theorem, the displacement w_{BC} at point B ($x = 3$ ft) produced by the 4,000-lb load at point C ($x = 0$) is equal to the deflection w_{CB} at point C produced by applying the 4,000 lb load at point B. Let w_B and θ_B denote the deflection and slope, respectively, at point B owing to load $F = 4,000$ lb applied at point B. Then, the deflection at point B ($x = 3$ ft)

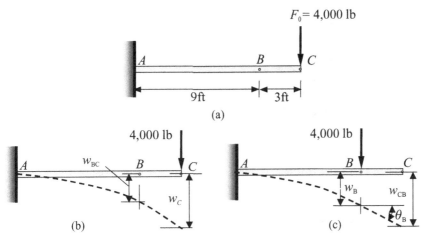

Figure 7.6.7. The cantilever beam of Example 7.6.3.

caused by load $F_0 = 4,000$ lb at point C ($x = 0$) is ($w_B = FL^3/3EI$ and $\theta_B = FL^2/2EI$)

$$w_{BC} = w_{CB} = w_B + (3 \times 12)\theta_B$$

$$= \frac{4000(9 \times 12)^3}{3EI} + \frac{(3 \times 12)4000(9 \times 12)^2}{2EI}$$

$$= \frac{243 \times 6000 \times (12)^3}{24 \times 10^6 \times 120} = 0.8748 \text{ in.}$$

2. Consider a circular plate of radius a with an axisymmetric boundary condition and subjected to an asymmetric loading of the type (see Figure 7.6.8)

$$q(r, \theta) = q_0 + q_1 \frac{r}{a} \cos\theta, \qquad (7.6.20)$$

where q_0 represents the uniform part of the load for which the solution can be determined for various axisymmetric boundary conditions [see Reddy (2007)]. In particular, the deflection of a clamped circular plate under a point load Q_0 at the center is given by

$$w(r) = \frac{Q_0 a^2}{16\pi D} \left[1 - \frac{r^2}{a^2} + 2\frac{r^2}{a^2} \log\left(\frac{r}{a}\right) \right]. \qquad (7.6.21)$$

Use the Betti–Maxwell's reciprocity theorem to determine the center deflection of a clamped plate under asymmetric distributed load.

SOLUTION: By Maxwell's theorem, the work done by a point load ($Q_0 = 1$) at the center of the plate due to the deflection (at the center) w_c caused by the distributed load $q(r, \theta)$ is equal to the work done by the distributed load $q(r, \theta)$ in moving through the displacement $w_0(r)$ caused by the point load

$q(r, \theta) = q_0 + q_1 \dfrac{r}{a} \cos\theta$

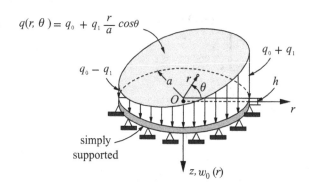

Figure 7.6.8. A circular plate subjected to an asymmetric loading.

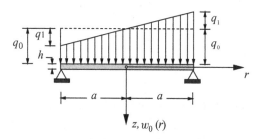

at the center. Hence, the center deflection of a clamped circular plate under asymmetric load (7.6.20) is

$$w_c = \frac{a^2}{16\pi D} \int_0^{2\pi} \int_0^a q(r, \theta) \left[1 - \frac{r^2}{a^2} \left(1 - 2\log\frac{r}{a} \right) \right] r \, dr \, d\theta = \frac{q_0 a^4}{64 D}. \quad (7.6.22)$$

7.7 Solution Methods

7.7.1 Types of Solution Methods

Analytical solution of a problem is one that satisfies the governing differential equation at every point of the domain as well as the boundary conditions exactly. In general, finding analytical solutions of elasticity problems is not simple due to complicated geometries and boundary conditions. *Approximate solution* is one that satisfies governing differential equations as well as the boundary conditions approximately. *Numerical solutions* are approximate solutions that are developed using a numerical method, such as finite difference methods, the finite element method, the boundary element method, and so on. Often one seeks approximate solutions of practical problems using numerical methods. In this section, we discuss methods for finding solutions, exact as well approximate.

The solutions of elasticity problems are developed using one of the following methods (see Slaughter, 2002):

1. The *inverse method* is one in which one finds the solution for displacement, strain, and stress fields that satisfy the governing equations of elasticity and then tries to find a problem with boundary conditions to which the fields correspond.

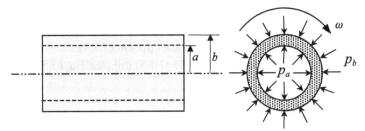

Figure 7.7.1. Rotating cylindrical pressure vessel.

2. The *semi-inverse method* is one in which the solution form in terms of unknown functions is arrived with the help of a qualitative understanding of the problem characteristics, and then the unknown functions are determined to satisfy the governing equations.
3. The *method of potentials* is one in which some of the governing equations are trivially satisfied by the choice of potential functions from which stresses or displacements are derived. The potential functions are determined by finding solutions to remaining equations.
4. The *variational methods* are those which make use of extremum (i.e., minimum or maximum) and stationary principles. The principles are often cast in terms of energies of the system.

In the remainder of this chapter, we consider mostly the semi-inverse method and the method of potentials to formulate and solve certain problems of elasticity.

7.7.2 An Example: Rotating Thick-Walled Cylinder

Consider an isotropic, hollow circular cylinder of internal radius a and outside radius b. The cylinder is pressurized at $r = a$ and/or at $r = b$, and rotating with a uniform speed of ω about its axis (z-axis). Under these applied loads, stresses are developed in the cylinder. Define a cylindrical coordinate system (r, θ, z), as shown in Figure 7.7.1. We assume that body force vector is $\mathbf{f} = \rho\omega^2 r\,\hat{\mathbf{e}}_r$.

For this problem, we have only stress boundary conditions (BVP Type II). We have

$$\text{At } r = a: \quad \hat{\mathbf{n}} = -\hat{\mathbf{e}}_r, \quad \mathbf{t} = p_a\,\hat{\mathbf{e}}_r \quad \text{or} \quad \sigma_{rr} = -p_a, \quad \sigma_{r\theta} = 0 \qquad (7.7.1)$$

$$\text{At } r = b: \quad \hat{\mathbf{n}} = \hat{\mathbf{e}}_r, \quad \mathbf{t} = -p_b\,\hat{\mathbf{e}}_r \quad \text{or} \quad \sigma_{rr} = -p_b, \quad \sigma_{r\theta} = 0. \qquad (7.7.2)$$

We wish to determine the displacements, strains, and stresses in the cylinder using the semi-inverse method.

Because of the symmetry about the z-axis, we assume that the displacement field is of the form

$$u_r = U(r), \quad u_\theta = u_z = 0, \qquad (7.7.3)$$

where $U(r)$ is an unknown function to be determined such that the equations of elasticity and boundary conditions are satisfied. If we cannot find $U(r)$ that satisfies the governing equations, then we must abandon the assumption (7.7.3).

The strains associated with the displacement field (7.7.3) are [see Eq. (3.5.21)]

$$\varepsilon_{rr} = \frac{dU}{dr}, \quad \varepsilon_{\theta\theta} = \frac{U}{r}, \quad \varepsilon_{zz} = 0,$$

$$\varepsilon_{r\theta} = 0, \quad \varepsilon_{z\theta} = 0, \quad \varepsilon_{rz} = 0. \tag{7.7.4}$$

The stresses are given by

$$\sigma_{rr} = 2\mu\varepsilon_{rr} + \lambda\left(\varepsilon_{rr} + \varepsilon_{\theta\theta}\right) = (2\mu + \lambda)\frac{dU}{dr} + \lambda\frac{U}{r},$$

$$\sigma_{\theta\theta} = 2\mu\varepsilon_{\theta\theta} + \lambda\left(\varepsilon_{rr} + \varepsilon_{\theta\theta}\right) = (2\mu + \lambda)\frac{U}{r} + \lambda\frac{dU}{dr},$$

$$\sigma_{zz} = 2\mu\varepsilon_{zz} + \lambda\left(\varepsilon_{rr} + \varepsilon_{\theta\theta}\right) = \lambda\left(\frac{dU}{dr} + \frac{U}{r}\right), \tag{7.7.5}$$

$$\sigma_{r\theta} = 0, \quad \sigma_{rz} = 0, \quad \sigma_{\theta z} = 0.$$

Substituting the stresses from Eq. (7.7.5) into the equations of equilibrium (5.3.17), we note that the last two equations are trivially satisfied, and the first equation reduces to

$$\frac{d\sigma_{rr}}{dr} + \frac{1}{r}\left(\sigma_{rr} - \sigma_{\theta\theta}\right) = -\rho\omega^2 r,$$

$$(2\mu + \lambda)\frac{d^2U}{dr^2} + \lambda\frac{d}{dr}\left(\frac{U}{r}\right) + \frac{2\mu}{r}\left(\frac{dU}{dr} - \frac{U}{r}\right) = -\rho\omega^2 r. \tag{7.7.6}$$

Simplifying the expression, we obtain

$$r^2\frac{d^2U}{dr^2} + r\frac{dU}{dr} - U = -\alpha r^3, \quad \alpha = \frac{\rho\omega^2}{2\mu + \lambda}. \tag{7.7.7}$$

The linear ordinary differential equation (7.7.7) can be transformed to one with constant coefficients by a change of independent variable, $r = e^\xi$ (or $\xi = \ln r$). Using the chain rule of differentiation, we obtain

$$\frac{dU}{dr} = \frac{dU}{d\xi}\frac{d\xi}{dr} = \frac{1}{r}\frac{dU}{d\xi}, \quad \frac{d^2U}{dr^2} = \frac{d}{dr}\left(\frac{1}{r}\frac{dU}{d\xi}\right) = \frac{1}{r^2}\left(-\frac{dU}{dr} + \frac{d^2U}{d\xi^2}\right). \tag{7.7.8}$$

Substituting the above expressions into (7.7.7), we obtain

$$\frac{d^2U}{d\xi^2} - U = -\alpha e^{3\xi}. \tag{7.7.9}$$

Seeking solution in the form $U(\xi) = e^{m\xi}$, we obtain the following general solution to the problem:

$$U_h(\xi) = c_1 e^\xi + c_2 e^{-\xi} - \frac{\alpha}{8}e^{3\xi}. \tag{7.7.10}$$

Changing back to the original independent variable r, we have

$$U(r) = c_1 r + \frac{c_2}{r} - \frac{\alpha}{8} r^3. \tag{7.7.11}$$

The stress σ_{rr} is given by

$$\sigma_{rr} = (2\mu + \lambda)\left(c_1 - \frac{c_2}{r^2} - \frac{3\alpha}{8} r^2\right) + \lambda\left(c_1 + \frac{c_2}{r^2} - \frac{\alpha}{8} r^2\right)$$

$$= 2(\mu + \lambda)c_1 - 2\mu\frac{c_2}{r^2} - \frac{(3\mu + 2\lambda)\alpha}{4} r^2. \tag{7.7.12}$$

Applying the stress boundary conditions in Eqs. (7.7.1) and (7.7.2), we obtain

$$2(\mu + \lambda)c_1 - 2\mu\frac{c_2}{a^2} - \frac{(3\mu + 2\lambda)\alpha}{4} a^2 = -p_a,$$

$$2(\mu + \lambda)c_1 - 2\mu\frac{c_2}{b^2} - \frac{(3\mu + 2\lambda)\alpha}{4} b^2 = -p_b. \tag{7.7.13}$$

Solving for the constants c_1 and c_2,

$$c_1 = \frac{1}{2(\mu + \lambda)}\left[\left(\frac{p_a a^2 - p_b b^2}{b^2 - a^2}\right) + (b^2 + a^2)\frac{(3\mu + 2\lambda)}{(2\mu + \lambda)}\frac{\rho\omega^2}{4}\right],$$

$$c_2 = \frac{a^2 b^2}{2\mu}\left[\left(\frac{p_a - p_b}{b^2 - a^2}\right) + \frac{(3\mu + 2\lambda)}{(2\mu + \lambda)}\frac{\rho\omega^2}{4}\right]. \tag{7.7.14}$$

Finally, the displacement u_r and stress σ_{rr} in the cylinder are given by

$$u_r = \frac{1}{2(\mu + \lambda)}\left[\left(\frac{p_a a^2 - p_b b^2}{b^2 - a^2}\right) + (b^2 + a^2)\frac{(3\mu + 2\lambda)}{(2\mu + \lambda)}\frac{\rho\omega^2}{4}\right] r$$

$$+ \frac{a^2 b^2}{2\mu}\left[\left(\frac{p_a - p_b}{b^2 - a^2}\right) + \frac{(3\mu + 2\lambda)}{(2\mu + \lambda)}\frac{\rho\omega^2}{4}\right]\frac{1}{r} - \frac{\rho\omega^2}{8(2\mu + \lambda)} r^3, \tag{7.7.15}$$

$$\sigma_{rr} = \left[\left(\frac{p_a a^2 - p_b b^2}{b^2 - a^2}\right) + (b^2 + a^2)\frac{(3\mu + 2\lambda)}{(2\mu + \lambda)}\frac{\rho\omega^2}{4}\right]$$

$$- \frac{a^2 b^2}{r^2}\left[\left(\frac{p_a - p_b}{b^2 - a^2}\right) + \frac{(3\mu + 2\lambda)}{(2\mu + \lambda)}\frac{\rho\omega^2}{4}\right] - \frac{(3\mu + 2\lambda)\alpha}{4} r^2. \tag{7.7.16}$$

7.7.3 Two-Dimensional Problems

A class of problems in elasticity, due to geometry, boundary conditions, and external applied loads, have their solutions (i.e., displacements and stresses) not dependent on one of the coordinates. Such problems are called *plane elasticity problems*. The plane elasticity problems considered here are grouped into *plane strain* and *plane stress* problems. Both classes of problems are described by a set of two *coupled* partial differential equations expressed in terms of two dependent variables that represent the two components of the displacement vector. The governing equations of plane strain problems differ from those of the plane stress problems only in the

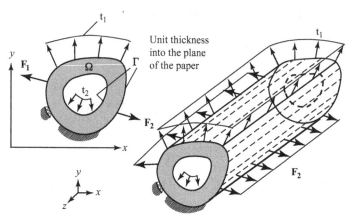

Figure 7.7.2. Examples of plane strain problems.

coefficients of the differential equations. The discussion here is limited to isotropic materials.

7.7.3.1 Plane Strain Problems

Plane strain problems are characterized by the displacement field

$$u_x = u_x(x, y), \quad u_y = u_y(x, y), \quad u_z = 0, \tag{7.7.17}$$

where (u_x, u_y, u_z) denote the components of the displacement vector **u** in the (x, y, z) coordinate system. An example of a plane strain problem is provided by the long cylindrical member under external loads that are independent of z, as shown in Figure 7.7.2. For cross sections sufficiently far from the ends, it is clear that the displacement u_z is zero and that u_x and u_y are independent of z, that is, a state of plane strain exists.

The displacement field (7.7.17) results in the following strain field:

$$\varepsilon_{xz} = \varepsilon_{yz} = \varepsilon_{zz} = 0,$$
$$\varepsilon_{xx} = \frac{\partial u_x}{\partial x}, \quad 2\varepsilon_{xy} = \frac{\partial u_x}{\partial y} + \frac{\partial u_y}{\partial x}, \quad \varepsilon_{yy} = \frac{\partial u_y}{\partial y}. \tag{7.7.18}$$

Clearly, the body is in a state of plane strain.

For an isotropic material, the stress components are given by [see Eq. (6.2.29)]

$$\sigma_{xz} = \sigma_{yz} = 0, \quad \sigma_{zz} = \nu \left(\sigma_{xx} + \sigma_{yy}\right), \tag{7.7.19}$$

$$\left\{ \begin{array}{c} \sigma_{xx} \\ \sigma_{yy} \\ \sigma_{xy} \end{array} \right\} = \frac{E}{(1+\nu)(1-2\nu)} \left[\begin{array}{ccc} 1-\nu & \nu & 0 \\ \nu & 1-\nu & 0 \\ 0 & 0 & \frac{(1-2\nu)}{2} \end{array} \right] \left\{ \begin{array}{c} \varepsilon_{xx} \\ \varepsilon_{yy} \\ 2\varepsilon_{xy} \end{array} \right\}. \tag{7.7.20}$$

The equations of equilibrium of three-dimensional linear elasticity, with the body-force components

$$f_3 = f_z = 0, \quad f_1 = f_x = f_x(x, y), \quad f_2 = f_y = f_y(x, y), \tag{7.7.21}$$

reduce to the following two plane-strain equations

$$\frac{\partial \sigma_{xx}}{\partial x} + \frac{\partial \sigma_{xy}}{\partial y} + f_x = 0, \tag{7.7.22}$$

$$\frac{\partial \sigma_{xy}}{\partial x} + \frac{\partial \sigma_{yy}}{\partial y} + f_y = 0. \tag{7.7.23}$$

The boundary conditions are either the stress type

$$\left. \begin{array}{l} t_x \equiv \sigma_{xx} n_x + \sigma_{xy} n_y = \hat{t}_x \\ t_y \equiv \sigma_{xy} n_x + \sigma_{yy} n_y = \hat{t}_y \end{array} \right\} \quad \text{on } \Gamma_\sigma, \tag{7.7.24}$$

or the displacement type

$$u_x = \hat{u}_x, \quad u_y = \hat{u}_y \text{ on } \Gamma_u. \tag{7.7.25}$$

Here (n_x, n_y) denote the components (or direction cosines) of the unit normal vector on the boundary Γ, Γ_σ, and Γ_u are disjoint portions of the boundary Γ, $\hat{t}_x$, and $\hat{t}_y$ are the components of the specified traction vector, and $\hat{u}_x$ and $\hat{u}_y$ are the components of specified displacement vector. Only one element of each pair, (u_x, t_x) and (u_y, t_y), may be specified at a boundary point.

7.7.3.2 Plane Stress Problems

A state of *plane stress* is defined as one in which the following stress field exists:

$$\sigma_{xz} = \sigma_{yz} = \sigma_{zz} = 0,$$

$$\sigma_{xx} = \sigma_{xx}(x, y), \quad \sigma_{xy} = \sigma_{xy}(x, y), \quad \sigma_{yy} = \sigma_{yy}(x, y). \tag{7.7.26}$$

An example of a plane stress problem is provided by a thin plate under external loads applied in the xy plane (or parallel to it) that are independent of z, as shown in Figure 7.7.3. The top and bottom surfaces of the plate are assumed to be traction-free, and the specified boundary forces are in the xy-plane so that $f_z = 0$ and $u_z = 0$.

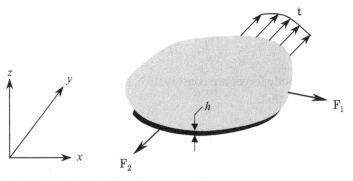

Figure 7.7.3. A thin plate in a state of plane stress.

The stress-strain relations of a plane stress state are

$$
\begin{Bmatrix} \sigma_{xx} \\ \sigma_{yy} \\ \sigma_{xy} \end{Bmatrix} = \frac{E}{1-\nu^2} \begin{bmatrix} 1 & \nu & 0 \\ \nu & 1 & 0 \\ 0 & 0 & \frac{(1+\nu)}{2} \end{bmatrix} \begin{Bmatrix} \varepsilon_{xx} \\ \varepsilon_{yy} \\ 2\varepsilon_{xy} \end{Bmatrix}.
\tag{7.7.27}
$$

The equations of equilibrium as well as boundary conditions of a plane stress problem are the same as those listed in Eqs. (7.7.22)–(7.7.25). The equilibrium equations (7.7.22) and (7.7.23) can be written in index notation as

$$
\sigma_{\beta\alpha,\beta} + f_\alpha = 0,
\tag{7.7.28}
$$

where α and β take the values of 1 and 2. The governing equations of plane stress and plane strain differ from each other only on account of the difference in the constitutive equations for the two cases. To unify the formulation for plane strain and plane stress, we introduce the parameter s

$$
s = \begin{cases} \frac{1}{1-\nu}, & \text{for plane strain,} \\ 1+\nu, & \text{for plane stress.} \end{cases}
\tag{7.7.29}
$$

Then the constitutive equations of plane stress as well as plane strain can be expressed as

$$
\sigma_{\alpha\beta} = 2\mu \left[\varepsilon_{\alpha\beta} + \left(\frac{s-1}{2-s} \right) \varepsilon_{\gamma\gamma} \delta_{\alpha\beta} \right],
\tag{7.7.30}
$$

$$
\varepsilon_{\alpha\beta} = \frac{1}{2\mu} \left[\sigma_{\alpha\beta} - \left(\frac{s-1}{s} \right) \sigma_{\gamma\gamma} \delta_{\alpha\beta} \right],
\tag{7.7.31}
$$

where α, β, and γ take values of 1 and 2. The compatibility equations (7.4.9) for plane stress and plane strain now take the form

$$
\nabla^2 \sigma_{\alpha\alpha} = -s \, f_{\alpha,\alpha}.
\tag{7.7.32}
$$

7.7.4 Airy Stress Function

Airy stress function is a potential function introduced to identically satisfy the equations of equilibrium, Eqs. (7.7.22) and (7.7.23). First, we assume that the body force vector $\mathbf{f}$ is derivable from a scalar potential V_f such that

$$
\mathbf{f} = -\nabla V_f \quad \text{or} \quad f_x = -\frac{\partial V_f}{\partial x}, \quad f_y = -\frac{\partial V_f}{\partial y}.
\tag{7.7.33}
$$

This amounts to assuming that body forces are conservative. Next, we introduce the Airy stress function $\Phi(x, y)$ such that

$$
\sigma_{xx} = \frac{\partial^2 \Phi}{\partial y^2} + V_f, \quad \sigma_{yy} = \frac{\partial^2 \Phi}{\partial x^2} + V_f, \quad \sigma_{xy} = -\frac{\partial^2 \Phi}{\partial x \partial y}.
\tag{7.7.34}
$$

This definition of $\Phi(x, y)$ automatically satisfies the equations of equilibrium (7.7.22) and (7.7.23).

The stresses derived from (7.7.34) are subject to the compatibility conditions (7.7.32). Substituting for $\sigma_{\alpha\beta}$ in terms of Φ from Eq. (7.7.34) into Eq. (7.7.32), we obtain

$$\nabla^4\Phi + (2-s)\nabla^2 V_f = 0, \tag{7.7.35}$$

where $\nabla^4 = \nabla^2\nabla^2$ is the *biharmonic operator*, which, in two dimensions, has the form

$$\nabla^4 = \frac{\partial^4}{\partial x^4} + 2\frac{\partial^4}{\partial x^2 \partial y^2} + \frac{\partial^4}{\partial y^4}. \tag{7.7.36}$$

If the body forces are zero, we have $V_f = 0$ and Eq. (7.7.35) reduces to the *biharmonic equation*

$$\nabla^4\Phi = 0. \tag{7.7.37}$$

In cylindrical coordinate system, Eqs. (7.7.33) and (7.7.34) take the form

$$f_r = -\frac{\partial V_f}{\partial r}, \qquad f_\theta = -\frac{1}{r}\frac{\partial V_f}{\partial \theta}, \tag{7.7.38}$$

$$\sigma_{rr} = \frac{1}{r}\frac{\partial\Phi}{\partial r} + \frac{1}{r^2}\frac{\partial^2\Phi}{\partial\theta^2} + V_f,$$

$$\sigma_{\theta\theta} = \frac{\partial^2\Phi}{\partial r^2} + V_f, \tag{7.7.39}$$

$$\sigma_{r\theta} = -\frac{\partial}{\partial r}\left(\frac{1}{r}\frac{\partial\Phi}{\partial\theta}\right).$$

The biharmonic operator $\nabla^4 = \nabla^2\nabla^2$ can be expressed using the definition of ∇^2 in a cylindrical coordinate system

$$\nabla^2 = \frac{\partial^2}{\partial r^2} + \frac{1}{r}\frac{\partial}{\partial r} + \frac{1}{r^2}\frac{\partial^2}{\partial\theta^2}. \tag{7.7.40}$$

In summary, solution to a plane elastic problem using the Airy stress function involves finding the solution to Eq. (7.7.35) and satisfying the boundary conditions of the problem. The most difficult part is finding solution to the fourth-order equation (7.7.35) over a given domain. Often the form of the Airy stress function is obtained by either the inverse method or semi-inverse method. Next, we consider some examples of the Airy function approach [see Timoshenko and Goodier (1970), Slaughter (2002), and Mase and Mase (1999) for additional examples].

EXAMPLE 7.7.1:

1. Suppose that the Airy stress function is a second-order polynomial (which is the lowest order that gives a nonzero stress field) of the form

$$\Phi(x, y) = c_1 xy + c_2 x^2 + c_3 y^2. \tag{7.7.41}$$

Determine constants c_1, c_2, and c_3 such that Φ satisfies the biharmonic equation $\nabla^4\Phi = 0$ (body force field is zero, $V_f = 0$) and corresponds to a possible state of stress for some boundary value problem (inverse method).

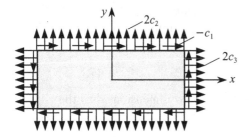

Figure 7.7.4. A problem with uniform stress field.

SOLUTION: Clearly, the biharmonic equation is satisfied by Φ in Eq. (7.7.41). The corresponding stress field is

$$\sigma_{xx} = \frac{\partial^2 \Phi}{\partial y^2} = 2c_3, \quad \sigma_{yy} = \frac{\partial^2 \Phi}{\partial x^2} = 2c_2, \quad \sigma_{xy} = -\frac{\partial^2 \Phi}{\partial x \partial y} = -c_1. \quad (7.7.42)$$

Thus, the state of stress is uniform (i.e., constant) throughout the body, and it is independent of the geometry. Thus, there are infinite number of problems for which the stress field is a solution. In particular, the rectangular domain shown in Figure 7.7.4 is one such problem.

2. Take the Airy stress function to be a third-order polynomial of the form

$$\Phi(x, y) = c_1 xy + c_2 x^2 + c_3 y^2 + c_4 x^2 y + c_5 xy^2 + c_6 x^3 + c_7 y^3. \quad (7.7.43)$$

Determine the stress field and identify various possible boundary-value problems.

SOLUTION: We note that $\nabla^4 \Phi = 0$ for any c_i (body force field is zero). The corresponding stress field is

$$\sigma_{xx} = 2c_3 + 2c_5 x + 6c_7 y, \quad \sigma_{yy} = 2c_2 + 2c_4 y + 6c_6 y, \quad \sigma_{xy} = -c_1 - 2c_4 x - 2c_5 y. \quad (7.7.44)$$

Again, there are infinite number of problems for which the stress field is a solution. In particular, for $c_1 = c_2 = c_3 = c_4 = c_5 = c_6 = 0$, the solution corresponds to a thin beam in pure bending (see Figure 7.7.5).

3. Last, take the Airy stress function to be a fourth-order polynomial of the form (omit terms that were already considered in the last two cases)

$$\Phi(x, y) = c_8 x^2 y^2 + c_9 x^3 y + c_{10} xy^3 + c_{11} x^4 + c_{12} y^4. \quad (7.7.45)$$

Determine the stress field and associated boundary-value problems.

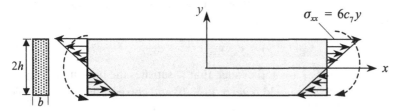

Figure 7.7.5. A thin beam in pure bending.

SOLUTION: Computing $\nabla^4 \Phi$ and equating it to zero (body-force field is zero) we find that

$$c_8 + 3(c_{11} + c_{12}) = 0.$$

Thus out of five constants only four of them are independent. The corresponding stress field is

$$\sigma_{xx} = 2c_8 x^2 + 6c_{10}xy + 12c_{12}y^2 = -6c_{11}x^2 + 6c_{10}xy + 6c_{12}(2y^2 - x^2),$$

$$\sigma_{yy} = 2c_8 y^2 + 6c_9 xy + 12c_{11}x^2 = 6c_9 xy + 2c_{11}(2x^2 - y^2) - 6c_{12}y^2,$$

$$\sigma_{xy} = -4c_8 xy - 3c_9 x^2 - 3c_{10}y^2 = 12c_{11}xy + 12c_{12}xy - 3c_9 x^2 - 3c_{10}y^2.$$

$$(7.7.46)$$

By suitable adjustment of the constants, we can obtain various loads on rectangular plates. For instance, taking all coefficients except c_{10} equal to zero, we obtain (see Problem 7.17).

$$\sigma_{xx} = 6c_{10}xy, \quad \sigma_{yy} = 0, \quad \sigma_{xy} = -3c_{10}y^2.$$

7.7.5 End Effects: Saint–Venant's Principle

A boundary-value problem of elasticity requires the boundary conditions to be known in the form of displacements or stresses [see Eqs. (7.7.24) and (7.7.25)] at *every point* of the boundary. As shown in Example 7.7.1, the boundary forces are distributed as a function of the distance along the boundary. If the boundary forces (and moments) are distributed in any other form (than per unit surface area), the boundary conditions cannot be expressed as point-wise quantities.

For example, consider the cantilevered beam with an end load, as shown in Figure 7.7.6. At $x = 0$, we are required to specify σ_{xx} and σ_{xy} (because u_x and u_y are clearly not zero there). There is no problem in stating that $\sigma_{xx}(0, z) = 0$; but we only know that the integral of σ_{xz} over the beam cross section must be equal to P:

$$\int_A \sigma_{xz}(0, z)\, dA = P,$$

which is not equal to specifying σ_{xz} point-wise. If we state that $\sigma_{xy}(0, z) = P/A$, where A is the cross-sectional area of the beam, then we have a stress singularity at points $(x, z) = (0, \pm h)$, because σ_{xz} is zero at $z = \pm h$ [we also have a different type of singularity at points $(x, z) = (L, \pm h)$].

Analytical solutions for such problems, when exist, show that a change in the distribution of the load on the end, without change of the resultant, alters the stress significantly only near the end. *Saint–Venant's principle* says that the effect of the change in the boundary condition to a statically equivalent condition is local; that is, the solutions obtained with the two sets of boundary conditions are approximately the same at points sufficiently far from the points where the elasticity boundary conditions are replaced with statically equivalent boundary conditions. Of course, "sufficiently far" is rather ambiguous and problem dependent. It is often taken to

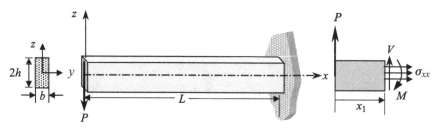

Figure 7.7.6. A cantilevered beam under an end load.

be equal to or greater than the length scale of the portion of the boundary where the boundary conditions are replaced. In the case of the beam shown in Figure 7.7.6, the distance is $2h$ (height of the beam).

EXAMPLE 7.7.2: Here, we consider the problem of a cantilevered beam with an end load, as shown in Figure 7.7.6. The problem can be treated as a plane stress if the beam is of small thickness b compared to the height, $b \ll h$ (of course, $h << L$ to call it a beam). If the beam is a portion of a very long slab, in the thickness direction, it can be treated as a plane strain problem. Write the boundary conditions and determine the Airy stress function, stresses, and displacements of the problem.

SOLUTION: The boundary conditions are of mixed type (see Figure 7.7.6): the tractions are specified on the boundaries $x = 0$ and $z = \pm h$, while the displacements are specified on the boundary $x = L$. However, boundary conditions of plane elasticity can be written only on $x = L$ and $z = \pm h$. On $x = 0$, we only know the total force in the z-direction and not the associated stress. Hence, it must be written as an integral condition on stress $\sigma_{xz}(0, z)$. Thus, we have

$$\sigma_{xx}(0, z) = 0, \quad \sigma_{xz}(x, -h) = 0, \quad \sigma_{zz}(x, -h) = 0,$$
$$\sigma_{xz}(x, h) = 0, \quad \sigma_{zz}(x, h) = 0, \tag{7.7.47}$$

$$u_x(L, z) = 0, \quad u_z(L, z) = 0, \tag{7.7.48}$$

$$\int_{-h}^{h} \sigma_{xz}(0, z) \, dz = \frac{P}{b}. \tag{7.7.49}$$

Because of the boundary condition in Eq. (7.7.49), the resulting boundary-value problem is not an exact elasticity problem in the sense that boundary values are not specified point-wise. If P is replaced with a shear stress condition $\sigma_{xz}(0, z) = \tau_0$, it is a proper elasticity boundary condition, but this creates another problem as discussed next.

This problem is discussed in most elasticity and continuum mechanics books, despite the fact that it is not a well-posed problem due to point singularities at the corners of the domain. For example, at points $(x, z) = (L, \pm h)$ we have both traction and displacement boundary condition, which is not allowed.

Similarly, at points $(x, z) = (0, \pm h)$ we have $\sigma_{xz} = \tau_0$ from one side and $\sigma_{xz} = 0$ from the other sides, which violates symmetry of stress tensor. Therefore, the solution being sought is an approximate solution, which is a reasonable one, by Saint–Venant's principle, away from the isolated points of singularity.

The semi-inverse method allows us to identify the form of the Airy stress function. The knowledge of the stress distributions from the elementary theory of beams provides the needed clue to identify the terms in the Airy stress function. Recall the following stress field from the Euler–Bernoulli beam theory:

$$\sigma_{xx} = \frac{M(x)z}{EI}, \quad \sigma_{zz} = 0, \quad \sigma_{xz} = \frac{V(x)Q(z)}{Ib}, \tag{7.7.50}$$

where M is the bending moment and V is the shear force,

$$M = \int_A z\sigma_{xx}\, dA, \quad V = \int_A \sigma_{xz}\, dA, \tag{7.7.51}$$

I is the second moment of area about the axis (y) of bending, and Q is the first moment of area

$$I = \int_A z^2\, dA, \quad Q(z) = \int_{\bar{A}} z\, dA. \tag{7.7.52}$$

Here $\bar{A}$ denotes the cross-sectional area between line z and the top of the beam. Clearly, Q is a quadratic function of z. We also note that $M(x)$ is a linear function of x while V is a constant for the problem at hand. Using this qualitative information and definitions (7.7.34), in the absence of body forces ($V_f = 0$), we take the Airy stress function to be

$$\Phi(x, z) = x(c_1 z + c_2 z^2 + c_3 z^3). \tag{7.7.53}$$

Only the first and third terms are dictated by the stress field in a beam. The second term is added to make it complete quadratic polynomial in z. The nonzero stresses are

$$\sigma_{xx} = 6c_3 xz, \quad \sigma_{xz} = -(c_1 + 2c_2 z + 3c_3 z^2). \tag{7.7.54}$$

The choice in (7.7.53) satisfies the biharmonic equation for any values of c_1, c_2, and c_3. We determine their values using the stress boundary conditions in Eqs. (7.7.47) and (7.7.49). The stress boundary conditions $\sigma_{xx}(0, z) = 0$ and $\sigma_{zz}(x, \pm h) = 0$ are trivially satisfied. From the remaining stress boundary conditions, we obtain

$$\sigma_{xz}(x, \pm h) = 0 \rightarrow c_1 - 2c_2 h + 3c_3 h^2 = 0 \quad \text{and} \quad c_1 + 2c_2 h + 3c_3 h^2 = 0,$$

$$\text{which yield} \quad c_2 = 0, \quad c_1 = -3h^2 c_3, \tag{7.7.55}$$

$$\int_{-h}^{h} \sigma_{xz}(0, z)\, dz = \frac{P}{b} \rightarrow -2h(c_1 + h^2 c_3) = \frac{P}{b}. \tag{7.7.56}$$

We have $c_3 = P/4bh^3$ and $c_1 = -3P/4bh$. Hence, the stress field becomes

$$\sigma_{xx} = \frac{6Pxz}{4bh^3} = \frac{Pxz}{I},$$

$$\sigma_{xz} = \left(\frac{3P}{4bh} - \frac{3Pz^2}{4bh^3}\right) = \frac{P}{2I}\left(h^2 - z^2\right), \tag{7.7.57}$$

where $I = 2bh^3/3$. The Airy stress function is

$$\Phi(x, z) = \frac{Pxz}{6I}\left(z^2 - 3h^2\right). \tag{7.7.58}$$

The computed stresses are exactly those predicted by the classical (i.e., Euler–Bernoulli) beam theory, where $M(x) = Px$ and $V = P$. This is not surprising because our choice of terms in the Airy stress function was dictated by the form of the stress field in the classical beam theory. This also indicates that we cannot obtain any better stress field than the elementary beam theory for the boundary conditions (7.7.47)–(7.7.49). This stress field can be used to determine the displacements in the beam.

The strain field associated with the stress field (7.7.57) is

$$\varepsilon_{xx} = \frac{Pxz}{EI},$$

$$\varepsilon_{zz} = -\nu\varepsilon_{xx} = -\nu\frac{Pxz}{EI}, \tag{7.7.59}$$

$$\varepsilon_{xz} = \frac{(1+\nu)P}{2EI}(h^2 - z^2),$$

where ν is the Poisson ratio and E is Young's modulus. We wish to determine the two-dimensional displacement field (u_x, u_z) in the beam.

Using the strain-displacement relations and integrating the strains in Eq. (7.7.59), we obtain (see Example 3.8.2 and note the change in the coordinate system used)

$$\varepsilon_{xx} = \frac{\partial u_x}{\partial x} = \frac{Pxz}{EI} \quad \text{or} \quad u_x = \frac{Px^2 z}{2EI} + f_1(z),$$

$$\varepsilon_{zz} = \frac{\partial u_z}{\partial z} = -\frac{\nu Pxz}{EI} \quad \text{or} \quad u_z = -\frac{\nu Pxz^2}{2EI} + f_2(x), \tag{7.7.60}$$

where (f_1, f_2) are functions of integration. Substituting u_x and u_z into the definition of the shear strain $2\varepsilon_{xz}$, we obtain

$$2\varepsilon_{xz} = \frac{\partial u_x}{\partial z} + \frac{\partial u_z}{\partial x} = \frac{Px^2}{2EI} + \frac{df_1}{dz} - \frac{\nu Pz^2}{2EI} + \frac{df_2}{dx}.$$

But this must be equal to the strain value given in Eq. (7.7.59):

$$\frac{Px^2}{2EI} + \frac{df_1}{dz} - \frac{\nu Pz^2}{2EI} + \frac{df_2}{dx} = \frac{(1+\nu)}{EI}P(h^2 - z^2).$$

Separating the x and z terms, we obtain

$$-\frac{df_2}{dx} - \frac{Px^2}{2EI} + \frac{(1+v)Ph^2}{EI} = \frac{df_1}{dz} + \frac{(2+v)Pz^2}{2EI}.$$

Since the left side depends only on x and the right side depends only on z, and yet the equality must hold, it follows that both sides should be equal to a constant, say c_0:

$$\frac{df_1}{dz} + \frac{(2+v)Pz^2}{2EI} = c_0, \quad -\frac{df_2}{dx} - \frac{Px^2}{2EI} + \frac{(1+v)Ph^2}{EI} = c_0.$$

Integrating the expressions for f_1 and f_2, we obtain

$$f_1(z) = -\frac{(2+v)Pz^3}{6EI} + c_0 z + c_1,$$

$$f_2(x) = -\frac{Px^3}{6EI} + \frac{(1+v)Ph^2 x}{EI} - c_0 x + c_2,$$

where c_1 and c_2 are constants of integration that are to be determined. The displacements (u_x, u_z) are now given by

$$u_x(x, z) = \frac{Px^2 z}{2EI} - \frac{(2+v)Pz^3}{6EI} + c_0 z + c_1, \qquad (7.7.61)$$

$$u_z(x, z) = \frac{(1+v)Ph^2 x}{EI} - \frac{v Pxz^2}{2EI} - \frac{Px^3}{6EI} - c_0 x + c_2. \qquad (7.7.62)$$

The constants c_0, c_1, and c_2 can be evaluated using the boundary conditions on the displacements. The displacement boundary conditions $u_x(L, z) = 0$ and $u_z(L, z) = 0$ cannot be satisfied (why?) by the displacement field in Eqs. (7.7.61) and (7.7.62). Therefore, we impose alternate boundary conditions that admit a meaningful solution. We set u_x, u_z, and rotation θ_3 to zero at $(x, z) = (L, 0)$:

$$u_x(L, 0) = 0, \quad u_z(L, 0) = 0, \quad \left(\frac{\partial u_x}{\partial z} - \frac{\partial u_z}{\partial x}\right)\bigg|_{(L,0)} = 0. \qquad (7.7.63)$$

Substituting the expressions for u_x and u_z into the boundary conditions (7.7.63), we obtain [note that the rotation condition used here is different from that used in Eq. (3.8.21)].

$$u_x(L, 0) = 0 \rightarrow c_1 = 0,$$

$$u_z(L, 0) = 0 \rightarrow c_0 L - c_2 = \frac{(1+v)Ph^2 L}{EI} - \frac{PL^3}{6EI},$$

$$\left(\frac{\partial u_x}{\partial z} - \frac{\partial u_z}{\partial x}\right)\bigg|_{(L,0)} = 0 \rightarrow 2c_0 = \frac{(1+v)Ph^2}{EI} - \frac{PL^2}{EI}.$$

Thus, the constants of integration are

$$c_0 = -\frac{PL^2}{2EI} + \frac{(1+v)Ph^2}{2EI}, \quad c_1 = 0, \quad c_2 = -\frac{PL^3}{3EI} - \frac{(1+v)Ph^2 L}{2EI}. \qquad (7.7.64)$$

Finally, the displacement field of two-dimensional elasticity theory becomes

$$u_x(x, z) = \frac{PL^2z}{6EI}\left[-3\left(1 - \frac{x^2}{L^2}\right) - (2 + v)\frac{z^2}{L^2} + 3(1 + v)\frac{h^2}{L^2}\right], \quad (7.7.65)$$

$$u_z(x, z) = -\frac{PL^3}{6EI}\left[2 - 3\frac{x}{L}\left(1 - v\frac{z^2}{L^2}\right) + \frac{x^3}{L^3} + 3(1 + v)\frac{h^2}{L^2}\left(1 - \frac{x}{L}\right)\right].$$

$$(7.7.66)$$

The displacement field in the beam according to the Euler–Bernoulli beam theory is given by

$$u_x(x, z) = -\frac{PL^2z}{2EI}\left(1 - \frac{x^2}{L^2}\right), \quad (7.7.67)$$

$$u_z(x, z) = \frac{PL^3}{6EI}\left(2 - 3\frac{x}{L} + \frac{x^3}{L^3}\right), \quad (7.7.68)$$

and, according to the Timoshenko beam theory, it is [see Reddy (2007, 2002)]

$$u_x(x, z) = \frac{PL^2z}{2EI}\left(1 - \frac{x^2}{L^2}\right), \quad (7.7.69)$$

$$u_z(x, z) = \frac{PL^3}{6EI}\left(2 - 3\frac{x}{L} + \frac{x^3}{L^3}\right) + 1.6\frac{(1 + v)Ph^2L}{EI}\left(1 - \frac{x}{L}\right), \quad (7.7.70)$$

where a shear correction factor of 5/6 is used in the Timoshhenko beam solution. Clearly, the Timoshenko beam solution is closer to the elasticity solution than the Euler–Bernoulli beam solution. Also, the elasticity solution indicates that plane sections perpendicular to the x-axis before deformation do not remain plane after deformation, as suggested by the quadratic and cubic terms in z.

EXAMPLE 7.7.3: Consider a thin rectangular plate of length $2a$, width $2b$, and thickness h, and has a circular hole of radius R at the center of the plate. A uniform traction of magnitude σ_0 is applied to the ends of the plate, as shown in Figure 7.7.7. Determine the stress field in the plate under the assumption that $R << b$.

SOLUTION: The boundary conditions of the problem are

$$\sigma_{xx}(\pm a, y) = \sigma_0, \quad \sigma_{xy}(\pm a, y) = 0, \quad \sigma_{yy}(x, \pm b) = 0, \quad \sigma_{xy}(x, \pm b) = 0, \quad (7.7.71)$$

$$\sigma_{rr}(R, \theta) = 0, \quad \sigma_{r\theta}(R, \theta) = 0. \quad (7.7.72)$$

Since the hole is assumed to be very small compared with the height of the plate, we can solve the problem for stress field inside a circular region of radius $c > R$, as shown in Figure 7.7.7. The stresses at the radius c are essentially the same as in the plate without the hole (a consequence of Saint–Venant's principle). We now set up the cylindrical coordinate system and solve the problem in that system.

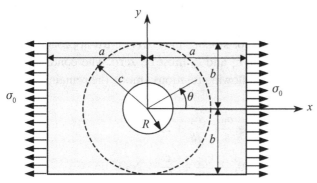

Figure 7.7.7. A thin rectangular plate with a central hole.

Using the transformation equations in Eq. (6.2.37), we can write the boundary conditions at $r = c$ for any θ as

$$\sigma_{rr}(c, \theta) = \sigma_0 \cos^2 \theta = \frac{\sigma_0}{2}\left(1 + \cos 2\theta\right),$$

$$\sigma_{\theta\theta}(c, \theta) = \sigma_0 \sin^2 \theta = \frac{\sigma_0}{2}\left(1 - \cos 2\theta\right), \qquad (7.7.73)$$

$$\sigma_{r\theta}(c, \theta) = -\frac{\sigma_0}{2} \cos 2\theta.$$

Using the definition of the Airy stress function for a cylindrical coordinate system and the conditions in Eq. (7.7.73), we see that $\Phi(r, \theta)$ must be of the form

$$\Phi(r, \theta) = G(r) - F(r) \cos 2\theta, \qquad (7.7.74)$$

with $G(r)$ and $F(r)$ satisfying

$$\left(\frac{d^2}{dr^2} + \frac{1}{r}\frac{d}{dr}\right)^2 G(r) = 0, \quad \left(\frac{d^2}{dr^2} + \frac{1}{r}\frac{d}{dr} - \frac{4}{r^2}\right)^2 F(r) = 0. \qquad (7.7.75)$$

The general solutions to the equations in (7.7.75) are of the form

$$F(r) = \frac{c_1}{r^2} + c_2 + c_3 r^2 + c_4 r^4, \quad G(r) = c_5 + c_6 \ln r + c_7 r^2 + c_8 r^2 \ln r. \quad (7.7.76)$$

Substituting the expressions from Eq. (7.7.76) into Eq. (7.7.74) and using the definition of the stress components, we obtain

$$\sigma_{rr} = \frac{1}{r}\frac{\partial \Phi}{\partial r} + \frac{1}{r^2}\frac{\partial^2 \Phi}{\partial \theta^2} = \frac{c_6}{r^2} + 2c_7 + c_8(1 + 2\ln r) - \left(\frac{6c_1}{r^4} + \frac{4c_2}{r^2} + 2c_3\right)\cos 2\theta,$$

$$\sigma_{\theta\theta} = \frac{\partial^2 \Phi}{\partial r^2} = -\frac{c_6}{r^2} + 2c_7 + c_8(3 + 2\ln r) - \left(\frac{6c_1}{r^4} + 2c_3 + 12c_4 r^2\right)\cos 2\theta,$$

$$\sigma_{r\theta} = -\frac{\partial}{\partial r}\left(\frac{1}{r}\frac{\partial \Phi}{\partial \theta}\right) = \left(-\frac{6c_1}{r^4} - \frac{2c_2}{r^2} + 2c_3 + 6c_4 r^2\right)\sin 2\theta.$$

$$(7.7.77)$$

Note that c_5 does not enter the calculation of stresses. The boundary conditions in Eqs. (7.7.72) and (7.7.73) are used to determine the remaining constants. As $r \to \infty$, the expressions for stresses in (7.7.77) must approach those
in Eq. (7.7.73). For this to happen, c_4 and c_8 must be zero. The conditions in
Eq. (7.7.72) and (7.7.73) yield the following relations among the remaining constants:

$$c_7 = \frac{\sigma_0}{4}, \quad c_3 = -\frac{\sigma_0}{4}, \quad \frac{c_6}{R^2} + 2c_7 = 0,$$

$$\frac{6c_1}{R^4} + \frac{4c_2}{R^2} + 2c_3 = 0, \quad -\frac{6c_1}{R^4} + \frac{2c_2}{R^2} + 2c_3 = 0. \tag{7.7.78}$$

The solution of these equations yield the values

$$c_1 = -\frac{\sigma_0 R^4}{4} \quad c_2 = \frac{\sigma_0 R^2}{2} \quad c_3 = -\frac{\sigma_0}{4} \quad c_6 = -\frac{\sigma_0 R^2}{2} \quad c_7 = \frac{\sigma_0}{4}. \tag{7.7.79}$$

Substituting these values into Eqs. (7.7.77), we obtain

$$\sigma_{rr} = \frac{\sigma_0}{2} \left[\left(1 - \frac{R^2}{r^2} \right) + \left(1 + \frac{3R^4}{r^4} - \frac{4R^2}{r^2} \right) \cos 2\theta \right],$$

$$\sigma_{\theta\theta} = \frac{\sigma_0}{2} \left[\left(1 + \frac{R^2}{r^2} \right) - \left(1 + \frac{3R^4}{r^4} \right) \cos 2\theta \right], \tag{7.7.80}$$

$$\sigma_{r\theta} = -\frac{\sigma_0}{2} \left(1 - \frac{3R^4}{r^4} + \frac{2R^2}{r^2} \right) \sin 2\theta.$$

The maximum stress occurs at $r = R$ and $\theta = \pm 90°$

$$\sigma_{max} = \sigma_{\theta\theta}(R, \pm 90) = 3\sigma_0. \tag{7.7.81}$$

7.7.6 Torsion of Noncircular Cylinders

The stress function approach used to study a number of plane elasticity problems
in Section 7.7.5 is also useful in studying torsion of noncircular members (see Figure 7.7.8). However, we cannot use the Airy stress function here because the present
problem does not quite fall into the category of plane elasticity problems. The governing equations for this problem must be developed from basic principles. The
problem was first studied by Saint–Venant using the semi-inverse method.

Consider a cylindrical member of noncircular cross section and length L sub-
jected to an end torque $\mathbf{T} = T\hat{\mathbf{e}}_3$, as shown in Figure 7.7.8. The lateral surface of the
cylinder is free of tractions. We assume that the magnitude of the applied torque is
small so that (a) all cross sections rotate about a single axis in proportion to their
distance from the end of the cylinder and (b) the out-of-plane distortion of each
cross section is the same. Therefore, attention is focused on a typical cross section,
denoted Ω. There are two different formulations to study the problem. One is based
on the warping function and the other on Prandtl stress function. Here we consider
the latter approach.

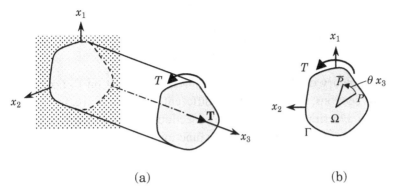

Figure 7.7.8. Torsion of a cylindrical member.

First, we note that the following stresses are identically zero for the problem:

$$\sigma_{11} = \sigma_{22} = \sigma_{33} = \sigma_{12} = 0. \tag{7.7.82}$$

Therefore, only stress equilibrium equation left to be satisfied is

$$\frac{\partial \sigma_{31}}{\partial x_1} + \frac{\partial \sigma_{32}}{\partial x_2} = 0. \tag{7.7.83}$$

Since $\sigma_{33} = 0$, it follows that σ_{31} and σ_{32} are independent of x_3. By introducing a stress function $\Psi(x_1, x_2)$, called the *Prandtl stress function*,

$$\sigma_{31} = \frac{\partial \Psi}{\partial x_2}, \qquad \sigma_{32} = -\frac{\partial \Psi}{\partial x_1}, \tag{7.7.84}$$

we trivially satisfy the equilibrium equation in (7.7.83). Now Ψ is to be determined such that it satisfies the compatibility conditions ($\sigma_{3\alpha, \beta\beta} = 0$, for $\alpha, \beta = 1, 2$)

$$\frac{\partial^2 \sigma_{31}}{\partial x_1^2} + \frac{\partial^2 \sigma_{31}}{\partial x_2^2} = 0 \Rightarrow \frac{\partial}{\partial x_2}\left(\frac{\partial^2 \Psi}{\partial x_1^2} + \frac{\partial^2 \Psi}{\partial x_2^2}\right) = 0,$$

$$\frac{\partial^2 \sigma_{32}}{\partial x_1^2} + \frac{\partial^2 \sigma_{32}}{\partial x_2^2} = 0 \Rightarrow \frac{\partial}{\partial x_1}\left(\frac{\partial^2 \Psi}{\partial x_1^2} + \frac{\partial^2 \Psi}{\partial x_2^2}\right) = 0. \tag{7.7.85}$$

From these two equations, it follows that Ψ is governed by the equation

$$\nabla^2 \Psi = C, \tag{7.7.86}$$

where C is a constant. Equation (7.7.86) must be solved subject to the traction-free boundary condition on the lateral surface

$$\sigma_{31} n_1 + \sigma_{32} n_2 = 0, \tag{7.7.87}$$

where the components n_1 and n_2 of the unit outward normal to the lateral surface $\hat{\mathbf{n}}$ can be expressed as

$$\hat{\mathbf{n}} = \frac{dx_2}{ds}\hat{\mathbf{e}}_1 - \frac{dx_1}{ds}\hat{\mathbf{e}}_2. \tag{7.7.88}$$

Then the boundary condition (7.7.87) takes the form

$$\frac{\partial \Psi}{\partial x_2} \frac{dx_2}{ds} + \frac{\partial \Psi}{\partial x_1} \frac{dx_1}{ds} \equiv \frac{d\Psi}{ds} = 0 \text{ on } \Gamma, \tag{7.7.89}$$

with Γ being the boundary of Ω. In other words, Ψ is a constant on Γ. For multiply connected cross sections, the constants on different boundaries are, in general, have different values. For simply connected cross sections, we can arbitrarily set the constant to zero. In summary, the Prandtl stress function is determined from

$$\nabla^2 \Psi = C \text{ in } \Omega, \quad \Psi = 0 \text{ on } \Gamma. \tag{7.7.90}$$

Although not discussed in detail here, the so-called warping function $\Phi(x_1, x_2)$ is related to the Prandtl stress function by

$$\frac{\partial \Phi}{\partial x_1} = \frac{1}{\mu \theta} \frac{\partial \Psi}{\partial x_2} + x_2, \quad \frac{\partial \Phi}{\partial x_2} = -\frac{1}{\mu \theta} \frac{\partial \Psi}{\partial x_1} - x_1, \tag{7.7.91}$$

where θ is the *twist per unit length* (assumed to be small) and μ is the Lamé constant (same as the shear modulus, G). Equations in (7.7.91) can be combined (by differentiating the first one with respect to x_2 and the second one with respect to x_1 and eliminating Φ) to obtain

$$-\nabla^2 \Psi = 2G\theta \text{ in } \Omega, \quad \Psi = 0 \text{ on } \Gamma. \tag{7.7.92}$$

Once Ψ is known, the displacements (u_1, u_2, u_3) are calculated from the equation (see Figure 7.7.8)

$$u_1 = -\theta x_2 x_3, \quad u_2 = \theta x_1 x_3, \quad u_3 = \theta \Phi(x_1, x_2), \tag{7.7.93}$$

and the stresses are calculated form Eq. (7.7.84).

Exact solutions of the torsion problem (7.7.92) are possible for few cases. Typically, one assumes Ψ to be in the form $\Psi = Af(x_1, x_2)$, where A is a constant and f is a sufficiently differentiable (i.e., $\nabla^2 f \neq 0$) function that is identically zero on the boundary. If $-\nabla^2 f$ is a non-zero constant c (so that Ac can be equated to $2\mu\theta$), then we solve for A and obtain the complete solution. If $\nabla^2 f$ is not a constant, exact solution is not possible, although an approximate solution can be obtained, for example, using the Galerkin method. Here, we consider an example.

EXAMPLE 7.7.4: Consider a cylindrical shaft of elliptical cross section, Ω. The boundary Γ is the ellipse with semiaxes a and b:

$$\Gamma = \left\{ (x_1, x_2) : \frac{x_1^2}{a^2} + \frac{x_2^2}{b^2} = 1 \right\}. \tag{7.7.94}$$

Determine the Prandtl stress function and the shear stresses.

SOLUTION: We select Ψ to be

$$\Psi(x_1, x_2) = A\left(\frac{x_1^2}{a^2} + \frac{x_2^2}{b^2} - 1\right),\tag{7.7.95}$$

where A is a constant to be determined such that Eq. (7.7.92) is satisfied. Since the boundary condition $\Psi = 0$ on Γ is satisfied, we substitute Ψ from Eq. (7.7.95) into $-\nabla^2\Psi = 2G\theta$ and obtain

$$-2A\left(\frac{1}{a^2} + \frac{1}{b^2}\right) = 2\mu\theta \Rightarrow A = -\frac{\mu\theta a^2 b^2}{a^2 + b^2}.\tag{7.7.96}$$

The Prandtl stress function is then given by

$$\Psi(x_1, x_2) = -\frac{\mu\theta a^2 b^2}{a^2 + b^2}\left(\frac{x_1^2}{a^2} + \frac{x_2^2}{b^2} - 1\right).\tag{7.7.97}$$

For solid cylinders, the twist per unit length θ can be related to the applied torque T by

$$T = 2\int_\Omega \Psi(x_1, x_2)\, dx_1\, dx_2.\tag{7.7.98}$$

For the problem at hand we obtain

$$\theta = \frac{(a^2 + b^2)T}{\mu\pi a^3 b^3}.\tag{7.7.99}$$

The stresses σ_{31} and σ_{32} are calculated using Eq. (7.7.84) to be

$$\sigma_{31} = -\frac{2a^2}{a^2 + b^2}\mu\theta x_2, \quad \sigma_{32} = -\frac{2b^2}{a^2 + b^2}\mu\theta x_1.\tag{7.7.100}$$

7.8 Principle of Minimum Total Potential Energy

7.8.1 Introduction

In Chapter 5 of this book, laws of physics (or conservation principles) and vector mechanics are used to derive the equations governing continua. These equations, as applied to solid bodies, can also be formulated by means of variational principles. Variational principles have played an important role in solid mechanics. The principle of minimum total potential energy, for example, can be regarded as a substitute to the equations of equilibrium of elastic bodies. Similarly, Hamilton's principle can be used in lieu of the equations governing dynamical systems, and the variational forms presented by Biot replace certain equations in linear continuum thermodynamics.

The use of variational principles makes it possible to concentrate in a single functional all of the intrinsic features of the problem at hand: the governing equations, the boundary conditions, initial conditions, constraint conditions, even jump conditions. Variational principles can serve not only to derive the governing equations but they also suggest new theories. Finally, and perhaps most importantly, variational principle provide a natural means for seeking approximate solutions; they are at the heart of the most powerful approximate methods in use in mechanics (e.g., traditional Ritz, Galerkin, least-squares, weighted-residual, and the finite element method). In many cases, they can also be used to establish upper and lower bounds on approximate solutions.

This section is devoted to the study of the principle of minimum total potential energy and its applications. To keep the scope of the chapter within reasonable limits, only key elements of the principle are presented here. Additional information can be found in the books by the author [see, for example, Reddy (2002)].

7.8.2 Total Potential Energy Principle

Recall from Sections 6.2 and 7.6 that for elastic bodies (in the absence of temperature variations) there exists a strain energy density function U_0 such that [see Eq. (6.2.2)]

$$\sigma_{ij} = \frac{\partial U_0}{\partial \varepsilon_{ij}}. \tag{7.8.1}$$

The strain energy density U_0 is a function of strains at a point and is assumed to be positive definite. For linear elastic bodies (i.e., obeying the generalized Hooke's law), the strain energy density is given by [see Eq. (7.6.1)],

$$U_0 = \frac{1}{2}\boldsymbol{\sigma} : \boldsymbol{\varepsilon} = \frac{1}{2}\sigma_{ij}\varepsilon_{ij} = \frac{1}{2}c_{ijkl}\varepsilon_{ij}\varepsilon_{kl}. \tag{7.8.2}$$

Hence, the total strain energy of the body $\mathcal{B}$ occupying the volume Ω is given by [see Eq. (7.6.2)]

$$U = \int_\Omega U_0(\varepsilon_{ij})\, d\mathbf{x} = \frac{1}{2}\int_\Omega \boldsymbol{\sigma} : \boldsymbol{\varepsilon}\, d\mathbf{x} = \frac{1}{2}\int_\Omega \sigma_{ij}\,\varepsilon_{ij}\, d\mathbf{x}. \tag{7.8.3}$$

The total work done by applied body force $\mathbf{f}$ and surface force $\mathbf{t}$ is given by [see Eq. (7.6.3)]

$$V = -\left[\int_\Omega \mathbf{f}\cdot\mathbf{u}\, d\mathbf{x} + \oint_\Gamma \mathbf{t}\cdot\mathbf{u}\, ds\right], \tag{7.8.4}$$

the minus sign in V indicates that the work is expended, whereas U is the available energy stored in the body $\mathcal{B}$. The total potential energy (functional) of the body $\mathcal{B}$ is the sum of the strain energy stored in the body and the work done by external forces

$$\Pi = U + V = \frac{1}{2}\int_\Omega \boldsymbol{\sigma} : \boldsymbol{\varepsilon}\, d\mathbf{x} - \left[\int_\Omega \mathbf{f}\cdot\mathbf{u}\, d\mathbf{x} + \oint_\Gamma \mathbf{t}\cdot\mathbf{u}\, ds\right]$$

$$= \frac{1}{2}\int_\Omega \sigma_{ij}\,\varepsilon_{ij}\, d\mathbf{x} - \left[\int_\Omega f_i u_i\, d\mathbf{x} + \oint_\Gamma t_i u_i\, ds\right]. \tag{7.8.5}$$

The *principle of minimum total potential energy* states that if a body is in equilibrium, of all *admissible* displacement fields $\mathbf{u}$, the one $\mathbf{u}_0$ that makes the total potential energy a minimum corresponds to the equilibrium solution:

$$\Pi(\mathbf{u}_0) \leq \Pi(\mathbf{u}). \tag{7.8.6}$$

An admissible displacement is the one that satisfies the geometric constraints (or boundary conditions) of the problem. The statement in Eq. (7.8.6) can be expressed (based on the conditions of calculus of variations for the minimum of a functional) as

$$\delta I = 0 \quad \text{(necessary condition)}, \tag{7.8.7}$$

$$\delta^2 I > 0 \quad \text{(sufficient condition)}, \tag{7.8.8}$$

where δ is the *variational operator*.

The variational operator δ is much like a total differential operator d, except that it operates with respect to the dependent variable $\mathbf{u}$ rather than the independent variable $\mathbf{x}$. Indeed, the laws of *variation* of sums, products, ratios, and powers of functions of a dependent variable u are completely analogous to the corresponding laws of differentiation; that is, the variational calculus resembles the differential calculus. For example, if $F_1 = F_1(u)$ and $F_2 = F_2(u)$, we have

$$
\begin{aligned}
&(1) && \delta(F_1 \pm F_2) = \delta F_1 \pm \delta F_2. \\
&(2) && \delta(F_1 \, F_2) = \delta F_1 \, F_2 + F_1 \delta F_2. \\
&(3) && \delta\left(\frac{F_1}{F_2}\right) = \frac{\delta F_1 \, F_2 - F_1 \, \delta F_2}{F_2^2}. \\
&(4) && \delta(F_1)^n = n(F_1)^{n-1}\delta F_1.
\end{aligned}
\tag{7.8.9}
$$

If $G = G(u, v, w)$ is a function of several dependent variables u, v, and w, and possibly their derivatives, the total variation is the sum of partial variations:

$$\delta G = \delta_u G + \delta_v G + \delta_w G, \tag{7.8.10}$$

where, for example, δ_u denotes the partial variation with respect to u. The variational operator can be interchanged with differential and integral operators:

$$
\begin{aligned}
&(1) && \delta(\nabla u) = \nabla(\delta u). \tag{7.8.11}
\end{aligned}
$$

$$
\begin{aligned}
&(2) && \delta\left(\int_\Omega u \, d\mathbf{x}\right) = \int_\Omega \delta u \, d\mathbf{x}. \tag{7.8.12}
\end{aligned}
$$

All of the above relations are valid in multidimensions and for functions that depend on more than one dependent variable.

The necessary condition (7.8.7) yields the governing equations of the problem, which are equivalent to those derived from the principle of linear momentum.

However, Eq. (7.8.7) also gives the boundary conditions on the forces of the problem. The equations obtained in Ω from the necessary condition (7.8.7) are known as the *Euler equations* and those obtained on Γ (or on a portion of Γ) are known as the *natural boundary conditions*.

7.8.3 Derivation of Navier's Equations

Here, we illustrate how the Navier equations of elasticity (7.3.3) and (7.3.4) can be derived using the principle of minimum total potential energy. Consider a linear elastic body B occupying volume Ω with boundary Γ and subjected to body force $\mathbf{f}$ (measured per unit volume) and surface traction $\hat{\mathbf{t}}$ on portion Γ_σ of the surface. We assume that the displacement vector $\mathbf{u}$ is specified to be $\hat{\mathbf{u}}$ on the remaining portion, Γ_u, of the boundary ($\Gamma = \Gamma_u \cup \Gamma_\sigma$). Therefore, $\delta\mathbf{u} = \mathbf{0}$ on Γ_u.

The total potential energy functional is given by (summation on repeated indices is implied throughout this discussion)

$$\Pi(\mathbf{u}) = \int_\Omega \left(\frac{1}{2}\sigma_{ij}\varepsilon_{ij} - f_i u_i \right) dx - \int_{\Gamma_\sigma} \hat{t}_i u_i \, ds. \tag{7.8.13}$$

The first term under the volume integral represents the strain energy density of the elastic body, the second term represents the work done by the body force $\mathbf{f}$, and the third term represents the work done by the specified traction $\hat{\mathbf{t}}$.

The strain-displacement relations and stress–strain relations for an isotropic elastic body are given by Eqs. (7.2.1) and (7.2.3), respectively. Substituting Eqs. (7.2.1) and (7.2.3) into Eq. (7.8.13), we obtain

$$\Pi(\mathbf{u}) = \int_\Omega \left[\frac{\mu}{4} (u_{i,j} + u_{j,i})(u_{i,j} + u_{j,i}) + \frac{\lambda}{2} u_{i,i} u_{k,k} - f_i u_i \right] dx - \int_{\Gamma_\sigma} \hat{t}_i u_i \, ds.$$
$$\tag{7.8.14}$$

Setting the first variation of Π to zero (i.e., using the principle of minimum total potential energy), we obtain

$$0 = \int_\Omega \left[\frac{\mu}{2} (\delta u_{i,j} + \delta u_{j,i})(u_{i,j} + u_{j,i}) + \lambda \delta u_{i,i} u_{k,k} - f_i \delta u_i \right] dx - \int_{\Gamma_\sigma} \hat{t}_i \delta u_i \, ds,$$
$$\tag{7.8.15}$$

wherein the product rule of variation is used and similar terms are combined. Next, we use the component form of the gradient theorem to relieve δu_i of any derivative so that we can use the fundamental lemma of variational calculus to the coefficients of δu_i to zero in Ω and on the portion of Γ where δu_i is arbitrary. Using the gradient theorem, we can write

$$\int_\Omega \delta u_{i,j} (u_{i,j} + u_{j,i}) \, dx = -\int_\Omega \delta u_i (u_{i,j} + u_{j,i})_{,j} \, dx + \oint_\Gamma \delta u_i (u_{i,j} + u_{j,i}) n_j \, ds,$$

where n_j denotes the jth direction cosine of the unit normal vector to the surface $\hat{\mathbf{n}}$. Using this result in Eq. (7.8.15), we arrive at

$$0 = \int_\Omega \left[-\frac{\mu}{2} \left(u_{i,j} + u_{j,i} \right)_{,j} \delta u_i - \frac{\mu}{2} \left(u_{i,j} + u_{j,i} \right)_{,i} \delta u_j - \lambda u_{k,ki} \delta u_i - f_i \delta u_i \right] d\mathbf{x}$$

$$+ \oint_\Gamma \left[\frac{\mu}{2} \left(u_{i,j} + u_{j,i} \right) \left(n_j \delta u_i + n_i \delta u_j \right) + \lambda u_{k,k} n_i \delta u_i \right] ds - \int_{\Gamma_\sigma} \delta u_i \hat{t}_i \, ds$$

$$= \int_\Omega \left[-\mu \left(u_{i,j} + u_{j,i} \right)_{,j} - \lambda u_{k,ki} - f_i \right] \delta u_i \, d\mathbf{x}$$

$$+ \oint_\Gamma \left[\mu \left(u_{i,j} + u_{j,i} \right) + \lambda u_{k,k} \delta_{ij} \right] n_j \delta u_i \, ds - \int_{\Gamma_\sigma} \delta u_i \hat{t}_i \, ds. \tag{7.8.16}$$

In arriving at the last step, change of dummy indices is made to combine terms.

Recognizing that the expression inside the square brackets of the closed surface integral is nothing but σ_{ij} and $\sigma_{ij} n_j = t_i$ by Cauchy's formula, we can write

$$\oint_\Gamma \left[\mu \left(u_{i,j} + u_{j,i} \right) + \lambda u_{k,k} \delta_{ij} \right] n_j \delta u_i \, ds = \oint_\Gamma t_i \delta u_i \, ds.$$

This boundary expression resulting from the "integration-by-parts" to relieve $\delta\mathbf{u}$ of any derivatives is used to classify the variables of the problem. The coefficient of δu_i is called the *secondary variable*, and the varied quantity itself (without the variational symbol) is called the *primary variable*. Thus, u_i is the primary variable and t_i is the corresponding secondary variable. They always appear in pairs, and only one element of the pair may be specified at any boundary point. Specification of a primary variable is called *essential boundary condition* and specification of a secondary variable is termed *natural boundary condition*. They are also known as the geometric and force boundary conditions, respectively. In applied mathematics field, they are known as the *Dirichlet boundary condition* and *Neumann boundary condition*, respectively.

Returning to the boundary integral, it can be expressed as the sum of integrals on Γ_u and Γ_σ:

$$\oint_\Gamma t_i \delta u_i \, ds = \int_{\Gamma_u} t_i \delta u_i \, ds + \int_{\Gamma_\sigma} t_i \delta u_i \, ds = \int_{\Gamma_\sigma} t_i \delta u_i \, ds.$$

The integral over Γ_u is set to zero because of the fact that $\mathbf{u}$ is specified there, i.e., $\delta\mathbf{u} = \mathbf{0}$. Hence, Eq. (7.8.16) becomes

$$0 = \int_\Omega \left[-\mu \left(u_{i,j} + u_{j,i} \right)_{,j} - \lambda u_{k,ki} - f_i \right] \delta u_i \, d\mathbf{x} + \int_{\Gamma_\sigma} \delta u_i \left(t_i - \hat{t}_i \right) ds. \tag{7.8.17}$$

Using the fundamental lemma of calculus of variations we set the coefficients of δu_i in Ω and δu_i on Γ_σ from Eq. (7.8.17) to zero separately and obtain

$$\mu(u_{i,jj} + u_{j,ij}) + \lambda u_{k,ki} + f_i = 0 \text{ in } \Omega, \tag{7.8.18}$$

$$t_i - \hat{t}_i = 0 \text{ on } \Gamma_\sigma, \tag{7.8.19}$$

for $i = 1, 2, 3$. Equation (7.8.18) represents the Navier's equations of elasticity (7.3.3), and the natural boundary conditions (7.8.19) are the same as those listed in Eq. (7.3.4).

To show that the total potential energy of a linear elasticity body is the indeed the minimum at its equilibrium configuration, we consider the total potential energy functional [more general than the one considered in Eq. (7.8.14); see Reddy (2002)]:

$$\Pi(\mathbf{u}) = \int_\Omega \left(\frac{1}{2} C_{ijk\ell} \varepsilon_{k\ell} \varepsilon_{ij} - f_i u_i \right) d\mathbf{x} - \int_{\Gamma_\sigma} \hat{t}_i u_i \, ds, \tag{7.8.20}$$

where $C_{ijk\ell}$ are the components of the fourth-order elasticity tensor.

Let $\mathbf{u}$ be the true displacement field and $\bar{\mathbf{u}}$ be an arbitrary but admissible displacement field. Then $\bar{\mathbf{u}}$ is of the form

$$\bar{\mathbf{u}} = \mathbf{u} + \alpha \mathbf{v},$$

where α is a real number and $\mathbf{v}$ is a sufficiently differentiable function that satisfies the homogeneous form of the essential boundary condition $\mathbf{v} = \mathbf{0}$ on Γ_u. Then $\Pi(\bar{\mathbf{u}})$ is given by

$$\Pi(\mathbf{u} + \alpha \mathbf{v}) = \int_\Omega \left[\frac{1}{2} C_{ijk\ell} (\varepsilon_{k\ell} + \alpha g_{k\ell})(\varepsilon_{ij} + \alpha g_{ij}) - f_i (u_i + \alpha v_i) \right] d\mathbf{x}$$
$$- \int_{\Gamma_\sigma} \hat{t}_i (u_i + \alpha v_i) ds,$$

where

$$g_{ij} = \frac{1}{2}(v_{i,j} + v_{j,i}).$$

Collecting the terms, we obtain (because $C_{ijk\ell} = C_{k\ell ij}$)

$$\Pi(\bar{\mathbf{u}}) = \Pi(\mathbf{u}) + \alpha \left[\int_\Omega \left(-f_i v_i + C_{ijk\ell} \varepsilon_{k\ell} g_{ij} + \frac{\alpha}{2} C_{ijk\ell} g_{ij} g_{k\ell} \right) d\mathbf{x} - \int_{\Gamma_\sigma} \hat{t}_i v_i \, ds \right].$$
$$\tag{7.8.21}$$

Using the equilibrium equations (7.2.2) and the generalized Hooke's law $\sigma_{ij} = C_{ijk\ell} \varepsilon_{k\ell}$, we can write

$$- \int_\Omega f_i v_i \, d\mathbf{x} = \int_\Omega \sigma_{ij,j} v_i \, d\mathbf{x} = \int_\Omega C_{ijk\ell} \varepsilon_{k\ell,j} v_i \, d\mathbf{x}$$
$$= - \int_\Omega C_{ijk\ell} \varepsilon_{k\ell} v_{i,j} \, d\mathbf{x} + \int_{\Gamma_\sigma} C_{ijk\ell} \varepsilon_{k\ell} v_i n_j \, ds$$
$$= - \int_\Omega C_{ijk\ell} \varepsilon_{k\ell} g_{ij} \, d\mathbf{x} + \int_{\Gamma_\sigma} \hat{t}_i v_i \, ds, \tag{7.8.22}$$

where the condition $v_i = 0$ on Γ_u is used in arriving at the last step. Substituting Eq. (7.8.22) into Eq. (7.8.21), we arrive at

$$\Pi(\bar{\mathbf{u}}) = \Pi(\mathbf{u}) + \frac{\alpha^2}{2} \int_\Omega C_{ijk\ell} g_{ij} g_{k\ell} \, d\mathbf{x}. \tag{7.8.23}$$

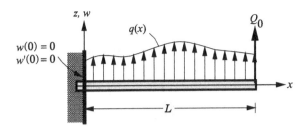

Figure 7.8.1. A beam with applied loads.

In view of the nonnegative nature of the second term on the right-hand side of Eq. (7.8.23), it follows that

$$\Pi(\bar{\mathbf{u}}) \geq \Pi(\mathbf{u}), \qquad (7.8.24)$$

and $\Pi(\bar{\mathbf{u}}) = \Pi(\mathbf{u})$ only if the quadratic expression $\frac{1}{2}C_{ijk\ell}g_{ij}g_{k\ell}$ is zero. Because of the positive definiteness of the strain energy density, the quadratic expression is zero only if $v_i = 0$, which in turn implies $\bar{u}_i = u_i$. Thus, Eq. (7.8.24) implies that, of all admissible displacement fields the body can assume, the true one is that which makes the total potential energy a minimum.

In the following, we consider an example to illustrate the use of the principle of minimum total potential energy for the bending of beams [see Reddy (2002)].

EXAMPLE 7.8.1: Consider the bending of a beam according to the Euler–Bernoulli beam theory (see Part 3 of Example 7.6.1). We wish to construct the total potential energy functional and then determine the governing equation and boundary conditions of the problem.

From Part 3 of Example 7.6.1 the total potential energy of a cantilevered beam bent by distributed transverse force $q(x)$ and point load Q_0 (see Figure 7.8.1), under the assumption of small strains and displacements for the linear elastic case (i.e., obeys Hooke's law), is given by [see the right-hand side of Eq. (7.6.8)]

$$\Pi(w) = \frac{1}{2} \int_0^L \left[EI \left(\frac{d^2 w}{dx^2} \right)^2 \right] dx - \left[\int_0^L q(x)w(x)\, dx + Q_0 w(L) \right], \quad (7.8.25)$$

where L is the length, A is the cross-sectional area, I is the second moment of area about the axis (y) of bending, and E is the Young's modulus of the beam. The first term represents the strain energy U, the second term represents the work done by the applied distributed load $q(x)$ in moving through the deflection $w(x)$, and the last expression represents the work done by the point load Q_0 in moving through the displacement $w(L)$.

Applying the principle of minimum total potential energy, $\delta\Pi = 0$, we obtain

$$0 = \delta\Pi = \int_0^L EI \frac{d^2 w}{dx^2} \frac{d^2 \delta w}{dx^2}\, dx - \left[\int_0^L q \delta w\, dx + Q_0 \delta w(L) \right]. \qquad (7.8.26)$$

Next, we carry out integration-by-parts on the first term to relieve δw of any derivative so that we can use the fundamental lemma of variational calculus to obtain the Euler equation. We obtain

$$
0 = \int_0^L \frac{d^2}{dx^2}\left(EI\frac{d^2w}{dx^2}\right)\delta w\,dx + \left[EI\frac{d^2w}{dx^2}\frac{d\delta w}{dx} - \frac{d}{dx}\left(EI\frac{d^2w}{dx^2}\right)\delta w\right]_0^L
$$
$$
- \left[\int_0^L q\delta w\,dx + Q_0\delta w(L)\right]. \tag{7.8.27}
$$

The boundary terms resulting from integration-by-parts allows us to classify the boundary conditions of the problem. The expressions that are coefficients of δw and $\delta(dw/dx)$ are the secondary variables:

$$
\delta w: \quad \frac{d}{dx}\left(EI\frac{d^2w}{dx^2}\right); \quad \delta\left(\frac{dw}{dx}\right): \quad EI\frac{d^2w}{dx^2}. \tag{7.8.28}
$$

It is clear that the secondary variables are nothing but the shear force $V(x) = dM/dx$ and bending moment $M(x)$ [see Eq. (7.6.9)]

$$
V(x) = -\frac{d}{dx}\left(EI\frac{d^2w}{dx^2}\right); \quad M(x) = -EI\frac{d^2w}{dx^2}. \tag{7.8.29}
$$

The respective primary variables are the varied quantities appearing in the boundary terms (just remove the variational operator from the varied quantities):

$$
\delta w \Rightarrow w; \quad \delta\left(\frac{dw}{dx}\right) \Rightarrow \frac{dw}{dx}. \tag{7.8.30}
$$

Thus, the deflection w and slope (or rotation) dw/dx are the primary variables of the problem. Only one element of each of the pairs (w, V) and $(dw/dx, M)$ may be specified at a point. Note that the definitions of the primary and secondary variables is unique and there should be no confusion in identifying them.

Returning to the expression in Eq. (7.8.27), first we collect the coefficients of δw in $(0, L)$ together and set them to zero, because δw is arbitrary in $(0, L)$,

$$
\frac{d^2}{dx^2}\left(EI\frac{d^2w}{dx^2}\right) - q(x) = 0, \quad 0 < x < L. \tag{7.8.31}
$$

Equation (7.8.31) is the Euler equation, which can also be derived from vector mechanics by considering an element of the beam and summing the forces and moments, and then relating the bending moment M to the deflection w, as given in Eq. (7.6.9). The summation of forces in the z-direction and moments about the y-axis give (the reader should verify these equations)

$$
\frac{dV}{dx} + q(x) = 0, \quad \frac{dM}{dx} - V(x) = 0. \tag{7.8.32}
$$

Combining Eqs. (7.6.9) and (7.8.32), we arrive at the equation in (7.8.31).

Now consider all boundary terms in Eq. (7.8.27), we conclude that

$$\left[\frac{d}{dx}\left(EI\frac{d^2w}{dx^2}\right)\right]_{x=0}\delta w(0) = 0, \quad \left[-\frac{d}{dx}\left(EI\frac{d^2w}{dx^2}\right) - Q_0\right]_{x=L}\delta w(L) = 0,$$

$$\left(EI\frac{d^2w}{dx^2}\right)_{x=0}\left(\frac{d\delta w}{dx}\right)_{x=0} = 0, \quad \left(EI\frac{d^2w}{dx^2}\right)_{x=L}\left(\frac{d\delta w}{dx}\right)_{x=L} = 0.$$

$$(7.8.33)$$

If either of the quantities δw and $d\delta w/dx$ are zero at $x = 0$ or $x = L$, because of specified geometric boundary conditions there, the corresponding expressions vanish because a specified quantity cannot be varied; the vanishing of the coefficients of δw and $d\delta w/dx$ at points where the geometric boundary conditions are not specified provides the natural boundary conditions. For example, suppose that the beam is clamped at $x = 0$ and free at $x = L$. Then, $\delta w(0) = 0$ and $d\delta w(0)/dx = 0$, and the specified natural boundary conditions of a cantilevered beam with an end load become

$$\left[-\frac{d}{dx}\left(EI\frac{d^2w}{dx^2}\right) - Q_0\right]_{x=L} = 0, \quad \left(EI\frac{d^2w}{dx^2}\right)_{x=L} = 0. \quad (7.8.34)$$

7.8.4 Castigliano's Theorem I

Suppose that the displacement field of a solid body can be expressed in terms of the displacements of a finite number of points x_i $(i = 1, 2, \ldots N)$ as

$$\mathbf{u}(\mathbf{x}) = \sum_{i=1}^{N} \mathbf{u}_i \phi_i(\mathbf{x}), \quad (7.8.35)$$

where $\mathbf{u}_i$ are unknown displacement parameters, called *generalized displacements*, and ϕ_i are known functions of position, called *interpolation functions* with the property that ϕ_i is unity at the ith point (i.e., $\mathbf{x} = \mathbf{x}_i$) and zero at all other points $(\mathbf{x}_j, j \neq i)$. Then it is possible to represent the strain energy U and potential energy V due to applied loads in terms of the generalized displacements $\mathbf{u}_i$. Then the principle of minimum total potential energy can be written as

$$\delta\Pi = \delta U + \delta V = 0 \Rightarrow \delta U = -\delta V \quad \text{or} \quad \frac{\partial U}{\partial \mathbf{u}_i}\cdot\delta\mathbf{u}_i = -\frac{\partial V}{\partial \mathbf{u}_i}\cdot\delta\mathbf{u}_i, \quad (7.8.36)$$

where sum on repeated indices is implied. Since

$$\frac{\partial V}{\partial \mathbf{u}_i} = -\mathbf{F}_i,$$

it follows, since $\delta\mathbf{u}_i$ are arbitrary, that

$$\left(\frac{\partial U}{\partial \mathbf{u}_i} - \mathbf{F}_i\right)\cdot\delta\mathbf{u}_i = 0 \quad \text{or} \quad \frac{\partial U}{\partial \mathbf{u}_i} = \mathbf{F}_i. \quad (7.8.37)$$

Equation (7.8.37) is known as Castigliano's Theorem I.

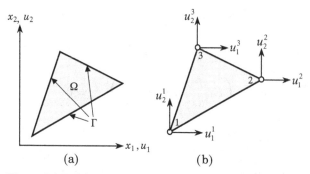

Figure 7.8.2. (*a*) A plane elastic triangular domain. (*b*) Domain with vertex displacement components.

When applied to a structure with point loads F_i (or moment M_i) moving through displacements u_i (or rotation θ_i), both having the same sense, Castigliano's Theorem I states that

$$\frac{\partial U}{\partial u_i} = F_i, \quad \text{or} \quad \frac{\partial U}{\partial \theta_i} = M_i. \tag{7.8.38}$$

It is clear from the derivation that Castigliano's Theorem I is a special case of the principle of minimum total potential energy.

Application of Castigliano's Theorem I to structural members (trusses and frames) can be found in many books [see Example 7.8.2 below and the book by Reddy (2002) and references therein]. In the following paragraphs, application to a plane elastic problem is illustrated.

Consider an arbitrary triangular, plane elastic domain Ω of thickness h and made of orthotropic material, as shown in Figure 7.8.2(a). Suppose that the body is free of body forces but subjected to tractions on its sides.

The strain energy and potential energy due to applied loads are

$$U = \frac{h}{2} \int_\Omega \sigma_{ij} \varepsilon_{ij} \, d\mathbf{x}, \quad V = -\oint_\Gamma \hat{t}_i u_i \, ds, \tag{7.8.39}$$

where Γ represents the collection of line segments enclosing the domain Ω and $\hat{t}_i(s)$ are the components of the boundary stresses. It is convenient to express the expressions for U and V in matrix form. The strain–displacement relations and constitutive equations can be written in matrix form as

$$\begin{Bmatrix} \varepsilon_{11} \\ \varepsilon_{22} \\ 2\varepsilon_{12} \end{Bmatrix} = \begin{bmatrix} \partial/\partial x_1 & 0 \\ 0 & \partial/\partial x_2 \\ \partial/\partial x_2 & \partial/\partial x_1 \end{bmatrix} \begin{Bmatrix} u_1 \\ u_2 \end{Bmatrix} \quad \text{or} \quad \{\varepsilon\} = [D]\{u\}, \tag{7.8.40}$$

$$\begin{Bmatrix} \sigma_{11} \\ \sigma_{22} \\ \sigma_{12} \end{Bmatrix} = \begin{bmatrix} c_{11} & c_{12} & 0 \\ c_{12} & c_{22} & 0 \\ 0 & 0 & c_{66} \end{bmatrix} \begin{Bmatrix} \varepsilon_{11} \\ \varepsilon_{22} \\ 2\varepsilon_{12} \end{Bmatrix} \quad \text{or} \quad \{\sigma\} = [C]\{\varepsilon\}. \tag{7.8.41}$$

Now suppose that the displacements (u_1, u_2) in the body Ω can be expressed (often, it is an approximation) as a linear combination of unknown values of the displacement vector at the vertices of the triangle and known functions of position $\psi_j(x_1, x_2)$

$$\begin{Bmatrix} u_1(x_1, x_2) \\ u_2(x_1, x_2) \end{Bmatrix} = \begin{Bmatrix} \sum_{j=1}^{3} u_1^j \, \psi_j(x_1, x_2) \\ \sum_{j=1}^{3} u_2^j \, \psi_j(x_1, x_2) \end{Bmatrix} \quad \text{or} \quad \{u\} = [\Psi]\{\Delta\}, \qquad (7.8.42)$$

where u_i^j denotes the value displacement component u_i at the jth vertex of the domain [see Figure 7.8.2(b)], and

$$[\Psi] = \begin{bmatrix} \psi_1 & 0 & \psi_2 & 0 & \psi_3 & 0 \\ 0 & \psi_1 & 0 & \psi_2 & 0 & \psi_3 \end{bmatrix} \quad (2 \times 6),$$

$$\{\Delta\} = \begin{Bmatrix} u_1^1 & u_2^1 & u_1^2 & u_2^2 & u_1^3 & u_2^3 \end{Bmatrix}^{\mathrm{T}} \quad (6 \times 1). \qquad (7.8.43)$$

Substituting (7.8.42) for $\{u\}$ into Eqs. (7.8.40) and (7.8.41), we obtain

$$\{\varepsilon\} = [D]\{u\} \Rightarrow \{\varepsilon\} = [B]\{\Delta\}; \quad \{\sigma\} = [C]\{\varepsilon\} \Rightarrow \{\sigma\} = [C][B]\{\Delta\}, \qquad (7.8.44)$$

where

$$[B] = [D][\Psi] = \begin{bmatrix} \frac{\partial \psi_1}{\partial x_1} & 0 & \frac{\partial \psi_2}{\partial x_1} & 0 & \frac{\partial \psi_3}{\partial x_1} & 0 \\ 0 & \frac{\partial \psi_1}{\partial x_2} & 0 & \frac{\partial \psi_2}{\partial x_2} & 0 & \frac{\partial \psi_3}{\partial x_2} \\ \frac{\partial \psi_1}{\partial x_2} & \frac{\partial \psi_1}{\partial x_1} & \frac{\partial \psi_2}{\partial x_2} & \frac{\partial \psi_2}{\partial x_1} & \frac{\partial \psi_3}{\partial x_2} & \frac{\partial \psi_3}{\partial x_1} \end{bmatrix} \quad (3 \times 6). \qquad (7.8.45)$$

Substituting these expressions into Eq. (7.8.39), we obtain

$$U = \frac{h}{2} \int_{\Omega} \{\Delta\}^{\mathrm{T}}[B]^{\mathrm{T}}[C][B]\{\Delta\}\, d\mathbf{x}, \quad V = -\oint_{\Gamma} \{\Delta\}^{\mathrm{T}}[\Psi]\{\hat{t}\}\, ds. \qquad (7.8.46)$$

Now applying Castigliano's Theorem II, we obtain

$$\frac{\partial U}{\partial \{\Delta\}} = -\frac{\partial V}{\partial \{\Delta\}} \Rightarrow [K]\{\Delta\} = \{F\}, \qquad (7.8.47)$$

where

$$[K] = h \int_{\Omega} [B]^{\mathrm{T}}[C][B]\, d\mathbf{x} \quad (6 \times 6), \quad \{F\} = \oint_{\Gamma} [\Psi]^{\mathrm{T}}\{\hat{t}\}\, ds \quad (6 \times 1). \qquad (7.8.48)$$

Equation (7.8.47) provides the necessary algebraic equations to solve for the unknown displacement components. However, the matrix $[K]$, known as the *stiffness matrix*, is singular to begin with. After imposing the necessary boundary conditions to eliminate the rigid body translations and the rotation, the condensed matrix becomes nonsingular. Equation (7.8.47) is not exact unless the representation in Eq. (7.8.42) is exact, which is most often not the case. The procedure described in Eqs. (7.8.42)–(7.8.48) is nothing but the finite element development for a typical domain Ω. This particular triangular element is known as the *constant strain triangle*, because the functions ψ_i are linear in x_1 and x_2 for a triangle with three vertex points where the displacement degrees of freedom are identified. Consequently, the strains are constant within the domain Ω. For more details, the reader may consult a finite element book [e.g., see Reddy (2006)].

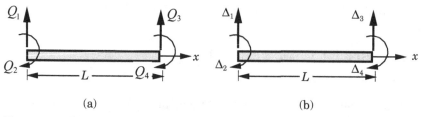

Figure 7.8.3. (a) Beam with end forces and moments (or generalized forces). (b) Generalized displacements.

EXAMPLE 7.8.2: To further illustrate the use of Castigliano's Theorem I, consider a straight beam of length L and constant bending stiffness EI (modulus E and second moment of area I about bending axis y) and subjected to point loads and moments, as shown in Figure 7.8.3(a). The equilibrium equation of the beam according to the Euler–Bernoulli beam theory (see Example 7.8.1) is

$$EI\frac{d^4w}{dx^4} = 0. \tag{7.8.49}$$

The exact solution to this fourth-order equation is

$$w_0(x) = c_1 + c_2 x + c_3 x^2 + c_4 x^3, \tag{7.8.50}$$

where c_i $(i = 1, 2, 3, 4)$ are constants of integration, which we express in terms of the deflections and rotations at the two ends of the beam. Let

$$\Delta_1 \equiv w(0) = c_1,$$

$$\Delta_2 \equiv \left(-\frac{dw}{dx}\right)_{x=0} = -c_2,$$

$$\Delta_3 \equiv w(L) = c_1 + c_2 L + c_3 L^2 + c_4 L^3, \tag{7.8.51}$$

$$\Delta_4 \equiv \left(-\frac{dw}{dx}\right)_{x=L} = -c_2 - 2c_3 L - 3c_4 L^2.$$

Clearly, Δ_1 and Δ_3 are the values of the transverse deflection w at $x = 0$ and $x = L$, respectively, and Δ_2 and Δ_4 are the rotations $-dw/dx$, measured positive clockwise, at $x = 0$ and $x = L$, respectively [see Figure 7.8.3(b)].

The reason for picking two deflection values and two rotations, as opposed four deflections at four points of the beam needs to be understood. From Example 7.8.1, it is clear that both w and dw/dx are the primary (kinematic) variables, which must be continuous at every point of the beam. If we were to join two such beams (possibly made of different bending stiffness EI), the kinematic variables can be made continuous by equating the like degrees of freedom at the common node.

The four equations in Eq. (7.8.51) can be solved for c_i in terms of Δ_i, called *generalized displacements*, which will serve as the generalized coordinates for the application of Castigliano's Theorem I. Then substituting the result into Eq. (7.8.50) yields

$$w(x) = \phi_1(x)\Delta_1 + \phi_2(x)\Delta_2 + \phi_3(x)\Delta_3 + \phi_4(x)\Delta_4 = \sum_{i=1}^{4} \phi_i(x)\Delta_i, \quad (7.8.52)$$

where $\phi_i(x)$ are the Hermite cubic polynomials

$$\phi_1(x) = 1 - 3\left(\frac{x}{L}\right)^2 + 2\left(\frac{x}{L}\right)^3,$$

$$\phi_2(x) = -x\left[1 - 2\left(\frac{x}{L}\right) + \left(\frac{x}{L}\right)^2\right],$$

$$\phi_3(x) = \left(\frac{x}{L}\right)^2\left(3 - 2\frac{x}{L}\right), \quad (7.8.53)$$

$$\phi_4(x) = x\frac{x}{L}\left(1 - \frac{x}{L}\right).$$

Equation (7.8.52) is analogous to Eq. (7.8.35) used (but not derived; the derivation can be found in any book on finite element analysis) in the plane elasticity problem discussed before.

The strain energy of the beam [see Eq. (7.8.25)] now can be expressed in terms of the generalized coordinates Δ_i $(i = 1, 2, 3, 4)$ as

$$U = \frac{EI}{2}\int_0^L \left(\frac{d^2w}{dx^2}\right)^2 dx = \frac{EI}{2}\int_0^L \left(\sum_{i=1}^{4} \Delta_i \frac{d^2\phi_i}{dx^2}\right)\left(\sum_{j=1}^{4} \Delta_j \frac{d^2\phi_j}{dx^2}\right) dx$$

$$= \frac{1}{2}\sum_{i=1}^{4}\sum_{j=1}^{4} K_{ij}\Delta_i\Delta_j, \quad (7.8.54)$$

where

$$K_{ij} = EI \int_0^L \frac{d^2\phi_i}{dx^2}\frac{d^2\phi_j}{dx^2}\, dx. \quad (7.8.55)$$

The stiffness coefficients K_{ij} are symmetric $(K_{ij} = K_{ji})$. By carrying out the indicated integration, K_{ij} can be evaluated [see Eq. (7.8.59)].

The work done by applied forces (q is taken as acting downward) is given by

$$V = -\left[-\int_0^L q(x)w(x)dx + \sum_{i=1}^{4} Q_i u_i\right]$$

$$= -\sum_{i=1}^{4}(q_i u_i + Q_i u_i), \quad (7.8.56)$$

where

$$q_i = -\int_0^L q(x)\phi_i(x)\, dx, \quad (7.8.57)$$

and Q_i are the generalized forces associated with the generalized displacements Δ_i. Thus, Q_1 and Q_3 are the transverse forces at $x = 0$ and $x = L$, respectively, and Q_2 and Q_4 are the bending moments at $x = 0$ and $x = L$, respectively, as shown in Figure 7.8.3(a). The transverse forces q_1 and q_3 and bending moments q_2 and q_4 together are statically equivalent (i.e., satisfy the force and moment equilibrium conditions of the beam) to the distributed load $q(x)$ on the beam. Using the Castigliano's Theorem I, we obtain

$$\frac{\partial U}{\partial \Delta_i} = -\frac{\partial V}{\partial \Delta_i} \Rightarrow \sum_{j=1}^{4} K_{ij}\Delta_j = Q_i + q_i \tag{7.8.58}$$

or, in explicit matrix form,

$$\frac{2EI}{L^3}\begin{bmatrix} 6 & -3L & -6 & -3L \\ -3L & 2L^2 & 3L & L^2 \\ -6 & 3L & 6 & 3L \\ -3L & L^2 & 3L & 2L^2 \end{bmatrix}\begin{Bmatrix} \Delta_1 \\ \Delta_2 \\ \Delta_3 \\ \Delta_4 \end{Bmatrix} = \begin{Bmatrix} q_1 \\ q_2 \\ q_3 \\ q_4 \end{Bmatrix} + \begin{Bmatrix} Q_1 \\ Q_2 \\ Q_3 \\ Q_4 \end{Bmatrix}. \tag{7.8.59}$$

It can be verified that the stiffness matrix $[K]$ is singular. For uniformly distributed load acting downward, $q(x) = q_0$, the load vector $\{q\}$ is given by

$$\begin{Bmatrix} q_1 \\ q_2 \\ q_3 \\ q_4 \end{Bmatrix} = -\frac{q_0 L}{12}\begin{Bmatrix} 6 \\ -L \\ 6 \\ L \end{Bmatrix}. \tag{7.8.60}$$

As a specific example, consider a beam fixed at $x = 0$, supported at $x = L$ by a linear elastic spring with spring constant k, subjected to uniformly distributed load of intensity q_0, and clockwise bending moment M_0 at $x = L$ (see Figure 7.8.4). We wish to determine the compression in the spring, that is, determine $w(L)$.

The geometric boundary conditions at $x = 0$ require $\Delta_1 = \Delta_2 = 0$. These conditions remove the rigid body modes of vertical translation and rotation about the y-axis. The force boundary conditions at $x = L$ require $Q_3 = -F_s = -kw(L) = -k\Delta_3$ and $Q_4 = M_0$. Thus, we have

$$\frac{2EI}{L^3}\begin{bmatrix} 6 & -3L & -6 & -3L \\ -3L & 2L^2 & 3L & L^2 \\ -6 & 3L & 6 & 3L \\ -3L & L^2 & 3L & 2L^2 \end{bmatrix}\begin{Bmatrix} 0 \\ 0 \\ u_3 \\ u_4 \end{Bmatrix} = -\frac{q_0 L}{12}\begin{Bmatrix} 6 \\ -L \\ 6 \\ L \end{Bmatrix} + \begin{Bmatrix} Q_1 \\ Q_2 \\ -k\Delta_3 \\ M_0 \end{Bmatrix}. \tag{7.8.61}$$

Thus, there are four equations in four unknowns, $(Q_1, Q_2, \Delta_3, \Delta_4)$. Since the last two equations have only the displacement unknowns Δ_3 and Δ_4, we can write

$$\begin{bmatrix} \frac{12EI}{L^3} + k & \frac{6EI}{L^2} \\ \frac{6EI}{L^2} & \frac{4EI}{L} \end{bmatrix}\begin{Bmatrix} \Delta_3 \\ \Delta_4 \end{Bmatrix} = -\frac{q_0 L}{12}\begin{Bmatrix} 6 \\ L \end{Bmatrix} + \begin{Bmatrix} 0 \\ M_0 \end{Bmatrix}. \tag{7.8.62}$$

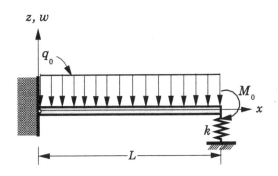

Figure 7.8.4. A beam fixed at $x = 0$ and supported by a spring at $x = L$.

Solving for $\Delta_3 = w(L)$ and $\Delta_4 = -(dw/dx)(L)$ by Cramer's rule, we obtain

$$\Delta_3 = -\left(q_0 L^2 + 4M_0\right) \frac{3L^2}{8\left(3EI + kL^3\right)} ,$$

$$\Delta_4 = \frac{q_0 L^3}{48EI} \frac{\left(24EI - kL^3\right)}{\left(3EI + kL^3\right)} + \frac{M_0 L}{4EI} \frac{\left(12EI + kL^3\right)}{\left(3EI + kL^3\right)} . \tag{7.8.63}$$

The solution obtained is exact because the representation in Eq. (7.8.52) is exact when EI is a constant.

7.9 Hamilton's Principle

7.9.1 Introduction

The principle of total potential energy discussed in the previous section can be generalized to dynamics of solid bodies, and it is known as Hamilton's principle. In Hamilton's principle, the system under consideration is assumed to be characterized by two energy functions: a *kinetic energy K* and a *potential energy* Π. For *discrete* systems (i.e., systems with a finite number of degrees of freedom), these energies can be described in terms of a finite number of generalized coordinates and their derivatives with respect to time t. For *continuous* systems (i.e., systems that cannot be described by a finite number of generalized coordinates), the energies can be expressed in terms of the dependent variables of the problem that are functions of position and time.

7.9.2 Hamilton's Principle for a Rigid Body

To gain a simple understanding of Hamilton's principle, consider a single particle or a rigid body (which is a collection of particles, distance between which is unaltered at all times) of mass m moving under the influence of a force $\mathbf{F} = \mathbf{F}(\mathbf{r})$ [see Reddy (2002)]. The path $\mathbf{r}(t)$ followed by the particle is related to the force $\mathbf{F}$ and mass m by the principle of conservation of linear momentum (i.e., Newton's second law of motion)

$$\mathbf{F}(\mathbf{r}) = m\frac{d^2\mathbf{r}}{dt^2}. \tag{7.9.1}$$

A path that differs from the actual path is expressed as $\mathbf{r} + \delta\mathbf{r}$, where $\delta\mathbf{r}$ is the variation of the path for any *fixed* time t. We suppose that the actual path $\mathbf{r}$ and the *varied* path differ except at two distinct times t_1 and t_2, that is, $\delta\mathbf{r}(t_1) = \delta\mathbf{r}(t_2) = \mathbf{0}$. Taking the scalar product of Eq. (7.9.1) with the variation $\delta\mathbf{r}$, and integrating with respect to time between t_1 and t_2, we obtain

$$\int_{t_1}^{t_2} \left[m\frac{d^2\mathbf{r}}{dt^2} - \mathbf{F}(\mathbf{r}) \right] \cdot \delta\mathbf{r}\, dt = 0. \tag{7.9.2}$$

Integration-by-parts of the first term in Eq. (7.9.2) yields

$$-\int_{t_1}^{t_2} \left(m\frac{d\mathbf{r}}{dt} \cdot \frac{d\delta\mathbf{r}}{dt} + \mathbf{F}(\mathbf{r}) \cdot \delta\mathbf{r} \right) dt + \left(m\frac{d\mathbf{r}}{dt} \cdot \delta\mathbf{r} \right)\Bigg|_{t_1}^{t_2} = 0. \tag{7.9.3}$$

The last term in Eq. (7.9.3) vanishes because $\delta\mathbf{r}(t_1) = \delta\mathbf{r}(t_2) = \mathbf{0}$. Also, note that

$$m\frac{d\mathbf{r}}{dt} \cdot \frac{d\delta\mathbf{r}}{dt} = \delta\left[\frac{m}{2}\frac{d\mathbf{r}}{dt} \cdot \frac{d\mathbf{r}}{dt} \right] \equiv \delta K, \tag{7.9.4}$$

where K is the kinetic energy of the particle or rigid body

$$K = \frac{1}{2}m\frac{d\mathbf{r}}{dt} \cdot \frac{d\mathbf{r}}{dt} = \frac{1}{2}m\mathbf{v} \cdot \mathbf{v}, \tag{7.9.5}$$

and δK is called the *virtual kinetic energy*. The expression $\mathbf{F}(\mathbf{r}) \cdot \delta\mathbf{r}$ is called the *virtual work done by external forces* and denoted by

$$\delta W_E = -\mathbf{F}(\mathbf{r}) \cdot \delta\mathbf{r}. \tag{7.9.6}$$

The minus sign indicates that the work is done by external force $\mathbf{F}$ *on* the body in moving through the displacement $\delta\mathbf{r}$. Equation (7.9.3) now takes the form

$$\int_{t_1}^{t_2} (\delta K - \delta W_E)dt = 0, \tag{7.9.7}$$

which is known as the *general form of Hamilton's principle* for a single particle or rigid body. Note that a particle or a rigid body has no strain energy Π because the distance between the particles is unaltered.

Suppose that the force $\mathbf{F}$ is conservative (that is, the sum of the potential and kinetic energies is conserved) such that it can be replaced by the gradient of a potential

$$\mathbf{F} = -\text{grad } V, \tag{7.9.8}$$

where $V = V(\mathbf{r})$ is the *potential energy due to the loads* on the body. Then Eq. (7.9.7) can be expressed in the form

$$\delta \int_{t_1}^{t_2} (K - V)\, dt = 0, \tag{7.9.9}$$

because $(\mathbf{r} = x_i\hat{\mathbf{e}}_i)$

$$\text{grad } V \cdot \delta\mathbf{r} = \frac{\partial V}{\partial x_i}\delta x_i = \delta V(\mathbf{x}).$$

The difference between the kinetic and potential energies is called the *Lagrangian* function

$$L \equiv K - V. \tag{7.9.10}$$

Equation (7.9.9) is known as Hamilton's principle for the conservative motion of a particle (or a rigid body). The principle can be stated as: *the motion of a particle acted on by conservative forces between two arbitrary instants of time t_1 and t_2 is such that the line integral over the Lagrangian function is an extremum for the path motion.* Stated in other words, of all possible paths that the particle could travel from its position at time t_1 to its position at time t_2, its actual path will be one for which the integral

$$I \equiv \int_{t_1}^{t_2} L \, dt \tag{7.9.11}$$

is an extremum (i.e., a minimum, maximum, or an inflection).

If the path $\mathbf{r}$ can be expressed in terms of the generalized coordinates $q_i (i = 1, 2, 3)$, the Lagrangian function can be written in terms of q_i and their time derivatives

$$L = L(q_1, q_2, q_3, \dot{q}_1, \dot{q}_2, \dot{q}_3). \tag{7.9.12}$$

Then the condition for the extremum of I in Eq. (7.9.11) results in the equation ($\delta q_i = 0$ at t_1 and t_2)

$$\delta I = \delta \int_{t_1}^{t_2} L(q_1, q_2, q_3, \dot{q}_1, \dot{q}_2, \dot{q}_3) dt = 0$$

$$= \int_{t_1}^{t_2} \sum_{i=1}^{3} \left[\frac{\partial L}{\partial q_i} - \frac{d}{dt} \left(\frac{\partial L}{\partial \dot{q}_i} \right) \right] \delta q_i \, dt. \tag{7.9.13}$$

When all q_i are linearly independent (i.e., no constraints among q_i), the variations δq_i are independent for all t, except $\delta q_i = 0$ at t_1 and t_2. Therefore, the coefficients of $\delta q_1, \delta q_2,$ and δq_3 vanish separately:

$$\frac{\partial L}{\partial q_i} - \frac{d}{dt} \left(\frac{\partial L}{\partial \dot{q}_i} \right) = 0, \quad i = 1, 2, 3. \tag{7.9.14}$$

These equations are called the *Lagrange equations of motion*. Recall that in Section 7.8 (for a static case) these equations were also called the Euler equations. For the dynamic case, these equations will be called the *Euler–Lagrange equations*.

When the forces are not conservative, we must deal with the general form of Hamilton's principle in Eq. (7.9.7). In this case, there exists no functional I that must be an extremum. If the virtual work can be expressed in terms of the generalized coordinates q_i by

$$\delta W_E = -(F_1 \delta q_1 + F_2 \delta q_2 + F_3 \delta q_3), \tag{7.9.15}$$

where F_i are the *generalized forces*, then we can write Eq. (7.9.13) as

$$\int_{t_1}^{t_2} \sum_{i=1}^{3} \left[\frac{\partial K}{\partial q_i} - \frac{d}{dt}\left(\frac{\partial K}{\partial \dot{q}_i}\right) + F_i \right] \delta q_i \, dt = 0, \tag{7.9.16}$$

and the Euler–Lagrange equations for the nonconservative forces are given by

$$\delta q_i : \quad \frac{\partial K}{\partial q_i} - \frac{d}{dt}\left(\frac{\partial K}{\partial \dot{q}_i}\right) + F_i = 0, \quad i = 1, 2, 3. \tag{7.9.17}$$

EXAMPLE 7.9.1: Consider the planar motion of a pendulum that consists of a mass m attached at the end of a rigid massless rod of length L that pivots about a fixed point O, as shown in Figure 7.9.1. Determine the equation of motion.

SOLUTION: The position of the mass can be expressed in terms of the generalized coordinates $q_1 = l$ and $q_2 = \theta$, measured from the vertical position. Since l is a constant, we have $\dot{q}_1 = 0$ and θ is the only independent generalized coordinate. The force **F** acting on the mass m is the component of the gravitational force,

$$\mathbf{F} = mg\,(\cos\theta\,\hat{\mathbf{e}}_r - \sin\theta\,\hat{\mathbf{e}}_\theta) \equiv F_r\hat{\mathbf{e}}_r + F_\theta\hat{\mathbf{e}}_\theta. \tag{7.9.18}$$

The component along $\hat{\mathbf{e}}_r$ does no work because $q_1 = l$ is a constant. The second component, F_θ, is derivable from the potential $(\nabla V = -F_\theta\hat{\mathbf{e}}_\theta)$

$$V = -[-mgl(1 - \cos\theta)] = mgl(1 - \cos\theta), \tag{7.9.19}$$

where V represents the potential energy of the mass m at any instant of time with respect to the static equilibrium position $\theta = 0$, and ∇ is the gradient operator in the polar coordinate system

$$\nabla = \hat{\mathbf{e}}_r\frac{\partial}{\partial r} + \frac{\hat{\mathbf{e}}_\theta}{r}\frac{\partial}{\partial\theta}. \tag{7.9.20}$$

Thus, the kinetic energy and the potential energy due to external load are given by

$$K = \frac{m}{2}(l\dot{\theta})^2, \quad V = mgl(1 - \cos\theta),$$

$$\delta K = ml^2\dot{\theta}\,\delta\dot{\theta}, \quad \delta V = mgl\sin\theta\,\delta\theta = -F_\theta(l\,\delta\theta). \tag{7.9.21}$$

Therefore, the Lagrangian function L is a function of θ and $\dot{\theta}$. The Euler–Lagrange equation is given by

$$\delta q_2 = \delta\theta : \quad \frac{\partial L}{\partial\theta} - \frac{d}{dt}\left(\frac{\partial L}{\partial\dot{\theta}}\right) = 0,$$

which yields

$$-mgl\sin\theta - \frac{d}{dt}(ml^2\dot{\theta}) = 0 \quad \text{or} \quad \ddot{\theta} + \frac{g}{l}\sin\theta = 0 \quad (F_\theta = ml\ddot{\theta}). \tag{7.9.22}$$

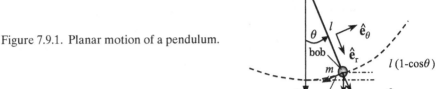

Figure 7.9.1. Planar motion of a pendulum.

Equation (7.9.22) represents a second-order nonlinear differential equation governing θ. For small angular motions, Eq. (7.9.22) can be linearized by replacing $\sin \theta \approx \theta$:

$$\ddot{\theta} + \frac{g}{\ell}\theta = 0. \tag{7.9.23}$$

Now suppose that the mass experiences a resistance force $\mathbf{F}^*$ proportional to its speed (e.g., the mass m is suspended in a medium with viscosity μ). According to Stoke's law,

$$\mathbf{F}^* = -6\pi \mu a l \dot{\theta} \hat{\mathbf{e}}_\theta, \tag{7.9.24}$$

where μ is the viscosity of the surrounding medium, a is the radius of the bob, and $\hat{\mathbf{e}}_\theta$ is the unit vector tangential to the circular path. The resistance of the massless rod supporting the bob is neglected. The force $\mathbf{F}^*$ is not derivable from a potential function (i.e., nonconservative). Thus, we have one part of the force (i.e., gravitational force) conservative and the other (i.e., viscous force) nonconservative. Hence, we use Hamilton's principle expressed by Eq. (7.9.14) or Eq. (7.9.17) with

$$\delta W_E = \delta V - \mathbf{F}^* \cdot (l\delta\theta\hat{\mathbf{e}}_\theta) = \left(mgl \sin \theta + 6\pi \mu a l^2 \dot{\theta}\right) \delta\theta \equiv -F_\theta l \delta\theta.$$

Then the equation of motion is given by $[K = K(\dot{\theta})]$

$$-\frac{d}{dt}\left(\frac{\partial K}{\partial \dot{\theta}}\right) + F_\theta l = 0 \quad \text{or} \quad \ddot{\theta} + \frac{g}{l}\sin \theta + \frac{6\pi a \mu}{m}\dot{\theta} = 0. \tag{7.9.25}$$

The coefficient $c = 6\pi a \mu / m$ is called the *damping* coefficient.

7.9.3 Hamilton's Principle for a Continuum

Hamilton's principle for a continuous body $\mathcal{B}$ occupying configuration κ with volume Ω with boundary Γ can be derived following essentially the same ideas as discussed for a particle or a rigid body. In contrast to a rigid body, a continuum is characterized by the kinetic energy K as well the strain (or internal) energy U. Newton's second law of motion for a continuous body can be written in general terms as

$$\mathbf{F} - m\mathbf{a} = \mathbf{0}, \tag{7.9.26}$$

where m is the mass, $\mathbf{a}$ the acceleration vector, and $\mathbf{F}$ is the resultant of *all* forces acting on the body $\mathcal{B}$. The actual path $\mathbf{u} = \mathbf{u}(\mathbf{x}, t)$ followed by a material particle in position $\mathbf{x}$ in the body is varied, consistent with kinematic (essential) boundary conditions on Γ, to $\mathbf{u} + \delta\mathbf{u}$, where $\delta\mathbf{u}$ is the admissible variation (or virtual displacement) of the path. We assume that the varied path differs from the actual path except at initial and final times, t_1 and t_2, respectively. Thus, an admissible variation $\delta\mathbf{u}$ satisfies the conditions,

$$\delta\mathbf{u} = \mathbf{0} \text{ on } \Gamma_u \text{ for all } t,$$
$$\delta\mathbf{u}(\mathbf{x}, t_1) = \delta\mathbf{u}(\mathbf{x}, t_2) = \mathbf{0} \text{ for all } \mathbf{x},$$

$$(7.9.27)$$

where Γ_u denotes the portion of the boundary Γ of the body where the displacement vector $\mathbf{u}$ is specified.

The work done on the body $\mathcal{B}$ at time t by the resultant force $\mathbf{F}$, which consists of body force $\mathbf{f}$ and specified surface traction $\hat{\mathbf{t}}$ in moving through respective virtual displacements $\delta\mathbf{u}$ is given by

$$\int_\Omega \mathbf{f} \cdot \delta\mathbf{u} \, dx + \int_{\Gamma_\sigma} \hat{\mathbf{t}} \cdot \delta\mathbf{u} \, ds - \int_\Omega \sigma : \delta\varepsilon \, dx, \qquad (7.9.28)$$

where σ and ε are the stress and strain tensors, and Γ_σ is the portion of the boundary Γ on which tractions are specified ($\Gamma = \Gamma_u \cup \Gamma_\sigma$). The last term in Eq. (7.9.28) is known as the *virtual work stored in the body* $\mathcal{B}$ due to deformation. The strains $\delta\varepsilon$ are assumed to be compatible in the sense that the strain-displacement relations (7.2.1) are satisfied. The work done by the inertia force $m\mathbf{a}$ in moving through the virtual displacement $\delta\mathbf{u}$ is given by

$$\int_\Omega \rho \frac{\partial^2 \mathbf{u}}{\partial t^2} \cdot \delta\mathbf{u} \, dx, \qquad (7.9.29)$$

where ρ is the mass density of the medium. We have, analogous to Eq. (7.9.2) for a rigid body, the result

$$\int_{t_1}^{t_2} \left\{ \int_\Omega \rho \frac{\partial^2 \mathbf{u}}{\partial t^2} \cdot \delta\mathbf{u} \, dx - \left[\int_\Omega (\mathbf{f} \cdot \delta\mathbf{u} - \sigma : \delta\varepsilon) \, dx + \int_{\Gamma_\sigma} \hat{\mathbf{t}} \cdot \delta\mathbf{u} \, ds \right] \right\} dt = 0$$

or

$$-\int_{t_1}^{t_2} \left[\int_\Omega \rho \frac{\partial \mathbf{u}}{\partial t} \cdot \frac{\partial \delta\mathbf{u}}{\partial t} \, dx + \int_\Omega (\mathbf{f} \cdot \delta\mathbf{u} - \sigma : \delta\varepsilon) \, dx + \int_{\Gamma_\sigma} \hat{\mathbf{t}} \cdot \delta\mathbf{u} \, ds \right] dt = 0. \quad (7.9.30)$$

In arriving at the expression in Eq. (7.9.30), integration-by-parts is used on the first term; the integrated terms vanish because of the initial and final conditions in Eq. (7.9.27). Equation (7.9.30) is known as the general form of Hamilton's principle for a continuous medium – conservative or not and elastic or not.

For an ideal elastic body, we recall from the previous sections that the forces $\mathbf{f}$ and $\mathbf{t}$ are conservative,

$$\delta V = -\left(\int_\Omega \mathbf{f} \cdot \delta\mathbf{u} \, dx + \int_{\Gamma_\sigma} \hat{\mathbf{t}} \cdot \delta\mathbf{u} \, ds \right), \qquad (7.9.31)$$

and that there exists a strain energy density function $U_0 = U_0(\varepsilon)$ such that

$$\sigma = \frac{\partial U_0}{\partial \varepsilon}. \tag{7.9.32}$$

Substituting Eqs. (7.9.31) and (7.9.32) into Eq. (7.9.30), we obtain

$$\delta \int_{t_1}^{t_2} [K - (V + U)]dt = 0, \tag{7.9.33}$$

where K and U are the kinetic and strain energies:

$$K = \int_{\Omega} \frac{\rho}{2} \frac{\partial \mathbf{u}}{\partial t} \cdot \frac{\partial \mathbf{u}}{\partial t} \, d\mathbf{x}, \quad U = \int_{\Omega} U_0 \, d\mathbf{x}. \tag{7.9.34}$$

Equation (7.9.33) represents Hamilton's principle for an elastic body. Recall that the sum of the strain energy and potential energy of external forces, $U + V$, is called the *total potential energy*, Π, of the body. For bodies involving no motion (i.e., forces are applied sufficiently slowly such that the motion is independent of time and the inertia forces are negligible), Hamilton's principle (7.9.33) reduces to the *principle of virtual displacements*. Equation (7.9.33) may be viewed as the dynamics version of the principle of virtual displacements.

The Euler–Lagrange equations associated with the Lagrangian, $L = K - \Pi$, can be obtained from Eq. (7.9.33):

$$0 = \delta \int_{t_1}^{t_2} L(\mathbf{u}, \nabla \mathbf{u}, \dot{\mathbf{u}}) \, dt$$

$$= \int_{t_1}^{t_2} \left[\int_{\Omega} \left(\rho \frac{\partial^2 \mathbf{u}}{\partial t^2} - \operatorname{div} \sigma - \mathbf{f} \right) \cdot \delta \mathbf{u} \, d\mathbf{x} + \int_{\Gamma_\sigma} (\mathbf{t} - \hat{\mathbf{t}}) \cdot \delta \mathbf{u} \, ds \right] dt, \tag{7.9.35}$$

where integration-by-parts, gradient theorems, and Eqs. (7.9.27) were used in arriving at Eq. (7.9.35) from Eq. (7.9.33). Because $\delta \mathbf{u}$ is arbitrary for t, $t_1 < t < t_2$, and for $\mathbf{x}$ in Ω and also on Γ_σ, it follows that

$$\rho \frac{\partial^2 \mathbf{u}}{\partial t^2} - \operatorname{div} \sigma - \mathbf{f} = \mathbf{0} \quad \text{in } \Omega, \tag{7.9.36}$$

$$\mathbf{t} - \hat{\mathbf{t}} = \mathbf{0} \quad \text{on } \Gamma_\sigma.$$

Equations (7.9.36) are the Euler–Lagrange equations for an elastic body.

EXAMPLE 7.9.2: The displacement field for pure bending of the Euler–Bernoulli beam theory is

$$u_1 = -z \frac{\partial w}{\partial x}, \quad u_2 = 0, \quad u_3 = w(x, t). \tag{7.9.37}$$

The Lagrange function associated with the dynamics of the Euler–Bernoulli beam is given by $L = K - (U + V)$, where

$$K = \int_0^L \int_A \left[\frac{\rho}{2} \left(-z \frac{\partial^2 w}{\partial x \partial t} \right)^2 + \frac{\rho}{2} \left(\frac{\partial w}{\partial t} \right)^2 \right] dA \, dx$$

$$= \int_0^L \left[\frac{\rho I}{2} \left(\frac{\partial^2 w}{\partial x \partial t} \right)^2 + \frac{\rho A}{2} \left(\frac{\partial w}{\partial t} \right)^2 \right] dx,$$

$$U = \int_0^L \int_A \frac{E}{2} \left(-z \frac{\partial^2 w}{\partial x^2} \right)^2 dA \, dx \tag{7.9.38}$$

$$= \int_0^L \frac{EI}{2} \left(\frac{\partial^2 w}{\partial x^2} \right)^2 dx,$$

$$V = - \int_0^L q(x, t) w \, dx.$$

Here w denotes the transverse displacement, which is a function of x and t, and q is the transverse distributed load. In arriving at the expressions for K and U, we have used the fact that the x-axis coincides with the geometric centroidal axis, $\int_A z \, dA = 0$.

The Hamilton principle gives

$$0 = \int_0^T (\delta K - \delta U - \delta V) \, dt$$

$$= \int_0^T \int_0^L \left[\rho I \frac{\partial \dot{w}}{\partial x} \frac{\partial \delta \dot{w}}{\partial x} + \rho A \dot{w} \delta \dot{w} - EI \frac{\partial^2 w}{\partial x^2} \frac{\partial^2 \delta w}{\partial x^2} + q \delta w \right] dx \, dt. \tag{7.9.39}$$

The Euler–Lagrange equation obtained from the above statement is the equation of motion governing the Euler–Bernoulli beam theory

$$\frac{\partial^2}{\partial x \partial t} \left(\rho I \frac{\partial^2 w}{\partial x \partial t} \right) - \frac{\partial}{\partial t} \left(\rho A \frac{\partial w}{\partial t} \right) - \frac{\partial^2}{\partial x^2} \left(EI \frac{\partial^2 w}{\partial x^2} \right) + q = 0. \tag{7.9.40}$$

The first term is known as the contribution due to *rotary inertia*.

Now suppose that the beam experiences two types of viscous (velocity-dependent) damping: (1) viscous resistance to transverse displacement of the beam and (2) a viscous resistance to straining of the beam material. If the resistance to transverse velocity is denoted by $c(x)$, the corresponding damping force is given by $q_D(x, t) = c(x) \dot{w}_0$. If the resistance to strain velocity is c_s, the damping stress is $\sigma_{xx}^D = c_s \dot{\varepsilon}_{xx}$. We wish to derive the equations of motion of the beam with both types of damping.

We must add the following terms due to damping to the expression in Eq. (7.9.39):

$$-\int_0^T \left[\int_\Omega \sigma_D \delta\varepsilon \, dx + \int_0^L q_D \delta w \, dx \right] dt$$

$$= -\int_0^T \left[\int_0^L \int_A c_s \left(-z\frac{\partial^3 w}{\partial x^2 \partial t} \right)\left(-z\frac{\partial^2 \delta w}{\partial x^2} \right) dA dx + \int_0^L q_D \delta w \, dx \right] dt$$

$$= -\int_0^T \int_0^L \left(Ic_s \frac{\partial^3 w}{\partial x^2 \partial t}\frac{\partial^2 \delta w}{\partial x^2} + c\frac{\partial w}{\partial t}\delta w \right) dx dt. \tag{7.9.41}$$

The Euler–Lagrange equations of the statement

$$0 = \int_0^T \int_0^L \left[\rho I \frac{\partial \dot{w}}{\partial x}\frac{\partial \delta\dot{w}}{\partial x} + \rho A\dot{w}\delta\dot{w} - EI\frac{\partial^2 w}{\partial x^2}\frac{\partial^2 \delta w}{\partial x^2} + q\delta w \right] dx dt$$

$$-\int_0^T \int_0^L \left(Ic_s \frac{\partial^3 w}{\partial x^2 \partial t}\frac{\partial^2 \delta w}{\partial x^2} + c\frac{\partial w}{\partial t}\delta w \right) dx dt \tag{7.9.42}$$

are

$$\frac{\partial^2}{\partial x \partial t}\left(\rho I \frac{\partial^2 w}{\partial x \partial t} \right) - \frac{\partial}{\partial t}\left(\rho A \frac{\partial w}{\partial t} \right) - \frac{\partial^2}{\partial x^2}\left(EI\frac{\partial^2 w}{\partial x^2} \right)$$

$$- \frac{\partial^2}{\partial x^2}\left(Ic_s \frac{\partial^3 w}{\partial x^2 \partial t} \right) - c\frac{\partial w}{\partial t} + q = 0. \tag{7.9.43}$$

7.10 Summary

This is a very comprehensive chapter on linearized elasticity. Beginning with a summary of the linearized elasticity equations that include the Navier equations and the Beltrami–Michell equations of elasticity, types of boundary value problems and principle of superposition were discussed. The Clapeyron theorem and Betti and Maxwell reciprocity theorems and their applications were also presented. Analytical solutions of a number of examples of standard boundary-value problems of elasticity using the Airy stress function are developed. Then, the principle of minimum total potential energy and its derivative the Castigliano Theorem I are discussed. Last, Hamilton's principle for problems of dynamics is presented.

PROBLEMS

7.1 An isotropic body ($E = 210$ GPa and $v = 0.3$) with two-dimensional state of stress experiences the following displacement field (in mm)

$$u_1 = 3x_1^2 - x_1^3 x_2 + 2x_2^3, \quad u_2 = x_1^3 + 2x_1 x_2,$$

where x_i are in meters. Determine the stresses and rotation of the body at point $(x_1, x_2) = (0.05, 0.02)$ m. Is the displacement field compatible (pulling your legs)?

7.2 A two-dimensional state of stress exists in a body with the following components of stress:

$$\sigma_{11} = c_1 x_2^3 + c_2 x_1^2 x_2 - c_3 x_1, \quad \sigma_{22} = c_4 x_2^3 - c_5, \quad \sigma_{12} = c_6 x_1 x_2^2 + c_7 x_1^2 x_2 - c_8,$$

where c_i are constants. Assuming that the body forces are zero, determine the conditions on the constants so that the stress field is in equilibrium and satisfies the compatibility equations.

7.3 For the plane elasticity problems shown in Figs. P7.3(a)–(d), write the boundary conditions and classify them into Type I, Type II, or Type III.

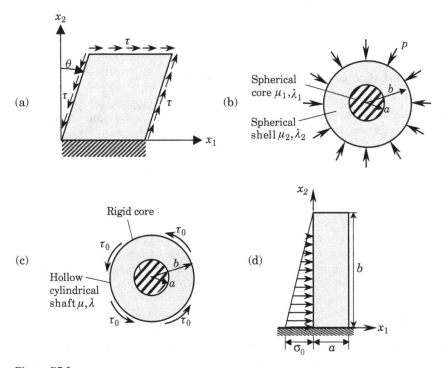

Figure P7.3.

7.4 Determine the deflection at the midspan of a cantilever beam subjected to uniformly distributed load q_0 throughout the span and a point load F_0 at the free end. Use Maxwell's theorem and superposition.

7.5 Consider a simply supported beam of length L subjected to a concentrated load F_B at the midspan and a bending moment M_A at the left end, as shown in Figure P7.5. Verify that Betti's theorem holds.

Figure P7.5.

7.6 A load $P = 4,000$ lb acting at a point A of a beam produces 0.25 in at point B and 0.75 in at point C of the beam. Find the deflection of point A produced by loads 4,500 lb and 2,000 lb acting at points B and C, respectively.

7.7 Use the reciprocity theorem to determine the deflection at the center of a simply supported circular plate under asymmetric loading (see Figure 7.6.8)

$$q(r, \theta) = q_0 + q_1 \frac{r}{a} \cos \theta.$$

The deflection due to a point load Q_0 at the center of a simply supported circular plate is

$$w(r) = \frac{Q_0 a^2}{16\pi D} \left[\left(\frac{3+\nu}{1+\nu} \right) \left(1 - \frac{r^2}{a^2} \right) + 2 \left(\frac{r}{a} \right)^2 \log \left(\frac{r}{a} \right) \right],$$

where $D = Eh^3/[12(1 - \nu^2)]$ and h is the plate thickness.

7.8 Use the reciprocity theorem to determine the center deflection of a simply supported circular plate under hydrostatic loading $q(r) = q_0(1 - r/a)$. See Problem 7.7.

7.9 Use the reciprocity theorem to determine the center deflection of a clamped circular plate under hydrostatic loading $q(r) = q_0(1 - r/a)$. The deflection due to a point load Q_0 at the center of a clamped circular plate is given in Eq. (7.6.21).

7.10 Determine the center deflection of a clamped circular plate subjected to a point load Q_0 at a distance b from the center (and for some θ) using the reciprocity theorem.

7.11 The lateral surface of a homogeneous, isotropic, solid circular cylinder of radius a, length L, and mass density ρ is bonded to a rigid surface. Assuming that the ends of the cylinder at $z = 0$ and $z = L$ are traction-free (see Figure P7.11), determine the displacement and stress fields in the cylinder due to its own weight.

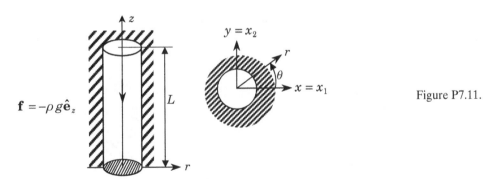

Figure P7.11.

7.12 A solid circular cylindrical body of radius a and height h is placed between two rigid plates, as shown in Figure P7.12. The plate at B is held stationary and the plate at A is subjected to a downward displacement of δ. Using a suitable coordinate system, write the boundary conditions for the following two cases:

(a) When the cylindrical object is bonded to the plates at A and B.

(b) When the plates at A and B are frictionless.

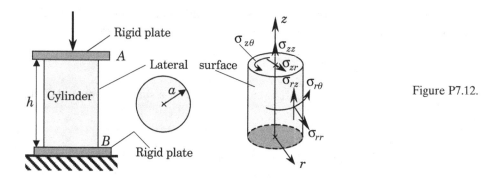

Figure P7.12.

7.13 An external hydrostatic pressure of magnitude p is applied to the surface of a spherical body of radius b with a concentric rigid spherical inclusion of radius a, as shown in Figure P7.13. Determine the displacement and stress fields in the spherical body.

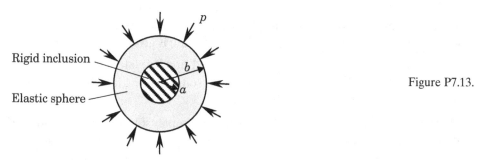

Rigid inclusion

Elastic sphere

Figure P7.13.

7.14 Reconsider the concentric spheres of Problem 7.13. As opposed to the rigid core in Problem 7.13, suppose that the core is elastic and the outer shell is subjected to external pressure p (both are linearly elastic). Assuming Lamé constants of μ_1 and λ_1 for the core and μ_2 and λ_2 for the outer shell (see Figure P7.14), and that the interface is perfectly bonded at $r = a$, determine the displacements of the core as well as for the shell.

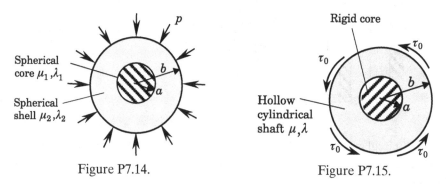

Spherical core μ_1, λ_1

Spherical shell μ_2, λ_2

Figure P7.14.

Rigid core

Hollow cylindrical shaft μ, λ

Figure P7.15.

7.15 Consider a long hollow circular shaft with a rigid internal core (a cross section of the shaft is shown in Figure P7.15). Assuming that the inner surface of the shaft at $r = a$ is perfectly bonded to the rigid core and the outer boundary at $r = b$ is subjected to a uniform *shearing* traction of magnitude τ_0, find the displacement and stress fields in the problem.

7.16 Interpret the stress field obtained with the Airy stress function in Eq. (7.7.43) when all constants except c_3 are zero. Use Figure 7.7.4 to sketch the stress field.

7.17 Interpret the following stress field obtained in Case 3 of Example 7.7.1 using Figure 7.7.4:

$$\sigma_{xx} = 6c_{10}xy, \quad \sigma_{yy} = 0, \quad \sigma_{xy} = -3c_{10}y^2.$$

Assume that c_{10} is a positive constant.

7.18 Compute the stress field associated with the Airy stress function

$$\Phi(x, y) = Ax^5 + Bx^4y + Cx^3y^2 + Dx^2y^3 + Exy^4 + Fy^5.$$

Interpret the stress field for the case in which all constants except D are zero. Use Figure 7.7.4 to sketch the stress field.

7.19 Investigate what problem is solved by the Airy stress function

$$\Phi = \frac{3A}{4b}\left(xy - \frac{xy^3}{3b^2}\right) + \frac{B}{4b}y^2.$$

7.20 Show that the Airy stress function

$$\Phi(x, y) = \frac{q_0}{8b^3}\left[x^2\left(y^3 - 3b^2y + 2b^3\right) - \frac{1}{5}y^3\left(y^2 - 2b^2\right)\right]$$

satisfies the compatibility condition. Determine the stress field and find what problem it corresponds to when applied to the region $-b \le y \le b$ and $x = 0, a$ (see Figure P7.20).

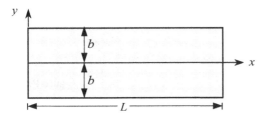

Figure P7.20.

7.21 Determine the Airy stress function for the stress field of the domain shown in Figure P7.21 and evaluate the stress field.

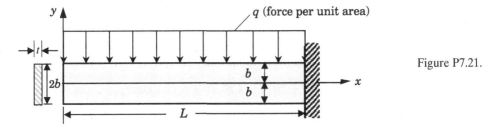

Figure P7.21.

7.22 The thin cantilever beam shown in Figure P7.22 is subjected to a uniform shearing traction of magnitude τ_0 along its upper surface. Determine whether the Airy stress function

$$\Phi(x, y) = \frac{\tau_0}{4}\left(xy - \frac{xy^2}{b} - \frac{xy^3}{b^2} + \frac{ay^2}{b} + \frac{ay^3}{b^2}\right)$$

satisfies the compatibility condition and stress boundary conditions of the problem.

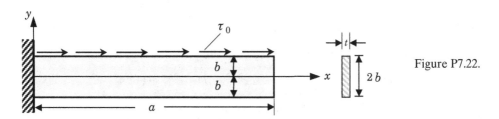

Figure P7.22.

7.23 The curved beam shown in Figure P7.23 is curved along a circular arc. The beam is fixed at the upper end, and it is subjected at the lower end to a distribution of tractions statically equivalent to a force per unit thickness $\mathbf{P} = -P\hat{\mathbf{e}}_1$. Assume that the beam is in a state of plane strain/stress. Show that an Airy stress function of the form

$$\Phi(r) = \left(Ar^3 + \frac{B}{r} + Cr\log r\right)\sin\theta$$

provides an approximate solution to this problem and solve for the values of the constants A, B, and C.

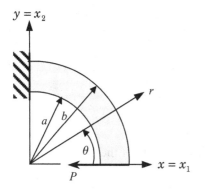

Figure P7.23.

7.24 Determine the stress field in a semi-infinite plate due to a normal load, f_0 force/unit length, acting on its edge, as shown in Figure P7.24. Use the following Airy stress function (that satisfies the compatibility condition $\nabla^4\Phi = 0$)

$$\Phi(r, \theta) = A\theta + Br^2\theta + Cr\theta\sin\theta + Dr\theta\cos\theta,$$

where A, B, C, and D are constants [see Eq. (7.7.39) for the definition of stress components in terms of the Airy stress function Φ]. Neglect the body forces (i.e., $V_f = 0$). *Hint:* Stresses must be single-valued. Determine the constants using the boundary conditions of the problem.

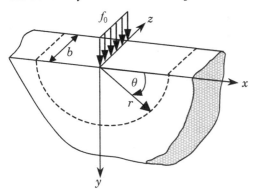

Figure P7.24.

7.25 Consider a cylindrical member with the equilateral triangular cross section shown in Figure P7.25. The equations of various sides of the triangle are

$$\text{side 1:}\quad x_1 - \sqrt{3}x_2 + 2a = 0,$$

$$\text{side 2:}\quad x_1 + \sqrt{3}x_2 + 2a = 0,$$

$$\text{side 3:}\quad x_1 - a = 0.$$

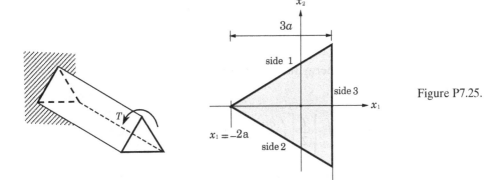

Figure P7.25.

Show that the exact solution for the problem can be obtained and that the twist per unit length θ and stresses σ_{31} and σ_{32} are given by

$$\theta = \frac{5\sqrt{3}T}{27\mu a^4}, \quad \sigma_{31} = \frac{\mu\theta}{a}(x_1 - a)x_2, \quad \sigma_{32} = \frac{\mu\theta}{2a}(x_1^2 + 2ax_1 - x_2^2).$$

7.26 Consider torsion of a cylindrical member with the rectangular cross section shown in Figure P7.26. Determine if a function of the form

$$\Psi = A\left(\frac{x_1^2}{a^2} - 1\right)\left(\frac{x_2^2}{b^2} - 1\right),$$

where A is a constant, can be used as a Prandtl stress function.

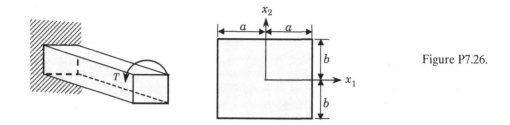

Figure P7.26.

7.27 Use the principle of minimum total potential energy to derive the Euler equations associated with the *Timoshenko beam theory*, which is based on the displacement field

$$u_1(x, z) = z\phi(x), \quad u_2(x, z) = w(x). \tag{1}$$

Use the cantilevered beam in Figure 7.8.1. *Hints:* Follow Part 3 of Example 7.6.1 and Example 7.8.1 to develop the total potential energy functional in terms of the dependent variables w and ϕ. Also the nonzero strains are

$$\varepsilon_{xx} = z\frac{d\phi}{dx}, \quad 2\varepsilon_{xz} = \phi + \frac{dw}{dx}. \tag{2}$$

Assume the following one-dimensional constitutive equations:

$$\sigma_{xx} = E\varepsilon_{xx}, \quad \sigma_{xz} = 2G\varepsilon_{xz}. \tag{3}$$

7.28 The total potential energy functional for a membrane stretched over domain $\Omega \in \mathfrak{R}^2$ is given by

$$\Pi(u) = \int_\Omega \left\{ \frac{T}{2}\left[\left(\frac{\partial u}{\partial x_1}\right)^2 + \left(\frac{\partial u}{\partial x_2}\right)^2\right] - fu \right\} d\mathbf{x},$$

where $u = u(x_1, x_2)$ denotes the transverse deflection of the membrane, T is the tension in the membrane, and $f = f(x_1, x_2)$ is the transversely distributed load on the membrane. Determine the governing differential equation and the permissible boundary conditions for the problem (i.e., identify the essential and natural boundary conditions of the problem) using the principle of minimum total potential energy.

7.29 Use the results of Example 7.8.2 to obtain the deflection at the center of a clamped-clamped beam ($EI = $ constant) under uniform load of intensity q_0 and supported at the center by a linear elastic spring (k).

7.30 Use the results of Example 7.8.2 to obtain the deflection $w(L)$ and slopes $(-dw/dx)(L)$ and $(-dw/dx)(2L)$ under a point load Q_0 for the beam shown in Figure P7.30. It is sufficient to set up the three equations for the three unknowns.

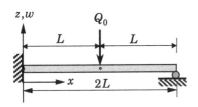

Figure P7.30.

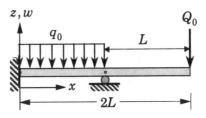

Figure P7.31.

7.31 Use the results of Example 7.8.2 to obtain the deflection $w(2L)$ and slopes at $x = L$ and $x = 2L$ for the beam shown in Figure P7.31. It is sufficient to set up the three equations for the three unknowns.

7.32 Consider a pendulum of mass m_1 with a flexible suspension, as shown in Figure P7.32. The hinge of the pendulum is in a block of mass m_2, which can move up and down between the frictionless guides. The block is connected by a linear spring (of spring constant k) to an immovable support. The coordinate x is measured from the position of the block in which the system remains stationary. Derive the Euler–Lagrange equations of motion for the system.

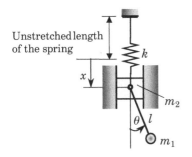

Figure P7.32.

7.33 A chain of total length L and mass m per unit length slides down from the edge of smooth table. Assuming that the chain is rigid, find the equation of motion governing the chain (see Example 5.2.2).

7.34 Consider a cantilever beam supporting a lumped mass M at its end (J is the mass moment of inertia), as shown in Figure P7.34. Derive the equations of motion and natural boundary conditions for the problem using the Euler–Bernoulli beam theory.

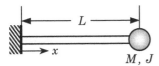

Figure P7.34.

7.35 Derive the equations of motion of the system shown in Figure P7.35. Assume that the mass moment of inertia of the link about its mass center is $J = m\Omega^2$, where Ω is the radius of gyration.

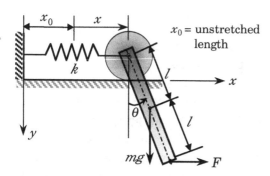

7.36 Derive the equations of motion of the Timoshenko beam theory, starting with the displacement field

$$u_1(x, z, t) = u(x, t) + z\phi(x, t), \quad u_2 = 0, \quad u_3 = w(x, t).$$

Assume that the beam is subjected to distributed axial load $f(x, t)$ and transverse load $q(x, t)$ and that the x-axis coincides with the geometric centroidal axis.

8 Fluid Mechanics and Heat Transfer Problems

> The only solid piece of scientific truth about which I feel totally confident is that we are profoundly ignorant about nature. It is this sudden confrontation with the depth and scope of ignorance that represents the most significant contribution of twentieth-century science to the human intellect.
>
> Lewis Thomas

8.1 Governing Equations

8.1.1 Preliminary Comments

Matter exists only in two states: solid and fluid. The difference between the two is that a solid can resist shear force in static deformation, whereas a fluid cannot. Shear force acting on a fluid causes it to deform continuously. Thus, a fluid at rest can only take hydrostatic pressure and no shear stress. Therefore, the stress vector at a point in a fluid at rest can be expressed as

$$\mathbf{t}^{(\hat{\mathbf{n}})} = \hat{\mathbf{n}} \cdot \boldsymbol{\sigma} = -p\hat{\mathbf{n}} \quad \text{or} \quad \boldsymbol{\sigma} = -p\mathbf{I}, \qquad (8.1.1)$$

where $\hat{\mathbf{n}}$ is unit vector normal to the surface and p is called the *hydrostatic pressure*. It is clear from Eq. (8.1.1) that hydrostatic pressure is equal to the negative of the mean stress

$$p = -\frac{1}{3}\sigma_{ii} = -\tilde{\sigma}. \qquad (8.1.2)$$

In general, p is related to temperature θ and density ρ by equation of the form

$$F(p, \rho, \theta) = 0. \qquad (8.1.3)$$

This equation is called the *equation of state*. Recall from Section 6.3.3, that the hydrostatic pressure p is not equal, in general, to the thermodynamic pressure P appearing in the constitutive equation of a fluid in motion [see Eqs. (6.3.15) and (6.3.16)]

$$\boldsymbol{\sigma} = \mathbf{F}(\mathbf{D}) - P\mathbf{I} = \boldsymbol{\tau} - P\mathbf{I}, \qquad (8.1.4)$$

where τ is the viscous stress tensor, which is a function of the motion, namely, the rate of deformation tensor $\mathbf{D}$; τ vanishes when fluid is at rest.

Fluid mechanics is a branch of mechanics that deals with the effects of fluids at rest (statics) or in motion (dynamics) on surfaces they come in contact. Fluids do not have the so-called natural state to which they return upon removal of forces causing deformation. Therefore, we use spatial (or Eulerian) description to write the governing equations. Pertinent equations are summarized next for an isotropic, Newtonian fluid. *Heat transfer* is a branch of engineering that deals with the transfer of thermal energy within a medium or from one medium to another due to a temperature difference. In this chapter, we study some typical problems of fluid mechanics and heat transfer.

8.1.2 Summary of Equations

The basic equations of viscous fluids are listed here. The number of equations (N_{eq}) and number of new dependent variables (N_{var}) for three-dimensional problems are listed in parenthesis. (in the equations that follow, Q denotes internal heat generation per unit mass and T denotes temperature).

Continuity equation ($N_{eq} = 1$, $N_{var} = 4$)

$$\frac{\partial \rho}{\partial t} + \text{div}(\rho \mathbf{v}) = 0, \quad \frac{D\rho}{Dt} + \rho \frac{\partial v_i}{\partial x_i} = 0. \tag{8.1.5}$$

Equations of motion ($N_{eq} = 3$, $N_{var} = 6$)

$$\nabla \cdot \sigma + \rho \mathbf{f} = \rho \left(\frac{\partial \mathbf{v}}{\partial t} + \mathbf{v} \cdot \nabla \mathbf{v} \right), \quad \frac{\partial \sigma_{ji}}{\partial x_j} + \rho f_i = \rho \frac{Dv_i}{Dt}. \tag{8.1.6}$$

Energy equation ($N_{eq} = 1$, $N_{var} = 4$)

$$\rho \frac{De}{Dt} = \sigma : \mathbf{D} - \nabla \cdot \mathbf{q} + \rho Q, \quad \rho \frac{De}{Dt} = \sigma_{ij} D_{ij} - \frac{\partial q_i}{\partial x_i} + \rho Q. \tag{8.1.7}$$

Constitutive equation ($N_{eq} = 6$, $N_{var} = 7$)

$$\sigma = 2\mu \mathbf{D} + \lambda (\text{tr} \, \mathbf{D})\mathbf{I} - P\mathbf{I}, \quad \sigma_{ij} = 2\mu D_{ij} + \lambda D_{kk} \delta_{ij} - P\delta_{ij}. \tag{8.1.8}$$

Heat conduction equation ($N_{eq} = 3$, $N_{var} = 1$)

$$\mathbf{q} = -k\nabla T, \quad q_i = -k\frac{\partial T}{\partial x_i}. \tag{8.1.9}$$

Kinetic equation of state ($N_{eq} = 1$, $N_{var} = 0$)

$$P = P(\rho, T). \tag{8.1.10}$$

Caloric equation of state ($N_{eq} = 1$, $N_{var} = 0$)

$$e = e(\rho, T). \tag{8.1.11}$$

Rate of deformation-velocity equations ($N_{eq} = 6$, $N_{var} = 0$)

$$\mathbf{D} = \frac{1}{2}[\nabla\mathbf{v} + (\nabla\mathbf{v})^{\mathrm{T}}], \quad D_{ij} = \frac{1}{2}\left(\frac{\partial v_i}{\partial x_j} + \frac{\partial v_j}{\partial x_i}\right). \tag{8.1.12}$$

Material time derivative

$$\frac{D}{Dt} \equiv \frac{\partial}{\partial t} + \mathbf{v} \cdot \nabla, \quad \frac{D}{Dt} \equiv \frac{\partial}{\partial t} + v_i \frac{\partial}{\partial x_i}. \tag{8.1.13}$$

Thus, there are 22 equations and 22 variables.

8.1.3 Viscous Incompressible Fluids

Here, we summarize the governing equations of fluid flows for the isothermal case. Like in elasticity, the number of equations of fluid flow can be combined to obtain a smaller number of equations in as many unknowns. For instance, Eqs. (8.1.5), (8.1.6), (8.1.8), and (8.1.12) can be combined to yield the following equations:

$$\frac{\partial \rho}{\partial t} + \nabla \cdot (\rho\mathbf{v}) = 0, \quad \frac{\partial \rho}{\partial t} + \frac{\partial(\rho v_i)}{\partial x_i} = 0. \tag{8.1.14}$$

$$\mu\nabla^2\mathbf{v} + (\mu + \lambda)\nabla(\nabla \cdot \mathbf{v}) - \nabla P + \rho\mathbf{f} = \rho\left(\frac{\partial \mathbf{v}}{\partial t} + \mathbf{v} \cdot \nabla\mathbf{v}\right),$$

$$\mu v_{i,jj} + (\mu + \lambda)v_{j,ji} - \frac{\partial P}{\partial x_i} + \rho f_i = \rho\left(\frac{\partial v_i}{\partial t} + v_j\frac{\partial v_i}{\partial x_j}\right). \tag{8.1.15}$$

Equations (8.1.14) and (8.1.15) are known as the *Navier–Stokes equations*.

Equations (8.1.14) and (8.1.15) together contain four equations in five unknowns (v_1, v_2, v_3, ρ, P). For compressible fluids, Eqs. (8.1.14) and (8.1.15) are appended with Eqs. (8.1.7) and (8.1.9)–(8.1.11). For the isothermal case, Eqs. (8.1.14) and (8.1.15) are appended with Eq. (8.1.10), where $P = P(I)$.

For incompressible fluids, ρ is constant and is a known variable, and thus we have four equations in four unknowns,

$$\nabla \cdot \mathbf{v} = 0, \quad \frac{\partial v_i}{\partial x_i} = 0. \tag{8.1.16}$$

$$\mu\nabla^2\mathbf{v} - \nabla P + \rho\mathbf{f} = \rho\left(\frac{\partial \mathbf{v}}{\partial t} + \mathbf{v} \cdot \nabla\mathbf{v}\right),$$

$$\mu v_{i,jj} - \frac{\partial P}{\partial x_i} + \rho f_i = \rho\left(\frac{\partial v_i}{\partial t} + v_j\frac{\partial v_i}{\partial x_j}\right). \tag{8.1.17}$$

The expanded forms of these four equations in rectangular Cartesian system and orthogonal curvilinear (i.e., cylindrical and spherical) coordinate systems are given below.

Cartesian coordinate system: (x, y, z); $v_1 = v_x$, $v_2 = v_y$, and $v_3 = v_z$.

$$\frac{\partial v_x}{\partial x} + \frac{\partial v_y}{\partial y} + \frac{\partial v_z}{\partial z} = 0, \tag{8.1.18}$$

$$\mu\left(\frac{\partial^2 v_x}{\partial x^2} + \frac{\partial^2 v_x}{\partial y^2} + \frac{\partial^2 v_x}{\partial z^2}\right) - \frac{\partial P}{\partial x} + \rho f_x = \rho\left(\frac{\partial v_x}{\partial t} + v_x\frac{\partial v_x}{\partial x} + v_y\frac{\partial v_x}{\partial y} + v_z\frac{\partial v_x}{\partial z}\right),$$

(8.1.19)

$$\mu\left(\frac{\partial^2 v_y}{\partial x^2} + \frac{\partial^2 v_y}{\partial y^2} + \frac{\partial^2 v_y}{\partial z^2}\right) - \frac{\partial P}{\partial y} + \rho f_y = \rho\left(\frac{\partial v_y}{\partial t} + v_x\frac{\partial v_y}{\partial x} + v_y\frac{\partial v_y}{\partial y} + v_z\frac{\partial v_y}{\partial z}\right),$$

(8.1.20)

$$\mu\left(\frac{\partial^2 v_z}{\partial x^2} + \frac{\partial^2 v_z}{\partial y^2} + \frac{\partial^2 v_z}{\partial z^2}\right) - \frac{\partial P}{\partial z} + \rho f_z = \rho\left(\frac{\partial v_z}{\partial t} + v_x\frac{\partial v_z}{\partial x} + v_y\frac{\partial v_z}{\partial y} + v_z\frac{\partial v_z}{\partial z}\right).$$

(8.1.21)

Cylindrical coordinate system: (r, θ, z); $v_1 = v_r$, $v_2 = v_\theta$, and $v_3 = v_z$.

$$\frac{1}{r}\frac{\partial(r v_r)}{\partial r} + \frac{1}{r}\frac{\partial v_\theta}{\partial \theta} + \frac{\partial v_z}{\partial z} = 0,$$

(8.1.22)

$$\mu\left[\frac{\partial}{\partial r}\left(\frac{1}{r}\frac{\partial}{\partial r}(r v_r)\right) + \frac{1}{r^2}\left(\frac{\partial^2 v_r}{\partial \theta^2} - 2\frac{\partial v_\theta}{\partial \theta}\right) + \frac{\partial^2 v_r}{\partial z^2}\right] - \frac{\partial P}{\partial r} + \rho f_r$$

$$= \rho\left(\frac{\partial v_r}{\partial t} + v_r\frac{\partial v_r}{\partial r} + \frac{v_\theta}{r}\frac{\partial v_r}{\partial \theta} - \frac{v_\theta^2}{r} + v_z\frac{\partial v_z}{\partial z}\right),$$

(8.1.23)

$$\mu\left[\frac{\partial}{\partial r}\left(\frac{1}{r}\frac{\partial}{\partial r}(r v_\theta)\right) + \frac{1}{r^2}\left(\frac{\partial^2 v_\theta}{\partial \theta^2} + 2\frac{\partial v_r}{\partial \theta}\right) + \frac{\partial^2 v_\theta}{\partial z^2}\right] - \frac{\partial P}{\partial \theta} + \rho f_\theta$$

$$= \rho\left(\frac{\partial v_\theta}{\partial t} + v_r\frac{\partial v_\theta}{\partial r} + \frac{v_\theta}{r}\frac{\partial v_\theta}{\partial \theta} + \frac{v_r v_\theta}{r} + v_z\frac{\partial v_\theta}{\partial z}\right),$$

(8.1.24)

$$\mu\left[\frac{1}{r}\frac{\partial}{\partial r}\left(r\frac{\partial v_z}{\partial r}\right) + \frac{1}{r^2}\frac{\partial^2 v_z}{\partial \theta^2} + \frac{\partial^2 v_z}{\partial z^2}\right] - \frac{\partial P}{\partial z} + \rho f_z$$

$$= \rho\left(\frac{\partial v_z}{\partial t} + v_r\frac{\partial v_z}{\partial r} + \frac{v_\theta}{r}\frac{\partial v_z}{\partial \theta} + v_z\frac{\partial v_z}{\partial z}\right).$$

(8.1.25)

Spherical coordinate system: (r, ϕ, θ); $v_1 = v_r$, $v_2 = v_\phi$, and $v_3 = v_\theta$.

$$2\frac{v_r}{r} + \frac{\partial v_r}{\partial r} + \frac{1}{r\sin\phi}\frac{\partial(v_\phi\sin\phi)}{\partial\phi} + \frac{1}{r\sin\phi}\frac{\partial v_\theta}{\partial\theta} = 0,$$

(8.1.26)

$$\mu\left[\frac{1}{r^2}\frac{\partial^2}{\partial r^2}(r^2 v_r) + \frac{1}{r^2\sin\phi}\frac{\partial}{\partial\phi}\left(\sin\phi\frac{\partial v_r}{\partial\phi}\right) + \frac{1}{r^2\sin^2\phi}\frac{\partial^2 v_r}{\partial\theta^2}\right] - \frac{\partial P}{\partial r} + \rho f_r$$

$$= \rho\left[\frac{\partial v_r}{\partial t} + v_r\frac{\partial v_r}{\partial r} + \frac{v_\phi}{r}\frac{\partial v_r}{\partial\phi} + \frac{v_\theta}{r\sin\phi}\frac{\partial v_r}{\partial\theta} - \left(\frac{v_\phi^2 + v_\theta^2}{r}\right)\right],$$

(8.1.27)

$$\mu \left[\frac{1}{r^2} \frac{\partial}{\partial r} \left(r^2 \frac{\partial v_\phi}{\partial r} \right) + \frac{1}{r^2} \frac{\partial}{\partial \phi} \left(\frac{1}{\sin \phi} \frac{\partial}{\partial \phi} (v_\phi \sin \phi) \right) + \frac{1}{r^2 \sin^2 \phi} \frac{\partial^2 v_\phi}{\partial \theta^2} \right.$$

$$\left. + \frac{2}{r^2} \left(\frac{\partial v_r}{\partial \phi} - \frac{\cos \phi}{\sin^2 \phi} \frac{\partial v_\theta}{\partial \theta} \right) \right] - \frac{1}{r} \frac{\partial P}{\partial \phi} + \rho f_\phi$$

$$= \rho \left(\frac{\partial v_\phi}{\partial t} + v_r \frac{\partial v_\phi}{\partial r} + \frac{v_\phi}{r} \frac{\partial v_\phi}{\partial \phi} + \frac{v_\theta}{r \sin \phi} \frac{\partial v_\phi}{\partial \theta} + \frac{v_r v_\phi}{r} - \frac{v_\theta^2 \cot \phi}{r} \right), \qquad (8.1.28)$$

$$\mu \left[\frac{1}{r^2} \frac{\partial}{\partial r} \left(r^2 \frac{\partial v_\theta}{\partial r} \right) + \frac{1}{r^2} \frac{\partial}{\partial \phi} \left(\frac{1}{\sin \phi} \frac{\partial}{\partial \phi} (v_\theta \sin \phi) \right) + \frac{1}{r^2 \sin^2 \phi} \frac{\partial^2 v_\theta}{\partial \theta^2} \right.$$

$$\left. + \frac{2}{r^2 \sin \phi} \left(\frac{\partial v_r}{\partial \theta} + \cot \phi \frac{\partial v_\phi}{\partial \theta} \right) \right] - \frac{1}{r \sin \phi} \frac{\partial P}{\partial \theta} + \rho f_\theta$$

$$= \rho \left(\frac{\partial v_\theta}{\partial t} + v_r \frac{\partial v_\theta}{\partial r} + \frac{v_\phi}{r} \frac{\partial v_\theta}{\partial \phi} + \frac{v_\theta}{r \sin \phi} \frac{\partial v_\theta}{\partial \theta} + \frac{v_\theta v_r}{r} + \frac{v_\theta v_\phi}{r} \cot \phi \right). \qquad (8.1.29)$$

In general, finding exact solutions of the Navier–Stokes equations is an impossible task. The principal reason is the nonlinearity of the equations, and consequently, the principle of superposition is not valid. In the following sections, we shall find exact solutions of Eqs. (8.1.16) and (8.1.17) for certain flow problems for which the convective terms (i.e., $\mathbf{v} \cdot \nabla \mathbf{v}$) vanish and problems become linear. Of course, even for linear problems, flow geometry must be simple to be able to determine the exact solution. The books by Bird et al. (1960) and Schlichting (1979) contains a number of such problems, and we discuss a few of them here (also, see Problems 8.1–8.5). Like in linearized elasticity, often the semi-inverse method is used to obtain the solutions.

For several classes of flows with constant density and viscosity, the differential equations are expressed in terms of a potential function called *stream function, ψ*. For two-dimensional planar problems (where $v_z = 0$ and data as well as solution does not depend on z), the stream function is defined by

$$v_x = -\frac{\partial \psi}{\partial y}, \quad v_y = \frac{\partial \psi}{\partial x}. \qquad (8.1.30)$$

This definition of ψ automatically satisfies the continuity equation (8.1.18):

$$\frac{\partial v_x}{\partial x} + \frac{\partial v_y}{\partial y} = -\frac{\partial^2 \psi}{\partial x \partial y} + \frac{\partial^2 \psi}{\partial x \partial y} = 0.$$

Next, we determine the governing equation of ψ. Recall the definition of the vorticity $\boldsymbol{\omega} = \nabla \times v$. In two dimensions, the only nonzero component of the vorticity vector is ζ ($\boldsymbol{\omega} = \zeta \, \hat{\mathbf{e}}_z$)

$$\boldsymbol{\omega} = \nabla \times v, \quad \zeta = \frac{\partial v_y}{\partial x} - \frac{\partial v_x}{\partial y}. \qquad (8.1.31)$$

Substituting the definition (8.1.30) into Eq. (8.1.31), we obtain

$$\mathbf{w} = \zeta\,\hat{\mathbf{e}}_z = \left(\frac{\partial^2\psi}{\partial x^2} + \frac{\partial^2\psi}{\partial y^2}\right)\hat{\mathbf{e}}_z = \nabla^2\psi\,\hat{\mathbf{e}}_z. \tag{8.1.32}$$

Next, recall the vorticity equation from Problem 5.15:

$$\frac{\partial\mathbf{w}}{\partial t} + (\mathbf{v}\cdot\nabla)\mathbf{w} = (\mathbf{w}\cdot\nabla)\mathbf{v} + \nu\nabla^2\mathbf{w}, \quad \nu = \frac{\mu}{\rho}. \tag{8.1.33}$$

Since for two-dimensional flows the vorticity vector $\mathbf{w}$ is perpendicular to the plane of the flow, $(\mathbf{w}\cdot\nabla)\mathbf{v}$ is zero. Then

$$\frac{\partial\mathbf{w}}{\partial t} + (\mathbf{v}\cdot\nabla)\mathbf{w} = \nu\nabla^2\mathbf{w}. \tag{8.1.34}$$

Substituting Eq. (8.1.32) into the vorticity equation (8.1.34), we obtain

$$\frac{\partial\nabla^2\psi}{\partial t} + (\mathbf{v}\cdot\nabla)(\nabla^2\psi) = \nu\nabla^4\psi. \tag{8.1.35}$$

In the rectangular Cartesian coordinate system, Eq. (8.1.35) has the form

$$\frac{\partial\nabla^2\psi}{\partial t} + \left(-\frac{\partial\psi}{\partial y}\frac{\partial\nabla^2\psi}{\partial x} + \frac{\partial\psi}{\partial x}\frac{\partial\nabla^2\psi}{\partial y}\right) = \nu\nabla^4\psi. \tag{8.1.36}$$

In the cylindrical coordinate system, the stress function is related to the velocities

$$v_r = -\frac{1}{r}\frac{\partial\psi}{\partial\theta}, \quad v_\theta = \frac{\partial\psi}{\partial r}, \tag{8.1.37}$$

and the governing equation (8.1.36) takes the form

$$\frac{\partial\nabla^2\psi}{\partial t} + \frac{1}{r}\left(-\frac{\partial\psi}{\partial\theta}\frac{\partial\nabla^2\psi}{\partial r} + \frac{\partial\psi}{\partial r}\frac{\partial\nabla^2\psi}{\partial\theta}\right) = \nu\nabla^4\psi, \tag{8.1.38}$$

where ∇^2 is given in Table 2.4.2 for the cylindrical coordinate system.

In the spherical coordinate system, the stress function is defined by

$$v_r = -\frac{1}{r^2\sin\phi}\frac{\partial\psi}{\partial\phi}, \quad v_\phi = \frac{1}{r\sin\phi}\frac{\partial\psi}{\partial r}, \tag{8.1.39}$$

and Eq. (8.1.36) has the form

$$\frac{\partial\tilde{\nabla}^2\psi}{\partial t} + \frac{1}{r^2\sin\phi}\left(-\frac{\partial\psi}{\partial\phi}\frac{\partial\tilde{\nabla}^2\psi}{\partial r} + \frac{\partial\psi}{\partial r}\frac{\partial\tilde{\nabla}^2\psi}{\partial\phi}\right) = \nu\tilde{\nabla}^4\psi,$$
$$\tilde{\nabla}^2 = \frac{\partial^2}{\partial r^2} + \frac{\sin\phi}{r^2}\frac{\partial}{\partial\phi}\left(\frac{1}{\sin\phi}\frac{\partial}{\partial\phi}\right). \tag{8.1.40}$$

8.1.4 Heat Transfer

Recall from Section 6.4 that the balance of energy is given by [see Eqs. (5.4.10) and (6.4.3) with $\mathcal{E}$ replaced by Q and θ by T]

$$\rho c_P\frac{DT}{Dt} = \Phi - P\nabla\cdot\mathbf{v} + \nabla\cdot(k\nabla T) + \rho Q, \tag{8.1.41}$$

where Q is the internal heat generation per unit mass. P is the pressure, T is the temperature, and Φ is the dissipation function

$$\Phi = \boldsymbol{\tau} : \mathbf{D}. \tag{8.1.42}$$

For an incompressible fluid, Eq. (8.1.41) takes the simpler form

$$\rho c_P \frac{DT}{Dt} = \Phi + \nabla \cdot (k\nabla T) + \rho Q. \tag{8.1.43}$$

The expanded form of Eq. (8.1.43) in rectangular Cartesian system and orthogonal curvilinear (i.e., cylindrical and spherical) coordinate systems are given below for the case in which k and μ are constants. For heat transfer in a solid medium, all of the velocity components should be set to zero.

Cartesian coordinate system (x, y, z):

$$\rho c_P \left(\frac{\partial T}{\partial t} + v_x \frac{\partial T}{\partial x} + v_y \frac{\partial T}{\partial y} + v_z \frac{\partial T}{\partial z} \right) = k \left(\frac{\partial^2 T}{\partial x^2} + \frac{\partial^2 T}{\partial y^2} + \frac{\partial^2 T}{\partial z^2} \right)$$

$$+ 2\mu \left[\left(\frac{\partial v_x}{\partial x} \right)^2 + \left(\frac{\partial v_y}{\partial y} \right)^2 + \left(\frac{\partial v_z}{\partial z} \right)^2 \right] + \mu \left[\left(\frac{\partial v_x}{\partial y} + \frac{\partial v_y}{\partial x} \right)^2 \right.$$

$$\left. + \left(\frac{\partial v_x}{\partial z} + \frac{\partial v_z}{\partial x} \right)^2 + \left(\frac{\partial v_y}{\partial z} + \frac{\partial v_z}{\partial y} \right)^2 \right] + \rho Q. \tag{8.1.44}$$

Cylindrical coordinate system (r, θ, z):

$$\rho c_P \left(\frac{\partial T}{\partial t} + v_r \frac{\partial T}{\partial r} + \frac{v_\theta}{r} \frac{\partial T}{\partial \theta} + v_z \frac{\partial T}{\partial z} \right) = k \left[\frac{1}{r} \frac{\partial}{\partial r} \left(r \frac{\partial T}{\partial r} \right) + \frac{1}{r^2} \frac{\partial^2 T}{\partial \theta^2} + \frac{\partial^2 T}{\partial z^2} \right]$$

$$+ 2\mu \left\{ \left(\frac{\partial v_r}{\partial r} \right)^2 + \left[\frac{1}{r} \left(\frac{\partial v_\theta}{\partial \theta} + v_r \right) \right]^2 + \left(\frac{\partial v_z}{\partial z} \right)^2 \right\}$$

$$+ \mu \left\{ \left(\frac{\partial v_\theta}{\partial z} + \frac{1}{r} \frac{\partial v_z}{\partial \theta} \right)^2 + \left(\frac{\partial v_z}{\partial r} + \frac{\partial v_r}{\partial z} \right)^2 + \left[\frac{1}{r} \frac{\partial v_r}{\partial \theta} + r \frac{\partial}{\partial r} \left(\frac{v_\theta}{r} \right) \right]^2 \right\} + \rho Q.$$

$$\tag{8.1.45}$$

Spherical coordinate system (r, ϕ, θ):

$$\rho c_P \left(\frac{\partial T}{\partial t} + v_r \frac{\partial T}{\partial r} + \frac{v_\phi}{r} \frac{\partial T}{\partial \phi} + \frac{v_\theta}{r \sin \phi} \frac{\partial T}{\partial \theta} \right) = k \left[\frac{1}{r^2} \frac{\partial}{\partial r} \left(r^2 \frac{\partial T}{\partial r} \right) \right.$$

$$\left. + \frac{1}{r^2 \sin \phi} \frac{\partial}{\partial \phi} \left(\sin \phi \frac{\partial T}{\partial \phi} \right) + \frac{1}{r^2 \sin^2 \phi} \frac{\partial^2 T}{\partial \theta^2} \right]$$

$$+ 2\mu \left\{ \left(\frac{\partial v_r}{\partial r} \right)^2 + \left(\frac{1}{r} \frac{\partial v_\phi}{\partial \phi} + \frac{v_r}{r} \right)^2 + \left(\frac{1}{r \sin \phi} \frac{\partial v_\theta}{\partial \theta} + \frac{v_r}{r} + \frac{v_\phi \cot \phi}{r} \right)^2 \right\}$$

$$+ \mu \left\{ \left[r \frac{\partial}{\partial r} \left(\frac{v_\phi}{r} \right) + \frac{1}{r} \frac{\partial v_r}{\partial \phi} \right]^2 + \left[\frac{1}{r \sin \phi} \frac{\partial v_r}{\partial \theta} + r \frac{\partial}{\partial r} \left(\frac{v_\theta}{r} \right) \right]^2 \right.$$

$$\left. + \left[\frac{\sin \phi}{r} \frac{\partial}{\partial \phi} \left(\frac{v_\theta}{\sin \phi} \right) + \frac{1}{r \sin \phi} \frac{\partial v_\phi}{\partial \theta} \right]^2 \right\} + \rho Q. \qquad (8.1.46)$$

8.2 Fluid Mechanics Problems

8.2.1 Inviscid Fluid Statics

For inviscid fluids (i.e., fluids with zero viscosity), the constitutive equation for stress is [see Eq. (6.3.16)]

$$\sigma = -P\mathbf{I} \quad (\sigma_{ij} = -P\delta_{ij}),$$

where P is the hydrostatic pressure, the equations of motion (8.1.6) reduces to

$$- \operatorname{grad} P + \rho \mathbf{f} = \rho \frac{D\mathbf{v}}{Dt} . \qquad (8.2.1)$$

The body force in hydrostatics problem often represents the gravitational force, $\rho \mathbf{f} = -\rho g \, \hat{\mathbf{e}}_3$, where the positive x_3-axis is taken positive upward. Consequently, the equations of motion reduce to

$$- \frac{\partial P}{\partial x_1} = \rho a_1, \quad - \frac{\partial P}{\partial x_2} = \rho a_2, \quad - \frac{\partial P}{\partial x_3} = \rho g + \rho a_3, \qquad (8.2.2)$$

where $a_i = \dot{v}_i$ is the ith component of acceleration.

For steady flows with constant velocity field, equations in (8.2.2) simplify to

$$- \frac{\partial P}{\partial x_1} = 0, \quad - \frac{\partial P}{\partial x_2} = 0, \quad - \frac{\partial P}{\partial x_3} = \rho g. \qquad (8.2.3)$$

The first two equations in (8.2.3) imply that $P = P(x_3)$. Integrating the third equation with respect to x_3, we obtain

$$P(x_3) = -\rho g x_3 + c_1,$$

where c_1 is the constant of integration, which can be evaluated using the pressure boundary condition at $x_3 = H$, where H is the height of the column of liquid [see Fig. 8.2.1(a)]. On the free surface, we have $P = P_0$, where P_0 is the atmospheric pressure. Then the constant of integration is $c_1 = P_0 + \rho g H$, and we have

$$P(x_3) = \rho g (H - x_3) + P_0. \qquad (8.2.4)$$

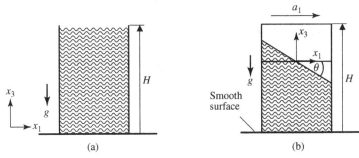

Figure 8.2.1. (a) Column of liquid of height H. (b) A container of fluid moving with a constant acceleration, $\mathbf{a} = a_1 \hat{\mathbf{e}}_1$.

For the unsteady case in which the fluid (i.e., a rectangular container with the fluid) moves at a constant acceleration a_1 in the x_1-direction, the equations of motion in Eq. (8.2.2) become

$$-\frac{\partial P}{\partial x_1} = \rho a_1, \qquad -\frac{\partial P}{\partial x_2} = 0, \qquad -\frac{\partial P}{\partial x_3} = \rho g, \qquad (8.2.5)$$

From the second equation, it follows that $P = P(x_1, x_3)$. Integrating the first equation with respect to x_1, we obtain

$$P(x_1, x_3) = -\rho a_1 x_1 + f(x_3),$$

where $f(x_3)$ is a function of x_3 alone. Substituting the above equation for P into the third equation in Eq. (8.2.5), and integrating with respect to x_3, we arrive at

$$f(x_3) = \rho g x_3 + c_2, \qquad P(x_1, x_3) = -\rho a_1 x_1 + \rho g x_3 + c_2,$$

where c_2 is a constant of integration. If $x_3 = 0$ is taken on the free surface of the fluid in the container, then $P = P_0$ at $x_1 = x_3 = 0$, giving $c_2 = P_0$. Thus,

$$P(x_1, x_3) = P_0 - \rho a_1 x_1 + \rho g x_3. \qquad (8.2.6)$$

Equation (8.2.6) suggests that the free surface (which is a plane), where $P = P_0$, is given by the equation $a_1 x_1 = g x_3$. The orientation of the plane is given by the angle θ as shown in Fig. 8.2.1(b), where

$$\tan \theta = \frac{dx_3}{dx_1} = \frac{a_1}{g}. \qquad (8.2.7)$$

When the fluid is a perfect gas, the constitutive equation for pressure is the equation of state

$$P = \rho RT, \qquad (8.2.8)$$

where T is the absolute temperature (in degree Kelvin) and R is the gas constant (m·N/kg·K). If the perfect gas is at rest at a constant temperature, then we have

$$\frac{P}{P_0} = \frac{\rho}{\rho_0}, \qquad (8.2.9)$$

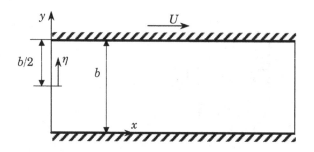

Figure 8.2.2. Parallel flow through a straight channel.

where ρ_0 is the density at pressure P_0. From the third equation in (8.2.3), we have

$$dx_3 = -\frac{1}{\rho g} dP = -\frac{P_0}{\rho_0 g} \frac{dP}{P}.$$

Integrating from $x_3 = x_3^0$ to x_3, we obtain

$$x_3 - x_3^0 = -\frac{P_0}{\rho_0 g} \ln\left(\frac{P}{P_0}\right) \quad \text{or} \quad P = P_0 \exp\left(-\frac{x_3 - x_3^0}{P_0/\rho_0 g}\right). \qquad (8.2.10)$$

8.2.2 Parallel Flow (Navier–Stokes Equations)

A flow is called *parallel* if only one velocity component is nonzero (i.e., all fluid particles moving in the same direction). Suppose that $v_2 = v_3 = 0$ and that the body forces are negligible. Then, from Eq. (8.1.16), it follows that

$$\frac{\partial v_1}{\partial x_1} = 0 \rightarrow v_1 = v_1(x_2, x_3, t). \qquad (8.2.11)$$

Thus, for a parallel flow, we have

$$v_1 = v_1(x_2, x_3, t), \quad v_2 = v_3 = 0. \qquad (8.2.12)$$

Consequently, the three equations of motion in (8.1.17) simplify to the following linear differential equations

$$-\frac{\partial P}{\partial x_1} + \mu\left(\frac{\partial^2 v_1}{\partial x_2^2} + \frac{\partial^2 v_1}{\partial x_3^2}\right) = \rho \frac{\partial v_1}{\partial t}, \quad \frac{\partial P}{\partial x_2} = 0, \quad \frac{\partial P}{\partial x_3} = 0. \qquad (8.2.13)$$

The last two equations in (8.2.13) imply that P is only a function of x_1. Thus, given the pressure gradient dP/dx_1, the first equation in (8.2.13) can be used to determine v_1.

8.2.2.1 Steady Flow of Viscous Incompressible Fluid between Parallel Plates

Consider a steady flow (i.e., $\partial v_1/\partial t = 0$) in a channel with two parallel flat walls (see Figure 8.2.2). Let the distance between the two walls be b. Using the alternative

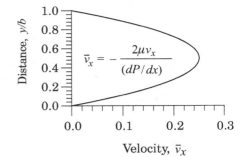

Figure 8.2.3. Velocity distributions for Poiseuille flow.

notation, $v_1 = v_x$, $x_1 = x$, $x_2 = y$, Eq. (8.2.13) can be reduced to the boundary value problem:

$$\mu \frac{d^2 v_x}{dy^2} = \frac{dP}{dx}, \quad 0 < y < b$$

$$v_x(0) = 0, \quad v_x(b) = U. \tag{8.2.14}$$

When $U \neq 0$, the problem is known as the *Couette flow*. The solution of Eq. (8.2.14) is given by

$$v_x(y) = \frac{y}{b} U - \frac{b^2}{2\mu} \frac{dP}{dx} \frac{y}{b} \left(1 - \frac{y}{b}\right), \quad 0 < y < b, \tag{8.2.15}$$

$$\bar{v}_x(\bar{y}) = \bar{y} + f\bar{y}(1 - \bar{y}), \quad \bar{v}_x = \frac{v_x}{U}, \quad \bar{y} = \frac{y}{b}, \quad f = -\frac{b^2}{2\mu U} \frac{dP}{dx}. \tag{8.2.16}$$

When $U = 0$, the flow is known as the *Poiseuille flow*. In this case, the solution (8.2.16) reduces to

$$v_x(y) = -\frac{b^2}{2\mu} \frac{dP}{dx} \frac{y}{b} \left(1 - \frac{y}{b}\right), \quad 0 < y < b, \tag{8.2.17}$$

$$v_x(\eta) = -\frac{1}{2\mu} \frac{dP}{dx} \left(\frac{b^2}{4} - \eta^2\right), \quad \eta = y - \frac{b}{2}, \quad -\frac{b}{2} < \eta < \frac{b}{2}. \tag{8.2.18}$$

Figures 8.2.3 and 8.2.4 show the velocity distributions for cases $U = 0$ and $U \neq 0$ (Couette flow).

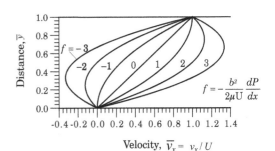

Figure 8.2.4. Velocity distributions for the Couette flow.

8.2.2.2 Steady Flow of a Viscous Incompressible Fluid through a Pipe

The steady flow through a long, straight, horizontal circular pipe is another problem that admits exact solution to the Navier–Stokes equations. We use the cylindrical coordinate system with r being the radial coordinate and the z-coordinate is taken along the axis of the pipe. The velocity components v_r and v_θ in the radial and tangential directions, respectively, are zero. Then the continuity equation (8.1.22) coupled with the axisymmetric flow situation (i.e., the flow field is independent of θ) implies that the velocity component parallel to the axis of the pipe, v_z, is only a function of r.

Equations (8.1.23) and (8.1.24) yield $(\partial P/\partial r) = 0$ and $(\partial P/\partial \theta) = 0$, implying that P is only a function of z (or P is a constant in every cross section). Equation (8.1.25) simplifies to

$$\frac{\mu}{r} \frac{d}{dr}\left(r \frac{dv_z}{dr}\right) = \frac{dP}{dz}, \tag{8.2.19}$$

whose solution is given by

$$v_z(r) = \frac{r^2}{4\mu} \frac{dP}{dz} + Ar(\log r - 1) + B, \tag{8.2.20}$$

where the constants of integration, A and B, are determined using the boundary conditions

$$r\tau_{rz} \equiv r\mu\left(\frac{\partial v_x}{\partial z} + \frac{\partial v_z}{\partial r}\right) = 0 \quad \text{at} \quad r = 0; \quad \text{and} \quad v_z = 0 \quad \text{at} \quad r = R, \tag{8.2.21}$$

where R is the radius of the pipe. We find that

$$A = 0, \quad B = -\frac{R^2}{4\mu} \frac{dP}{dz}, \tag{8.2.22}$$

and the solution becomes

$$v_z(r) = -\frac{1}{4\mu} \frac{dP}{dz}(R^2 - r^2). \tag{8.2.23}$$

Thus, the velocity over the cross section of the pipe varies as a paraboloid of revolution. The maximum velocity occurs along the axis of the pipe and it is equal to

$$(v_z)_{max} = v_z(0) = -\frac{R^2}{4\mu} \frac{dP}{dz}, \tag{8.2.24}$$

The volume rate of flow through the pipe is

$$Q = \int_0^{2\pi} \int_0^R v_z(r)\, r\, dr\, d\theta = \frac{\pi R^4}{8\mu}\left(-\frac{dP}{dz}\right). \tag{8.2.25}$$

The wall shear stress is

$$\tau_w = -\mu\left(\frac{dv_z}{dr}\right)_{r=R} = \frac{R}{2} \frac{dP}{dz}. \tag{8.2.26}$$

8.2.2.3 Unsteady Flow of a Viscous Incompressible Fluid through a Pipe

Here, we consider unsteady flow of a viscous fluid of constant ρ and μ through a long, horizontal circular pipe of length L and radius R. Assume that the fluid is initially at rest. At $t = 0$, a pressure gradient dP/dz (assumed to be independent of t) is applied to the system. We wish to determine the velocity profile as a function of time for $t > 0$.

For this case, Eq. (8.1.25) takes the form

$$\rho \frac{\partial v_z}{\partial t} = \frac{\mu}{r} \frac{\partial}{\partial r}\left(r \frac{\partial v_z}{\partial r}\right) - \frac{dP}{dz}. \tag{8.2.27}$$

The boundary conditions in Eq. (8.2.21) are still valid for this problem. The initial condition is

$$v_z(r, 0) = 0, \quad 0 < r \le R. \tag{8.2.28}$$

To solve the problem, we introduce the following dimensionless variables

$$\bar{v}_z = -\frac{4\mu L}{(dP/dx)R^2}v_z; \quad \xi = \frac{r}{R}; \quad \tau = \frac{\mu}{\rho R^2}t. \tag{8.2.29}$$

Then Eqs. (8.2.27), (8.2.21), and (8.2.28) become, respectively,

$$\frac{\partial \bar{v}_z}{\partial \tau} = \frac{1}{\xi}\frac{\partial}{\partial \xi}\left(\xi \frac{\partial \bar{v}_z}{\partial \xi}\right) + 4, \tag{8.2.30}$$

B.C.: $\bar{v}_z(1, \tau) = 0$, $\bar{v}_z(0, \tau)$ is finite;   I.C.: $\bar{v}_z(\xi, 0) = 0$.

Next, we seek the solution $\bar{v}_z(\xi, \tau)$ as the sum of steady-state solution $\bar{v}_z(\xi, \tau) \to (\bar{v}_z)_\infty(\xi)$ as $\tau \to \infty$ and transient solution $(\bar{v}_z)_\tau(\xi, \tau)$ such that

$$-4 = \frac{1}{\xi}\frac{d}{d\xi}\left(\xi \frac{d(\bar{v}_z)_\infty}{d\xi}\right), \tag{8.2.31}$$

$$\frac{\partial(\bar{v}_z)_\tau}{\partial \tau} = \frac{1}{\xi}\frac{\partial}{\partial \xi}\left(\xi \frac{\partial(\bar{v}_z)_\tau}{\partial \xi}\right). \tag{8.2.32}$$

Equation (8.2.31) is subjected to the conditions

$$(\bar{v}_z)_\infty(1) = 0, \quad (\bar{v}_z)_\infty(0) \text{ is finite}; \tag{8.2.33}$$

Equation (8.2.32) is to be solved with the boundary and initial conditions

B.C.: $(\bar{v}_z)_\tau(1, \tau) = 0$, $(\bar{v}_z)_\tau(0, \tau)$ is finite;   I.C.: $(\bar{v}_z)_\tau(\xi, 0) = -(\bar{v}_z)_\infty$. (8.2.34)

The solution of Eqs. (8.2.31) and (8.2.33) is

$$(\bar{v}_z)_\infty(\xi) = 1 - \xi^2. \tag{8.2.35}$$

The solution to Eqs. (8.2.32) and (8.2.34) can be obtained using the separation of variables technique. We assume solution in the form

$$(\bar{v}_z)_\tau(\xi, \tau) = X(\xi)T(\tau) \tag{8.2.36}$$

and substitute into Eq. (8.2.32) to obtain

$$\frac{1}{T}\frac{dT}{d\tau} = \frac{1}{X}\frac{1}{\xi}\frac{d}{d\xi}\left(\xi\frac{dX}{d\xi}\right). \tag{8.2.37}$$

Since the left side is a function of τ alone and the right side is a function of ξ alone, it follows that both sides must be equal to a constant, which we choose to designate as $-\alpha^2$ (because the solution must be a decay type in τ and periodic in ξ). Thus, we have

$$\frac{dT}{d\tau} + \alpha^2 T = 0 \rightarrow T(\tau) = Ae^{-\alpha^2\tau}, \tag{8.2.38}$$

and

$$\frac{1}{\xi}\frac{d}{d\xi}\left(\xi\frac{dX}{d\xi}\right) + \alpha^2 X = 0 \rightarrow X(\xi) = C_1 J_0(\alpha\xi) + C_2 Y_0(\alpha\xi), \tag{8.2.39}$$

where J_0 and Y_0 are the zero order Bessel functions of the first and second kind, respectively. The constants A, C_1, and C_2 must be determined such that the initial conditions in Eq. (8.2.34) are satisfied. The condition that $(\bar{v}_z)_\tau(0, \tau)$ be finite requires $X(0)$ to be finite. Since $Y_0(0) = -\infty$, it follows that $C_2 = 0$. The boundary condition $(\bar{v}_z)_\tau(1, \tau) = 0$ requires $X(1) = J_0(\alpha) = 0$. Since $J_0(\alpha)$ is an oscillating function, it has the following zeros (i.e., the roots of $J_0(\alpha)$ are):

$$\alpha_1 = 2.4048, \quad \alpha_1 = 5.5201, \quad \alpha_3 = 8.6537, \quad \alpha_4 = 11.7915, \quad \alpha_5 = 14.9309, \cdots.$$

Thus the total solution can be written as

$$(\bar{v}_z)_\tau(\xi, \tau) = \sum_{n=1}^{\infty} C_n e^{-\alpha_n^2\tau} J_0(\alpha_n\xi). \tag{8.2.40}$$

The constants $C_n = AC_{1n}$ are determined using the initial condition in Eq. (8.2.34). We have

$$(\bar{v}_z)_\tau(\xi, 0) = -(\bar{v}_z)_\infty = -(1 - \xi^2) = \sum_{n=1}^{\infty} C_n J_0(\alpha_n\xi). \tag{8.2.41}$$

Using the orthogonality of $J_0(\alpha_n)$

$$\int_0^1 J_0(\alpha_n\xi) J_0(\alpha_m\xi) \xi \, d\xi = \begin{cases} 0, & m \neq n \\ \beta_n, & m = n \end{cases}, \tag{8.2.42}$$

where β_n is given by

$$\beta_n = \int_0^1 [J_0(\alpha_n\xi)]^2 \xi \, d\xi = \frac{1}{2}[J_1(\alpha_m)]^2,$$

$$\int_0^1 J_0(\alpha_n\xi)(1 - \xi^2)\xi \, d\xi = \frac{4J_1(\alpha_n)}{\alpha_n^3}. \tag{8.2.43}$$

The above integrals are evaluated using some standard relations for the Bessel functions. Thus, we obtain

$$C_n = -\frac{8}{\alpha_n^3 J_1(\alpha_n)}. \tag{8.2.44}$$

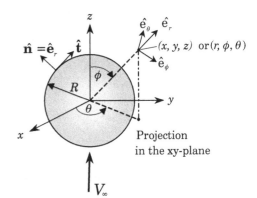

Figure 8.2.5. Creeping flow around a sphere.

The final expression for the velocity $W(\xi)$ is

$$(\bar{v}_z)(\xi, \tau) = (1 - \xi^2) - 8 \sum_{n=1}^{\infty} \frac{J_0(\alpha_n \xi)}{\alpha_n^3 J_1(\alpha_n)} e^{-\alpha_n^2 \tau}. \qquad (8.2.45)$$

8.2.3 Problems with Negligible Convective Terms

The exact solution of the Navier–Stokes equations is made difficult by the presence of the convective, nonlinear terms, $\mathbf{v} \cdot \nabla \mathbf{v}$. When the motion is assumed to be very slow, the convective terms are very small compared to the viscous terms $\mu \nabla^2 \mathbf{v}$ and can be neglected, resulting in linear equations of motion. Such flows are called *creeping flows* and the Navier–Stokes equations without the convective terms are often called the *Stokes equations*. For creeping flows, the governing equations reduce to

$$\nabla \cdot \mathbf{v} = 0, \quad \frac{\partial v_i}{\partial x_i} = 0, \qquad (8.2.46)$$

$$\rho \frac{\partial \mathbf{v}}{\partial t} = \mu \nabla^2 \mathbf{v} - \nabla P + \rho \mathbf{f}, \quad \rho \frac{\partial v_i}{\partial t} = \mu u_{i,jj} - \frac{\partial P}{\partial x_i} + \rho f_i. \qquad (8.2.47)$$

Equations (8.1.19)–(8.1.21), (8.1.23)–(8.1.25), and (8.1.27)–(8.1.29) can be simplified by omitting the convective terms.

8.2.3.1 Flow of a Viscous Incompressible Fluid around a Sphere

Here we consider the steady slow flow of a viscous fluid around a sphere of radius R. The fluid approaches the sphere in the z direction at a velocity V_∞, as shown in Figure 8.2.5. Neglecting the convective terms in Eq. (8.1.40), the governing equation (with no θ dependence and omitting v_θ terms) in terms of the stream function is $\tilde{\nabla}^4 \psi = 0$:

$$\left[\frac{\partial^2}{\partial r^2} + \frac{\sin \phi}{r^2} \frac{\partial}{\partial \phi} \left(\frac{1}{\sin \phi} \frac{\partial}{\partial \phi} \right) \right]^2 \psi = 0. \qquad (8.2.48)$$

This equation must be solved subjected to the boundary conditions

$$v_r(R, \phi) = -\frac{1}{R^2 \sin \phi} \frac{\partial \psi}{\partial \phi}\bigg|_{r=R} = 0,$$

$$v_\phi(R, \phi) = \frac{1}{R \sin \phi} \frac{\partial \psi}{\partial r}\bigg|_{r=R} = 0, \qquad (8.2.49)$$

$$\psi \to -\frac{1}{2} V_\infty r^2 \sin^2 \phi \text{ as } r \to \infty.$$

The first two conditions reflect the attachment of the viscous fluid to the surface of the sphere. The third condition implies that $v_r = V_\infty$ far from the sphere,

$$r \xrightarrow{\text{lim}} \infty \, v_r(r, \phi) = V_\infty. \qquad (8.2.50)$$

Assuming solution of the form

$$\psi(r, \phi) = f(r) \sin^2 \phi, \qquad (8.2.51)$$

and substituting into Eq. (8.2.48) gives

$$\left(\frac{d^2}{dr^2} - \frac{2}{r^2}\right)\left(\frac{d^2}{dr^2} - \frac{2}{r^2}\right) f(r) = 0, \qquad (8.2.52)$$

whose general solution is

$$f(r) = \frac{c_1}{r} + c_2 r + c_3 r^2 + c_4 r^4. \qquad (8.2.53)$$

Satisfaction of the third boundary condition in Eq. (8.2.49) requires $c_4 = 0$ and c_3 to be equal to $-V_\infty/2$. Hence, the solution is

$$\psi(r, \phi) = \left(\frac{c_1}{r} + c_2 r - \frac{V_\infty}{2} r^2\right) \sin^2 \phi. \qquad (8.2.54)$$

The velocity components are

$$v_r = -\frac{1}{r^2 \sin \phi} \frac{\partial \psi}{\partial \phi} = \left(V_\infty - 2\frac{c_1}{r^3} - 2\frac{c_2}{r}\right) \cos \phi,$$

$$\qquad (8.2.55)$$

$$v_\phi = \frac{1}{r \sin \phi} \frac{\partial \psi}{\partial r} = \left(-V_\infty - 2\frac{c_1}{r^3} + \frac{c_2}{r}\right) \sin \phi.$$

The boundary conditions in Eq. (8.2.49) gives $c_1 = -V_\infty R^3/4$ and $c_2 = 3V_\infty R/4$ so that the velocity distributions are

$$v_r = V_\infty \left[1 - \frac{3}{2}\left(\frac{R}{r}\right) + \frac{1}{2}\left(\frac{R}{r}\right)^3\right] \cos \phi,$$

$$\qquad (8.2.56)$$

$$v_\phi = -V_\infty \left[1 - \frac{3}{4}\left(\frac{R}{r}\right) - \frac{1}{4}\left(\frac{R}{r}\right)^3\right] \sin \phi.$$

See Problem 8.15 for the shear stress and pressure distributions on the sphere.

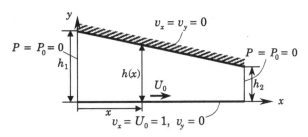

Figure 8.2.6. Schematic of a slider bearing.

8.2.3.2 Flow of a Viscous Incompressible Lubricant in a Bearing

A slider (or slipper) bearing consists of a short sliding pad moving at a velocity $v_x = U_0$ relative to a stationary pad inclined at a small angle with respect to the stationary pad, and the small gap between the two pads is filled with a lubricant, as shown in Figure 8.2.6. Since the ends of the bearing are generally open, the pressure there is atmospheric, say $P = P_0$. When the upper pad is parallel to the base plate, the pressure everywhere in the gap will be atmospheric, and the bearing cannot support any transverse load. If the upper pad is inclined to the base pad, a pressure distribution is set up in the gap. For large values of U_0, the pressure generated can be of sufficient magnitude to support heavy loads normal to the base pad.

When the width of the gap and the angle of inclination are small, one may assume that $v_y = 0$ and $v_z = 0$ and the pressure is only a function of x. Assuming a two-dimensional state of flow in the xy plane and a small angle of inclination, and neglecting the normal stress gradient (in comparison with the shear stress gradient), the equations governing the flow of the lubricant between the pads can be reduced to [see Schlichting (1979) for details]

$$\mu \frac{\partial^2 v_x}{\partial y^2} = \frac{dP}{dx}, \quad \frac{dP}{dx} = \frac{6\mu U_0}{h^2} \left(1 - \frac{H}{h}\right), \quad 0 < x < L, \tag{8.2.57}$$

where

$$h(x) = h_1 + \frac{h_2 - h_1}{L} x, \quad H = \frac{2h_1 h_2}{h_1 + h_2}. \tag{8.2.58}$$

The solution of Eq. (8.2.57), subject to the boundary conditions $v_x(x, 0) = U_0$ and $v_x(x, h) = 0$ is

$$v_x(x, y) = \left(U_0 - \frac{h^2}{2\mu} \frac{dP}{dx} \frac{y}{h}\right)\left(1 - \frac{y}{h}\right), \tag{8.2.59}$$

$$P(x) = \frac{6\mu U_0 L (h_1 - h)(h - h_2)}{h^2 (h_1^2 - h_2^2)}, \tag{8.2.60}$$

$$\sigma_{xy}(x, y) = \mu \frac{\partial v_x}{\partial y} = \frac{dP}{dx}\left(y - \frac{h}{2}\right) - \mu \frac{U_0}{h}. \tag{8.2.61}$$

Table 8.2.1. *Comparison of finite element solutions velocities with the analytical solutions for viscous fluid in a slider bearing*

$\bar{y}$	$v_x(0, y)$	$\bar{y}$	$v_x(0.18, y)$	$\bar{y}$	$v_x(0.36, y)$	x	$\bar{P}(x, 0)$	$-\sigma_{xy}(x, 0)$
0.0	30.000	0.00	30.000	0.00	30.000	0.01	7.50	59.99
1.0	22.969	0.75	25.156	0.50	29.531	0.03	22.46	59.89
2.0	16.875	1.50	20.625	1.00	28.125	0.05	37.29	59.67
3.0	11.719	2.25	16.406	1.50	25.781	0.07	51.89	59.30
4.0	7.500	3.00	12.500	2.00	22.500	0.09	66.12	58.77
5.0	4.219	3.75	8.906	2.50	18.281	0.27	129.60	38.40
6.0	1.875	4.50	5.625	3.00	13.125	0.29	118.57	32.71
7.0	0.469	5.25	2.656	3.50	7.031	0.31	99.58	25.70
8.0	0.000	6.00	0.000	4.00	0.000	0.33	70.30	17.04

$\bar{x} = 10x$, $\quad \bar{y} = y \times 10^4$, $\quad \bar{P} = P \times 10^{-2}$.

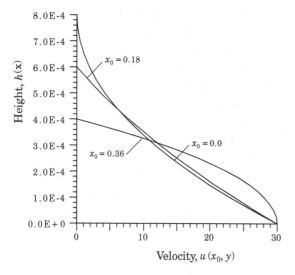

Figure 8.2.7. Velocity distributions for the slider bearing problem.

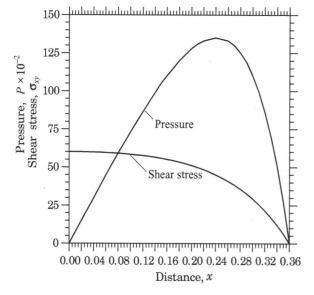

Figure 8.2.8. Pressure and shear stress distributions for the slider bearing problem.

Numerical results are obtained using the following parameters:

$$h_1 = 2h_2 = 8 \times 10^{-4} \text{ ft}, \quad L = 0.36 \text{ ft}, \quad \mu = 8 \times 10^{-4} \text{ lb/ft}^2, \quad U_0 = 30 \text{ ft} \quad (8.2.62)$$

Table 8.2.1 contains numerical values of the velocity, pressure, and shear stress as a function of position. Figure 8.2.7 contains plots of the horizontal velocity v_x at $x = 0$ ft, $x = 0.18$ ft, and $x = 0.36$ ft, while Figure 8.2.8 contains plots of pressure and shear stress as a function of x at $y = 0$.

8.3 Heat Transfer Problems

8.3.1 Heat Conduction in a Cooling Fin

Heat transfer from a surface to the surrounding fluid medium can be increased by attaching thin strips, called *fins*, of conducting material to the surface (see Figure 8.3.1). We assume that the fins are very long in the y-direction, and heat conducts only along the x-direction and convects through the lateral surface, that is, $T = T(x, t)$. This assumption reduces the three-dimensional problem to a one-dimensional problem. By setting the velocity components to zero in Eq. (8.1.44) and noting that $T = T(x, t)$, we obtain

$$\rho c_p \frac{\partial T}{\partial t} = k \frac{\partial^2 T}{\partial x^2} + \rho Q. \tag{8.3.1}$$

Equation (8.3.1) does not account for the cross-sectional area of the fin and convective heat transfer through the surface. Therefore, we derive the governing equation from the first principles. We assume steady heat conduction.

Consider an element of length Δx at a distance x in the fin. The balance of energy in the element requires that

$$(qA)_x - (qA)_{x+\Delta x} - hP\Delta x(T - T_\infty) + \rho Q \left(\frac{A_x + A_{x+\Delta x}}{2} \right) \Delta x = 0, \tag{8.3.2}$$

where q is the heat flux, A is the area of cross section (which can be a function of x), P is the perimeter, h is the film conductance, and Q is internal heat generation per unit mass (which is zero in the case of fins). Dividing throughout by Δx and taking the limit $\Delta x \to 0$, we obtain

$$-\frac{d}{dx}(qA) + Ph(T - T_\infty) + \rho QA = 0. \tag{8.3.3}$$

Using Fourier's law, $q = -k(dT/dx)$, where k is thermal conductivity of the fin, we obtain

$$\frac{d}{dx}\left(kA\frac{dT}{dx} \right) + Ph(T - T_\infty) + \rho QA = 0. \tag{8.3.4}$$

Equation (8.3.4) must be solved subject to the boundary conditions

$$T(0) = T_0, \quad \left[kA\frac{dT}{dx} + hA(T - T_\infty) \right]_{x=a} = 0. \tag{8.3.5}$$

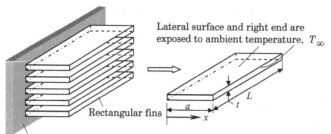

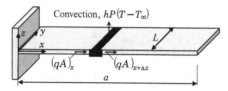

Figure 8.3.1. Heat transfer in a cooling fin.

The second boundary condition is a statement of the balance of energy (conductive and convective) at $x = a$.

We introduce the following nondimensional quantities for convenience of solving the problem (assume that k and A are constant):

$$\theta = \frac{T - T_\infty}{T_0 - T_\infty}, \quad \xi = \frac{x}{a}, \quad m^2 = \frac{hPa^2}{kA}, \quad N = \frac{ha}{k}. \tag{8.3.6}$$

Then Eqs. (8.3.4) and (8.3.5) take the form

$$\frac{d^2\theta}{d\xi^2} - m^2\theta = 0, \quad \theta(0) = 1, \quad \left[\frac{d\theta}{d\xi} + N\theta\right]_{\xi=1} = 0. \tag{8.3.7}$$

The general solution to the differential equation in (8.3.7) is

$$\theta(\xi) = C_1 \cosh m\xi + C_2 \sinh m\xi, \quad 0 < \xi < a,$$

where the constants C_1 and C_2 are determined using the boundary conditions. We obtain

$$C_1 = 1; \quad C_2 = -\frac{m \sinh m + N \cosh m}{m \cosh m + N \sinh m}, \tag{8.3.8}$$

and the solution becomes

$$\theta(\xi) = \frac{\cosh m\xi \, (m \cosh m + N \sinh m) - (m \sinh m + N \cosh m) \sinh m\xi}{m \cosh m + N \sinh m}$$

$$= \frac{m \cosh m(1 - \xi) + N \sinh m(1 - \xi)}{m \cosh m + N \sinh m}. \tag{8.3.9}$$

The *effectiveness of a fin* is defined by (omitting the end effects)

$$\mathcal{E} = \frac{\text{Actual heat convected by the fin surface}}{\text{Heat that would be convected if the fin surface were held at } T_0}$$

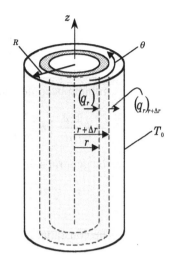

Figure 8.3.2. Heat conduction in a circular cylinder.

$$= \frac{\int_0^L \int_0^a h(T - T_\infty)dxdy}{\int_0^L \int_0^a h(T_0 - T_\infty)dxdy} = \int_0^1 \theta(\xi)\,d\xi$$

$$= \int_0^1 \frac{m\cosh m(1-\xi) + N\sinh m(1-\xi)}{m\cosh m + N\sinh m}\,d\xi$$

$$= \frac{1}{m}\frac{m\sinh m + N(\cosh m - 1)}{m\cosh m + N\sinh m}. \tag{8.3.10}$$

8.3.2 Axisymmetric Heat Conduction in a Circular Cylinder

Here we consider heat transfer in a long circular cylinder (see Figure 8.3.2). If the boundary conditions and material of the cylinder are axisymmetric, that is, independent of the circumferential coordinate θ, it is sufficient to consider a typical rz-plane, where r is the radial coordinate and z is the axial coordinate. Further, if the cylinder is very long, say 10 diameters length, then heat transfer along typical radial line is all we need to determine; thus, the problem is reduced to one dimension.

The governing equation for this one-dimensional problem can be obtained from Eq. (8.1.45) as

$$\rho c_P \frac{\partial T}{\partial t} = \frac{1}{r}\frac{\partial}{\partial r}\left(kr\frac{\partial T}{\partial r}\right) + \rho Q(r), \tag{8.3.11}$$

where ρQ is internal heat generation per unit volume. For example, in the case of an electric wire of circular cross section and electrical conductivity k_e (1/Ohm/m) heat is produced at the rate of

$$\rho Q = \frac{I^2}{k_e}, \tag{8.3.12}$$

where I is electric current density (amps/m^2) passing through the wire.

Equation (8.3.11) is to be solved subjected to appropriate initial condition and boundary conditions at $r = 0$ and $r = R$, where R is the radius of the cylinder. Here, we consider a steady heat transfer when there is an internal heat generation of $\rho Q = g$ and the surface of the cylinder is subjected to a temperature $T(R) = T_0$. Then the problem becomes one of solving the equation

$$k\frac{1}{r}\frac{d}{dr}\left(r\frac{dT}{dr}\right) + g = 0, \quad (rq_r)_{r=0} = \left[-kr\frac{dT}{dr}\right]_{r=0} = 0, \quad T(R) = T_0. \quad (8.3.13)$$

The general solution is given by

$$T(r) = -\frac{gr^2}{4k} + A\log r + B. \quad (8.3.14)$$

The constants A and B are determined using the boundary conditions:

$$(rq_r)(0) = 0 \rightarrow A = 0; \quad T(R) = T_0 \rightarrow B = T_0 + \frac{gR^2}{4k}.$$

The final solution is given by

$$T(r) = T_0 + \frac{gR^2}{4k}\left[1 - \left(\frac{r}{R}\right)^2\right], \quad (8.3.15)$$

which is a parabolic function of the distance r. The heat flux is given by

$$q(r) = -k\frac{dT}{dr} = \frac{gr}{2}, \quad (8.3.16)$$

and the total heat flow at the surface is

$$Q = 2\pi RL\, q(R) = \pi R^2 L\, g.$$

The problem of solving Eq. (8.3.11) subjected to the initial condition and boundary conditions

I.C.: $T(r, 0) = 0$

B.C.: $T(R, t) = 0, \quad (rq_r)_{r=0} = \left[-kr\frac{\partial T}{\partial r}\right]_{r=0} = 0. \quad (8.3.17)$

is equivalent to solving the problem described by Eqs. (8.2.27), (8.2.21), and (8.2.28). In particular, we take

$$\theta = \frac{4kL}{gR^2}T; \quad \xi = \frac{r}{R}; \quad \tau = \frac{k}{\rho c_P R^2}t. \quad (8.3.18)$$

Then the transient solution is given by

$$\theta(\xi, \tau) = (1 - \xi^2) - 8\sum_{n=1}^{\infty}\frac{J_0(\alpha_n\xi)}{\alpha_n^3 J_1(\alpha_n)}e^{-\alpha_n^2\tau}. \quad (8.3.19)$$

8.3.3 Two-Dimensional Heat Transfer

Here, we consider steady heat conduction in a rectangular plate with sinusoidal temperature distribution on one edge (see Figure 8.3.3). The governing equation is a special case of Eq. (8.1.44). Taking $T = T(x, y)$, and setting the time derivative term and velocity components to zero, we obtain

$$k \left(\frac{\partial^2 T}{\partial x^2} + \frac{\partial^2 T}{\partial y^2} \right) = 0. \tag{8.3.20}$$

The boundary conditions are

$$T(x, 0) = 0, \quad T(0, y) = 0, \quad T(a, y) = 0, \quad T(x, b) = T_0 \sin \frac{\pi x}{a}. \tag{8.3.21}$$

Once the temperature $T(x, y)$ is known, we can determine the components of heat flux, q_x and q_y, from Fourier's law

$$q_x = -k \frac{\partial T}{\partial x}, \quad q_y = -k \frac{\partial T}{\partial y}. \tag{8.3.22}$$

The classical approach to an analytical solution of the Laplace or Poisson equation over a regular (i.e., rectangular or circular) domain is the separation-of-variables technique. In this technique, we assume the temperature $T(x, y)$ to be of the form

$$T(x, y) = X(x)Y(y), \tag{8.3.23}$$

where X is a function of x alone and Y is a function of y alone. Substituting Eq. (8.3.23) into Eq. (8.3.20) and rearranging the terms, we obtain

$$\frac{1}{X} \frac{d^2 X}{dx^2} = -\frac{1}{Y} \frac{d^2 Y}{dy^2}. \tag{8.3.24}$$

Since the left side is a function of x alone and the right side is a function of y alone, it follows that both sides must be equal to a constant, which we choose to be $-\lambda^2$ (because the solution must be periodic in x so as to satisfy the boundary condition on the edge $y = b$). Thus, we have

$$\frac{d^2 X}{dx^2} + \lambda^2 X = 0, \quad \frac{d^2 Y}{dy^2} - \lambda^2 Y = 0, \tag{8.3.25}$$

Figure 8.3.3. Heat conduction in a rectangular plate.

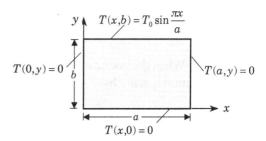

whose general solutions are

$$X(x) = C_1 \cos \lambda x + C_2 \sin \lambda x, \quad Y(y) = C_3 e^{-\lambda y} + C_4 e^{\lambda y}. \tag{8.3.26}$$

The solution $T(x, y)$ is given by

$$T(x, y) = (C_1 \cos \lambda x + C_2 \sin \lambda x)(C_3 e^{-\lambda y} + C_4 e^{\lambda y}). \tag{8.3.27}$$

The constants C_i $(i = 1, 2, 3, 4)$ are determined using the boundary conditions in Eq. (8.3.21). We obtain

$$T(x, 0) = 0 \rightarrow (C_1 \cos \lambda x + C_2 \sin \lambda x)(C_3 + C_4) = 0 \rightarrow C_3 = -C_4,$$

$$T(0, y) = 0 \rightarrow C_1 (C_3 e^{-\lambda y} + C_4 e^{\lambda y}) = 0 \rightarrow C_1 = 0,$$

$$T(a, y) = 0 \rightarrow C_2 \sin \lambda a (C_3 e^{-\lambda y} + C_4 e^{\lambda y}) = 0 \rightarrow \sin \lambda a = 0.$$

The last conclusion is reached because $C_2 = 0$ will make the whole solution trivial. We have

$$\sin \lambda a = 0 \rightarrow \lambda a = n\pi \quad \text{or} \quad \lambda_n = \frac{n\pi}{a}. \tag{8.3.28}$$

The solution in Eq. (8.3.27) now can be expressed as

$$T(x, y) = \sum_{n=1}^{\infty} A_n \sin \frac{n\pi x}{a} \sinh \frac{n\pi y}{a}. \tag{8.3.29}$$

The constants $A_n, n = 1, 2, \ldots$, are determined using the remaining boundary condition. We have

$$T(x, b) = T_0 \sin \frac{\pi x}{a} = \sum_{n=1}^{\infty} A_n \sin \frac{n\pi x}{a} \sinh \frac{n\pi b}{a}.$$

Multiplying both sides with $\sin(m\pi x/a)$ and integrating from 0 to a and using the orthogonality of the sine functions

$$\int_0^a \sin \frac{n\pi x}{a} \sin \frac{m\pi x}{a} dx = \begin{cases} 0, & m \neq n \\ \frac{a}{2}, & m = n \end{cases},$$

we obtain

$$A_1 = \frac{T_0}{\sinh \frac{n\pi b}{a}}, \quad A_n = 0 \quad \text{for} \quad n \neq 1.$$

Hence, the final solution is

$$T(x, y) = T_0 \frac{\sinh \frac{\pi y}{a}}{\sinh \frac{\pi b}{a}} \sin \left(\frac{\pi x}{a} \right). \tag{8.3.30}$$

When the boundary condition at $y = b$ is replaced with $T(x, b) = f(x)$, then the solution is given by

$$T(x, y) = \sum_{n=1}^{\infty} A_n \frac{\sinh \frac{n\pi y}{a}}{\sinh \frac{n\pi b}{a}} \sin \left(\frac{n\pi x}{a} \right), \tag{8.3.31}$$

with A_n given by

$$A_n = \frac{2}{a} \int_0^a f(x) \sin \frac{n\pi x}{a} \, dx. \qquad (8.3.32)$$

8.3.4 Coupled Fluid Flow and Heat Transfer

Next, we consider an example in which the fluid flow is coupled to heat transfer. Consider the fully developed, incompressible, steady Couette flow between parallel plates with zero pressure gradient (see Section 8.2). Suppose that the top plate moving with a velocity U and maintained at a temperature T_1 and the bottom plate is stationary and maintained at temperature T_0 (see Figure 8.3.4). Assuming fully developed temperature profile and zero internal heat generation, we wish to determine the temperature field.

For fully developed temperature field, we can assume that $T = T(y)$. Then the energy equation (8.1.44) reduces to

$$k\frac{d^2 T}{dy^2} + \mu \left(\frac{dv_x}{dy}\right)^2 = 0 \rightarrow \frac{d^2 T}{dy^2} = -\frac{\mu}{kb^2}U^2. \qquad (8.3.33)$$

The solution of this equation is

$$T(y) = -\frac{\mu U^2}{kb^2}\frac{y^2}{2} + Ay + B,$$

where the constants A and B are determined using the boundary conditions $T(0) = T_0$ and $T(b) = T_1$. We obtain

$$
\begin{aligned}
T(0) &= T_0: \quad B = T_0, \\
T(b) &= T_1: \quad A = \frac{T_1 - T_0}{b} + \frac{\mu U^2}{2kb}.
\end{aligned}
\qquad (8.3.34)
$$

Thus the temperature field in the channel is given by

$$\frac{T(y) - T_0}{T_1 - T_0} = \frac{y}{b} + \frac{\mu U^2}{2k(T_1 - T_0)}\frac{y}{b}\left(1 - \frac{y}{b}\right). \qquad (8.3.35)$$

Figure 8.3.4. Velocity and temperature distributions for the Couette flow.

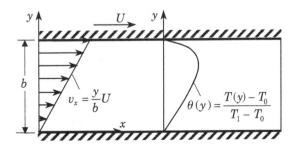

8.4 Summary

In this chapter, applications of the equations of continuum mechanics to fluid mechanics and heat transfer are presented. First, a summary of the equations as applied to decoupled viscous incompressible fluids and heat transfer are presented in Cartesian, cylindrical, and spherical coordinates systems. Then applications to some simple problems of fluid mechanics and heat transfer are discussed. The classes of problems of fluid mechanics that admit analytical solutions are rather limited.

PROBLEMS

8.1 Assume that the velocity components in an incompressible flow are independent of the x coordinate and $v_z = 0$ to simplify the continuity equation (8.1.18) and the equations of motion (8.1.19)–(8.1.21).

8.2 An engineer is to design a sea lab 4 m high, 5 m width, and 10 m long to withstand submersion to 120 m, measured from the surface of the sea to the top of the sea lab. Determine the (a) pressure on the top and (b) pressure variation on the side of the cubic structure. Assume a density of salt water to be $\rho = 1{,}020$ kg/m^3.

8.3 Compute the pressure and density at an elevation of 1,600 m for isothermal conditions. Assume $P_0 = 10^2$ kPa, $\rho_0 = 1.24$ kg/m^3 at sea level.

8.4 Derive the pressure-temperature and density-temperature relations for an ideal gas when temperature varies according to $\theta(x_3) = \theta_0 + mx_3$, where m is taken to be $m = -0.0065°$C/m up to the stratosphere, and x_3 is measured upward from sea level. *Hint:* Use Eq. (8.2.8) and the third equation in Eq. (8.2.3).

8.5 Consider the steady flow of a viscous incompressible Newtonian fluid down an inclined surface of slope α under the action of gravity (see Figure P8.5). The thickness of the fluid perpendicular to the plane is h and the pressure on the free surface is p_0, a constant. Use the semi-inverse method (i.e., assume the form of the velocity field) to determine the pressure and velocity field.

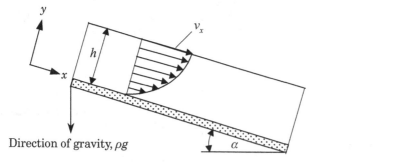

Figure P8.5.

Direction of gravity, ρg

8.6 Two immiscible fluids are flowing in the x-direction in a horizontal channel of length L and width $2b$ under the influence of a fixed pressure gradient. The fluid rates are adjusted such that the channel is half filled with Fluid I (denser phase) and half filled with Fluid II (less dense phase). Assuming that the gravity of the fluids is negligible, determine the velocity field. Use the geometry and coordinate system shown in Figure P8.6.

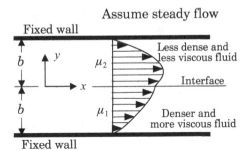

Figure P8.6.

8.7 Consider the steady flow of a viscous, incompressible fluid in the annular region between two coaxial circular cylinders of radii R and αR, $\alpha < 1$, as shown in Figure P8.7. Take $\bar{P} = P + \rho g z$. Determine the velocity and shear stress distributions in the annulus.

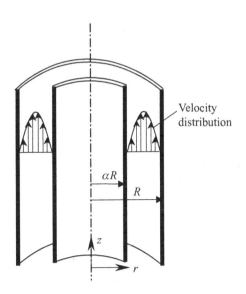

Figure P8.7.

8.8 Consider a steady, isothermal, incompressible fluid flowing between two vertical concentric long circular cylinders with radii r_1 and r_2. If the outer one rotating with an angular velocity Ω, show that the Navier–Stokes equations reduce to the following equations governing the circumferential velocity $v_\theta = v(r)$ and pressure P:

$$\rho \frac{v^2}{r} = \frac{\partial P}{\partial r}, \quad \mu \frac{d}{dr}\left(\frac{1}{r}\frac{d}{dr}(rv)\right) = 0, \quad 0 = -\frac{\partial P}{\partial z} + \rho g.$$

Determine the velocity v and shear stress $\tau_{r\theta}$ distributions.

8.9 Consider an isothermal, incompressible fluid flowing radially between two concentric porous spherical shells. Assume steady flow with $v_r = v(r)$. Simplify the continuity and momentum equations for the problem.

8.10 A fluid of constant density and viscosity is in a cylindrical container of radius R and the container is rotated about its axis with an angular velocity of Ω. Use the cylindrical coordinate system with the z-coordinate along the cylinder axis. Let the body force vector to be equal to $\rho \mathbf{f} = -g\hat{\mathbf{e}}_z$. Assume that $v_r = 0$ and $v_z = 0$, and $v_\theta = v(r)$ and simplify the governing equations. Determine $v(r)$ from the second momentum equation subject to the boundary condition $v(R) = \Omega r$. Then evaluate P from remaining equations.

8.11 Consider the *unsteady* parallel flow on a flat plate (or plane wall). Assume that the motion is started impulsively from rest. Take the x-coordinate along the plate and the y-coordinate perpendicular to the wall. Assume that only nonzero velocity component is $v_x = v_x(y, t)$ and that the pressure P is a constant. Show that the Navier–Stokes equations for this case are simplified to

$$\rho \frac{\partial v_x}{\partial t} = \mu \frac{\partial^2 v_x}{\partial y^2}, \quad 0 < y < \infty. \tag{1}$$

Solve the above equation for $v_x(y)$ using the following initial and boundary conditions:

$$\text{Initial condition} \qquad v_x(y, 0) = 0,$$
$$\text{Boundary conditions} \quad v_x(\infty, t) = 0. \tag{2}$$

Hint: Introduce a new coordinate η by assuming $\eta = y/(2\sqrt{vt})$, where v is the kinematic viscosity $v = \mu/\rho$, and seek solution in the form $v_x(\eta) = U_0\, f(\eta)$. The solution is obtained in terms of the *complementary error function*

$$\text{erfc}\,\eta = \frac{2}{\sqrt{\pi}} \int_\eta^\infty e^{-\eta^2}\, d\eta = 1 - \text{erf}\,\eta = 1 - \frac{2}{\sqrt{\pi}} \int_0^\eta e^{-\eta^2}\, d\eta, \tag{3}$$

where $\text{erf}\,\eta$ is the *error function*.

8.12 Solve Eq. (1) of Problem 8.11 for the following boundary conditions (i.e., flow near an oscillating flat plate)

$$\text{Initial condition} \qquad v_x(y, 0) = 0,$$
$$\text{Boundary conditions} \quad v_x(0, t) = U_0 \cos nt, \quad v_x(\infty, t) = 0. \tag{1}$$

In particular, obtain the solution

$$v_x(y, t) = U_0 e^{\lambda y} \cos(nt - \lambda y), \quad \lambda = \sqrt{\frac{\rho n}{2\mu}}. \tag{2}$$

8.13 Show that the components of the viscous stress tensor τ [see Eq. (6.3.7)] for an isotropic, viscous, Newtonian fluid in cylindrical coordinates are related to the velocity gradients by

$$\tau_{rr} = 2\mu \frac{\partial v_r}{\partial r} + \lambda \, \nabla \cdot \mathbf{v}, \quad \tau_{\theta\theta} = 2\mu \left(\frac{1}{r} \frac{\partial v_\theta}{\partial \theta} + \frac{v_r}{r} \right) + \lambda \, \nabla \cdot \mathbf{v},$$

$$\tau_{zz} = 2\mu \frac{\partial v_z}{\partial z} + \lambda \, \nabla \cdot \mathbf{v}, \quad \tau_{r\theta} = \mu \left[r \frac{\partial}{\partial r} \left(\frac{v_\theta}{r} \right) + \frac{1}{r} \frac{\partial v_r}{\partial \theta} \right],$$

$$\tau_{z\theta} = \mu \left(\frac{\partial v_\theta}{\partial z} + \frac{1}{r} \frac{\partial v_z}{\partial \theta} \right), \quad \tau_{zr} = \mu \left(\frac{\partial v_z}{\partial r} + \frac{\partial v_r}{\partial z} \right),$$

$$\nabla \cdot \mathbf{v} = \frac{1}{r} \frac{\partial (r v_r)}{\partial r} + \frac{1}{r} \frac{\partial v_\theta}{\partial \theta} + \frac{\partial v_z}{\partial z}.$$

8.14 Show that the components of the viscous stress tensor τ [see Eq. (6.3.7)] for an isotropic, viscous, Newtonian fluid in spherical coordinates are related to the velocity gradients by

$$\tau_{rr} = 2\mu \frac{\partial v_r}{\partial r} + \lambda \, \nabla \cdot \mathbf{v}, \quad \tau_{\phi\phi} = 2\mu \left(\frac{1}{r} \frac{\partial v_\theta}{\partial \theta} + \frac{v_r}{r} \right) + \lambda \, \nabla \cdot \mathbf{v},$$

$$\tau_{\theta\theta} = 2\mu \left(\frac{1}{r \sin\phi} \frac{\partial v_\phi}{\partial \phi} + \frac{v_r}{r} + \frac{v_\phi \cot\phi}{r} \right) + \lambda \, \nabla \cdot \mathbf{v},$$

$$\tau_{r\phi} = \mu \left[r \frac{\partial}{\partial r} \left(\frac{v_\phi}{r} \right) + \frac{1}{r} \frac{\partial v_r}{\partial \phi} \right], \quad \tau_{r\theta} = \mu \left[\frac{1}{r \sin\phi} \frac{\partial v_r}{\partial \theta} + r \frac{\partial}{\partial r} \left(\frac{v_\theta}{r} \right) \right],$$

$$\tau_{\phi\theta} = \mu \left[\frac{\sin\phi}{r} \frac{\partial}{\partial \phi} \left(\frac{v_\theta}{\sin\phi} \right) + \frac{1}{r \sin\phi} \frac{\partial v_\phi}{\partial \theta} \right],$$

$$\nabla \cdot \mathbf{v} = \frac{1}{r^2} \frac{\partial (r^2 v_r)}{\partial r} + \frac{1}{r \sin\phi} \frac{\partial}{\partial \phi} (v_\phi \sin\phi) + \frac{1}{r \sin\phi} \frac{\partial v_\theta}{\partial \theta}.$$

8.15 Use the velocity field in Eq. (8.2.56) to determine the shear stress component $\tau_{r\phi}$ and pressure P. *Ans:*

$$\tau_{r\phi} = \frac{3\mu V_\infty}{2R} \left(\frac{R}{r} \right)^4 \sin\phi, \quad P = P_0 - \rho g z - \frac{3\mu V_\infty}{2R} \left(\frac{R}{r} \right)^2 \cos\phi,$$

where P_0 is the pressure in the plane $z = 0$ far away from the sphere and $-\rho g z$ is the contribution of the fluid weight (hydrostatic effect).

8.16 Consider a long electric wire of length L and radius R and electrical conductivity k_e [1/(Ohm·m)]. An electric current with current density I (amps/m^2) is passing through the wire. The transmission of an electric current is an irreversible process in which some electrical energy is converted into thermal energy (heat). The rate of heat production per unit volume is given by

$$\rho Q_e = \frac{I^2}{k_e}.$$

Assuming that the temperature rise in the cylinder is small enough not to affect the thermal or electrical conductivities and heat transfer is one-dimensional along the radius of the cylinder, derive the governing equation using balance of energy.

8.17 Solve the equation derived in Problem 8.16 using the boundary conditions

$$q(0) = \text{finite}, \quad T(R) = T_0.$$

8.18 A slab of length L is initially at temperature $f(x)$. For times $t > 0$, the boundaries at $x = 0$ and $x = L$ are kept at temperatures T_0 and T_L, respectively. Obtain the temperature distribution in the slab as a function of position x and time t.

8.19 Obtain the steady-state temperature distribution $T(x, y)$ in a rectangular region, $0 \leq x \leq a, 0 \leq y \leq b$ for the boundary conditions

$$q_x(0, y) = 0, \quad q_y(x, b) = 0, \quad q_x(a, y) + hT(a, y) = 0, \quad T(x, 0) = f(x).$$

8.20 Consider the steady flow through a long, straight, horizontal circular pipe. The velocity field is given by [see Eq. (8.2.23)]

$$v_r = 0, \quad v_\theta = 0, \quad v_z(r) = -\frac{R^2}{4\mu}\frac{dP}{dz}\left(1 - \frac{r^2}{R^2}\right). \tag{1}$$

If the pipe is maintained at a temperature T_0 on the surface, determine the steady-state temperature distribution in the fluid.

8.21 Consider the free convection problem of flow between two parallel plates of different temperature. A fluid with density ρ and viscosity μ is placed between two *vertical* plates a distance $2a$ apart, as shown in Figure P8.21. Suppose that the plate at $x = a$ is maintained at a temperature T_1 and the plate at $x = -a$ is maintained at a temperature T_2, with $T_2 > T_1$. Assuming that the plates are very long in the y-direction and hence that the temperature and velocity fields are only a function of x, determine the temperature $T(x)$ and velocity $v_y(x)$. Assume that the volume rate of flow in the upward moving stream is the same as that in the downward moving stream and the pressure gradient is solely due to the weight of the fluid.

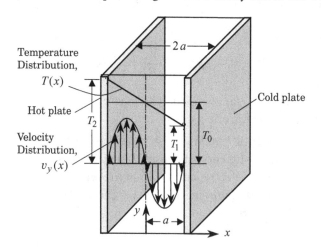

Figure P8.21.

In questions of science, the authority of a thousand is not worth the humble reasoning of a single individual.

Galileo Galilei

All truths are easy to understand once they are discovered; the point is to discover them.

Galileo Galilei

9.1 Introduction

9.1.1 Preliminary Comments

The class of materials that exhibit the characteristics of elastic as well as viscous materials are known as viscoelastic materials. Metals at elevated temperatures, concrete, and polymers provide examples of materials with viscoelastic behavior. In this section, we study mathematical models of linear viscoelastic behavior. The characteristics of a viscoelastic material are that they (a) have time-dependent behavior and (b) have permanent deformation (i.e., do not return to original geometry after the removal of forces causing the deformation).

The viscoelastic response characteristics of a material are determined often using (1) creep tests, (2) stress relaxation tests, or (3) dynamic response to loads varying sinusoidally with time. *Creep* test involves determining the strain response under constant stress, and it is done under uniaxial tensile stress due to its simplicity. Application of a constant stress σ_0 produces a strain, which, in general, contains three components: an instantaneous, a plastic, and a delayed reversible component

$$\varepsilon(t) = \left[J_\infty + \frac{t}{\eta_0} + \psi(t) \right] \sigma_0,$$

where $J_\infty \sigma_0$ is the instantaneous component of strain, η_0 is the Newtonian viscosity coefficient, and $\psi(t)$ the creep function such that $\psi(0) = 0$. *Relaxation* test involves determination of stress under constant strain. Application of a constant strain ε_0 produces a stress that contains two components

$$\sigma(t) = [E_0 + \phi(t)]\,\varepsilon_0,$$

where E_0 is the static elastic modulus and $\phi(t)$ is the relaxation function such that $\phi(0) = 0$.

A qualitative understanding of viscoelastic behavior of materials can be gained through spring-and-dashpot models. For linear responses, combinations of linear elastic springs and linear viscous dashpots are used. Two simple spring-and-dashpot models are the *Maxwell model* and *Kelvin–Voigt model*. The Maxwell model characterizes a viscoelastic fluid while Kelvin–Voigt model represents a viscoelastic solid. Other combinations of these models are also used. The mathematical models to be discussed here provide some insight into the creep and relaxation characteristics of viscoelastic responses, but they may not represent a satisfactory quantitative behavior of any real material. A combination of the Maxwell and Kelvin–Voigt models may represent the creep and relaxation responses of some materials.

9.1.2 Initial Value Problem, the Unit Impulse, and the Unit Step Function

The governing equations of the mathematical models involving springs and dashpots are ordinary differential equations in time, t. These equations relate stress σ to strain ε and they have the general form

$$P(\sigma) = Q(\varepsilon), \tag{9.1.1}$$

where P and Q are differential operators of order M and N, respectively,

$$P = \sum_{m=0}^{M} p_m \frac{d^m}{dt^m}, \quad Q = \sum_{n=0}^{N} q_n \frac{d^n}{dt^n}. \tag{9.1.2}$$

The coefficients p_m and q_n are known in terms of the spring constants k_i and dashpot constants η_i of the model. Equation (9.1.1) is solved either for $\varepsilon(t)$ for a specified $\sigma(t)$ (creep response) or for $\sigma(t)$ for a given $\varepsilon(t)$ (relaxation response). Since Eq. (9.1.1) is a Mth-order differential equation for the relaxation response (Nth-order equation for the creep response), we must know M (N) initial values, that is, values at time $t = 0$, of σ (ε):

$$\sigma(0) = \sigma_0, \quad \dot{\sigma}(0) = \dot{\sigma}_0, \ldots, \quad \left(\frac{d^{N-1}\sigma}{dt^{N-1}}\right)_{t=0} = \sigma_0^{(N-1)},$$

or   (9.1.3)

$$\varepsilon(0) = \varepsilon_0, \quad \dot{\varepsilon}(0) = \dot{\varepsilon}_0, \ldots, \quad \left(\frac{d^{M-1}\varepsilon}{dt^{M-1}}\right)_{t=0} = \varepsilon_0^{(M-1)},$$

where $\sigma_0^{(i)}$, for example, denotes the value of the ith time derivative of $\sigma(t)$ at time $t = 0$. Equation (9.1.1) together with (9.1.3) defines an initial value problem.

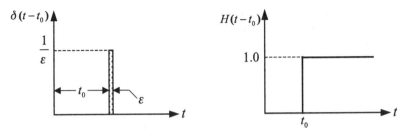

Figure 9.1.1. (a) The Dirac delta function. (b) Unit step function.

In the sequel, we will study the creep and relaxation responses of the discrete viscoelastic models under applied inputs. The applied stress or strain can be in the form of a *unit impulse* or *unit step function*. The unit impulse, also known as the *Dirac delta function*, is defined as

$$\delta(t - t_0) = 0, \quad \text{for } t \neq t_0,$$
$$\int_{-\infty}^{\infty} \delta(t - t_0)dt = 1. \tag{9.1.4}$$

The units of the Dirac delta function are $1/s = s^{-1}$. A plot of the Dirac delta function is shown in Figure 9.1.1(a). The time interval in which the Dirac delta function is nonzero is defined to be infinitely small, say ϵ. The Dirac delta function can be used to represent an arbitrary point value F_0 at $t = t_0$ as a function of time:

$$f(t) = F_0\delta(t - t_0); \quad \int_{-\infty}^{\infty} f(t)\, dt = \int_{-\infty}^{\infty} F_0\delta(t - t_0)dt = F_0, \tag{9.1.5}$$

where $f(t)$ has the units of F_0 per second.

The unit step function is defined as [see Figure 9.1.1(b)]

$$H(t - t_0) = \begin{cases} 0, & \text{for } t < t_0, \\ 1, & \text{for } t > t_0. \end{cases} \tag{9.1.6}$$

Clearly, the function $H(t)$ is discontinuous at $t = t_0$, where its value jumps from 0 to 1. The unit step function is dimensionless. The unit step function $H(t)$, when multiplies an arbitrary function $f(t)$, sets the portion of $f(t)$ corresponding to $t < 0$ to zero while leaving the portion corresponding to $t > 0$ unchanged.

The Dirac delta function is viewed as the derivative of the unit step function; conversely, the unit step function is the integral of the Dirac delta function

$$\delta(t) = \frac{dH(t)}{dt}; \quad H(t) = \int_{-\infty}^{t} \delta(\xi)d\xi. \tag{9.1.7}$$

9.1.3 The Laplace Transform Method

The Laplace transform method is widely used to solve linear differential equations, especially those governing initial-value problems. The significant feature of the method is that it allows in a natural way the use of singularity functions like the

Dirac delta function and the unit step function in the data of the problem. Here we review the method in the context of solving initial value problems.

The (one-sided) *Laplace transformation* of a function $f(t)$, denoted $\bar{f}(s)$, is defined as

$$\bar{f}(s) \equiv \mathcal{L}[f(t)] = \int_0^\infty e^{-st} f(t) \, dt, \tag{9.1.8}$$

where s is, in general, a complex quantity referred as a *subsidiary variable*, and the function e^{-st} is known as the *kernel* of the transformation. The Laplace transforms of some functions are given in Table 9.1.1. The table can also be used for inverse transforms. The following two examples illustrate the use of the Laplace transform method in the solution of differential equations.

EXAMPLE 9.1.1: Consider the first-order differential equation

$$b\frac{du}{dt} + cu = f_0, \tag{9.1.9}$$

where b, c, and f_0 are constants. Equation (9.1.9) is subjected to zero initial condition, $u(0) = 0$. Determine the solution using the Laplace transform method.

SOLUTION: The Laplace transform of the equation gives

$$(bs + c)\bar{u} = \frac{f_0}{s} \quad \text{or} \quad \bar{u}(s) = \frac{f_0}{bs\left(s + \frac{c}{b}\right)}. \tag{9.1.10}$$

To invert Eq. (9.1.10) to determine $u(t)$, we rewrite the expression as (i.e., split into partial fractions; see Problem 9.1 for an explanation of the *method of partial fractions*)

$$\bar{u}(s) = \frac{f_0}{c}\left(\frac{1}{s} - \frac{1}{s + \alpha}\right), \quad \alpha = \frac{c}{b}.$$

The inverse transform is given by (see Table 9.1.1)

$$u(t) = \frac{f_0}{c}\left(1 - e^{-\alpha t}\right). \tag{9.1.11}$$

When b and c are positive real numbers, $u(t)$ approaches f_0/c as $t \to \infty$.

EXAMPLE 9.1.2: Consider the second-order differential equation

$$a\frac{d^2u}{dt^2} + b\frac{du}{dt} + cu = f_0, \tag{9.1.12}$$

where a, b, c, and f_0 are constants. The equation is to be solved subjected to zero initial conditions, $u(0) = 0$ and $\dot{u}(0) = 0$. Determine the solution using the Laplace transform method.

SOLUTION: The Laplace transform of the equation gives

$$(as^2 + bs + c)\bar{u} = \frac{f_0}{s}$$

Table 9.1.1. *The Laplace transforms of some standard functions*

$f(t)$	$\bar{f}(s)$
$f(t)$	$\int_0^\infty e^{-st} f(t)\, dt$
$\dot{f} \equiv \frac{df}{dt}$	$s\bar{f}(s) - f(0)$
$\ddot{f} \equiv \frac{d^2 f}{dt^2}$	$s^2 \bar{f}(s) - sf(0) - \dot{f}(0)$
$f^{(n)}(t) \equiv \frac{d^n f}{dt^n}$	$s^n \bar{f}(s) - s^{n-1} f(0) - s^{n-2} \dot{f}(0)$
	$- \cdots - f^{(n-1)}(0)$
$\int_0^t f(\xi)\, d\xi$	$\frac{1}{s} \bar{f}(s)$
$\int_0^t f_1(t) f_2(t-\xi)\, d\xi$	$\bar{f}_1(s) \bar{f}_2(s)$
$H(t)$	$\frac{1}{s}$
$\delta(t) = \dot{H}(t)$	1
$\dot{\delta}(t) = \ddot{H}(t)$	s
$\delta^{(n)}(t)$	s^n
t	$\frac{1}{s^2}$
t^n	$\frac{n!}{s^{n+1}}$
$t f(t)$	$-\bar{f}'(s)$
$t^n f(t)$	$(-1)^n \bar{f}^{(n)}(s)$
$e^{at} f(t)$	$\bar{f}(s-a)\, dt$
e^{at}	$\frac{1}{s-a}$
$t e^{at}$	$\frac{1}{(s-a)^2}$
$t^n e^{at}$	$\frac{n!}{(s-a)^{n+1}}, \quad n = 0, 1, 2, \cdots$
$e^{at} - e^{bt}$	$\frac{a-b}{(s-a)(s-b)}$
$(a e^{at} - b e^{bt})$	$\frac{s(a-b)}{(s-a)(s-b)}$
$\sin at$	$\frac{a}{s^2+a^2}$
$\cos at$	$\frac{s}{s^2+a^2}$
$\sinh at$	$\frac{a}{s^2-a^2}$
$\cosh at$	$\frac{s}{s^2-a^2}$
$t \sin at$	$\frac{2as}{(s^2+a^2)^2}$
$t \cos at$	$\frac{s^2-a^2}{(s^2+a^2)^2}$
$e^{bt} \sin at$	$\frac{a}{(s-b)^2+a^2}$
$e^{bt} \cos at$	$\frac{s-b}{(s-b)^2+a^2}$
$1 - \cos at$	$\frac{a^2 s}{s(s^2+a^2)}$
$at - \sin at$	$\frac{a^3}{s^2(s^2+a^2)}$
$\sin at - at \cos at$	$\frac{2a^3}{(s^2+a^2)^2}$
$\sin at + at \cos at$	$\frac{2as^2}{(s^2+a^2)^2}$
$\cos at - \cos bt$	$\frac{(b^2-a^2)s}{(s^2+a^2)^2(s^2+b^2)}, \quad b^2 \neq a^2$
$\sin at \cosh at - \cos at \sinh at$	$\frac{4a^3 s}{s^4+4a^4}$
$\sin at \sinh at$	$\frac{2a^2 s}{s^4+4a^4}$
$\sinh at - \sin at$	$\frac{2a^3}{(s^4-a^4)}$
$\cosh at - \cos at$	$\frac{2a^2 s}{s^4-a^4}$
$\sqrt{t}$	$\frac{\sqrt{\pi}}{2} s^{-3/2}$
$\frac{1}{\sqrt{\pi t}}$	$\frac{1}{\sqrt{s}}$
$J_0(at)$	$\frac{1}{\sqrt{s^2+a^2}}$
$\frac{e^{bt} - e^{at}}{t}$	$\log \frac{s-a}{s-b}$
$\frac{1}{t}(1 - \cos at)$	$\frac{1}{2} \log \frac{s^2+a^2}{s^2}$
$\frac{1}{t}(1 - \cosh at)$	$\frac{1}{2} \log \frac{s^2-a^2}{s^2}$
$\frac{1}{t} \sin kt$	$\arctan \frac{k}{s}$

$J_0(at)$ is the Bessel function of the first kind.

or

$$\bar{u}(s) = \frac{f_0}{s(as^2 + bs + c)}.$$

To invert the above equation to determine $u(t)$, first we write $as^2 + bs + c$ as $a(s + \alpha)(s + \beta)$, where α and β are the roots of the equation $as^2 + bs + c = 0$:

$$\alpha = \frac{1}{2a}\left(b - \sqrt{b^2 - 4ac}\right), \quad \beta = \frac{1}{2a}\left(b + \sqrt{b^2 - 4ac}\right), \qquad (9.1.13)$$

so that

$$\bar{u}(s) = \frac{f_0}{as(s + \alpha)(s + \beta)}. \qquad (9.1.14)$$

The actual nature of the solution $u(t)$ depends on the nature of the roots α and β in Eq. (9.1.13). Three possible cases depend on whether $b^2 - 4ac > 0$, $b^2 - 4ac = 0$, or $b^2 - 4ac < 0$. We discuss them under the assumption that a, b, and c are positive real numbers.

CASE 1. When $b^2 - 4ac > 0$, the roots are real, positive, and unequal. Then, we can rewrite Eq. (9.1.14) as

$$\bar{u} = \frac{f_0}{a}\left[\frac{A}{s} + \frac{B}{s + \alpha} + \frac{C}{s + \beta}\right],$$

so that we can use the inverse Laplace transform to obtain $u(t)$. The constants A, B, and C satisfy the relations

$$A + B + C = 0, \quad (\alpha + \beta)A + \beta B + \alpha C = 0, \quad \alpha\beta A = 1.$$

The solution of these equations is

$$A = \frac{1}{\alpha\beta}, \quad B = \frac{1}{\alpha(\beta - \alpha)}, \quad C = \frac{1}{\beta(\beta - \alpha)}.$$

Thus, we have

$$\bar{u}(s) = \frac{f_0}{a}\left[\frac{1}{\alpha\beta s} - \frac{1}{\alpha(\beta - \alpha)(s + \alpha)} + \frac{1}{\beta(\beta - \alpha)(s + \beta)}\right]. \qquad (9.1.15)$$

The inverse transform is

$$u(t) = \frac{f_0}{a\alpha\beta}\left[1 - \frac{\beta}{\beta - \alpha}e^{-\alpha t} + \frac{\alpha}{\beta - \alpha}e^{-\beta t}\right]$$

$$= \frac{f_0}{a\alpha\beta(\beta - \alpha)}\left[\beta\left(1 - e^{-\alpha t}\right) - \alpha\left(1 - e^{-\beta t}\right)\right]. \qquad (9.1.16)$$

Hence, $u(t)$ approaches $f_0/a\alpha\beta$ as $t \to \infty$.

CASE 2. When $b^2 - 4ac = 0$, the roots are real, positive, and equal, $\alpha = \beta = b/2a$. Then Eq. (9.1.14) takes the form

$$\bar{u}(s) = \frac{f_0}{as(s + \alpha)^2} = \frac{f_0}{a\alpha}\left[\frac{1}{\alpha}\left(\frac{1}{s} - \frac{1}{s + \alpha}\right) - \frac{1}{(s + \alpha)^2}\right]. \qquad (9.1.17)$$

The inverse Laplace transform gives

$$u(t) = \frac{f_0}{a\alpha^2}\left[1 - (1 + \alpha t)e^{-\alpha t}\right]. \tag{9.1.18}$$

Hence, $u(t)$ approaches $4 f_0 a / b^2$ as $t \to \infty$.

CASE 3. When $b^2 - 4ac < 0$, the roots are complex, and they appear in complex conjugate pairs:

$$\alpha = \alpha_1 - i\alpha_2, \quad \beta = \alpha_1 + i\alpha_2; \qquad \alpha_1 = \frac{b}{2a}, \quad \alpha_2 = \sqrt{4ac - b^2}. \tag{9.1.19}$$

From Eq. (9.1.16), we obtain

$$u(t) = \frac{f_0}{a\alpha\beta(\beta - \alpha)}e^{-\alpha_1 t}\left[\beta\left(1 - e^{i\alpha_2 t}\right) - \alpha\left(1 - e^{-i\alpha_2 t}\right)\right]$$

$$= \frac{f_0}{a\left(\alpha_1^2 + \alpha_2^2\right)}e^{-\alpha_1 t}\left(1 - \cos\alpha_2 t - \frac{\alpha_1}{\alpha_2}\sin\alpha_2 t\right). \tag{9.1.20}$$

Hence, $u(t)$ approaches zero as $t \to \infty$.

9.2 Spring and Dashpot Models

9.2.1 Creep Compliance and Relaxation Modulus

The equations relating stress σ and strain ε in spring-dashpot models are ordinary differential equations, and they have the general form given in Eq. (9.1.1). The solution of Eq. (9.1.1) to determine $\sigma(t)$ for a given $\varepsilon(t)$ (relaxation response) or to determine $\varepsilon(t)$ for given $\sigma(t)$ (creep response) is made easy by the Laplace transform method. In this section, we shall study several standard spring-dashpot models for their constitutive models and creep and relaxation responses. First, we note certain features of the general constitutive equation (9.1.1). In general, the creep response and relaxation response are of the form

$$\varepsilon(t) = J(t)\sigma_0, \tag{9.2.1}$$

$$\sigma(t) = Y(t)\varepsilon_0, \tag{9.2.2}$$

where $J(t)$ is called the *creep compliance* and $Y(t)$ the *relaxation modulus* associated with (9.1.1). The function $J(t)$ is the strain per unit of applied stress, and $Y(t)$ is the stress per unit of applied strain. By definition, both $J(t)$ and $Y(t)$ are zero for all $t < 0$.

The Laplace transform of Eq. (9.1.1) for creep response and relaxation response have the forms

Creep response $\quad \bar{Q}_s\bar{\varepsilon}(s) = \bar{P}_s\bar{\sigma}(s) = \dfrac{1}{s}\bar{P}_s\sigma_0, \tag{9.2.3}$

Relaxation response $\quad \bar{P}_s\bar{\sigma}(s) = \bar{Q}_s\bar{\varepsilon}(s) = \dfrac{1}{s}\bar{Q}_s\varepsilon_0, \tag{9.2.4}$

where

$$P_s = \sum_{m=0}^{M} p_m s^m, \quad Q_s = \sum_{n=0}^{N} q_n s^n. \tag{9.2.5}$$

The Laplace transforms of Eqs. (9.2.1) and (9.2.2) are

$$\bar{\varepsilon}(s) = \bar{J}(s)\sigma_0, \tag{9.2.6}$$

$$\bar{\sigma}(s) = \bar{Y}(s)\varepsilon_0 \tag{9.2.7}$$

Comparing Eq. (9.2.3) with (9.2.6) and Eq. (9.2.4) with (9.2.7), we obtain

$$\bar{J}(s) = \frac{1}{s}\frac{\bar{P}_s}{\bar{Q}_s}, \quad \bar{Y}(s) = \frac{1}{s}\frac{\bar{Q}_s}{\bar{P}_s}. \tag{9.2.8}$$

It also follows that the Laplace transforms of the creep compliance and relaxation modulus are related by

$$\bar{J}(s)\,\bar{Y}(s) = \frac{1}{s^2} \quad \text{or} \quad t = \int_0^t Y(t-t')\,J(t')\,dt'. \tag{9.2.9}$$

Thus, once we know creep compliance $J(t)$, we can determine the relaxation modulus $Y(t)$ and vice versa

$$Y(t) = \mathcal{L}^{-1}\left[\frac{1}{s^2\bar{J}(s)}\right], \quad J(t) = \mathcal{L}^{-1}\left[\frac{1}{s^2\bar{Y}(s)}\right]. \tag{9.2.10}$$

Although creep and relaxation tests have the advantage of simplicity, there are also shortcomings. The first shortcoming is that uniaxial creep and relaxation test procedures assume the stress to be uniformly distributed through the specimen, with the lateral surfaces being free to expand and contract. This condition cannot be satisfied at the ends of a specimen that is attached to a test machine. The second shortcoming involves the dynamic effects which are encountered in obtaining data at short times. The relaxation and creep functions which are determined through Eqs. (9.2.1) and (9.2.2) are based on the assumption that all transients excited through the dynamic response of specimen and testing machine are neglected. Typically, this effect limits relaxation and creep data to times no less than 0.1 seconds.

9.2.2 Maxwell Element

The Maxwell element of Figure 9.2.1 consists of a linear elastic spring element in series with a dashpot element. The stress–strain relation for the model is developed using the following stress–strain relationships of individual elements:

$$\sigma = k\varepsilon, \quad \sigma = \eta\dot{\varepsilon}, \tag{9.2.11}$$

where k is the spring elastic constant, η is the dashpot viscous constant, and the superposed dot indicates time derivative. It is understood that the spring element responds instantly to a stress, while the dashpot cannot respond instantly (because its response is rate dependent). Let ε_1 be the strain in the spring and ε_2 be the strain

Figure 9.2.1. The Maxwell element.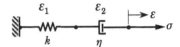

in the dashpot. When elements are connected in series, each element carries the same amount of stress while the strains are different in each element. We have

$$\dot{\varepsilon} = \dot{\varepsilon}_1 + \dot{\varepsilon}_2 = \frac{\dot{\sigma}}{k} + \frac{\sigma}{\eta}$$

or

$$\sigma + \frac{\eta}{k}\frac{d\sigma}{dt} = \eta\frac{d\varepsilon}{dt} \qquad [P(\sigma) = Q(\varepsilon)]. \qquad (9.2.12)$$

Thus, we have $M = N = 1$ [see Eqs. (9.1.1) and (9.1.2)] and $p_0 = 1$, $p_1 = \eta/k$, $q_0 = 0$ and $q_1 = \eta$.

9.2.2.1 Creep Response

Let $\sigma = \sigma_0\,H(t)$. Then differential equation in (9.2.12) simplifies to

$$q_1\frac{d\varepsilon}{dt} = p_1\sigma_0\delta(t) + p_0\sigma_0\,H(t). \qquad (9.2.13)$$

The Laplace transform of Eq. (9.2.13) is

$$q_1\left[s\bar{\varepsilon}(s) - \varepsilon(0)\right] = \sigma_0\left(p_1 + \frac{p_0}{s}\right).$$

Assuming that $\varepsilon(0) = 0$, we obtain

$$\bar{\varepsilon}(s) = \sigma_0\left(\frac{p_1}{q_1 s} + \frac{p_0}{q_1 s^2}\right).$$

The inverse transform gives the creep response

$$\varepsilon(t) = \frac{\sigma_0}{q_1}(p_1 + p_0 t) = \frac{\sigma_0}{k}\left(1 + \frac{t}{\tau}\right) \quad \text{for } t > 0, \qquad (9.2.14)$$

where τ is the *retardation time* or *relaxation time*,

$$\tau = \frac{\eta}{k}. \qquad (9.2.15)$$

Note that $\varepsilon(0^+) = \sigma_0/k$. The coefficient of σ_0 in Eq. (9.2.14) is called the *creep compliance*, denoted by $J(t)$

$$J(t) = \frac{1}{k}\left(1 + \frac{t}{\tau}\right). \qquad (9.2.16)$$

The creep response of the Maxwell model is shown in Figure 9.2.2(a).

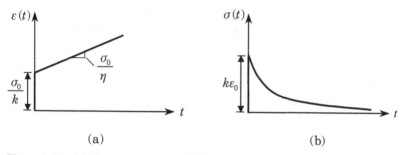

Figure 9.2.2. (a) Creep response and (b) relaxation response of the Maxwell element.

9.2.2.2 Relaxation Response

Let $\varepsilon = \varepsilon_0 H(t)$. Then Eq. (9.2.12) reduces to

$$p_1 \frac{d\sigma}{dt} + p_0 \sigma = q_1 \varepsilon_0 \delta(t). \tag{9.2.17}$$

The Laplace transform of the above equation is

$$p_1 (s\bar{\sigma} - \sigma(0)) + p_0 \bar{\sigma} = q_1 \varepsilon_0.$$

Using the initial condition $\sigma(0) = 0$, we write

$$\bar{\sigma}(s) = \varepsilon_0 \left(\frac{q_1}{p_0 + p_1 s} \right) = \frac{q_1}{p_1} \varepsilon_0 \left(\frac{1}{\frac{p_0}{p_1} + s} \right),$$

whose inverse transform is

$$\sigma(t) = \frac{q_1}{p_1} \varepsilon_0 e^{-p_0 t / p_1} = k \varepsilon_0 e^{-t/\tau}, \quad \text{for } t > 0. \tag{9.2.18}$$

The coefficient of ε_0 in Eq. (9.2.18) is called the *relaxation modulus*

$$Y(t) = k e^{-t/\tau}. \tag{9.2.19}$$

The relaxation response of the Maxwell model is shown in Figure 9.2.2(b).

Note that the relaxation modulus $Y(t)$ can also be obtained using Eq. (9.2.10). We have

$$\bar{J}(s) = \frac{1}{ks^2} \left(s + \frac{1}{\tau} \right),$$

and

$$Y(t) = \mathcal{L}^{-1} \left[\frac{1}{s^2 \bar{J}(s)} \right] = \mathcal{L}^{-1} \left[\frac{k}{(s + \frac{1}{\tau})} \right] = k e^{-t/\tau},$$

which is the same as that in Eq. (9.2.19).

Figure 9.2.3 shows the creep and relaxation responses of the Maxwell model in a standard test in which the stress and strain are monitored to see the creep and relaxation during the same test.

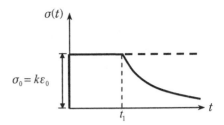

Figure 9.2.3. A standard test of a Maxwell fluid.

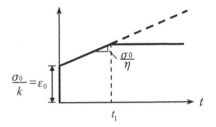

A *generalized Maxwell model* consists of N Maxwell elements in parallel and a free spring (k_0) in series. The relaxation response of the generalized Maxwell model is of the form [see Eq. (9.2.18)]

$$\sigma(t) = \varepsilon_0 \left[k_0 + \sum_{n=1}^{N} k_n e^{-\frac{t}{\tau_n}} \right], \quad \tau_n = \frac{\eta_n}{k_n}. \tag{9.2.20}$$

The relaxation modulus of the generalized Maxwell model is

$$Y(t) = k_0 + \sum_{n=1}^{N} k_n e^{-\frac{t}{\tau_n}}. \tag{9.2.21}$$

9.2.3 Kelvin–Voigt Element

The Kelvin–Voigt element of Figure 9.2.4 consists of a linear elastic spring element in parallel with a dashpot element. The stress–strain relation for the model is derived as follows. Let σ_1 be the stress in the spring and σ_2 be the stress in the dashpot. Each element carries the same amount of strain. Then

$$\sigma = \sigma_1 + \sigma_2 = k\varepsilon + \eta \frac{d\varepsilon}{dt}. \tag{9.2.22}$$

We have $p_0 = 1$, $q_0 = k$, and $q_1 = \eta$.

Figure 9.2.4. The Kelvin–Voigt solid element.

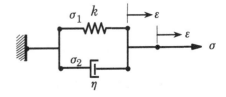

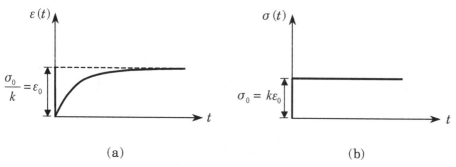

Figure 9.2.5. (a) Creep response and (b) relaxation response of the Kelvin–Voigt element.

9.2.3.1 Creep Response

Let $\sigma = \sigma_0 H(t)$. Then the differential equation in (9.2.22) becomes

$$q_1 \frac{d\varepsilon}{dt} + q_0 \varepsilon = p_0 \sigma_0 H(t). \tag{9.2.23}$$

The Laplace transform of the equation yields (with zero initial condition)

$$\bar{\varepsilon}(s) = \frac{p_0 \sigma_0}{q_1} \frac{1}{s\left(s + \frac{q_0}{q_1}\right)} = \frac{p_0 \sigma_0}{q_0} \left[\frac{1}{s} - \frac{1}{\left(s + \frac{q_0}{q_1}\right)}\right].$$

The inverse is

$$\varepsilon(t) = \frac{p_0 \sigma_0}{q_0}\left(1 - e^{-\frac{q_0}{q_1}t}\right) = \frac{\sigma_0}{k}\left(1 - e^{-\frac{t}{\tau}}\right). \tag{9.2.24}$$

The creep response of the Kelvin–Voigt model is shown in Figure 9.2.5(a). Note that in the limit $t \to \infty$, the strain attains the value $\varepsilon_\infty = \sigma_0/k$. The creep compliance of the Kelvin–Voigt model is

$$J(t) = \frac{1}{k}\left(1 - e^{-\frac{t}{\tau}}\right). \tag{9.2.25}$$

9.2.3.2 Relaxation Response

Let $\varepsilon(t) = \varepsilon_0 H(t)$ in Eq. (9.2.22). We obtain

$$\sigma(t) = \varepsilon_0 \left[q_0 H(t) + q_1 \delta(t)\right] = J(t)\varepsilon_0, \quad Y(t) = \left[k H(t) + \eta \delta(t)\right]. \tag{9.2.26}$$

Alternatively, we have

$$s^2 \bar{J}(s) = \frac{s}{\eta} \frac{1}{s + k/\eta}, \quad \bar{Y}(s) = \eta + \frac{k}{s},$$

from which we obtain $Y(t)$ as given in Eq. (9.2.26). The relaxation response of the Kelvin–Voigt model is shown in Figure 9.2.5(b). The creep and relaxation responses in the standard test of the Kelvin–Voigt model are shown in Figure 9.2.6.

A *generalized Kelvin–Voigt model* consists of N Kelvin–Voigt elements in series along with the Maxwell element, and it can be used to fit creep data to a

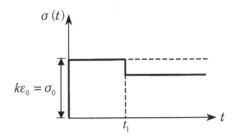

Figure 9.2.6. A standard test of a Kelvin–Voigt solid.

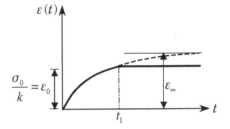

high degree. The creep compliance of the generalized Kelvin–Voigt model is [see Eq. (9.2.25)]

$$J(t) = \frac{1}{k_0} + \frac{t}{\eta_0} + \sum_{n=1}^{N} \frac{1}{k_n}\left(1 - e^{-\frac{t}{\tau_n}}\right), \quad \tau_n = \frac{\eta_n}{k_n}. \tag{9.2.27}$$

9.2.4 Three-Element Models

There are two three-element models, as shown in Figures 9.2.7(a) and 9.2.7(b). In the first one, an extra spring element is added in series to the Kelvin–Voigt element, and in the second one, a spring element is added in parallel to the Maxwell element. The constitutive equations for the two models are derived as follows.

For the three-element model in Figure 9.2.7(a), we have

$$\sigma = \sigma_1 + \sigma_2, \quad \varepsilon = \varepsilon_1 + \varepsilon_2, \quad \sigma_1 = k_2\varepsilon_2, \quad \sigma_2 = \eta\dot{\varepsilon}_2, \quad \varepsilon_1 = \frac{\sigma}{k_1}. \tag{9.2.28}$$

Using the relations in (9.2.28) we obtain

$$\frac{\eta}{k_1}\frac{d\sigma}{dt} + \left(1 + \frac{k_2}{k_1}\right)\sigma = k_2\varepsilon + \eta\frac{d\varepsilon}{dt}. \tag{9.2.29}$$

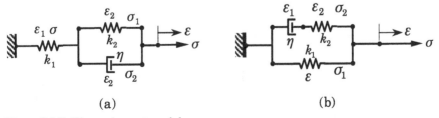

(a)   (b)

Figure 9.2.7. Three-element models.

Equation (9.2.29) is of the form $P(\sigma) = Q(\varepsilon)$

$$p_0\sigma + p_1\frac{d\sigma}{dt} = q_0\varepsilon + q_1\frac{d\varepsilon}{dt},$$

$$p_0 = 1 + \frac{k_2}{k_1}, \quad p_1 = \frac{\eta}{k_1}, \quad q_0 = k_2, \quad q_1 = \eta.$$

(9.2.30)

For the three-element model shown in Figure 9.2.7(b), we have

$$\sigma = \sigma_1 + \sigma_2, \quad \varepsilon = \varepsilon_1 + \varepsilon_2, \quad \varepsilon_1 = \frac{\sigma_2}{k_2}, \quad \dot{\varepsilon}_2 = \frac{\sigma_2}{\eta}, \quad \varepsilon = \frac{\sigma_1}{k_1}.$$

(9.2.31)

Combining the above relations, we arrive at

$$\frac{1}{\eta}\sigma + \frac{1}{k_2}\frac{d\sigma}{dt} = \frac{k_1}{\eta}\varepsilon + \left(1 + \frac{k_1}{k_2}\right)\frac{d\varepsilon}{dt},$$

or

$$p_0\sigma + p_1\frac{d\sigma}{dt} = q_0\varepsilon + q_1\frac{d\varepsilon}{dt},$$

$$p_0 = \frac{1}{\eta}, \quad p_1 = \frac{1}{k_2}, \quad q_0 = \frac{k_1}{\eta}, \quad q_1 = 1 + \frac{k_1}{k_2}.$$

(9.2.32)

Apparently, the three-element models represent the constitutive behavior of an ideal cross-linked polymer.

The creep and relaxation response of the three-element model shown in Figure 9.2.7(a) are studied next. Substituting $\sigma(t) = \sigma_0 H(t)$ into Eq. (9.2.30), we obtain

$$p_0\sigma_0 H(t) + p_1\sigma_0\delta(t) = q_0\varepsilon + q_1\frac{d\varepsilon}{dt}.$$

(9.2.33)

The Laplace transform of the above equation yields

$$(q_0 + q_1 s)\,\bar{\varepsilon}(s) = \sigma_0\left(\frac{p_0}{s} + p_1\right) \quad \text{or} \quad \bar{\varepsilon}(s) = \sigma_0\frac{(p_0 + p_1 s)}{s(q_0 + q_1 s)},$$

(9.2.34)

where zero initial conditions are used. We rewrite the above expression in a form suitable for inversion back to the time domain

$$\bar{\varepsilon}(s) = \sigma_0\left[\frac{p_0}{q_0}\left(\frac{1}{s} - \frac{1}{\frac{q_0}{q_1} + s}\right) + \frac{p_1}{q_1}\frac{1}{\left(\frac{q_0}{q_1} + s\right)}\right].$$

(9.2.35)

Using the inverse Laplace transform, we obtain

$$\varepsilon(t) = \sigma_0\left[\frac{p_0}{q_0}\left(1 - e^{-\frac{t}{\tau}}\right) + \frac{p_1}{q_1}e^{-\frac{t}{\tau}}\right], \quad \tau = \frac{q_1}{q_0}$$

$$= \sigma_0\left[\frac{k_1 + k_2}{k_1 k_2}\left(1 - e^{-\frac{t}{\tau}}\right) + \frac{1}{k_1}e^{-\frac{t}{\tau}}\right], \quad \tau = \frac{\eta}{k_2}.$$

(9.2.36)

Thus, the creep compliance is given by

$$J(t) = \left[\frac{k_1 + k_2}{k_1 k_2} \left(1 - e^{-\frac{t}{\tau}} \right) + \frac{1}{k_1} e^{-\frac{t}{\tau}} \right] = \frac{1}{k_1} + \frac{1}{k_2} \left(1 - e^{-\frac{t}{\tau}} \right). \tag{9.2.37}$$

For the relaxation response, let $\varepsilon(t) = \varepsilon_0 H(t)$ in Eq. (9.2.30) and obtain

$$p_0 \sigma + p_1 \frac{d\sigma}{dt} = q_0 \varepsilon_0 H(t) + q_1 \varepsilon_0 \delta(t). \tag{9.2.38}$$

The Laplace transform of the equation is

$$(p_0 + p_1 s) \bar{\sigma}(s) = \varepsilon_0 \left(\frac{q_0}{s} + q_1 \right) \quad \text{or} \quad \bar{\sigma}(s) = \varepsilon_0 \frac{(q_0 + q_1 s)}{s(p_0 + p_1 s)}, \tag{9.2.39}$$

where zero initial conditions are used. We rewrite the above expression in the form

$$\bar{\sigma}(s) = \varepsilon_0 \left[\frac{q_0}{p_0} \left(\frac{1}{s} - \frac{1}{\frac{p_0}{p_1} + s} \right) + \frac{q_1}{p_1} \frac{1}{\left(\frac{p_0}{p_1} + s \right)} \right]. \tag{9.2.40}$$

Using the inverse Laplace transform, we obtain

$$\sigma(t) = \varepsilon_0 \left[\frac{q_0}{p_0} \left(1 - e^{-\frac{t}{\tau}} \right) + \frac{q_1}{p_1} e^{-\frac{t}{\tau}} \right], \quad \tau = \frac{p_1}{p_0}$$

$$= \varepsilon_0 \left[\frac{k_1 k_2}{k_1 + k_2} \left(1 - e^{-\frac{t}{\tau}} \right) + k_1 e^{-\frac{t}{\tau}} \right], \quad \tau = \frac{\eta}{k_1 + k_2}. \tag{9.2.41}$$

Thus, the relaxation modulus is given by

$$Y(t) = \left[\frac{k_1 k_2}{k_1 + k_2} \left(1 - e^{-\frac{t}{\tau}} \right) + k_1 e^{-\frac{t}{\tau}} \right], \quad \tau = \frac{\eta}{k_1 + k_2}. \tag{9.2.42}$$

Determination of the creep and relaxation responses of the three-element model in Figure 9.2.7(b) will be considered in Example 9.2.3.

9.2.5 Four-Element Models

The four-element models, such as the ones shown in Figure 9.2.8, have constitutive relations that involve second-order derivatives of stress and strain. Here we discuss the creep response of such models in general terms. The relaxation response follows along similar lines to what is discussed for creep response.

Consider the second-order differential equation

$$p_0 \sigma + p_1 \dot{\sigma} + p_2 \ddot{\sigma} = q_0 \varepsilon + q_1 \dot{\varepsilon} + q_2 \ddot{\varepsilon}. \tag{9.2.43}$$

Let $\sigma(t) = \sigma_0 H(t)$. We have

$$p_0 \sigma_0 H(t) + p_1 \sigma_0 \delta(t) + p_2 \sigma_0 \dot{\delta}(t) = q_0 \varepsilon + q_1 \dot{\varepsilon} + q_2 \ddot{\varepsilon}. \tag{9.2.44}$$

Taking the Laplace transform and assuming homogeneous initial conditions, we obtain

$$\sigma_0 \left(\frac{p_0}{s} + p_1 + p_2 s \right) = \left(q_0 + q_1 s + q_2 s^2 \right) \bar{\varepsilon}(s) \tag{9.2.45}$$

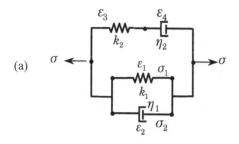

(a)

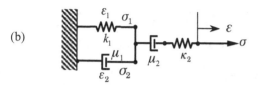

(b)

Figure 9.2.8. Four-element models.

or

$$\bar{\varepsilon}(s) = \sigma_0 \frac{p_0 + p_1 s + p_2 s^2}{s(q_0 + q_1 s + q_2 s^2)}. \tag{9.2.46}$$

To invert the above equation to determine $\varepsilon(t)$, first we write $q_2 s^2 + q_1 s + q_0$ as $q_2(s + \alpha)(s + \beta)$, where α and β are the roots of the equation $q_2 s^2 + q_1 s + q_0 = 0$:

$$\alpha = \frac{1}{2q_2}\left(q_1 - \sqrt{q_1^2 - 4q_2 q_0}\right), \quad \beta = \frac{1}{2q_2}\left(q_1 + \sqrt{q_1^2 - 4q_2 q_0}\right) \tag{9.2.47}$$

so that

$$\bar{\varepsilon}(s) = \sigma_0 \frac{p_0 + p_1 s + p_2 s^2}{q_2 s(s + \alpha)(s + \beta)}. \tag{9.2.48}$$

We write the solution in three parts for the case of real and unequal roots with $q_0 \neq 0$, $q_1 \neq 0$, and $q_2 \neq 0$:

$$\bar{\varepsilon}_1(s) = \sigma_0 \frac{p_0}{q_2}\left[\frac{1}{\alpha \beta s} - \frac{1}{\alpha(\beta - \alpha)(s + \alpha)} + \frac{1}{\beta(\beta - \alpha)(s + \beta)}\right], \tag{9.2.49}$$

$$\bar{\varepsilon}_2(s) = \sigma_0 \frac{p_1}{q_2}\left[\frac{1}{(\beta - \alpha)(s + \alpha)} - \frac{1}{(\beta - \alpha)(s + \beta)}\right], \tag{9.2.50}$$

$$\bar{\varepsilon}_3(s) = \sigma_0 \frac{p_2}{q_2}\left[-\frac{\alpha}{(\beta - \alpha)(s + \alpha)} + \frac{\beta}{(\beta - \alpha)(s + \beta)}\right]. \tag{9.2.51}$$

The solution is obtained by taking inverse Laplace transform

$$\varepsilon(t) = \frac{\sigma_0}{q_2}\left\{p_0\left[\frac{1}{\alpha \beta} - \frac{e^{-\alpha t}}{\alpha(\beta - \alpha)} + \frac{e^{-\beta t}}{\beta(\beta - \alpha)}\right]\right.$$

$$\left. + p_1\left[\frac{e^{-\alpha t}}{(\beta - \alpha)} - \frac{e^{-\beta t}}{(\beta - \alpha)}\right] + p_2\left[-\frac{\alpha e^{-\alpha t}}{(\beta - \alpha)} + \frac{\beta e^{-\beta t}}{(\beta - \alpha)}\right]\right\}. \tag{9.2.52}$$

When $q_2 = 0$, $q_1 \neq 0$, and $q_0 \neq 0$, Eq. (9.2.46) takes the form (with $\alpha = q_0/q_1$)

$$\bar{\varepsilon}(s) = \frac{\sigma_0}{q_1} \left[\frac{p_0}{\alpha} \left(\frac{1}{s} - \frac{1}{s+\alpha} \right) + \frac{p_1}{s+\alpha} + p_2 \left(1 - \frac{\alpha}{s+\alpha} \right) \right], \qquad (9.2.53)$$

and the solution is given by

$$\varepsilon(t) = \frac{\sigma_0}{q_1} \left[\frac{p_0}{\alpha} \left(1 - e^{-\alpha t} \right) + p_1 e^{-\alpha t} + p_2 \left(\delta(t) - \alpha e^{-\alpha t} \right) \right]. \qquad (9.2.54)$$

The Dirac delta function indicates that the model lacks impact response. That is, if a Dirac delta function appears in a relaxation function $Y(t)$, a finite stress is not sufficient to produce at once a finite strain, and an infinite one is needed.

When $q_0 = 0$, $q_1 \neq 0$, and $q_2 \neq 0$, Eq. (9.2.46) takes the form (with $\alpha = q_1/q_2$)

$$\bar{\varepsilon}(s) = \frac{\sigma_0}{q_2} \left[\frac{p_0}{\alpha^2} \left(\frac{\alpha}{s^2} - \frac{1}{s} + \frac{1}{s+\alpha} \right) + \frac{p_1}{\alpha} \left(\frac{1}{s} - \frac{1}{s+\alpha} \right) + \frac{p_2}{s+\alpha} \right], \qquad (9.2.55)$$

and the solution is given by

$$\varepsilon(t) = \frac{\sigma_0}{q_2} \left[\frac{p_0 t}{\alpha} + \frac{1}{\alpha} \left(p_1 - \frac{p_0}{\alpha} \right) \left(1 - e^{-\alpha t} \right) + p_2 e^{-\alpha t} \right]. \qquad (9.2.56)$$

This completes the general discussion of the creep response of four-element models. For the relaxation response the role of p's and q's is exchanged. Alternatively, we can use Eq. (9.2.10) to determine $Y(t)$.

EXAMPLE 9.2.3: Consider the differential equation in Eq. (9.2.32),

$$p_0 \sigma + p_1 \dot{\sigma} = q_0 \varepsilon + q_1 \dot{\varepsilon} \qquad (9.2.57)$$

with

$$p_0 = \frac{1}{\eta}, \quad p_1 = \frac{1}{k_2}, \quad p_2 = 0, \quad q_0 = \frac{k_1}{\eta}, \quad q_1 = \frac{k_1 + k_2}{k_2}, \quad q_2 = 0. \qquad (9.2.58)$$

Determine the creep and relaxation response.

SOLUTION: From Eq. (9.2.54), we have the creep response ($\alpha = q_0/q_1$)

$$\varepsilon(t) = \sigma_0 \frac{k_2}{k_1 + k_2} \left[\frac{1}{\alpha \eta} \left(1 - e^{-\alpha t} \right) + \frac{1}{k_2} e^{-\alpha t} \right]$$

$$= \sigma_0 \left[\frac{1}{k_1} \left(1 - e^{-\alpha t} \right) + \frac{1}{k_1 + k_2} e^{-\alpha t} \right], \quad \alpha = \frac{k_1 k_2}{\eta(k_1 + k_2)}. \qquad (9.2.59)$$

Thus, the creep compliance of the three-element model in Figure 9.2.7(b)

$$J(t) = \frac{1}{k_1} \left(1 - e^{-\alpha t} \right) + \frac{1}{k_1 + k_2} e^{-\alpha t}. \qquad (9.2.60)$$

The relaxation response is $\sigma(t) = Y(t)\varepsilon_0$ with $Y(t)$ computed as follows. We have

$$\bar{Y}(s) = \frac{1}{s^2 \bar{J}(s)}, \quad \bar{J}(s) = \frac{1}{k_1} \left(\frac{1}{s} - \frac{1}{s+\alpha} \right) + \frac{1}{k_1 + k_2} \frac{1}{s+\alpha}$$

and

$$s^2 \bar{J}(s) = \frac{s\left(s + \frac{k_2}{\eta}\right)}{(k_1 + k_2)(s + \alpha)}, \quad \frac{1}{s^2 \bar{J}(s)} = \frac{k_2}{s + \frac{k_2}{\eta}} + \frac{k_1}{s}. \tag{9.2.61}$$

Thus, the relaxation modulus is

$$Y(t) = k_1 + k_2 e^{-t/\tau}, \quad \tau = \frac{\eta}{k_2}. \tag{9.2.62}$$

EXAMPLE 9.2.4: Consider the differential equation

$$\ddot{\varepsilon} + \frac{k_2}{\eta_2}\dot{\varepsilon} = \frac{1}{k_1}\ddot{\sigma} + \left(\frac{1}{\eta_1} + \frac{1}{\eta_2} + \frac{k_2}{k_1 \eta_2}\right)\dot{\sigma} + \frac{k_2}{\eta_1 \eta_2}\sigma. \tag{9.2.63}$$

Thus, we have

$$q_0 = 0, \quad q_1 = \frac{k_2}{\eta_2}, \quad q_2 = 1, \quad p_0 = \frac{k_2}{\eta_1 \eta_2}, \quad p_1 = \frac{1}{\eta_1} + \frac{1}{\eta_2} + \frac{k_2}{k_1 \eta_2}, \quad p_2 = \frac{1}{k_1}. \tag{9.2.64}$$

Determine the creep and relaxation response of the model.

SOLUTION: The creep response is given by Eq. (9.2.56)

$$\varepsilon(t) = \frac{\sigma_0}{q_2}\left[\frac{p_0 t}{\alpha} + \frac{1}{\alpha}\left(p_1 - \frac{p_0}{\alpha}\right)(1 - e^{-\alpha t}) + p_2 e^{-\alpha t}\right]$$

$$= \sigma_0\left[\frac{1}{k_1} + \frac{t}{\eta_1} + \frac{1}{k_2}(1 - e^{-t/\tau})\right], \quad \tau = \frac{1}{\alpha} = \frac{\eta_2}{k_2}. \tag{9.2.65}$$

Thus, the creep compliance is

$$J(t) = \frac{1}{k_1} + \frac{t}{\eta_1} + \frac{1}{k_2}(1 - e^{-t/\tau}). \tag{9.2.66}$$

To determine the relaxation modulus, we compute

$$\bar{J}(s) = \frac{1}{k_1 s} + \frac{1}{\eta_1 s^2} + \frac{1}{k_2}\left(\frac{1}{s} - \frac{1}{s + \frac{1}{\tau}}\right),$$

$$s^2 \bar{J}(s) = \frac{s}{k_1} + \frac{1}{\eta_1} + \frac{1}{\eta_2}\left(\frac{s}{s + \frac{1}{\tau}}\right) = \frac{as^2 + bs + c}{d\left(s + \frac{1}{\tau}\right)}, \tag{9.2.67}$$

where

$$a = \eta_1 \eta_2, \quad b = (k_1 + k_2)\eta_1 + k_1 \eta_2, \quad c = k_1 k_2, \quad d = k_1 \eta_1 \eta_2. \tag{9.2.68}$$

Then

$$\bar{Y}(s) = \frac{1}{s^2 \bar{J}} = \frac{d\left(s + \frac{1}{\tau}\right)}{as^2 + bs + c} = \frac{d}{a}\left(\frac{A}{s + \alpha} + \frac{B}{s + \beta}\right) \tag{9.2.69}$$

where

$$\alpha = \frac{b}{2a} + \frac{1}{2a}\sqrt{b^2 - 4ac}, \quad \beta = \frac{b}{2a} - \frac{1}{2a}\sqrt{b^2 - 4ac},$$

$$A = -\frac{k_2 - \eta_2\alpha}{\eta_2(\alpha - \beta)}, \quad B = \frac{k_2 - \eta_2\beta}{\eta_2(\alpha - \beta)}. \tag{9.2.70}$$

It can be shown that $b^2 > 4ac$ and $\alpha > \beta > 0$ for $k_i > 0$ and $\eta_i > 0$. Hence, we have

$$Y(t) = \frac{k_1\eta_1}{\sqrt{b^2 - 4ac}} \left[-(k_2 - \eta_2\alpha)e^{-\alpha t} + (k_2 - \eta_2\beta)e^{-\beta t} \right]$$

$$= \frac{k_1\eta_1}{\sqrt{b^2 - 4ac}} \left[k_2 \left(e^{-\beta t} - e^{-\alpha t} \right) + \eta_2 \left(\alpha e^{-\alpha t} - \beta e^{-\beta t} \right) \right]. \tag{9.2.71}$$

9.3 Integral Constitutive Equations

9.3.1 Hereditary Integrals

The spring-and-dashpot elements are discrete models and are governed by differential equations. At $t = 0$, a stress σ_0 applied suddenly produces a strain $\varepsilon(t) = J(t)\sigma_0$ (see Figure 9.3.1). If the stress σ_0 is maintained unchanged, then $\varepsilon(t) = J(t)\sigma_0$ describes the strain for all $t > 0$. If we treat the material as linear, we can use the principle of linear superposition to calculate the strain produced in a given direction by the action of several loads of different magnitudes. If, at $t = t_1$, some more stress $\Delta\sigma_1$ is applied, then additional strain is produced which is proportional to $\Delta\sigma_1$ and

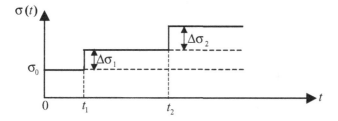

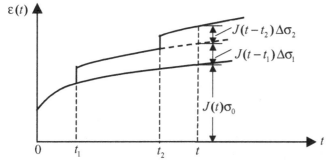

Figure 9.3.1. Strain response due to σ_0 and $\Delta\sigma_i$.

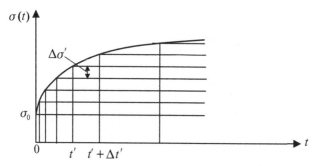

Figure 9.3.2. Linear superposition to derive hereditary integral.

depends on the same creep compliance. This additional strain is measured for $t > t'$. Hence, the total strain for $t > t_1$ is the sum of the strain due to σ_0 and that due to $\Delta\sigma_1$:

$$\varepsilon(t) = J(t)\sigma_0 + J(t - t_1)\Delta\sigma_1. \tag{9.3.1}$$

Similarly, if additional stress $\Delta\sigma_2$ is applied at time $t = t_2$, then the total strain for $t > t_2$ is

$$\varepsilon(t) = J(t)\sigma_0 + J(t - t_1)\Delta\sigma_1 + J(t - t_2)\Delta\sigma_2$$

$$= J(t)\sigma_0 + \sum_{i=1}^{2} J(t - t_i)\Delta\sigma_i. \tag{9.3.2}$$

If the stress applied is an arbitrary function of t, it can be divided into the first part $\sigma_0 H(t)$ and a sequence of infinitesimal stress increments $d\sigma(t')H(t - t')$ (see Fig. 9.3.2). The corresponding strain at time t can be written (using the Boltzmann's superposition principle)

$$\varepsilon(t) = J(t)\sigma_0 + \int_0^t J(t - t')d\sigma(t') = J(t)\sigma_0 + \int_0^t J(t - t')\frac{d\sigma(t')}{dt'} dt'. \tag{9.3.3}$$

Equation (9.3.3) indicates that the strain at any given time depends on all that has happened before, that is, on the entire stress history $\sigma(t')$ for $t' < t$.

This is in contrast to the elastic material whose strain only depends on the stress acting at that time only. Equation (9.3.3) is called a *hereditary integral*.

Equation (9.3.3) can be written in alternate form

$$\varepsilon(t) = J(t)\sigma(0) + [J(t - t')\sigma(t')]_0^t - \int_0^t \frac{dJ(t - t')}{dt'}\sigma(t') dt'$$

$$= J(0)\sigma(t) + \int_0^t \frac{dJ(t - t')}{d(t - t')}\sigma(t') dt' \tag{9.3.4}$$

$$= J(0)\sigma(t) + \int_0^t \frac{dJ(\tau)}{d\tau}\sigma(t - \tau) d\tau. \tag{9.3.5}$$

Equation (9.3.3) separates the strain caused by initial stress $\sigma(0)$ and that caused by stress increments. On the other hand, Eq. (9.3.5) separates the strain into the part that would occur if the total stress $\sigma(t)$ were applied at time t and additional strain produced due to creep.

It is possible to include the initial part due to σ_0 into the integral. For example, Eq. (9.3.3) can be written as

$$\varepsilon(t) = \int_{-\infty}^{t} J(t - t') \frac{d\sigma(t')}{dt'} dt'. \tag{9.3.6}$$

The fact that $J(t) = 0$ for $t < 0$ is used in writing the above integral, which is known as *Stieljes integral.*

Arguments similar to those presented for the creep compliance can be used to derive the hereditary integrals for the relaxation modulus $Y(t)$. If the strain history is known as a function of time, $\varepsilon(t)$, the stress is given by

$$\sigma(t) = Y(t)\varepsilon(0) + \int_{0}^{t} Y(t - t') \frac{d\varepsilon(t')}{dt'} dt' \tag{9.3.7}$$

$$= Y(0)\varepsilon(t) + \int_{0}^{t} \frac{dY(t')}{dt'} \varepsilon(t - t') dt' \tag{9.3.8}$$

$$= \int_{-\infty}^{t} Y(t - t') \frac{d\varepsilon(t')}{dt'} dt'. \tag{9.3.9}$$

EXAMPLE 9.3.1: Consider the stress history shown in Figure 9.3.3. Write the hereditary integral in Eq. (9.3.4) for the Maxwell model and Kelvin–Voigt model.

SOLUTION: The creep compliance of the Maxwell model is given in Eq. (9.2.16) as $J(t) = (1/k + t/\eta)$ with $J(0) = 1/k$. Then the strain response according to the hereditary integral in Eq. (9.3.4) is given by

For $t < t_1$: $\varepsilon(t) = \sigma_1 \dfrac{t}{t_1} \dfrac{1}{k} + \dfrac{\sigma_1}{t_1} \displaystyle\int_0^t t' \dfrac{1}{\eta} dt' = \dfrac{\sigma_1 t}{\eta t_1} \left(\dfrac{\eta}{k} + \dfrac{t}{2} \right).$ (9.3.10)

For $t > t_1$: $\varepsilon(t) = \sigma_1 \dfrac{1}{k} + \dfrac{\sigma_1}{t_1} \displaystyle\int_0^{t_1} t' \dfrac{1}{\eta} dt' + \sigma_1 \displaystyle\int_{t_1}^t 1 \cdot \dfrac{1}{\eta} dt'$

$$= \frac{\sigma_1}{\eta} \left(\frac{\eta}{k} + \frac{t_1}{2} + t \right). \tag{9.3.11}$$

By setting $t_1 = 0$, we obtain the same result as in Eq. (9.2.14).

The creep compliance of the Kelvin–Voigt model is given in Eq. (9.2.25). Then the strain response according to the hereditary integral in Eq. (9.3.4) is given by

For $t < t_1$: $\varepsilon(t) = \sigma_1 \dfrac{t}{t_1} \cdot 0 + \dfrac{\sigma_1}{\eta t_1} \displaystyle\int_0^t t' e^{-(t-t')/\tau} dt'$

$$= \frac{\sigma_1}{kt_1} \left[t - \frac{\eta}{k} \left(1 - e^{-t/\tau} \right) \right]. \tag{9.3.12}$$

For $t > t_1$: $\varepsilon(t) = \dfrac{\sigma_1}{\eta t_1} \displaystyle\int_0^{t_1} t' e^{-(t-t')/\tau} dt' + \dfrac{\sigma_1}{\eta} \displaystyle\int_{t_1}^t e^{-(t-t')/\tau} dt'$

$$= \frac{\sigma_1}{k} \left[1 + \frac{\eta}{kt_1} \left(1 - e^{t_1/\tau} \right) e^{-t/\tau} \right]. \tag{9.3.13}$$

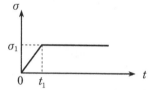

Figure 9.3.3. Stress history

By setting $t_1 = 0$ in Eq. (9.3.13), we obtain (use L'Hospital rule to deal with zero divided by zero condition) the same strain response as in Eq. (9.2.25). For $t \to \infty$, the strain goes to $\varepsilon = \sigma_1/k$, the same limit as if σ_1 were applied suddenly at $t = 0$ or $t = t_1$. This implies that the stress history is wiped out if sufficient time has elapsed. Thus, Kelvin–Voigt model represents the behavior of an elastic solid.

9.3.2 Hereditary Integrals for Deviatoric Components

The one-dimensional linear viscoelastic stress–strain relations developed in the previous sections can be extended in a straightforward manner to those relating the deviatoric stress components to the deviatoric strain components. Recall that the deviatoric stress and strain tensors are defined as

$$\text{deviatoric stress} \quad \boldsymbol{\sigma'} \equiv \boldsymbol{\sigma} - \tilde{\sigma}\mathbf{I}, \quad \left(\sigma'_{ij} = \sigma_{ij} - \frac{1}{3}\sigma_{kk}\delta_{ij} \right), \tag{9.3.14}$$

$$\text{deviatoric strain} \quad \boldsymbol{\varepsilon'} \equiv \boldsymbol{\varepsilon} - \frac{1}{3}\mathrm{tr}(\varepsilon), \quad \left(\varepsilon'_{ij} = \varepsilon_{ij} - \frac{1}{3}\varepsilon_{kk}\delta_{ij} \right), \tag{9.3.15}$$

where $\tilde{\sigma}$ is the mean stress and e is the dilatation

$$\text{mean stress} \quad \tilde{\sigma} \equiv \frac{1}{3}\sigma_{ii}, \quad \text{dilatation} \quad e \equiv \varepsilon_{ii}. \tag{9.3.16}$$

The constitutive equations between the deviatoric components of a linear elastic isotropic material are

$$\tilde{\sigma} = Ke, \quad \boldsymbol{\sigma'} = 2\mu\boldsymbol{\varepsilon'} \quad (\sigma'_{ij} = 2\mu\varepsilon'_{ij}). \tag{9.3.17}$$

Here K denotes the bulk modulus and μ is the Lamé constant (the same as the shear modulus), which are related to Young's modulus E and Poisson's ratio ν by

$$K = \frac{E}{3(1 - 2\nu)}, \quad \mu = G = \frac{E}{2(1 + \nu)}. \tag{9.3.18}$$

The linear viscoelastic strain–stress and stress–strain relations for the deviatoric components in Cartesian coordinates are

$$\varepsilon'_{ij}(t) = \int_{-\infty}^{t} J_s(t - t') \frac{d\sigma'_{ij}}{dt'} \, dt', \tag{9.3.19}$$

$$\varepsilon_{kk}(t) = \int_{-\infty}^{t} J_d(t - t') \frac{d\sigma_{kk}}{dt'} \, dt', \tag{9.3.20}$$

$$\sigma'_{ij}(t) = 2 \int_{-\infty}^{t} G(t - t') \frac{d\varepsilon'_{ij}}{dt'} dt', \qquad (9.3.21)$$

$$\sigma_{kk}(t) = 3 \int_{-\infty}^{t} K(t - t') \frac{d\varepsilon_{kk}}{dt'} dt' \qquad (9.3.22)$$

where $J_s(t)$ is the creep compliance in shear and J_d is the creep compliance in dilation. The general stress–strain relations may be written as

$$\sigma_{ij}(t) = 2 \int_{-\infty}^{t} G(t - t') \frac{d\varepsilon_{ij}(t')}{dt'} dt'$$
$$+ \delta_{ij} \int_{-\infty}^{t} \left[K(t - t') - \frac{2}{3} G(t - t') \right] \frac{d\varepsilon_{kk}(t')}{dt'} dt', \qquad (9.3.23)$$

$$\varepsilon_{ij}(t) = \int_{-\infty}^{t} J_s(t - t') \frac{d\sigma_{ij}(t')}{dt'} dt'$$
$$+ \frac{1}{3} \delta_{ij} \int_{-\infty}^{t} [J_d(t - t') - J_s(t - t')] \frac{d\sigma_{kk}(t')}{dt'} dt'. \qquad (9.3.24)$$

The Laplace transforms of Eqs. (9.3.19)–(9.3.22) are

$$\bar{\varepsilon}'_{ij}(s) = s \bar{J}_s(s) \bar{\sigma}'_{ij}(s), \quad \bar{\sigma}'_{ij}(s) = 2s \bar{G}(s) \bar{\varepsilon}'_{ij}(s), \qquad (9.3.25)$$

$$\bar{\varepsilon}_{kk}(s) = s \bar{J}_d(s) \bar{\sigma}_{kk}(s), \quad \bar{\sigma}_{kk}(s) = 3s \bar{K}(s) \bar{\varepsilon}_{kk}(s), \qquad (9.3.26)$$

from which it follows that

$$2\bar{G}(s) = \frac{1}{s^2 \bar{J}_s(s)}, \qquad (9.3.27)$$

$$3\bar{K}(s) = \frac{1}{s^2 \bar{J}_d(s)}. \qquad (9.3.28)$$

9.3.3 The Correspondence Principle

There exists certain correspondence between the elastic and viscoelastic solutions of a boundary value problem. The correspondence allows us to obtain solutions of a viscoelastic problem from that of the corresponding elastic problem.

Consider a one-dimensional elastic problem, such as a bar or beam, carrying certain applied loads F_i^0, $i = 1, 2, \cdots$. Suppose that the stress induced is σ^e. The strain is

$$\varepsilon^e = \sigma^e / E. \qquad (9.3.29)$$

Then consider the same structure but made of a viscoelastic material. Assume that the same loads are applied at time $t = 0$ and then held constant

$$F_i(t) = F_i^0 H(t).$$

The stress in the viscoelastic beam is $\sigma(t) = \sigma^e H(t)$. The strain in the viscoelastic structure is

$$\varepsilon(t) = J(t)\sigma^e. \tag{9.3.30}$$

For any time t, the strain in the viscoelastic structure is like the strain in an elastic beam of modulus $E = 1/J(t)$. Thus, we have the following *correspondence principle* (Part 1): If a viscoelastic structure is subjected to loads that are all applied simultaneously at $t = 0$ and then held constant, its stresses are the same as those in an elastic structure under the same loads, and its time-dependent strains and displacements are obtained from those of the elastic structure by replacing E by $1/J(t)$.

Next, consider an elastic structure in which the displacements are prescribed and held constant. Suppose that the displacement in the structure is u^e. The strain ε^e can be computed from the displacement u^e using an appropriate kinematic relation and stress σ using the constitutive equation

$$\sigma = E\varepsilon^e. \tag{9.3.31}$$

Then consider the same structure but made of a viscoelastic material. If we prescribe deflection $u(t) = u^e H(t)$, the strains produced are $\varepsilon(t) = \varepsilon^e H(t)$. The strain will produce a stress

$$\sigma(t) = Y(t)\varepsilon^e. \tag{9.3.32}$$

For any time t, the stress in the viscoelastic structure is like the stress in an elastic beam of modulus $E = Y(t)$. Thus, we have the second part of the *correspondence principle*: If a viscoelastic structure is subjected to displacements that are all imposed at $t = 0$ and then held constant, its displacements and strains are the same as those in the elastic structure under the same displacements, and its time-dependent stresses are obtained from those of the elastic structure by replacing E by $Y(t)$.

The ideas presented above for step loads or step displacements can be generalized to loads and displacements that are arbitrary functions of time. Let $w^e(\mathbf{x})$ be the deflection of a structure made of elastic material and subjected to a load $f_0(\mathbf{x})$. Then by the correspondence principle, the deflection of the same structure but made of viscoelastic material with creep compliance $J(t)$ and subjected to the step load $f(\mathbf{x}, t) = f_0(\mathbf{x})H(t)$ is

$$w(\mathbf{x}, t) = J(t)w^e(\mathbf{x}). \tag{9.3.33}$$

If the load history is of general type, $f(\mathbf{x}, t) = f_0(\mathbf{x})g(t)$, we can break the load history into a sequence of infinitesimal steps $dg(t')$, as shown in Figure 9.3.4. Then we can write the solution in the form of a hereditary integral

$$w(\mathbf{x}, t) = w^e(\mathbf{x})\left[g(0)J(t) + \int_0^t J(t - t')\frac{dg(t')}{dt'}\,dt'\right]. \tag{9.3.34}$$

Next we consider a number of examples to illustrate how to determine the viscoelastic response.

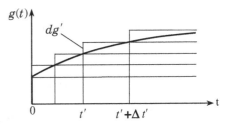

Figure 9.3.4. Load history as a sequence of infinitesimal load steps.

EXAMPLE 9.3.2: Consider a simply supported beam, as shown in Figure 9.3.5. At time $t = 0$, a point load P is placed at the center of the beam. Determine the viscoelastic center deflections using Maxwell's and Kelvin's models.

SOLUTION: The deflection at the center of the elastic beam is

$$w_0^e = \frac{PL^3}{48EI}. \tag{9.3.35}$$

For a viscoelastic beam, we replace $1/E$ with creep compliance $J(t)$ of a chosen viscoelastic material (e.g., Maxwell model or Kelvin model)

$$w_0^v(t) = J(t)\frac{PL^3}{48I}. \tag{9.3.36}$$

Using the Maxwell model, we can write [see Eq. (9.2.16)]

$$w_0^v(t) = \frac{1}{k}\left(1 + \frac{t}{\tau}\right)\frac{PL^3}{48I}, \quad \tau = \frac{\eta}{k}. \tag{9.3.37}$$

For the Kelvin model, we obtain [see Eq. (9.2.25)]

$$w_0^v(t) = \frac{1}{k}\left(1 - e^{-t/\tau}\right)\frac{PL^3}{48I}, \quad \tau = \frac{\eta}{k}. \tag{9.3.38}$$

Clearly, the response is quite different for the two materials.

EXAMPLE 9.3.3: Consider the simply supported beam of Figure 9.3.5 but with specified deflection w_0 at the center of the beam. Determine the viscoelastic center deflection.

SOLUTION: The force required to deflect the elastic beam at the center by w_0 is

$$P^e = \frac{48EIw_0}{L^3}. \tag{9.3.39}$$

Figure 9.3.5. A simply supported beam with a central point load.

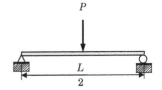

To obtain the load for a viscoelastic beam, we replace E with relaxation modulus $Y(t)$ of the viscoelastic material used

$$P^e(t) = Y(t)\frac{48Iw_0}{L^3}. \tag{9.3.40}$$

For the Maxwell model, we have the result [see Eq. (9.2.19)]

$$P_0^v(t) = ke^{-t/\tau}\frac{48Iw_0}{L^3}, \qquad \tau = \frac{\eta}{k}, \tag{9.3.41}$$

and for the Kelvin model, we obtain [see Eq. (9.2.26)]

$$P_0^v(t) = k\left[H(t) + \tau\,\delta(t)\right]\frac{48Iw_0}{L^3}, \qquad \tau = \frac{\eta}{k}. \tag{9.3.42}$$

EXAMPLE 9.3.4: Consider a simply supported beam with a uniformly distributed load of intensity q_0 as shown in Figure 9.3.6(a). Determine the viscoelastic deflection at the center.

SOLUTION: The elastic deflection of the beam is given by

$$w^e(x) = \frac{q_0L^4}{24EI}\left[\left(\frac{x}{L}\right) - 2\left(\frac{x}{L}\right)^3 + \left(\frac{x}{L}\right)^4\right]. \tag{9.3.43}$$

The midspan deflection is

$$w_0^e(L/2) = \frac{5q_0L^4}{384EI}. \tag{9.3.44}$$

For the load history shown in Figure 9.3.6(b), the midspan deflection of the viscoelastic beam is

$$w_0^v(L/2, t) = \frac{5q_0L^4}{384I}\frac{1}{t_1}\int_0^t J(t - t')\,dt', \quad 0 < t < t_1, \tag{9.3.45}$$

$$w_0^v(L/2, t) = \frac{5q_0L^4}{384I}\frac{1}{t_1}\int_0^{t_1} J(t - t')\,dt', \quad t > t_1. \tag{9.3.46}$$

For example, if we use the Kelvin–Voigt model, we obtain ($\tau = \eta/k$):

$$w_0^v(L/2, t) = \frac{5q_0L^4}{384I}\frac{1}{kt_1}\left[t - \frac{\eta}{k}\left(1 - e^{-t/\tau}\right)\right], \quad 0 < t < t_1, \tag{9.3.47}$$

$$w_0^v(L/2, t) = \frac{5q_0L^4}{384I}\frac{1}{k}\left[1 + \frac{\eta}{kt_1}\left(1 - e^{t_1/\tau}\right)e^{-t/\tau}\right], \quad t > t_1. \tag{9.3.48}$$

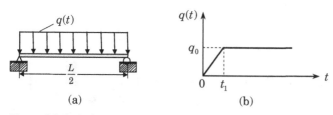

(a)   (b)

Figure 9.3.6. A simply supported beam with a uniform load.

Table 9.3.1. *Field equations of elastic and viscoelastic bodies*

Elasticity	Viscoelasticity
Equations of motion	
$\sigma_{ij,j} + f_i = \rho\ddot{u}_i$	$\sigma_{ij,j} + f_i = \rho\ddot{u}_i$
Strain–displacement equations	
$\varepsilon_{ij} = \frac{1}{2}(u_{i,j} + u_{j,i})$	$\varepsilon_{ij} = \frac{1}{2}(u_{i,j} + u_{j,i})$
Boundary conditions	
$u_i = \hat{u}_i$ on S_1	$u_i = \hat{u}_i$ on S_1
$t_i \equiv n_j\sigma_{ji} = \hat{t}_i$ on S_2	$t_i \equiv n_j\sigma_{ji} = \hat{t}_i$ on S_2
Constitutive equations	
$\sigma'_{ij} = 2G\varepsilon'_{ij}$	$\sigma'_{ij} = 2\int_{-\infty}^{t} G(t-t')\frac{d\varepsilon'_{ij}}{dt'}\,dt'$
$\sigma_{kk} = 3K\varepsilon_{kk}$	$\sigma_{kk} = 3\int_{-\infty}^{t} K(t-t')\frac{d\varepsilon_{kk}}{dt'}\,dt'$

9.3.4 Elastic and Viscoelastic Analogies

In this section, we examine the analogies between the field equations of elastic and viscoelastic bodies. These analogies help us to solve viscoelastic problems when solutions to the corresponding elastic problem are known. The field equations are summarized in Table 9.3.1 for the two cases. The Laplace transformed equations of elastic and viscoelastic bodies are summarized in Table 9.3.2. The correspondence is more apparent. A comparison of the Laplace transformed elastic and viscoelastic equations reveal the following correspondence

$$\sigma_{ij}^e(\mathbf{x}) \sim \bar{\sigma}_{ij}^v(\mathbf{x}, s), \quad \varepsilon_{ij}^e(\mathbf{x}) \sim \bar{\varepsilon}_{ij}^v(\mathbf{x}, s), \tag{9.3.49}$$

$$G^e(\mathbf{x}) \sim \bar{G}^*(\mathbf{x}, s) = s\bar{G}(\mathbf{x}, s) \quad K^e(\mathbf{x}) \sim \bar{K}^*(\mathbf{x}, s) = s\bar{K}(\mathbf{x}, s). \tag{9.3.50}$$

This correspondence allows us to use the solution of an elastic boundary value problem to obtain the transformed solution of the associated viscoelastic boundary-value

Table 9.3.2. *Field equations of elastic and Laplace transformed viscoelastic bodies for the quasi-static case*

Elasticity	Viscoelasticity
Equations of motion	
$\sigma_{ij,j} + f_i = 0$	$\bar{\sigma}_{ij,j} + \bar{f}_i = 0$
Strain-displacement equations	
$\varepsilon_{ij} = \frac{1}{2}(u_{i,j} + u_{j,i})$	$\bar{\varepsilon}_{ij} = \frac{1}{2}(\bar{u}_{i,j} + \bar{u}_{j,i})$
Boundary conditions	
$u_i = \hat{u}_i$ on S_1	$\bar{u}_i = \hat{\bar{u}}_i$ on S_1
$t_i \equiv n_j\sigma_{ji} = \hat{t}_i$ on S_2	$\hat{\bar{t}}_i \equiv n_j\sigma_{ji} = \hat{\bar{t}}_i$ on S_2
Constitutive equations	
$\sigma'_{ij} = 2G\varepsilon'_{ij}$	$\bar{\sigma}'_{ij} = s\bar{G}(s)\bar{\varepsilon}'_{ij} = G^*(s)\bar{\varepsilon}'_{ij}$
$\sigma_{kk} = 3K\varepsilon_{kk}$	$\bar{\sigma}_{kk} = 3s\bar{K}(s)\,\bar{\varepsilon}_{kk} = 3K^*(s)\,\bar{\varepsilon}_{kk}$

$G^*(s) = s\bar{G}(s), \quad K^*(s) = s\bar{K}(s).$

problem by simply replacing the elastic material properties G and K with G^* and K^*. One needs only to invert the solution to obtain the time-dependent viscoelastic solution. This analogy does not apply to problems for which the boundary conditions are time dependent.

The analogy also holds for the dynamic case, but it is between the Laplace transformed elastic variables and viscoelastic variables:

$$\bar{\sigma}_{ij}^e(\mathbf{x}, s) \sim \bar{\sigma}_{ij}^v(\mathbf{x}, s), \quad \bar{\varepsilon}_{ij}^e(\mathbf{x}, s) \sim \bar{\varepsilon}_{ij}^v(\mathbf{x}, s), \tag{9.3.51}$$

$$\bar{G}^e(\mathbf{x}, s) \sim \bar{G}^*(\mathbf{x}, s) = s\bar{G}(\mathbf{x}, s) \quad \bar{K}^e(\mathbf{x}, s) \sim \bar{K}^*(\mathbf{x}, s) = s\bar{K}(\mathbf{x}, s).$$

Next we consider an example of application of the elastic–viscoelastic analogy.

EXAMPLE 9.3.5: The structure shown in Figure 9.3.7 consists of a viscoelastic rod and elastic rod connected in parallel to a rigid bar. The areas of cross sections of the rods are the same. The modulus of the material of the rods are

Viscoelastic rod: $E(t) = 2\mu H(t) + 2\eta\delta(t).$

Elastic rod: $E =$ Young's modulus = constant. $\tag{9.3.52}$

If a load of $P(t) = P_0 H(t)$ acts on the rigid bar and the rigid bar is maintained horizontal, determine the resulting displacement of the rigid bar.

SOLUTION: Let u^e and $u^v(t)$ be the axial displacements in elastic and viscoelastic rods, respectively. Then the axial strains in elastic and viscoelastic rods are given by

$$\varepsilon^e = \frac{u^e}{L}, \quad \varepsilon^v(t) = \frac{u^v(t)}{L}. \tag{9.3.53}$$

The strain–stress relations for the two rods are

$$\varepsilon^e = \frac{\sigma^e}{E^e}, \quad \varepsilon^v(t) = \int_{-\infty}^{t} J(t - \tau)\frac{d\sigma^v}{d\tau}. \tag{9.3.54}$$

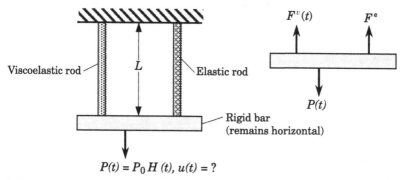

$P(t) = P_0 H (t),\ u(t) = ?$

Figure 9.3.7. Elastic–viscoelastic bar system.

The axial stresses in elastic and viscoelastic rods are given by

$$\sigma^e = \frac{F^e}{A}, \qquad \sigma^v(t) = \frac{F^v(t)}{A}. \tag{9.3.55}$$

From Eqs. (9.3.53)–(9.3.55) we have

$$u^e = \frac{F^e L}{E^e A}, \qquad u^v(t) = \frac{L}{A} \int_{-\infty}^{t} J(t-\tau) \frac{dF^v}{d\tau} \, d\tau, \tag{9.3.56}$$

where F^e and F^v are the axial forces in the elastic and viscoelastic rods, respectively. The geometric compatibility requires $u^e = u^v$, giving

$$\frac{F^e L}{A E^e} = \frac{L}{A} \int_{-\infty}^{t} J(t-\tau) \frac{dF^v}{d\tau} \, d\tau$$

or

$$F^e = E^e \int_{-\infty}^{t} J(t-\tau) \frac{dF^v}{d\tau} \, d\tau. \tag{9.3.57}$$

The force equilibrium requires

$$P(t) = F^v + F^e = F^v + E^e \int_{-\infty}^{t} J(t-\tau) \frac{dF^v}{d\tau} \, d\tau, \tag{9.3.58}$$

which is an integro-differential equation for $F^v(t)$.

Using the Laplace transform, we obtain

$$\frac{P_0}{s} = \left(1 + E^e s \bar{J}\right) \bar{F}^v. \tag{9.3.59}$$

Since $s\bar{J} = \frac{1}{s\bar{E}}$, we can write

$$\bar{J}(s) = \frac{1}{s^2 \bar{E}} = \frac{1}{s(2\eta s + 2\mu)} = \frac{1}{2\mu} \left(\frac{1}{s} - \frac{1}{s + \frac{\mu}{\eta}} \right), \tag{9.3.60}$$

and the inverse transform gives

$$J(t) = \frac{1}{2\mu} \left(1 - e^{-\frac{\mu t}{\eta}} \right). \tag{9.3.61}$$

Equation (9.3.59) takes the form

$$\bar{F}^v = \frac{P_0}{s} \left(\frac{s + \frac{\mu}{\eta}}{s + \alpha} \right), \qquad \alpha = \frac{2\mu + E^e}{2\eta},$$

$$= \frac{P_0}{2\mu + E^e} \left(\frac{2\mu}{s} - \frac{E^e}{s + \alpha} \right). \tag{9.3.62}$$

The inverse transform gives the force in the viscoelastic rod

$$F^v(t) = \frac{P_0}{2\mu + E^e} \left(2\mu - E^e e^{-\alpha t} \right). \tag{9.3.63}$$

Then from Eq. (9.3.56) we have

$$\bar{u}^v(s) = \frac{L}{A} s \bar{J} \bar{F}^v = \frac{P_0 L}{A s (s + \alpha)} = \frac{P_0 L}{A (2\mu + E^e)} \left[\frac{1}{s} - \frac{1}{s + \alpha} \right]. \tag{9.3.64}$$

The inverse transform yields the displacement

$$u^v(t) = \frac{P_0 L}{A(2\mu + E^e)}\left(1 - e^{-\alpha t}\right).$$ (9.3.65)

9.4 Summary

This chapter is dedicated to an introduction to linearized viscoelasticity. Beginning with simple spring-dashpot models of Maxwell and Kelvin–Voigt, three and four element models and integral constitutive models are discussed, and their creep and relaxation responses are derived. The discussion is then generalized to derive integral constitutive relations of viscoelastic materials. Analogies between elastic and viscoelastic solutions are discussed. Applications of the analogies to the solutions of some typical problems from mechanics of materials are presented. This chapter constitutes a good introduction to a course on theory of viscoelasticity.

PROBLEMS

9.1 *Method of partial fractions.* Suppose that we have a ratio of polynomials of the type

$$\frac{\bar{F}(s)}{\bar{G}(s)},$$

where $\bar{F}(s)$ is a polynomial of degree m and $\bar{G}(s)$ is a polynomial of degree n, with $n > m$. We wish to write in the form

$$\frac{\bar{F}(s)}{\bar{G}(s)} = \frac{c_1}{s + \alpha_1} + \frac{c_2}{s + \alpha_2} + \frac{c_3}{s + \alpha_3} + \cdots + \frac{c_n}{s + \alpha_n},$$

where c_i and α_i are constants to be determined using

$$c_i = \lim_{s \to -\alpha_i} \frac{(s + \alpha_i)\bar{F}(s)}{\bar{G}(s)}, \quad i = 1, 2, \ldots, n.$$

It is understood that $\bar{G}(s)$ is equal to the product $\bar{G}(s) = (s + \alpha_1)(s + \alpha_2)\ldots(s + \alpha_n)$. If

$$\bar{F}(s) = s^2 - 6, \quad \bar{G}(s) = s^3 + 4s^2 + 3s,$$

determine c_i.

9.2 Determine the creep and relaxation responses of the three-element model of Figure 9.2.7(b).

9.3 Derive the governing differential equation for the spring-dashpot model shown in Figure P9.3. Determine the creep compliance $J(t)$ and relaxation modulus $Y(t)$ associated with the model.

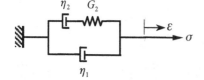

Figure P9.3.

9.4 Determine the relaxation modulus $Y(t)$ of the three-element model of Figure 9.2.7(a) using Eq. (9.2.10) and the creep compliance in Eq. (9.2.37) [i.e., verify the result in Eq. (9.2.42)].

9.5 Derive the governing differential equation for the mathematical model obtained by connecting the Maxwell element in *series* with the Kelvin–Voigt element (see Figure P9.5).

Figure P9.5.

9.6 Determine the creep compliance $J(t)$ and relaxation modulus $Y(t)$ of the four-element model of Problem 9.5.

9.7 Derive the governing differential equation for the mathematical model obtained by connecting the Maxwell element in *parallel* with the Kelvin–Voigt element (see Figure P9.7).

Figure P9.7.

9.8 Derive the governing differential equation of the four-parameter solid shown in Figure P9.8. Show that it degenerates into the Kelvin–Voigt solid when its components parts are made equal.

Figure P9.8.

9.9 Determine the creep compliance $J(t)$ and relaxation modulus $Y(t)$ of the four-element model of Problem 9.7.

9.10 If a strain of $\varepsilon(t) = \varepsilon_0 t$ is applied to the four-element model of Problem 9.7, determine the stress $\sigma(t)$ using a suitable hereditary integral [use $Y(t)$ from Problem 9.9].

9.11 For the three-element model of Figure 9.2.7(b), determine the stress $\sigma(t)$ when the applied strain is $\varepsilon(t) = \varepsilon_0 + \varepsilon_1 t$, where ε_0 and ε_1 are constants.

9.12 Determine expressions for the (Laplace) transformed modulus $\bar{E}(s)$ and Poisson's ratio $\bar{\nu}$ in terms of the transformed bulk modulus $\bar{K}(s)$ and transformed shear modulus $\bar{G}(s)$.

9.13 Evaluate the hereditary integral in Eq. (9.3.4) for the three-element model of Figure 9.2.7(a) and stress history shown in Figure 9.3.3.

9.14 Given that the shear creep compliance of a Kelvin–Voigt viscoelastic material is

$$J(t) = \frac{1}{2G_0}(1 - e^{-t/\tau}),$$

where G_0 and τ are material constants, determine the following properties for this material:

   (a) shear relaxation modulus, $2G(t)$,

   (b) the differential operators P and Q of Eq. (9.1.1),

   (c) integral form of the stress–strain relation, and

   (d) integral form of the strain–stress relation.

9.15 The strain in a uniaxial viscoelastic bar with viscoelastic modulus $E(t) = E_0/(1 + t/C)$ is $\varepsilon(t) = At$, where E_0, C, and A are constants. Determine the stress $\sigma(t)$ in the bar.

9.16 Determine the free end deflection $w^v(t)$ of a cantilever beam of length L, moment of inertia I, and subjected to a point load $P(t)$ at the free end, for the cases (a) $P(t) = P_0 H(t)$ and (b) $P(t) = P_0 e^{-\alpha t}$. The material of the beam has the relaxation modulus of $E(t) = Y(t) = A + B e^{-\alpha t}$.

9.17 A cantilever beam of length L is made of a viscoelastic material that can be represented by the three-parameter solid shown in Fig. 9.2.7(a). The beam carries a load of $P(t) = P_0 H(t)$ at its free end. Assuming that the second moment of area of the beam is I, determine the tip deflection.

9.18 A simply supported beam of length L, second moment of area I is made from the Kelvin–Voigt type viscoelastic material whose compliance constitutive response is

$$J(t) = \frac{1}{E_0}(1 - e^{-t/\tau}),$$

where E_0 and τ are material constants. The beam is loaded by a transverse distributed load

$$q(x, t) = q_0 \left(1 - \frac{x}{L}\right) t^2 = f(x) g(t),$$

where q_0 is the intensity of the distributed load at $x = 0$ and $g(t) = t^2$. Determine the deflection and stress in the viscoelastic beam using the Euler–Bernoulli beam theory.

9.19 The pin-connected structure shown in Figure P9.19 is made from an incompressible viscoelastic material whose shear response can be expressed as

$$P = 1 + \frac{\eta}{\mu} \frac{d}{dt}, \quad Q = \eta \frac{d}{dt},$$

where η and μ are material constants. The structure is subjected to a time-dependent vertical force $P(t)$, as shown in Figure P9.19. Determine the vertical load $P(t)$ required to produce this deflection history. Assume that member AB has an area of cross-section $A_1 = 9/16$ in.^{2} and member BC has an area of cross-section $A_2 = 125/48$ in.^{2}.

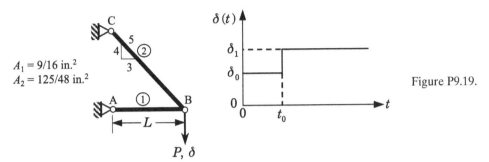

Figure P9.19.

9.20 Consider a hallow thick-walled spherical pressure vessel composed of two different viscoelastic materials, as shown in Figure P9.20. *Formulate* (you need not obtain complete solution to) the boundary value problem from which the stress and displacement fields may be determined.

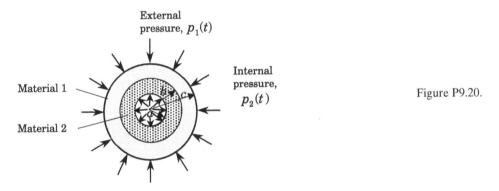

Figure P9.20.

References

[1] R. Aris, *Vectors, Tensors, and the Basic Equations in Fluid Mechanics,* Prentice Hall Englewood Cliffs, NJ (1962).

[2] J. P. Aubin, *Applied Functional Analysis,* John Wiley & Sons, New York (1979).

[3] E. Betti, "Teoria della' Elasticita," *Nuovo Cimento,* Serie 2, Tom VII and VIII (1872).

[4] R. B. Bird, W. E. Stewart, and E. N. Lightfoot, *Transport Phenomena,* John Wiley & Sons, New York (1960).

[5] R. B. Bird, R. C. Armstrong, and O. Hassager, *Dynamics of Polymeric Liquids, Vol. 1: Fluid Mechanics,* 2nd ed. John Wiley & Sons, New York (1971).

[6] J. Bonet and R. D. Wood, *Nonlinear Continuum Mechanics for Finite Element Analysis,* Cambridge University Press, New York (1997).

[7] P. Chadwick, *Continuum Mechanics: Concise Theory and Problems,* 2nd ed., Dover, Mineola, NY (1999).

[8] T. J. Chung, *Applied Continuum Mechanics,* Cambridge University Press, New York (1996).

[9] A. C. Eringen and G. W. Hanson, *Nonlocal Continuum Field Theories,* Springer-Verlag, New York (2002).

[10] W. Flügge, *Viscoelasticity,* 2nd ed., Springer-Verlag, New York (1975).

[11] Y. C. Fung, *First Course in Continuum Mechanics,* 3rd ed., Prentice Hall, Englewood Cliffs, NJ (1993).

[12] F. R. Gantmacher, *The Theory of Matrices,* Chelsea, New York (1959).

[13] M. E. Gurtin, *An Introduction to Continuum Mechanics,* Elsevier Science & Technology, San Diego, CA (1981).

[14] K. D. Hjelmstad, *Fundamentals of Structrual Mechanics,* 2nd ed., Springer, New York, 2005.

[15] H. Hochstadt, *Special Functions of Mathematical Physics,* Holt, New York (1961).

[16] W. F. Hughes and F. J. Young, *The Electromagnetodynamics of Fluids,* John Wiley & Sons, New York (1966).

[17] J. D. Jackson, *Classical Electrodynamics,* 2nd ed., John Wiley & Sons, New York (1975).

[18] W. M. Lai, D. Rubin, and E. Krempl, *Introduction to Continuum Mechanics,* 3rd ed., Elsevier Science & Technology, San Diego, CA (1997).

[19] E. H. Lee, "Viscoelasticity," in *Handbook of Engineering Mechanics,* McGraw-Hill, New York (1962).

[20] I.-S. Liu, *Continuum Mechanics,* Springer-Verlag, New York (2002).

[21] L. E. Malvern, *Introduction to the Mechanics of a Continuous Medium,* Prentice Hall, Englewood Cliffs, NJ (1997).

[22] G. T. Mase and G. E. Mase, *Continuum Mechanics for Engineers,* 2nd ed., CRC Press, Boca Raton, FL (1999).

[23] J. C. Maxwell, "On the Calculation of the Equilibrium and the Stiffness of Frames," *Philosophical Magazine Serial 4,* **27,** 294 (1864).

[24] N. I. Mushkelishvili, *Some Basic Problems of the Mathematical Theory of Elasticity,* Noordhoff, Gröningen, the Netherlands (1963).

[25] A. W. Naylor and G. R. Sell, *Linear Operator Theory in Engineering and Science,* Holt, Reinhart and Winston, New York (1971).

[26] W. Noll, *The Non-Linear Field Theories of Mechanics,* 3rd ed., Springer-Verlag, New York (2004).

[27] R. W. Ogden, *Non-Linear Elastic Deformations,* Halsted (John Wiley & Sons), New York (1984).

[28] W. Prager, *Introduction to Mechanics of Continua,* Dover, Mineola, NY (2004).

[29] J. N. Reddy, *Applied Functional Analysis and Variational Methods in Engineering,* McGraw-Hill, New York (1986); reprinted by Krieger, Malabar, FL (1991).

[30] J. N. Reddy, *Theory and Analysis of Elastic Plates and Shells,* 2nd ed., Taylor & Francis, Philadelphia (2007).

[31] J. N. Reddy, *Energy Principles and Variational Methods in Applied Mechanics,* 2nd ed., John Wiley & Sons, New York (2002).

[32] J. N. Reddy, *Mechanics of Laminated Composite Plates and Shells: Theory and Analysis,* 2nd ed., CRC Press, Boca Raton, FL (2004).

[33] J. N. Reddy, *An Introduction to the Finite Element Method,* 3rd ed., McGraw-Hill, New York (2006).

[34] J. N. Reddy and D. K. Gartling, *The Finite Element Method in Heat Transfer and Fluid Dynamics,* 2nd ed., CRC Press, Boca Raton, FL (2001).

[35] J. N. Reddy and M. L. Rasmussen, *Advanced Engineering Analysis,* John Wiley, New York (1982); reprinted by Krieger, Malabar, FL (1991).

[36] H. Schlichting, *Boundary Layer Theory* (translated from German by J. Kestin), 7th ed., McGraw-Hill, New York (1979).

[37] L. A. Segel, *Mathematics Applied to Continuum Mechanics,* Dover, Mineola, New York (1987).

[38] W. S. Slaughter, *The Linearized Theory of Elasticity,* Birkhäser, Boston (2002).

[39] D. R. Smith and C. Truesdell, *Introduction to Continuum Mechanics,* Kluwer, the Netherlands (1993).

[40] I. S. Sokolnikoff, *Mathematical Theory of Elasticity,* 2nd ed., McGraw-Hill, New York; reprinted by Krieger, Melbourne, FL (1956).

[41] A. J. M. Spencer, *Continuum Mechanics,* Dover, Mineola, NY (2004).

[42] V. L. Streeter, E. B. Wylie, and K. W. Bedford *Fluid Mechanics,* 9th ed., McGraw-Hill, New York (1998).

[43] R. M. Temam and A. M. Miranville, *Mathematical Modelling in Continuum Mechanics,* 2nd ed., Cambridge University Press, New York (2005).

[44] S. P. Timoshenko and J. N. Goodier, *Theory of Elasticity,* 3rd ed., McGraw-Hill, New York (1970).

[45] C. A. Truesdell, *The Elements of Continuum Mechanics,* Springer-Verlag, New York (1984).

[46] C. Truesdell and R. A. Toupin, "The Classical Field Theories," in *Encyclopedia of Physics, III/1,* S. Flügge (ed.), Springer-Verlag, Berlin (1965).

[47] C. Truesdell and W. Noll, "The Non-Linear Field Theories of Mechanics," in *Encyclopedia of Physics, III/3,* S. Flügge (ed.), Springer-Verlag, Berlin (1965).

Answers to Selected Problems

Chapter 1

1.1 The equation of motion is

$$\frac{dv}{dt} + \alpha v = g, \quad \alpha = \frac{c}{m}.$$

1.2 The energy balance gives

$$-\frac{d}{dx}(Aq) + \beta P(T_\infty - T) + Ag = 0.$$

1.4 The conservation of mass gives

$$\frac{d(Ah)}{dt} = q_i - q_0,$$

where A is the area of cross section of the tank $(A = \pi D^2/4)$.

Chapter 2

2.1 The equation of (or any multiple of it) the required line is

$$\mathbf{C} \cdot [\mathbf{A} - (\mathbf{A} \cdot \hat{\mathbf{e}}_B)\,\hat{\mathbf{e}}_B] = 0.$$

2.2 The equation for the required plane is

$$(\mathbf{A} - \mathbf{B}) \times (\mathbf{B} - \mathbf{C}) \cdot (\mathbf{A} - \mathbf{C}) = 0.$$

2.6 (a) $S_{ii} = 12$. (b) $S_{ij}S_{ji} = 240$. (e) $S_{ij}A_j = \{18\ 15\ 34\}^{\mathrm{T}}$.

2.8 The vectors are linearly dependent.

2.10 (a) The transformation is defined by

$$\left\{ \begin{array}{c} \hat{\mathbf{e}}_1' \\ \hat{\mathbf{e}}_2' \\ \hat{\mathbf{e}}_3' \end{array} \right\} = \left[\begin{array}{ccc} \frac{1}{\sqrt{3}} & \frac{-1}{\sqrt{3}} & \frac{1}{\sqrt{3}} \\ \frac{2}{\sqrt{14}} & \frac{3}{\sqrt{14}} & \frac{1}{\sqrt{14}} \\ \frac{-4}{\sqrt{42}} & \frac{1}{\sqrt{42}} & \frac{5}{\sqrt{42}} \end{array} \right] \left\{ \begin{array}{c} \hat{\mathbf{e}}_1 \\ \hat{\mathbf{e}}_2 \\ \hat{\mathbf{e}}_3 \end{array} \right\}.$$

2.12 Follows from the definition

$$[L] = \begin{bmatrix} \frac{1}{\sqrt{2}} & 0 & \frac{1}{\sqrt{2}} \\ \frac{1}{2} & \frac{1}{\sqrt{2}} & -\frac{1}{2} \\ -\frac{1}{2} & \frac{1}{\sqrt{2}} & \frac{1}{2} \end{bmatrix}.$$

2.17 Note that $\left(\frac{\partial r}{\partial x_i} = x_i / r \right)$

$$\text{grad}(r^2) = 2r\hat{e}_i \frac{\partial r}{\partial x_i} = 2\hat{e}_i x_i = 2\mathbf{r}.$$

Use of the divergence theorem gives the required result.

2.18 Use the divergence theorem to obtain the required result.

2.19 The integral relations are obvious.

(a) The identity is obtained by substituting $\mathbf{A} = \phi \nabla \psi$ for $\mathbf{A}$ into Eq. (2.4.34).

2.20 See Problem 2.10(a) for the basis vectors of the barred coordinate system in terms of the unbarred system; the matrix of direction cosines $[L]$ is given there. Then the components of the dyad in the barred coordinate system are

$$[\bar{S}] = \begin{bmatrix} 2 & -\frac{14}{\sqrt{42}} & 0 \\ -\frac{14}{\sqrt{42}} & \frac{15}{14} & -\frac{37}{14\sqrt{3}} \\ 0 & -\frac{37}{14\sqrt{3}} & \frac{13}{14} \end{bmatrix}.$$

2.24 Begin with

$$[(\mathbf{S} \cdot \mathbf{A}) \times (\mathbf{S} \cdot \mathbf{B})] \cdot (\mathbf{S} \cdot \mathbf{C}) = e_{ijk} S_{ip} S_{jq} S_{kr} A_p B_q C_r$$

obtain

$$|S|e_{pqr} - e_{ijk} S_{ip} S_{jq} S_{kr} = 0.$$

2.25 Use the del operator from Table 2.4.2 to compute the divergence of the tensor $\mathbf{S}$.

2.30 (a) $\lambda_1 = 3.0$, $\lambda_2 = 2(1 + \sqrt{5}) = 6.472$, $\lambda_3 = 2(1 - \sqrt{5}) = -2.472$. The eigenvector components A_i associated with λ_3 are $\hat{\mathbf{A}}^{(3)} = \pm(0.5257, 0.8507, 0)$.

(c) The eigenvalues are $\lambda_1 = 4$, $\lambda_2 = 2$, $\lambda_3 = 1$.

(d) The eigenvalues are $\lambda_1 = 3$, $\lambda_2 = 2$, $\lambda_3 = -1$. The eigenvector associated with λ_1 is $\hat{\mathbf{A}}^{(1)} = \pm\frac{1}{\sqrt{2}}(1, 0, 1)$.

(f) The eigenvalues are $\lambda_1 = 3.24698$, $\lambda_2 = 1.55496$, $\lambda_3 = 0.19806$. The eigenvectors are $\hat{\mathbf{A}}^{(1)} = \pm(0.328, -0.737, 0.591)$; $\hat{\mathbf{A}}^{(2)} = \pm(0.591, -0.328, -0.737)$; $\hat{\mathbf{A}}^{(3)} = \pm(0.737, 0.591, 0.328)$.

2.31 The inverse is

$$[A]^{-1} = \frac{1}{12} \begin{bmatrix} 7 & -2 & 1 \\ -2 & 4 & -2 \\ 1 & -2 & 7 \end{bmatrix}.$$

Chapter 3

3.1 $\mathbf{v} = \frac{\mathbf{x}}{1+t}$, $\mathbf{a} = \frac{t}{(1+t)^2}\mathbf{v}$.

3.3 (c) $\left\{\begin{matrix} 1 \\ 2 \\ 1 \end{matrix}\right\}$.

3.4 (b) $[C] = \begin{bmatrix} k_1^2 & 0 & 0 \\ 0 & k_2^2 & 0 \\ 0 & 0 & k_3^2 \end{bmatrix}$.

3.5 (a) $[F] = \begin{bmatrix} k_1 & e_0 k_2 & 0 \\ 0 & k_2 & 0 \\ 0 & 0 & k_3 \end{bmatrix}$.

3.6 (c) $[F] = \begin{bmatrix} \cos At & \sin At & 0 \\ -\sin At & \cos At & 0 \\ 0 & 0 & 1+Bt \end{bmatrix}$.

3.7 (a) $u_1(\mathbf{X}) = AX_2$, $u_2(\mathbf{X}) = BX_1$, $u_3(\mathbf{X}) = 0$.

(c) $2[E] = \begin{bmatrix} B^2 & A+B & 0 \\ A+B & A^2 & 0 \\ 0 & 0 & 0 \end{bmatrix}$.

3.9 (c) $[F] = \begin{bmatrix} \cosh t & \sinh t & 0 \\ \sinh t & \cosh t & 0 \\ 0 & 0 & 1 \end{bmatrix}$.

3.11 (b) The angle ABC after deformation is $90 - \beta$, where $\cos\beta = \frac{\mu}{\sqrt{1+\mu^2}}$.

3.12 (a) $[E] = \frac{1}{2}([C] - [I]) = \begin{bmatrix} 6 & 7 & 0 \\ 7 & 8 & 0 \\ 0 & 0 & 0 \end{bmatrix}$.

3.13 $u_1 = \frac{e_0}{b}X_2$, $u_2 = 0$.

3.14 $u_1 = \left(\frac{e_0}{b^2}\right)X_2^2$, $u_2 = 0$.

3.15 $E_{11} = \frac{e_1}{a}\frac{X_2}{b} + \frac{1}{2}X_2^2\left(\frac{e_1^2+e_2^2}{a^2b^2}\right)$, $E_{22} = \frac{e_2}{b}\frac{X_1}{a} + \frac{1}{2}X_1^2\left(\frac{e_1^2+e_2^2}{a^2b^2}\right)$, $2E_{12} = \frac{e_1}{b}\frac{X_1}{a} + \frac{e_2}{a}\frac{X_2}{b} + X_1 X_2\left(\frac{e_1^2+e_2^2}{a^2b^2}\right)$.

3.16 $u_1 = -0.2X_1 + 0.5X_2$, $u_2 = 0.2X_1 - 0.1X_2 + 0.1X_1 X_2$.

3.17 $\varepsilon_{rr} = A$, $\varepsilon_{r\theta} = 0$, $\varepsilon_{rz} = 0$, $\varepsilon_{\theta\theta} = A$, $\varepsilon_{z\theta} = \frac{1}{2}\left(Br + \frac{C}{r}\cos\theta\right)$, $\varepsilon_{zz} = 0$.

3.19 The linear components are given by $\varepsilon_{11} = 3x_1^2 x_2 + c_1\left(2c_2^3 + 3c_2^2 x_2 - x_2^3\right)$, $\varepsilon_{22} = -\left(2c_2^3 + 3c_2^2 x_2 - x_2^3 + 3c_1 x_1^2 x_2\right)$, $2\varepsilon_{12} = x_1\left[x_1^2 + c_1\left(3c_2^2 - 3x_2^2\right)\right] - 3c_1 x_1 x_2^2$.

3.20 (b) The strain field is *not* compatible.

3.21 (b) $E_{11}'(= E_{nn}) \approx \frac{ae_0}{a^2+b^2}$, $E_{12}'(= E_{ns}) \approx \frac{e_0}{2b}\left(\frac{a^2-b^2}{a^2+b^2}\right)$.

3.22 The principal strains are $\varepsilon_1 = 0$ and $\varepsilon = 10^{-3}$ in./in. The principal direction associated with $\varepsilon_1 = 0$ is $\mathbf{A}_1 = \hat{\mathbf{e}}_1 - 2\hat{\mathbf{e}}_2$ and that associated with $\varepsilon = 10^{-3}$ is $\mathbf{A}_2 = 2\hat{\mathbf{e}}_1 + \hat{\mathbf{e}}_2$.

3.23 (c) $u_1 = cX_1X_2^2$, $u_2 = cX_1^2X_2$.

3.26 Use the definition (3.6.3) and Eqs. (3.6.14) and (3.7.1) as well as the symmetry of $\mathbf{U}$ to establish the result.

3.29 The function $f(X_2, X_3)$ is of the form $f(X_2, X_3) = A + BX_2 + CX_3$, where A, B, and C are arbitrary constants.

3.35 $[C] = \begin{bmatrix} 5.0 & 0.40 \\ 0.4 & 1.16 \end{bmatrix}$, $[U] = \begin{bmatrix} 2.2313 & 0.1455 \\ 0.1455 & 1.0671 \end{bmatrix}$.

3.36 $[U] = \begin{bmatrix} 0.707 & 0.707 & 0 \\ 0.707 & 2.121 & 0 \\ 0 & 0 & 1.0 \end{bmatrix}$, $[V] = \begin{bmatrix} 2.121 & 0.707 & 0 \\ 0.707 & 0.707 & 0 \\ 0 & 0 & 1.0 \end{bmatrix}$.

Chapter 4

4.3 (i)(a) $\mathbf{t}^{\hat{n}} = 2(\hat{\mathbf{e}}_1 + 7\hat{\mathbf{e}}_2 + \hat{\mathbf{e}}_3)$. (c) $\sigma_n = -7.33$ MPa, $\sigma_s = 12.26$ MPa.

4.4 (a) $\mathbf{t}^{\hat{n}} = \frac{1}{\sqrt{3}}(5\hat{\mathbf{e}}_1 + 5\hat{\mathbf{e}}_2 + 9\hat{\mathbf{e}}_3)10^3$ psi. (b) $\sigma_n = 6,333.33$ psi, $\sigma_s = 1,885.62$ psi. (c) $\sigma_{p1} = 6,656.85$ psi, $\sigma_{p2} = 1,000$ psi, $\sigma_{p3} = -4,656.85$ psi.

4.5 $\sigma_n = -2.833$ MPa, $\sigma_s = 8.67$ MPa.

4.6 $\sigma_n = 0.3478$ MPa, $\sigma_s = 4.2955$ MPa.

4.9 $\sigma_n = 3.84$ MPa, $\sigma_s = -17.99$ MPa.

4.11 $\sigma_n = -76.60$ MPa, $\sigma_s = 32.68$ MPa.

4.12 $\sigma_s = 90$ MPa.

4.13 $\sigma_{p1} = 972.015$ kPa, $\sigma_{p2} = -72.015$ kPa.

4.14 $\sigma_{p1} = 121.98$ MPa, $\sigma_{p2} = -81.98$ MPa.

4.15 $\sigma_1 = 11.824 \times 10^6$ psi, $\mathbf{n}^{(1)} = \pm(1, 0.462, 0.814)$.

4.17 $\lambda_1' = \frac{2}{3}$, $\lambda_2' = \frac{5}{3}$, $\lambda_3' = -\frac{7}{3}$; $\hat{\mathbf{n}}^{(1)} = -0.577\hat{\mathbf{e}}_1 + 0.577\hat{\mathbf{e}}_2 + 0.577\hat{\mathbf{e}}_3$.

4.18 $\lambda_1 = 6.856$, $\hat{\mathbf{A}}^{(1)} = \pm(0.42, 0.0498, -0.906)$.

4.19 (b) $t_n = -16.67$ MPa, $t_s = 52.7$ MPa.

4.20 $\sigma_1 = 25$ MPa, $\sigma_2 = 50$ MPa, $\sigma_3 = 75$ MPa;
$\hat{\mathbf{n}}^{(1)} = \pm\left(\frac{3}{5}\hat{\mathbf{e}}_1 - \frac{4}{5}\hat{\mathbf{e}}_3\right)$, $\hat{\mathbf{n}}^{(2)} = \pm\hat{\mathbf{e}}_2$, $\hat{\mathbf{n}}^{(3)} = \pm\left(\frac{4}{5}\hat{\mathbf{e}}_1 + \frac{3}{5}\hat{\mathbf{e}}_3\right)$.

Chapter 5

5.6 (a) Satisfies. (b) Satisfies.

5.7 $Q = \frac{b}{6}(3v_0 - c)$ m³/(s.m).

5.8 (a) $F = 24.12$N. (b) $F = 12.06$ N. (c) $F_x = 45$ N.

5.9 $v(t) = \sqrt{\frac{g}{L}(x^2 - x_0^2)}$, $a(t) = \frac{g}{L}x(t)$;

$v(t_0) = \sqrt{\frac{g}{L}(L^2 - x_0^2)} \approx \sqrt{gL}$ when $L \gg x_0$.

5.14 The proof of this identity requires the following identities (here $\mathbf{A}$ is a vector and ϕ is a scalar function):

$$\nabla \cdot (\nabla \times \mathbf{A}) = 0. \tag{1}$$

$$\nabla \times (\nabla \phi) = 0. \tag{2}$$

$$\mathbf{v} \cdot \text{grad } \mathbf{v} = \nabla \left(\frac{v^2}{2}\right) - \mathbf{v} \times \nabla \times \mathbf{v}. \tag{3}$$

$$\nabla \times (\mathbf{A} \times \mathbf{B}) = \mathbf{B} \cdot \nabla \mathbf{A} - \mathbf{A} \cdot \nabla \mathbf{B} + \mathbf{A} \, \text{div} \mathbf{B} - \mathbf{B} \, \text{div} \mathbf{A}. \tag{4}$$

5.17 $v_2 = 9.9$ m/s, $Q = 19.45$ Liters/s.

5.18 $\rho f_1 = 0$, $\rho f_2 = a\left(b^2 + 2x_1 x_2 - x_2^2\right)$, $\rho f_3 = -4abx_3$.

5.20 $\sigma_{12} = -\frac{P}{2I_3}\left(h^2 - x_2^2\right)$, $\sigma_{22} = 0$ $\left(I_3 = \frac{2bh^3}{3}\right)$.

5.21 (a) $T = 0.15$ N-m. (b) When $T = 0$, $\omega_0 = 477.5$ rpm.

5.22 $\omega = 16.21$ rad/s $= 154.8$ rpm.

5.24 $v_1 = 0.69$ m/s, $v_2 = 2.76$ m/s, loss $= 5.3665$ N $\cdot$ m/kg.

Chapter 6

6.2 $\begin{Bmatrix} \sigma_{11} \\ \sigma_{22} \\ \sigma_{23} \\ \sigma_{13} \\ \sigma_{12} \end{Bmatrix} = 10^6 \begin{Bmatrix} 37.8 \\ 43.2 \\ 27.0 \\ 21.6 \\ 0.0 \\ 5.4 \end{Bmatrix}$ Pa.

6.3 $I_1 = 108$ MPa, $I_2 = 2,507.76$ MPa^2, $I_3 = 25,666.67$ MPa^3;
$J_1 = 500 \times 10^{-6}$, $J_2 = 235 \times 10^{-9}$, $J_3 = -32 \times 10^{-12}$.

6.4 $I_1 = 78.8$ MPa, $I_2 = 1,062.89$ MPa^2, $I_3 = 17,368.75$ MPa^3.

6.5 $J_1 = 66.65 \times 10^{-6}$, $J_2 = 63,883.2 \times 10^{-12}$, $J_3 = 244,236 \times 10^{-18}$.

6.6 $\tau_{11} = 0$, $\tau_{22} = \frac{2\mu k}{1+kt}$, $\tau_{12} = \mu\left(\frac{4tk}{(1+kt)^2}x_2\right)$.

6.8 (1) Physical admissibility, (2) determinism, (3) equipresence, (4) local action, (5) material frame indifference, (6) material symmetry, (7) dimensionality, (8) memory, and (9) causality.

Chapter 7

7.1 $\sigma_{11} = 96.88$ MPa, $\sigma_{22} = 64.597$ MPa, $\sigma_{33} = 48.443$ MPa, $\sigma_{12} = 4.02$ MPa,
$\sigma_{13} = 0$ MPa, $\sigma_{23} = 0$.

7.4 $w_0\left(\frac{L}{2}\right) = -\left(\frac{5F_0L^3}{48EI} + \frac{17q_0L^4}{384EI}\right)$.

7.6 $w_A = 0.656$ in.

7.7 $w_c = \frac{q_0a^4}{64D}\left(\frac{5+\nu}{1+\nu}\right)$.

7.8 $w_c = \frac{q_0a^4}{(1+\nu)D}\left(\frac{5+\nu}{64} - \frac{6+\nu}{150}\right)$.

7.9 $w_c = \frac{43}{4800}\frac{q_0a^4}{D}$.

7.10 $w_{cb} = \frac{b^2 Q_0}{16\pi D}\left(2\log\frac{b}{a} + \frac{a^2}{b^2} - 1\right)$.

7.11 $u_z(r) = -\frac{\rho g a^2}{4\mu}\left(1 - \frac{r^2}{a^2}\right)$, $\sigma_{\theta z} = 0$, $\sigma_{zr} = \frac{\rho g}{2}r$.

7.13 $\sigma_{rr} = -\left(1 + \frac{4\mu}{3K}\right)p$, $\sigma_{\theta\theta} = \sigma_{\phi\phi} = -\left(1 - \frac{2\mu}{3K}\right)p$.

7.15 $u_\theta(r) = \frac{\tau_0 b^2}{2\mu a}\left(\frac{r}{a} - \frac{a}{r}\right)$, $\sigma_{r\theta} = \frac{b^2\tau_0}{r^2}$.

7.18 $\sigma_{xx} = 2D\left(3x^2y - 2y^3\right)$, $\sigma_{yy} = 2Dy^3$, $\sigma_{xy} = -6Dxy^2$.

7.21

$$\sigma_{xx} = \frac{3q_0}{10}\left(\frac{y}{b} + \frac{5a^2}{2b^2}\frac{x^2}{a^2}\frac{y}{b} - \frac{5}{3}\frac{y^3}{b^3}\right),$$

$$\sigma_{yy} = \frac{q_0}{4}\left(-2 - 3\frac{y}{b} + \frac{y^3}{b^3}\right),$$

$$\sigma_{xy} = \frac{3q_0a}{4b}\frac{x}{a}\left(1 - \frac{y^2}{b^2}\right).$$

7.22

$$\sigma_{xx} = \frac{\partial^2\Phi}{\partial y^2} = \frac{\tau_0}{4}\left(-\frac{2x}{b} - \frac{6xy}{b^2} + \frac{2a}{b} + \frac{6ay}{b^2}\right), \sigma_{yy} = \frac{\partial^2\Phi}{\partial x^2} = 0,$$

$$\sigma_{xy} = -\frac{\partial^2\Phi}{\partial x\partial y} = -\frac{\tau_0}{4}\left(1 - \frac{2y}{b} - \frac{3y^2}{b^2}\right).$$

7.24 $\sigma_{rr} = -\frac{2f_0}{\pi r}\sin\theta$, $\sigma_{\theta\theta} = 0$, $\sigma_{r\theta} = 0$.

7.25

$$\sigma_{31} = \frac{\partial\Psi}{\partial x_2} = \frac{\mu\theta}{a}x_2(x_1 - a)$$

$$\sigma_{32} = -\frac{\partial\Psi}{\partial x_1} = \frac{\mu\theta}{2a}\left(x_1^2 + 2ax_1 - x_2^2\right).$$

The angle of twist is $\theta = \frac{5\sqrt{3}T}{27\mu a^4}$.

7.27 The Euler equations are

$$\delta w: \quad -\frac{d}{dx}\left[GA\left(\phi+\frac{dw}{dx}\right)\right] - q = 0$$

$$\delta\phi: \quad -\frac{d}{dx}\left(EI\frac{d\phi}{dx}\right) + GA\left(\phi+\frac{dw}{dx}\right) = 0.$$

7.29 $w(0) = u_1 = -\frac{q_0 L^4}{24EI+kL^3}.$

7.32

$$L = \frac{1}{2}m_1\left[l^2\dot\theta^2 + \dot x^2 - 2l\dot x\dot\theta\sin\theta\right] + \frac{1}{2}m_2\dot x^2$$

$$+ m_1g(x - l\cos\theta) + m_2gx + \frac{1}{2}k(x+h)^2$$

where h is the elongation in the spring due to the masses $h = \frac{g}{k}(m_1 + m_2)$.

7.33 $\rho l\ddot x - \rho x$ or $\ddot x - \frac{g}{l}x = 0.$

7.34

$$\frac{\partial}{\partial t}\left(\rho A\frac{\partial w}{\partial t}\right) + \frac{\partial^2}{\partial x^2}\left(EI\frac{\partial^2 w}{\partial x^2}\right) - \frac{\partial^2}{\partial x\partial t}\left(\rho I\frac{\partial^2 w}{\partial x\partial t}\right) = q.$$

7.35

$$m(\ddot x + l\ddot\theta\cos\theta - l\dot\theta^2\sin\theta) + kx = F,$$

$$m\left[l\ddot x\cos\theta + (l^2 + \Omega^2)\ddot\theta\right] + mgl\sin\theta = 2aF\cos\theta.$$

7.36

$$\delta u_0: \quad -\frac{\partial N_{xx}}{\partial x} - f + \frac{\partial}{\partial t}\left(m_0\frac{\partial u}{\partial t}\right) = 0,$$

$$\delta w: \quad -\frac{\partial Q_x}{\partial x} - q + \frac{\partial}{\partial t}\left(m_0\frac{\partial w}{\partial t}\right) = 0,$$

$$\delta\phi: \quad -\frac{\partial M_{xx}}{\partial x} + Q_x + \frac{\partial}{\partial t}\left(m_2\frac{\partial\phi}{\partial t}\right) = 0.$$

Chapter 8

8.2 (a) The pressure at the top of the sea lab is $P = 1.2$ MN/m^2.

8.3 $\rho = 1.02$ kg/m^3.

8.4 $P = P_0\left(1 + \frac{mx_3}{\theta_0}\right)^{-g/mR}, \quad \rho = \rho_0\left(1 + \frac{mx_3}{\theta_0}\right)^{-g/mR}.$

8.5 $P(y) = P_0 + \rho gh\cos\alpha\left(1 - \frac{y}{h}\right), \quad U(y) = \frac{\rho gh^2\sin\alpha}{2\mu}\left(2\frac{y}{h} - \frac{y^2}{h^2}\right).$

8.7 The shear stress is given by

$$\tau_{rz} = -\left(\frac{d\bar P}{dz}\frac{r}{2} + \frac{1}{r}c_1\right) = -\frac{d\bar P}{dz}\frac{R}{4}\left[2\left(\frac{r}{R}\right) + (1 - \alpha^2)\frac{1}{\log\alpha}\left(\frac{R}{r}\right)\right],$$

where $d\bar P/dz = \frac{dP}{dz} + \rho g$

8.8 The velocity field is

$$v_\theta(r) = \frac{\Omega r_1^2}{r_1^2 - r_2^2}\left(r - \frac{r_2^2}{r}\right).$$

If $r_1 = R$ and $r_2 = \alpha R$ with $0 < \alpha < 1$, we have

$$v_\theta(r) = \frac{\Omega R}{1 - \alpha^2}\left(\frac{r}{R} - \alpha^2\frac{R}{r}\right).$$

The shear stress distribution is given by $\tau_{r\theta} = -2\mu\Omega\frac{\alpha^2}{1-\alpha^2}\left(\frac{R}{r}\right)^2$.

8.10 $P = -\rho g z + \frac{1}{2}\rho\Omega^2 r^2 + c$, where $c = P_0 + \rho g z_0$.

8.12 $v_x(y, t) = U_0 e^{-\eta}\cos(nt - \eta)$.

8.15 $\tau_{r\phi} = \frac{3\mu V_\infty}{2R}\left(\frac{R}{r}\right)^4\sin\phi$, $\quad P = P_0 - \rho g z - \frac{3\mu V_\infty}{2R}\left(\frac{R}{r}\right)^2\cos\phi$, where P_0 is the pressure in the plane $z = 0$ far away from the sphere and $-\rho g z$ is the contribution of the fluid weight (hydrostatic effect).

8.16 $-\frac{d}{dr}(r q_r) + r\rho Q_e = 0$.

8.17 $T(r) = T_0 + \frac{\rho Q_e R^2}{4k}\left[1 - \left(\frac{r}{R}\right)^2\right]$.

8.18 $\theta(x, t) = \sum_{n=1}^{\infty} B_n \sin\lambda_n x\, e^{-\alpha\lambda_n^2 t}$, $\quad B_n = \frac{2}{L}\int_0^L f(x)\sin\lambda_n x\, dx$.

8.20 $T(r) = T_0 - \frac{\mu\alpha^2 R_0^3}{9k}\left[1 - \left(\frac{r}{R_0}\right)^3\right]$.

8.21 $v_y(x) = \frac{\rho_r\beta_r g a^2(T_2 - T_1)}{12\mu}\left[\left(\frac{x}{a}\right)^3 - \left(\frac{x}{a}\right)\right]$.

Chapter 9

9.1 $-2H(t) + 2.5e^{-t} + 0.5e^{-3t}$.

9.2 $J(t) = \frac{1}{k_1} - \frac{k_2}{k_1(k_1+k_2)}e^{-t/\tau}$, $\quad Y(t) = k_1 + k_2 e^{-t/\tau}$.

9.3 $J(t) = \left[\frac{t}{\eta_1+\eta_2} + \frac{1}{G_2}\left(\frac{\eta_2}{\eta_1+\eta_2}\right)^2(1 - e^{-\alpha_2 t})\right]$, $\quad Y(t) = \eta_1\delta(t) + G_2 e^{-t/\tau_2}$, $\quad \tau_2 = \frac{\eta_2}{G_2}$,
$\alpha_2 = \frac{G_2}{\eta_1} + \frac{G_2}{\eta_2}$.

9.4 $Y(t) = \frac{k_1 k_2}{k_1+k_2}\left(1 - e^{-\lambda t}\right) + k_1 e^{-\lambda t}$, $\quad \lambda = \frac{k_1+k_2}{\eta}$.

9.5 $q_1\dot{\varepsilon} + q_2\ddot{\varepsilon} = p_0\sigma + p_1\dot{\sigma} + p_2\ddot{\sigma}$, where $p_0 = \frac{k_1}{\mu_1\mu_2}$, $\quad p_1 = \frac{k_1}{k_2\mu_1} + \frac{1}{\mu_1} + \frac{1}{\mu_2}$,
$p_2 = \frac{1}{k_2}$ $\quad q_1 = \frac{k_1}{\mu_1}$ $\quad q_2 = 1$.

9.6 $Y(t) = \frac{k_1 k_2\mu_2}{\lambda_1-\lambda_2}\left[(\lambda_1 - \alpha)e^{-\lambda_1 t} - (\lambda_2 - \alpha)e^{-\lambda_2 t}\right]$.

9.7 $q_0\varepsilon + q_1\dot{\varepsilon} + q_2\ddot{\varepsilon} = p_0\sigma + p_1\dot{\sigma}$, where $p_0 = \frac{1}{\eta_2}$, $\quad p_1 = \frac{1}{k_2}$, $\quad q_0 = \frac{k_1}{\eta_2}$,
$q_1 = 1 + \frac{k_1}{k_2} + \frac{\eta_1}{\eta_2}$, $\quad q_2 = \frac{\eta_1}{k_2}$.

9.9
$$J(t) = \frac{1}{q_2}\left\{p_0\left[\frac{1}{\alpha\beta} - \frac{e^{-\alpha t}}{\alpha(\beta - \alpha)} + \frac{e^{-\beta t}}{\beta(\beta - \alpha)}\right]\right.$$

$$\left. + p_1\left[\frac{e^{-\alpha t}}{(\beta - \alpha)} - \frac{e^{-\beta t}}{(\beta - \alpha)}\right] + p_2\left[-\frac{\alpha e^{-\alpha t}}{(\beta - \alpha)} + \frac{\beta e^{-\beta t}}{(\beta - \alpha)}\right]\right\}.$$

The relaxation response is $Y(t) = k_1 + k_2 e^{-\alpha t} + \eta_1 \delta(t)$.

9.10 $\sigma(t) = \left[k_1 + k_2 e^{-\alpha t} + \eta_1 \delta(t)\right] \varepsilon_0 + \left[t k_1 + \frac{k_2}{\alpha}\left(1 - e^{-\alpha t}\right) + \eta_1 H(t)\right] \varepsilon_0$.

9.11 $Y(t) = k_1 + k_2 e^{-t/\tau}$, $\tau = \frac{\eta}{k_2}$.

9.12 $\bar{E}(s) = \frac{9\bar{K}(s)\bar{G}(s)}{3\bar{K}(s)+\bar{G}(s)}$, $s\bar{v}(s) = \frac{3\bar{K}(s)-2\bar{G}(s)}{2[3\bar{K}(s)+\bar{G}(s)]}$.

9.13 $\varepsilon(t) = \sigma_1 \left(\frac{t}{k_1} + \frac{1}{k_2} e^{-t/\tau}\right)$, for $t > t_0$.

9.14 (a) $2G(t) = 2G_0[H(t) + \tau\delta(t)]$.

(c) $\sigma'_{ij}(t) = 2G(t)\varepsilon'_{ij}(0) + 2\int_0^t G(t-t') \frac{d\varepsilon'_{ij}(t')}{dt'} dt'$.

9.15 $\sigma(t) = \ln(1 + t/C)$.

9.16 (a) $w^v(L, t) = \frac{P_0 L^3}{3E_0 I}\left[-\frac{B}{A} e^{-\frac{A\alpha}{E_0} t} + \frac{E_0}{A} H(t)\right]$. (b) $w^v(L, t) = \frac{P_0 L^3}{3E_0 I} e^{-\frac{A\alpha}{E_0} t}$.

9.17 $w^v(L, t) = \frac{P_0 L^3}{3I}\left[\frac{p_0}{q_0} H(t) + \left(\frac{q_0 p_1 - q_0 p_1}{q_1 q_0}\right) e^{-(q_1/p_1)t}\right]$.

9.18 $w^v(x, t) = \frac{q_0 L^4}{360 I}\left(1 - \frac{x}{L}\right)\left[7 - 10\left(1 - \frac{x}{L}\right)^2 + 3\left(1 - \frac{x}{L}\right)^4\right] h(t)$, where $h(t) = \frac{2\tau^2}{E_0}\left(1 - e^{-t/\tau}\right) + \frac{\tau^2}{E_0}\left(\frac{t}{\tau}\right)\left[\left(\frac{t}{\tau}\right) - 2\right]$, $\sigma(x, t) = -Ez\frac{\partial^2 w^v}{\partial x^2} = \frac{q_0 L^2 z}{60 I}\left(1 - \frac{x}{L}\right)\frac{x}{L} h(t)$.

9.19 $P(t) = \frac{1}{2L}\left[\delta_0 E(t) + (\delta_1 - \delta_0)E(t - t_0)\right]$.

9.20 The Laplace transformed viscoelastic solutions for the displacements and stresses are obtained from

$$\bar{u}_r(r, s) = \bar{A}_i(s)r + \frac{\bar{B}_i(s)}{r^2},$$

$$\bar{\sigma}_{rr}(r, s) = (2\mu + 3\lambda)\bar{A}_i(s) - \frac{4\mu}{r^3}\bar{B}_i(s),$$

$$\bar{\sigma}_{\theta\theta}(r, s) = \bar{\sigma}_{\phi\phi}(r, s) = [2s\bar{\mu}(s) + 3s\bar{\lambda}(s)]\bar{A}_i(s) + \frac{4s\bar{\mu}(s)}{r^3}\bar{B}_i(s),$$

where $\bar{A}_i(s)$ and $\bar{B}_i(s)$ are the same as A_i and B_i with v and E replaced by $s\bar{v}(s)$ and $s\bar{E}(s)$, respectively.

Index

Printed in the United States
By Bookmasters